ASTROPHYSICAL POLARIZED BACKGROUNDS

Related Titles from the AIP Conference Proceedings Subseries on Astronomy and Astrophysics

599 X-Ray Astronomy: Stellar Endpoints, AGN, and the Diffuse X-Ray Background
Edited by Nicholas E. White, Giuseppe Malaguti, and Giorgio G. C. Palumbo, December 2001, 0-7354-0043-1

598 Solar and Galactic Composition: A Joint SOHO/ACE Workshop
Edited by Robert F. Wimmer-Schweingruber, December 2001, CD-ROM included, 0-7354-0042-3

587 Gamma 2001: Gamma-Ray Astrophysics 2001
Edited by Steven Ritz, Neil Gehrels, and Chris R. Shrader, October 2001, 0-7354-0027-X; CD-ROM: 0-7354-0030-X

586 Relativistic Astrophysics: 20th Texas Symposium
Edited by J. Craig Wheeler and Hugo Martel, October 2001, 0-7354-0026-1

575 Astrophysical Sources for Ground-Based Gravitational Wave Detectors
Edited by Joan M. Centrella, July 2001, 0-7354-0014-8

566 Observing Ultrahigh Energy Cosmic Rays from Space and Earth: International Workshop
Edited by Humberto Salazar, Luis Villaseñor, and Arnulfo Zepeda, May 2001, 0-7354-0002-4

565 Young Supernova Remnants: Eleventh Astrophysics Conference
Edited by Stephen S. Holt and Una Hwang, May 2001, 0-7354-0001-6

558 High Energy Gamma-Ray Astronomy: International Symposium
Edited by Felix A. Aharonian and Heinz J. Völk, April 2001, 1-56396-990-4

556 Explosive Phenomena in Astrophysical Compact Objects: First KIAS Astrophysics Workshop
Edited by Heon-Young Chang, Chang-Hwan Lee, Mannque Rho, and Insu Yi, March 2001, 1-56396-987-4

476 3 K Cosmology: EC-TMR Conference
Edited by Luciano Maiani, Francesco Melchiorri, and Nicola Vittorio, May 1999, 1-56396-847-9

470 After the Dark Ages: When Galaxies Were Young (The Universe at 2<z<5)
Edited by Stephen S. Holt and Eric P. Smith, April 1999, 1-56396-855-X

To learn more about these titles, or the AIP Conference Proceedings Series, please visit the webpage **http://proceedings.aip.org**

ASTROPHYSICAL POLARIZED BACKGROUNDS

Workshop on Astrophysical Polarized Backgrounds

Bologna, Italy   *9-12 October 2001*

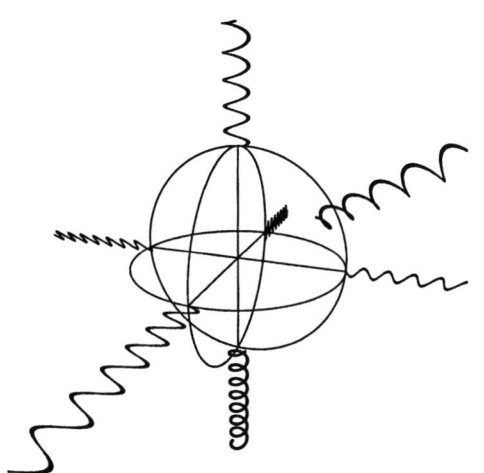

EDITORS
Stefano Cecchini
Stefano Cortiglioni
ITeSRE/CNR, Bologna, Italy
Robert Sault
Australia Telescope, CSIRO, Epping, Australia
Carla Sbarra
ITeSRE/CNR, Bologna, Italy

AMERICAN INSTITUTE OF PHYSICS

Melville, New York, 2002
AIP CONFERENCE PROCEEDINGS ■ VOLUME 609

Editors:

Stefano Cecchini
Stefano Cortiglioni
Istituto TeSRE/CNR
via P. Gobetti 101
40129 Bologna
ITALY

E-mail: cecchini@bo.infn.it
E-mail: cortiglioni@tesre.bo.cnr.it

Robert Sault
Australia Telescope, CSIRO
P.O. Box 76
Epping, New South Wales, 1710
AUSTRALIA

E-mail: rsault@atnf.csiro.au

Carla Sbarra
Istituto TeSRE/CNR
via P. Gobetti 101
40129 Bologna
ITALY

E-mail: sbarra@tesre.bo.cnr.it

L.C. Catalog Card No. 2002101276
ISBN 0-7354-0055-5
ISSN 0094-243X

Printed in the United States of America

CONTENTS

DEVICES, ANALYSIS TOOLS AND SYSTEMATICS

SPACE AND BALLOON FACILITIES AND AGENCIES

POSTER PRESENTATIONS

PREFACE

It is widely recognized that the polarization of the Cosmic Microwave Background (CMB) represents the new frontier for astrophysicists working on CMB and associated cosmology. The COBE satellite was the first to detect the anisotropy, opening a new observing era dedicated to the study of CMB temperature fluctuations in a wide range of angular scales. As is well known, in fact, the CMB can be studied by observing its spectrum, temperature anisotropies and polarization, the latter being generated by Thomson scattering of anisotropically distributed photons. The recent results of BOOMERanG, the great expectations from MAP, and the ultimate investigations that will be carried on in future by PLANCK, will probably close the era of anisotropies.

The detection of polarization in the CMB radiation will provide information about our universe which is not available in the temperature data, like that related to the ionisation history and to the inflation. Gravitational waves and vector modes can be studied as well, as they leave specific signatures in the CMB polarization pattern.

Similarly to both the anisotropy and the spectrum, the genuine CMB polarization can be probed only after subtraction of foregrounds, the polarization properties of which are not sufficiently known at present. In addition to the whole set of foregrounds, ground experiments must deal with the Earth atmosphere emission, while for space and balloon-borne experiments this problem is either not present or strongly reduced. At large angular scales, where the point source emission can be neglected, the Galactic emission is the most important foreground. Synchrotron, free-free and dust represent the potential sources of diffuse polarized emission distributed on different angular scales. However, such a Galactic foreground has been observed only at frequencies lower than 2.7 GHz and in limited sky regions, mostly near the Galactic plane. On the other hand the CMB measurements are generally concentrated at high galactic latitudes, where the sky emission is supposed to be lower.

This situation offers the opportunity for many investigations of polarization properties of the Galactic emission, which is definitely unknown at the level requested by the study of CMB polarization. It is necessary to remember, in fact, that the CMB is expected to show linear polarization at a very low level, in the order of 1 μK or less, depending on the angular scale. Such a level of polarization cannot be measured with "ordinary" instrumentation and techniques, but it rather requires a new kind of approach aimed specifically at the control of systematics. This scenario seems to provide one of the most exciting astrophysical environments, where scientists with different "backgrounds" may converge to face a common problem: the investigation of Astrophysical Polarized Backgrounds from radio to millimetric wavelengths.

The Italian Space Agency has demonstrated its understanding of this problem by allocating considerable resources to the field of polarization, including supporting the SPOrt (Sky Polarization Observatory) Program (a forerunner in polarimetric observations from space).

The Workshop on "Astrophysical Polarised Backgrounds" held in Bologna, also known as the Bologna APB Workshop, has been organised to review the current state in the field of searching, measuring, and modelling the diffuse polarised backgrounds in a very wide spectral range. The meeting has included scientists from many fields who deal with diffuse astrophysical polarization emission. They range from radioastronomers and cosmologists to engineers. Unfortunately, the APB Workshop took place only one month after the dramatic attack on the World Trade Centre in New York and most of the American scientists who were going to attend the meeting were forced to cancel. Scientists coming from other countries were hindered as well and the participation in the Workshop has been seriously affected by the severe situation throughout the world.

Despite this, the APB Workshop was successful. It provided the first opportunity for many scientists involved in parallel studies to gather in one forum. The panorama of ground and space observations and their related techniques was described by the contributions of 41 speakers and 10 poster presentations. There were also plenty of fruitful discussions. Some contributions from people who could not participate were presented by collegues, others have nevertheless been included in these Proceedings.

The Bologna APB Workshop also hosted a parallel round table on Balloon Facilities with representatives from the following Institutions:

- Italian Space Agency - ASI (Italy)
- Andoya Rocket Range (Norway)
- Swedish Space Corporation Esrange (Sweden)
- NASA-NSBF (USA)

We would like to thank the Area di Ricerca di Bologna (Consiglio Nazionale delle Ricerche) for allowing us to use its Conference Centre and facilities. Additional thanks is offered to all the young people (mainly students), who contributed to the local organization during the four days of the workshop. Of course, we also give warm thanks to all the speakers.

The APB Workshop has been made possible by the generous support of ASI, TILAB, LEYBOLD, ALENIA SPAZIO, LABEN, VARIAN and Banca Nazionale del Lavoro, which we gratefully acknowledge.

Stefano Cecchini
Stefano Cortiglioni
Robert Sault
Carla Sbarra

FIGURE 1. Participants to the APB Workshop in the garden of the Area della Ricerca of Bologna (CNR). The picture has been taken by Andrea Cremonini and Simona Righini, from the CNR Institute of Radioastronomy

DIFFUSE POLARIZED EMISSION AND MAGNETIC FIELDS

Polarization Surveys of the Galaxy

Wolfgang Reich, Ernst Fürst, Patricia Reich, Richard Wielebinski and
Maik Wolleben

Max-Planck-Institut für Radioastronomie, Auf dem Hügel 69, 53121 Bonn, Germany

Abstract. We report on sensitive $\lambda 21$ cm and $\lambda 11$ cm polarization surveys of the Galactic plane carried out with the Effelsberg 100-m telescope at arcmin angular resolution and some related work. Highly structured polarized emission is seen along the Galactic plane as well as up to very high Galactic latitudes. These observations reflect Faraday effects in the interstellar medium. Polarized foreground and background components along the line of sight, modified by Faraday rotation and depolarization, add in a complex way. The amplitudes of polarized emission features are highly frequency dependent. Small-scale components decrease in amplitude rapidly with increasing frequency. We stress the need for sensitive absolutely calibrated polarization data. These are essential for baseline setting and a correct interpretation of small-scale structures. Absolutely calibrated data are also needed to estimate the high-frequency polarized background. A recent study of polarized emission observed across the local Taurus-Auriga molecular cloud complexes indicates excessive synchrotron emission within a few hundred parsecs. These results suggest that possibly a large fraction of the Galactic high latitude total intensity and polarized emission is of local origin.

SURVEY HISTORY

Survey work has a tradition at the Max-Planck-Institut für Radioastronomie in Bonn. At first there are the all-sky surveys at 408 MHz ($\lambda 73$ cm, Haslam et al. [1]) and the recently completed $\lambda 21$ cm all-sky survey (Reich [2], Reich & Reich [3] and Reich et al. [4]). Both surveys have a comparable angular resolution of $0°8$ or $0°6$. Their sensitivities of 2 K or 0.05 K match for spectral studies of the Galactic synchroton emission. Higher angular resolution surveys at $\lambda 21$ cm and $\lambda 11$ cm were carried out with the Effelsberg 100-m telescope, but had to be limited to the Galactic plane because of the large amount of observing time needed. These and other survey data including all references are accessible via the internet at `http://www.mpifr-bonn.mpg.de/survey.html`.

Although a number of new total intensity surveys were carried out in the last two decades, polarization measurements were still just available from a number of long wavelengths surveys, where $\lambda 21$ cm is the shortest wavelength. These surveys were carried out in the sixties with the Dwingeloo 25-m telescope (Brouw & Spoelstra [5]). Although these data have rather moderate angular resolution they are quite carefully absolutely calibrated. The Dwingeloo surveys cover large sections of the northern sky, although they are not fully sampled. Large fractional polarizations are noted at high Galactic latitudes and a fairly smooth intensity and vector distribution is seen. Rotation Measures (RMs) are small in general. These results did not immediately trigger survey projects aiming for higher angular resolution.

The motivation for systematic polarization work at the Effelsberg telescope is mainly based on the unexpected $\lambda 11$ cm polarization results of Junkes et al. [6], which came out as a byproduct of the first section on the Effelsberg $\lambda 11$ cm Galactic plane survey (Reich et al. [7]). Beside polarized sources, like supernova remnants, these data show polarized emission patches with no apparent corresponding total intensity structure. These have to be associated to the unresolved diffuse background emission. It could be demonstrated by the anticorrelation of diffuse thermal emission and integrated polarized emission that a fraction of the emission must originate at a few kpc distance in the Galactic disk (Junkes et al. [8]). This indicates a highly structured interstellar medium, where "holes" with low Faraday effects allow a study of the Galactic magnetic field at large distances.

Sensitive high resolution polarization observations at $\lambda 90$ cm by Wieringa et al. [9] with the Westerbork telescope showed filamentary features on degree scales at high latitudes. Their origin has to be local. The total absence of enhanced synchrotron emission calls for an explanation by a local Faraday screen modifying the polarized background.

CP609, *Astrophysical Polarized Backgrounds*, edited by S. Cecchini et al.
© 2002 American Institute of Physics 0-7354-0055-5/02/$19.00

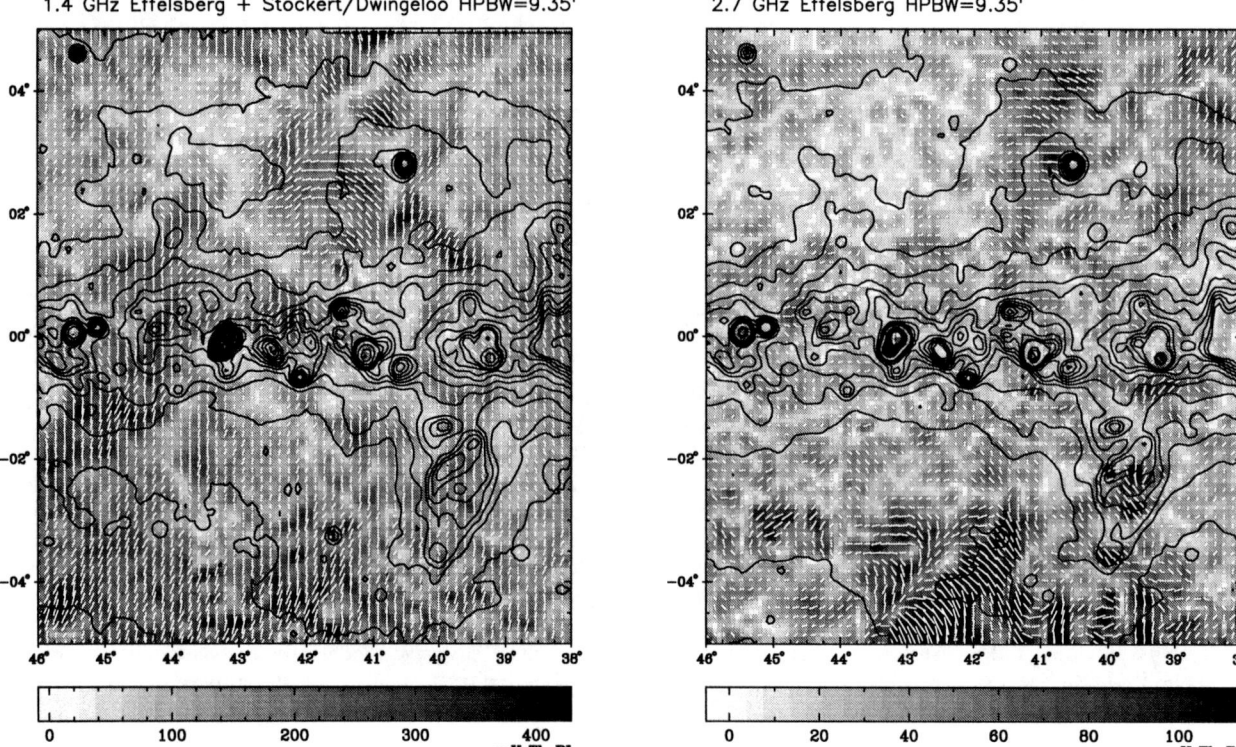

FIGURE 1. Section of the Galactic plane at λ21 cm and λ11 cm shown at the same angular resolution. The λ21 cm map was absolutely calibrated both for total intensity and polarization using Stockert (Reich & Reich [3]) and Dwingeloo (Brouw & Spoelstra [5]) data, while the λ11 cm map is not on an absolute scale. Polarized structures vary largely in intensity and angle.

However, a quantitative analysis has some difficulties. The detected filaments at λ90 cm disappear at shorter wavelengths, which again strengthens the case of a Faraday screen. Because of the λ^2 dependence of the Faraday rotation even small RMs have large effects at long wavelengths. We refer to Sokoloff et al. [10] for a detailed discussion of Faraday effects.

GALACTIC PLANE SURVEYS INCLUDING POLARIZATION

From the Junkes et al. [6] λ11 cm data it is obvious that low resolution data miss details of Galactic polarized emission. It is also clear that the fine structure is quite weak. The λ11 cm survey data need some smoothing to obtain a sufficient signal-to-noise. Since the distribution of polarized emission is not predictable from total intensity we started in 1994 the "λ21 cm Medium Galactic Latitude Survey" to map the Galactic plane within ±20° at 9.′35 (HPBW) with a sensitivity equivalent to the total intensity confusion limit of 7 mJy/beam or ~ 15 mK T_b. This sensitivity was reached with an integration time of 2 sec per 4′ pixel. Thus the surveyed area of ~ 7600 deg² needs ~ 1000h of net integration time. Observations were exclusively done at night time otherwise polarized solar emission shows up, which was picked up by far sidelobes. Varying interference sometimes forces to stop observations for several months. Nevertheless, in September 2001 the survey is observed to ~ 95%. Of course, the completion of data reduction needs some more time. Basic data of the survey are listed in Table 1.

The observing and calibration methods of the "λ21 cm Medium Galactic Latitude Survey" were already described in detail by Uyanıker et al. [11]. These include some additional steps to the standard Effelsberg reduction, in particular reduction of spurious polarization from cross-talk and absolute calibration using Dwingeloo survey data (Brouw & Spoelstra [5]), where available. First maps of typical regions ranging from the first quadrant towards the Galactic anticentre region are shown by Uyanıker et al. [12]. The characteristics of the total intensity are a smooth dominating background with faint ridges, arcs and complex emission regions superimposed. Diffuse polarized emission is seen

TABLE 1. Basic observational parameters of the Effelsberg λ21 cm and λ11 cm polarization surveys of the Galactic plane

Frequency [GHz]	1.4	2.695
HPBW	9.'35	5.'1
RMS PI [mK]	8	11
L-Coverage	$\sim 30° - 220°$	$4.°9 - 76°$
B-Coverage	+/-20°	+/-5°

everywhere. There are numerous polarized features without any associated total intensity structures. In particular remarkable depolarized loops, arcs and straight filaments are seen. These are most pronounced towards the anticentre direction, where the absolute intensities are low and the line of sight out of the Galaxy is short. The standard interpretation is the assumption of a highly polarized smooth background, which is seen through spatially varying Faraday screens. We refer to Uyanıker et al. [12] for maps illustrating these findings. A section of the Galactic plane is shown in Fig. 1.

In addition the work of Junkes et al. [6] was continued by analysing more polarization data of the Effelsberg λ11 cm survey of the first quadrant, which was extended from the initial latitude coverage of ±1.°5 to ±5° (Reich et al. [13]). The results of the analysis of the polarization data were published by Duncan et al. [14]. Figure 1 shows some data from that work at the angular resolution of the λ21 cm survey. Duncan et al. [14] made also some comparison with the Parkes λ13 cm polarization survey of the fourth quadrant (Duncan et al. [15]). The λ11 cm maps show a clear increase of polarized emission with Galactic latitude where depolarization is smaller. This result strengthens the conclusion of Junkes et al. [8] on the kpc-origin of some polarized emission. Duncan et al. [14] noted an anticorrelation of polarized emission at longitudes between 20° and 45° with H I gas at kinematic distances in the range 2 kpc – 2.5 kpc, which requires the polarized emission to originate at the same or larger distances.

SOME FOLLOW-UP WORK

Meanwhile some follow-up observations of λ21 cm and λ11 cm polarization features have been started at λ6 cm, although the required typical sensitivities of 1 mK and below limit observations to a few fields only. Some first results have been described by Reich et al. [16]. In brief: towards the first Galactic quadrant the situation appears rather complex. λ21 cm, λ11 cm (see Fig. 1), but also λ6 cm polarization maps might differ largely in the sense that features visible at one wavelength disappear at the other, while new features show up. This indicates a superposition of numerous emission layers along the line of sight across the Galaxy, which is quite plausible when looking at the structured polarized emission towards the anticentre at λ21 cm (Uyanıker et al. [12]), where the line of sight is much shorter. For a decomposition of these components observations at a denser sampling in wavelengths are required, but these are not yet available. Of course, narrow-band polarimetry within each available band will be quite helpful as well.

Towards the Galactic anticentre Reich et al. [16] find the situation to be less complex. For example: a ring-like structure of $\sim 1°$ in diameter centered at $\ell, b = 192.°2, 9.°4$ was also visible in polarization at λ11 cm and λ6 cm, although it becomes very faint. A RM of $\sim 120 \, \mathrm{rad\,m^{-2}}$ was derived. RM is calculated from: RM $[\mathrm{rad\,m^{-2}}] = 0.81 n_e [\mathrm{cm^{-3}}] \, B_\| [\mu G] \, l [\mathrm{pc}]$. Assuming distances between 0.5 kpc and 3 kpc for this feature, its size l is between 9 pc and 52 pc. Here a spherical shape is assumed. Upper limits for the electron density are $1.5 \, \mathrm{cm^{-3}}$ or $0.6 \, \mathrm{cm^{-3}}$ from the emission measure of less than $20 \, \mathrm{pc\,cm^{-6}}$ set by the noise of the λ6 cm total intensity signal. The magnetic field component along the line of sight calculates lower limits between $11 \, \mu G$ and $5 \, \mu G$. It is not clear what processes create such unusual magnetic-ionic structures high above the Galactic plane. We note that for this and similar features the spectrum of the polarized signal is steeper than the total intensity spectrum.

The Taurus-Auriga region and the local synchrotron emissivity

The λ21 cm Medium Galactic Latitude Survey data from longitudes 150° to 190° and latitudes from −4° to −20° have recently been reduced and analysed by Wolleben [17]. This region includes the well studied Taurus-Auriga and Perseus molecular cloud complexes, which are located at distances of ~ 140 pc and 350 pc. The molecular clouds are partly seen in superposition. CO survey data and IRAS data show highly structured emission across this area. Molecular gas and dust clouds are partly well correlated. There is an enhancement of polarized emission in the direction of these molecular cloud complexes. However, the total intensity smoothly increasing towards the Galactic plane but there is no indication of enhanced synchrotron emission. A number of discrete polarized features ranging between $\sim 0.3°$ and $1°$ in diameter could be identified adjacent to molecular and dust clouds. This association suggests the existence of a Faraday screen at the same distance. Polarized emission originating behind the molecular material gets modified and adds with the foreground polarization in a different way than outside the Faraday screen. Wolleben [17] noticed for a number of polarized structures a clear systematic dependence of the polarization angle with polarized intensity. Nine objects including their surroundings were modelled assuming regular foreground and background polarization. The background emission is then subject to Faraday modulation towards the polarized feature. Pure Faraday rotation as well as Faraday rotation with correlated depolarization were considered and fitted to the polarization angle–polarized intensity relation. RMs of up to $\sim 30 \, \mathrm{rad \, m^{-2}}$ were derived, quite much for clouds of a maximum extent between 0.7 pc and 2.4 pc (distance 140 pc) or 1.8 pc to 6 pc (distance 350 pc). For a number of polarized features both models fit the data quite well and a second frequency is needed for a distinction between both models. The average of $\lambda 21$ cm polarized intensities in front and behind the nine the Faraday screens near the molecular clouds have values of ~ 220 mK at polarization angle $25°$ and 290 mK at $-35°$ for the foreground and background components, respectively. Both components are related to a regular magnetic field. Both values are uncertain by ~ 100 mK and $\sim 20°$, respectively. Interestingly, a rather similar jump in polarization angle is seen from stellar polarization data (Heiles [18]) in the range up to 200 pc and 200 pc to 500 pc. The E-vectors agree on average within $20°$ with those from the radio data, where the internal scatter is about the same.

Polarized intensity of ~ 0.5 K from a regular magnetic field originating within 0.5 kpc requires a total intensity synchrotron component of at least 0.7 K. This is a lower limit from the intrinsically polarized regular field. Estimates for the fraction of the regular magnetic field component range from 60% to 70% of the total field (see Beck [19] for a review and references). If these estimates also hold for the very local magnetic field the $\lambda 21$ cm synchrotron component raises to ~ 1 K within 0.5 kpc.

Beuermann et al. [20] have unfolded the $\lambda 73$ cm all-sky survey (Haslam et al. [1]) for a three-dimensional radio emission model of the Galaxy. They quote a local synchrotron emissivity of ~ 11.7 K/kpc, which they found to be in agreement with previous studies. This scales to ~ 0.3 K/kpc at $\lambda 21$ cm, which is significantly below the 1.4 K/kpc to 2 K/kpc for the local emissivity towards the Taurus-Auriga complex. In this direction the total Galactic emission above the cosmic microwave background is at the 2 K level. However, this very local synchrotron component depends strongly on the magnetic field direction and therefore is unlikely isotropic. It is interesting to note that the recently derived average local emissivity at 22 MHz by Roger et al. [21] is about three times higher than the value adopted by Beuermann et al. [20]. The Roger et al. [21] result is based on emission towards extended opaque H II regions at distances between 400 pc and 2.2 kpc.

Enhanced local emissivity implies that emission at high Galactic latitudes is more affected by small scale fluctuations in the interstellar medium. This is expected to be the case in particular for polarized emission. The thick disk shrinks in size and its smooth intensity distribution is less dominant. Cosmic microwave studies at high Galactic latitudes should take this result into account.

ABSOLUTE CALIBRATION

The Effelsberg survey maps have a relative offset resulting from the assumed "zero" at the edges of each individual observed field. Each field is observed twice in Galactic longitude and latitude direction. Just the strong emission of the Galactic ridge in the first quadrant requires to scan along latitude direction only. The fields at $\lambda 21$ cm and $\lambda 11$ cm have typical sizes of $\sim 10° \times 10°$ or $\sim 3° \times 3°$, respectively. The combined observations have an "average zero" at their boundaries. As described by Uyanıker et al. [11] the total intensity and polarization data were absolutely calibrated by available lower angular resolution data from the Stockert and the Dwingeloo 25-m telescopes. While the total $\lambda 21$ cm intensities could be completely corrected, the polarization data could not because of incomplete mapping, undersampling and low sensitivity. The correction effects at low latitude are largest for total intensities and low for polarization, while at high latitudes the situation is reverse. This is problematic, because the correct zero level in polarization is much less predictable than for total intensities. Stokes U and Q may have positive or negative values depending on the polarization angle $\phi = 0.5 \, \mathrm{atan}(U/Q)$ and the polarized intensity calculates from

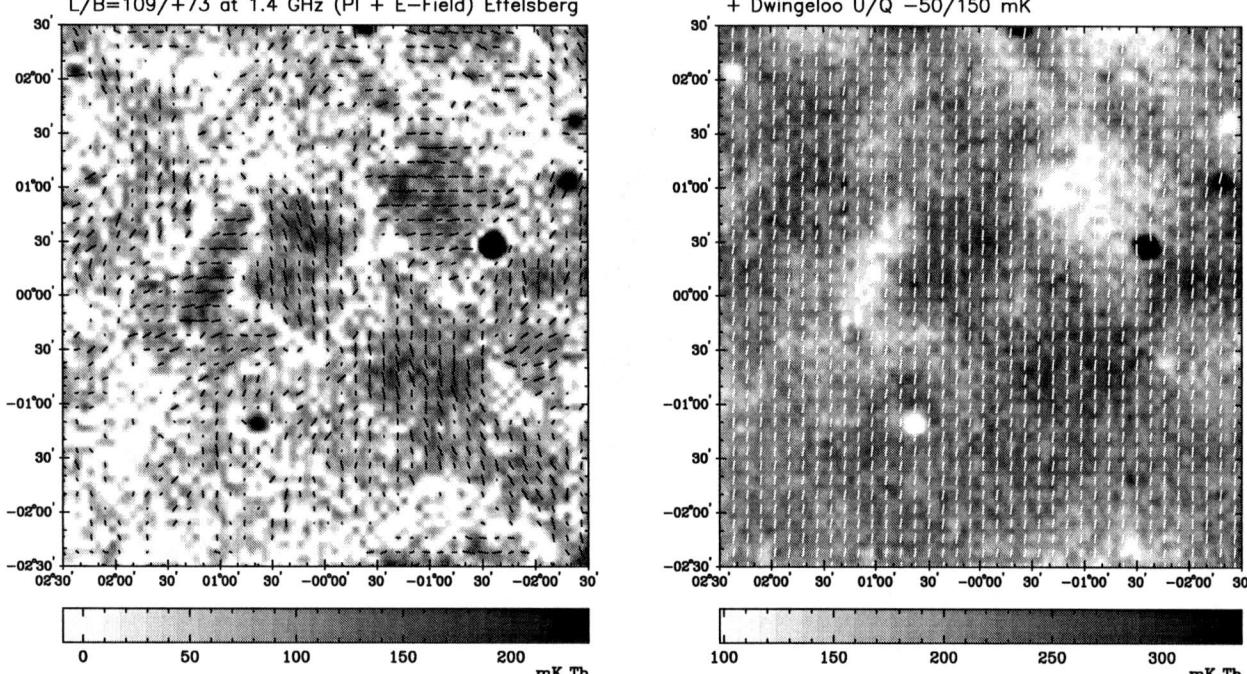

FIGURE 2. High latitude polarized emission at 1.4 GHz without (left) and with (right) large-scale structure added. Low-resolution absolutely calibrated Dwingeloo data were used to estimate the missing large-scale components of the Effelsberg data. Due to severe undersampling of the Dwingeloo data just constant offsets for U (-50 mK) and Q ($+150$ mK) have been added to the Effelsberg data. Depending on their polarization angle relative to the large-scale emission small-scale structures appear as enhancements or depressions superposed on the large-scale structure. A number of polarized sources are visible. The total intensity (not shown here) shows many compact sources in the field but no extended structures possibly related to the polarized features.

$(U^2 + Q^2)^{0.5}$. Adding large-scale dominating U and Q components with the same or the opposite sign converts a small-scale polarized emission feature into enhanced or reduced emission within the large-scale structure. Figure 2 shows an example of this effect: A high Galactic latitude $\lambda 21$ cm Effelsberg map with relative zero levels and with the large-scale components added. The large-scale structure entirely dominates. The average polarized emission from the Effelsberg map is about 8 mK, which is just 5% of the large-scale emission. However, the large-scale structure is estimated from a few data points near the observed field. These U and Q values were averaged and used to adopt the level of the Effelsberg U and Q maps. The uncertainty of the large-scale amplitudes is at least 30%. In addition, possible gradients in U and Q or fluctuations on scales of a degree or larger could not be taken into account.

The $\lambda 11$ cm total intensity data could be corrected for missing large-scale emission ($\geq 10°$), while the polarization data from all fields were adjusted relative to each other. In fact it was shown by Duncan et al. [14] that the $\lambda 13$ cm Parkes data contain emission from larger scales, because the survey was combined from larger individual fields. The only indirect method to adopt large-scale polarized structure at shorter wavelengths than $\lambda 21$ cm is to extrapolate by assuming a spectral index for the polarized emission and to adopt a long wavelength RM in part available from the work of Brouw & Spoelstra [5]. New absolutely calibrated data at $\lambda 21$ cm at mK-sensitivity are needed, which may use the already existing Dwingeloo data for adjustment purpose, as was discussed by Reich & Wielebinski [22]. Also plans for sensitive $\lambda 6$ cm observtions exist, where the angular resolution of a 25-m telescope matches that of the Effelsberg $\lambda 21$ cm observations. At $\lambda 6$ cm and shorter wavelengths extrapolation from the Brouw & Spoelstra [5] data seem irrelevant.

CONCLUDING REMARKS

We have discussed some work on polarized Galactic emission from the Effelsberg $\lambda 21$ cm and $\lambda 11$ cm surveys including some follow-up observations and analysis. Small-scale Faraday modulation causes highly fluctuating polarized or depolarized structures. The origin of the corresponding magneto-ionic structures – enhancement of thermal electron density or compression of the magnetic field – remains open so far. Small-scale emission becomes rather faint at shorter wavelengths. The dominating polarized background appears to be smooth on degree scales and larger. It is likely the dominating component at short wavelength. Its investigation requires very sensitive absolutely calibrated measurements. This needs substantial efforts and much telescope time, but can be done with small (25-m class) telescopes. At high latitudes the contribution from polarized synchrotron emission originating within a few hundred parsecs appears to be larger than previously assumed and fluctuations in polarization are also visible at high latitudes not much different from those seen at low latitudes.

ACKNOWLEDGMENTS

We like to thank A. R. Duncan, N. Junkes and B. Uyanıker for their contributions to the polarization survey work at Bonn.

REFERENCES

1. Haslam, C. G. T, Salter, C. J., Stoffel, H., and Wilson, W. E., *Astron. Astrophys. Suppl.*, **48**, 219 (1982).
2. Reich, W., *Astron. Astrophys. Suppl.*, **48**, 219 (1982).
3. Reich, P., and Reich, W., *Astron. Astrophys. Suppl.*, **48**, 219 (1986).
4. Reich, P., Testori, J. C., and Reich, W., *Astron. Astrophys.*, **376**, 861 (2001).
5. Brouw, W. N., and Spoelstra, T. A. Th., *Astron. Astrophys. Suppl.*, **26**, 129 (1976).
6. Junkes, N., Fürst, E., and Reich, W., *Astron. Astrophys. Suppl.*, **69**, 451 (1987).
7. Reich, W., Fürst, E., Steffen, P., Reif, K., and Haslam, C. G. T., *Astron. Astrophys. Suppl.*, **58**, 197 (1984).
8. Junkes, N., Fürst, E., and Reich, W., in *Interstellar Magnetic Fields*, edited by R. Beck and R. Gräve, Springer, Berlin, 1987, p. 115.
9. Wieringa, M. H., de Bruyn, A. G., Jansen. D., Brouw, W. N., and Katgert, P., *Astron. Astrophys.*, **268**, 215 (1993).
10. Sokoloff, D. D., Bykov, A. A., Shukurov, A., Berkhuijsen, E. M., Beck, R., and Poezd, A. D., *MNRAS*, **299**, 189 (1998).
11. Uyanıker, B., Fürst, E., Reich, W., Reich, P., and Wielebinski, R., *Astron. Astrophys. Suppl.*, **132**, 401 (1998).
12. Uyanıker, B., Fürst, E., Reich, W., Reich, P., and Wielebinski, R., *Astron. Astrophys. Suppl.*, **138**, 31 (1999).
13. Reich, W., Fürst, E., Reich, P., and Reif, K., *Astron. Astrophys. Suppl.*, **85**, 633 (1990)
14. Duncan, A. R., Reich, P., Reich, W., and Fürst, E., *Astron. Astrophys.*, **350**, 447 (1999).
15. Duncan, A. R., Haynes, R. F., Jones, K. L., and Stewart, R. T., *MNRAS*, **291**, 279 (1997).
16. Reich, W., Uyanıker, B., Fürst, E., Reich, P., and Wielebinski, R., in *Galactic Foreground Polarization*, edited by E. M. Berkhuijsen, Workshop Proceedings, MPIfR Bonn, 1999, p. 54.
17. Wolleben, M., Diploma Thesis, Bonn University (2001).
18. Heiles, C., *Astron. J.*, **119**, 923 (2000).
19. Beck, R., in *The Astrophysics of Galactic Cosmic Rays*, edited by R. Diehl et al., Space Sience Review, Kluwer Academic Publishers, Dordrecht, 2001, in press.
20. Beuermann, K., Kanbach, G., and Berkhuijsen, E. M., *Astron. Astrophys.*, **153**, 17 (1985).
21. Roger, R., Costain, C. H., Landecker, T. L., and Swerdlyk, C. M., *Astron. Astrophys. Suppl.*, **137**, 7 (1999).
22. Reich, W., and Wielebinski, R., in *"Radio Polarization: A New Probe of the Galaxy"*, edited by T. L. Landecker, Workshop Proceedings, DRAO, Penticton, 2001, p. 55.

Polarization Imaging of the Galactic Emission at 1420 MHz – the Canadian Galactic Plane Survey

T.L. Landecker, B. Uyanıker, R. Kothes

National Research Council Canada, Herzberg Institute of Astrophysics,Dominion Radio Astrophysical Observatory, Penticton, B.C., Canada

Abstract. The Canadian Galactic Plane Survey (CGPS) is providing images of total intensity and polarized emission at 1420 MHz with arcminute angular resolution over the region $75° < \ell < 145°$, $-3.5° < b < 5.5°$. Small-scale structure, attributed to Faraday-rotation effects, is detected over almost the entire survey area. Data are presented for the range $83° < \ell < 95°$. Here the polarization features lie in the Local arm, at distances less than 2 kpc.

INTRODUCTION

The Canadian Galactic Plane Survey, the CGPS, is a systematic mapping of the principal constituents of the interstellar medium (ISM) of the Galaxy with arcminute angular resolution [10]. Among those constituents are free electrons and magnetic fields forming a magneto-ionic medium (MIM), which is detected through the Faraday rotation it impresses upon the polarized emission from the Galactic synchrotron emission. The first phase of the CGPS covers a large region in the second Galactic quadrant ($75° < \ell < 145°$, $-3.5° < b < 5.5°$) and includes polarization imaging at 1420 MHz, with $1'$ resolution. The creation of a database of the major components of the ISM (atomic, molecular, ionized and relativistic gas, and dust) with high resolution is a significant step in the interpretation of the polarization observations.

This paper presents CGPS polarization data for a region in Cygnus, and shows that most of the features detected are in the Local arm. In other regions of the survey more distant polarization features are seen, attributable to a Perseus arm MIM.

PREPARATION OF SURVEY IMAGES

The DRAO Synthesis Telescope [6] produces a set of images in all four Stokes parameters I, Q, U, and V near 1420 MHz. The telescope receives both hands of circular polarization in four separate bands, each of width 7.5 MHz. For the observations presented here, the centre frequencies are 1406.89, 1414.39, 1426.89, and 1434.39 MHz. Visibilities from all four bands are separately gridded onto the $u - v$ plane and a combined I image is computed. Q and U images are made separately in each band to reduce bandwidth depolarization. The V maps contain flux which depends on errors in the telescope and its calibration, mostly arising from ellipticity of the feeds [7], and are usually of little value.

Calibration and image processing follows the practice conventional for aperture-synthesis data, to the point of image formation and deconvolution with CLEAN. Further processing [10] is based on routines developed especially for the DRAO Synthesis Telescope [12]. The essential processes are a visibility-based removal (using the routine MODCAL) of the effects of strong sources, both within and outside the main beam, and self-calibration. The effects of strong sources like Cas A and Cyg A are seen in Q and U images even when these sources are well outside the main beam.

Instrumental polarization varies across the field of view because of the effects of feed cross-polarization and aperture blockage by feed-support struts. The effects are corrected with empirical data obtained by observing the unpolarized sources 3C147 and 3C295 at a number of positions across the field of view out to a radius of $90'$. Individual images are assembled into mosaics, weighting data from individual fields to minimize the noise level in the final image.

CP609, *Astrophysical Polarized Backgrounds,* edited by S. Cecchini et al.

RESULTS AND THEIR INTERPRETATION

We present here a large mosaic, covering $82° < l < 95°$, $-3.5° < b < 5.5°$, in the Cygnus region. Figures 1 and 2 show I and U images, made from more than 40 telescope pointings. Note that single-antenna data have been incorporated into the I image, but not into the U image. The resolution ($\sim 1'$) in these $13° \times 9°$ images surpasses that of any previously available polarization data for this region. Here we aim to present a general conceptual framework within which the detailed interpretation of these results can proceed.

We expect to see polarized emission from extragalactic sources and from supernova remnants (SNRs). We know that the other polarized features result from small-scale structure in Faraday rotation acting on Galactic synchrotron emission, and we can expect interpretation to be complicated by the fact that synchrotron emission and Faraday rotation co-exist along most lines of sight.

In this region only about half the area shows detectable polarized emission. While we have a tendency to concentrate on regions of higher polarization, and to seek explanations of these "polarization features", we need to find a consistent interpretation which explains the regions of low polarization as well. Interpretation therefore requires a full consideration of depolarization mechanisms. A possible interpretation of the absence of polarized emission in our data is simply that the polarization structure is smooth, and there is no structure on the scales that the interferometer can detect, but this is unlikely. First, the Synthesis Telescope is sensitive to all scales from $1'$ to $\sim 45'$. Second, we see small-scale polarized features in almost every other direction in the Galactic plane. We prefer the interpretation that areas of low detected polarization are areas of intense depolarization.

There are three well-known depolarization mechanisms, bandwidth depolarization, depth (or front-back) depolarization, and beam depolarization [2, 8]. Bandwidth depolarization would only be a factor in the presence of rotation measures (RMs) of several thousand, and the typical RM of these regions is less than a few hundred. Under the conditions of magnetic field and electron density expected looking along the Local arm, total depth depolarization at 1420 MHz can result along lines of sight whose length is of order 1 to 4 kpc. Depth depolarization is expected to be a significant factor in our data, but cannot be the only factor at play. An effect which requires path lengths of order 1 kpc to function cannot produce arcminute structure in the Galactic emission. The presence of widespread small-scale polarization structure requires that there is small-scale structure in Faraday rotation. Beam depolarization must therefore have a significant role.

Inspection of polarization images suggests that the concept of "cells" is useful. A cell is loosely defined by the correlation scales of the magnetic field and of electron density. There is good evidence for actual cells in polarization observations of SNRs, and it is likely that they exist in the general ISM as well. When enough cells are present within the telescope beam, vector averaging reduces the observed fractional polarization to a very low level. Depth depolarization will still be a major factor in a "cellular" ISM, but beam depolarization will play a role as well. Without a detailed knowledge of the cellular structure of the ISM, exact predictions of depolarization are not possible [8].

We note that some SNRs in the field do not have detectable polarized emission, including some of dimensions large relative to the beam. All SNRs will generate polarized emission at source, and the only mechanism that can render it undetectable is beam depolarization occurring in the SNR and in the intervening medium. We detect polarized emission from the SNRs HB21 (G89.0+4.7) and CTB104A (G93.7−0.2). The distance to HB21 is 0.8 kpc [9] and to CTB104A 1.5 kpc [11]. We do not detect polarized emission from the SNR 3C434.1 (G94.0+1.0), whose distance is ~ 8 kpc [4] or from G84.2−0.8, whose distance is ~ 4 kpc [3] although both are bright SNRs with angular diameters of at least 20 beamwidths. Somewhere between 1.5 and 4 kpc beamwidth depolarization seems to take over for our telescope.

The separate or combined effects of depth depolarization and beam depolarization do not allow us to detect polarized features that originate beyond a certain distance. We refer to this distance as the *polarization horizon*. In this direction, the polarization horizon is within a few kpc, probably confined to the Local arm. While the concept of a polarization horizon applies only for one wavelength and one beamsize, we believe that it is a useful one. In the following we attempt to determine an approximate distance to the polarization horizon.

We interpret the regions where we have not detected significant polarized emission as regions where the properties of the medium along the line of sight are reasonably uniform all the way to the polarization horizon. The combined effects of depth depolarization and beam depolarization reduce the polarization fraction to a very low level in these directions. Large changes in the density of foreground ionized gas can change this situation, especially if the electron density rises to the point where we have an H II region. Electron densities in H II regions are much higher than in the surrounding ISM, and motions are turbulent, resulting in very tangled magnetic fields. Since magnetic fields are frozen into the ionized gas, and move with it, high field strengths can result from compression. Cell size is therefore very small, and Faraday rotation very high, leading to virtually total beam depolarization. A foreground H II region,

Galactic Latitude

−02° 00° 02° 04°

Canadian Galactic Plane Survey * DRAO * Stokes I

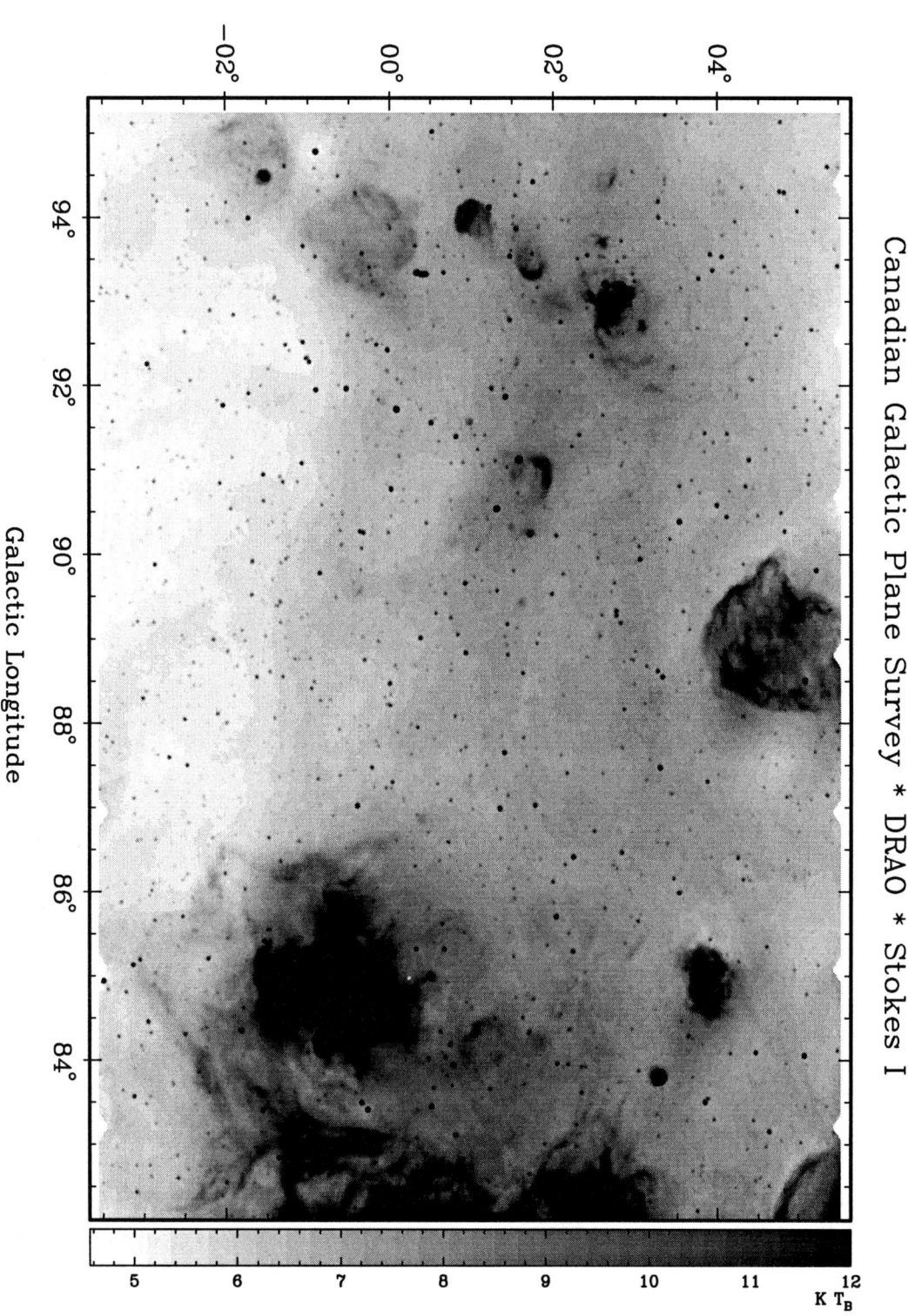

94° 92° 90° 88° 86° 84°

Galactic Longitude

5 6 7 8 9 10 11 12

K T$_B$

FIGURE 1. Total intensity, Stokes Parameter *I*

Galactic Latitude

−02° 00° 02° 04°

Canadian Galactic Plane Survey * DRAO * Stokes U

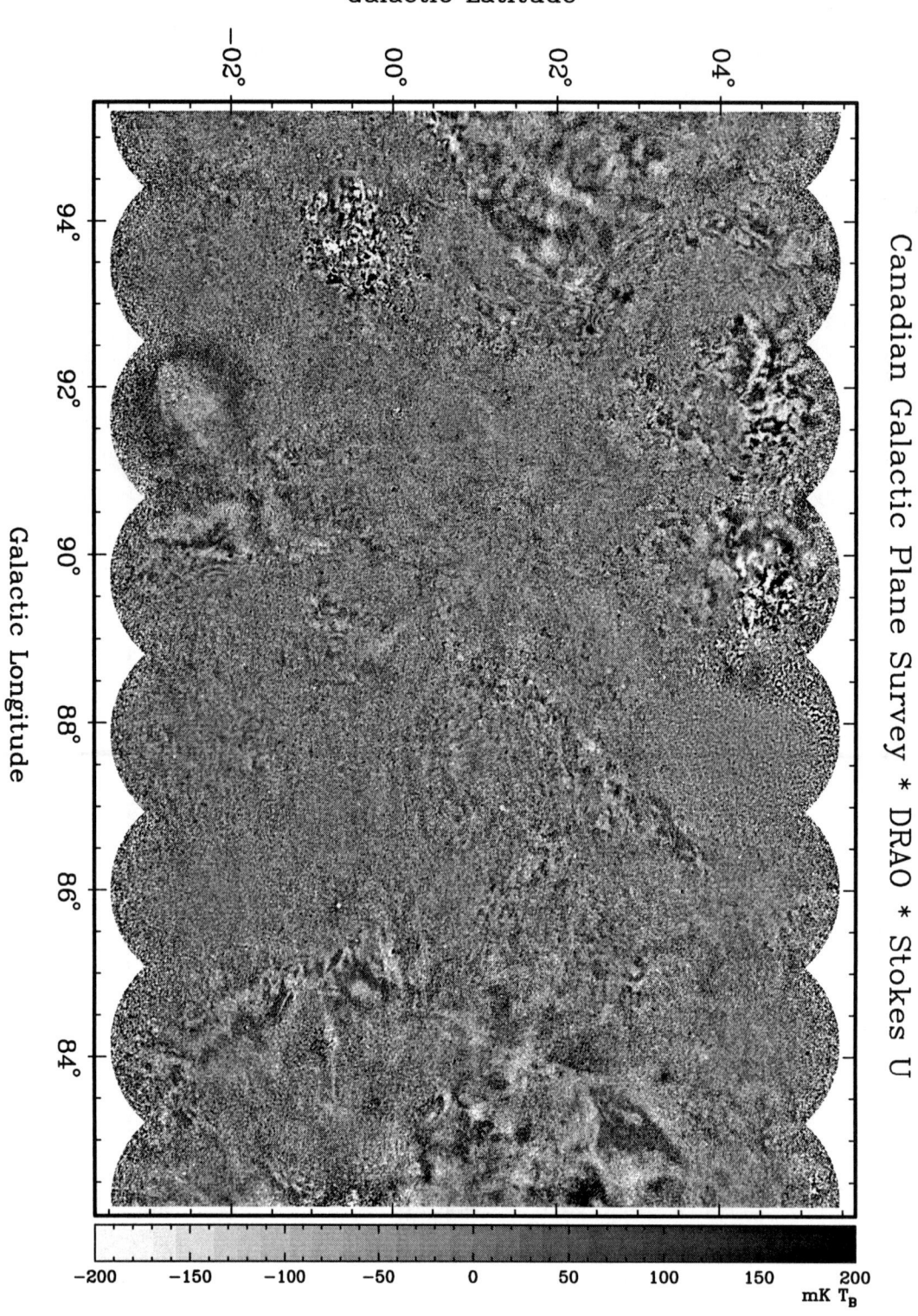

94° 92° 90° 88° 86° 84°

Galactic Longitude

−200 −150 −100 −50 0 50 100 150 200

mK T$_B$

FIGURE 2. Stokes Parameter U

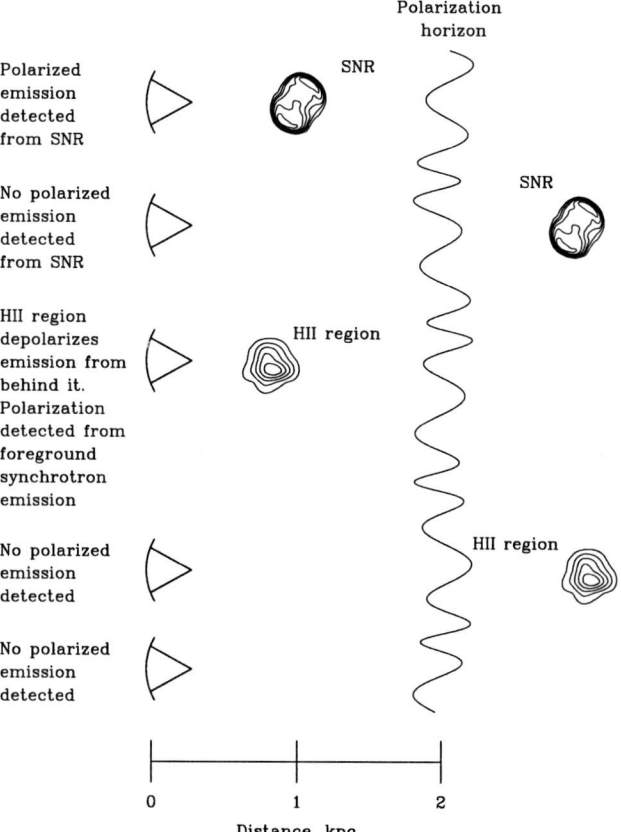

FIGURE 3. Polarized emission may or may not be detected by a radio telescope (at left) in the direction of polarized emitters (SNRs) and unpolarized emitters (H II regions) depending upon the location of the sources relative to the polarization horizon. The polarization horizon in the present data is estimated to be at a distance of ∼2 kpc.

while optically thin, is Faraday thick, and blocks access to any polarization originating at greater distances along the line of sight. If the H II region is closer than the polarization horizon, the emission generated between the H II region and the telescope will *not* be totally eliminated by depth or beam depolarization over the shorter path, and we may detect polarized emission superimposed on the H II region. This leads to the result, which at first seems puzzling, that we are seeing polarized emission where there are H II regions. These concepts are illustrated in Figure 3.

We can test this hypothesis against our data. We see polarization in front of W80 ($\ell = 85°, b = -1°$) whose distance is 0.5 kpc, but not in front of S112 ($\ell = 83.8°, b = 3.3° - 2.1$ kpc), S115 ($\ell = 84.8°, b = 3.9° - 3.0$ kpc), S124 ($\ell = 94.6°, b = -1.5° - 2.6$ kpc), or CTB102 ($\ell = 92.9°, b = 2.7° - \sim8$ kpc), apart from some small effects due to imperfect correction for instrumental polarization. Again, we have clear evidence that the distance to the polarization horizon is no more than a few kpc.

We conclude that in this direction the polarization horizon is no more than about 2 kpc distant, well within the Local arm. All the polarization features seen in Figure 2 are probaly generated within the Local arm.

COMPARISON WITH OTHER REGIONS

In the range $130° < \ell < 140°$ polarization features can be associated with H II regions at known distances, leading to the conclusion that the Faraday "screen" is in the Perseus arm at distances of 2 to 3 kpc. We have at least two examples. The area near W4 shows extra Faraday rotation in the diffuse outer envelope of the H II region, and depolarization by the H II region itself [5]. In a second field (at $\ell = 141°$) we see similar effects due to another Perseus arm H II region.

Measurements of the Faraday rotation of compact extragalactic sources seen through the Galactic plane [1] have shown that Rotation Measure (RM) declines between $\ell = 75°$ and $\ell = 145°$ in a manner consistent with a magnetic field aligned with the arm, locally directed towards $\ell = 84° \pm 4°$. Towards $\ell \approx 140°$ the line of sight is roughly at $60°$ to the field direction in both the Local and Perseus arms, and we expect RM to be lower. Consequently the polarization horizon is more distant.

ACKNOWLEDGEMENTS

The Dominion Radio Astrophysical Observatory is a National Facility operated by the National Research Council. The Canadian Galactic Plane Survey is a Canadian project with international partners, and is supported by the Natural Sciences and Engineering Research Council.

REFERENCES

1. Brown, J.C., and Taylor, A.R., ApJ, **563**, L31 (2001)
2. Burn, B.J. MNRAS, **133**, 67 (1066)
3. Feldt, C., and Green, D.A. A&A, **274**, 421 (1993)
4. Foster, T. 2000, thesis, University of Alberta
5. Gray, A.D., Landecker, T.L., Dewdney, P.E., Taylor, A.R., Willis, A.G., and Normandeau, M., ApJ, **514**, 221 (1998)
6. Landecker, T.L., Dewdney, P.E., Burgess, T.A., Gray, A.D., Higgs, L.A., Hoffmann, A.P., Hovey, G.J., Karpa, D.R., Lacey, J.D., Prowse, N., Purton, C.R., Roger, R.S., Willis, A.G., Wyslouzil, W., Routledge, D., and Vaneldik, J.F., A&AS, **145**, 509 (2000)
7. Smegal, R.J., Landecker, T.L., Vaneldik, J.F., Routledge, D., and Dewdney, P.E., Radio Science, **32**, 643 (1997)
8. Sokoloff, D.D, Bykov, A.A., Shukurov, A., Berjhuijsen, E.M., Beck, R., and Poezd, A.D., MNRAS, **299**, 189 (1998)
9. Tatematsu, K., Fukui, Y., Landecker, T.L., and Roger, R.S., A&A, **237**, 189 (1990)
10. Taylor, A.R., Dougherty, S.M., Gibson, S.J., Peracaula, M., Landecker, T.L., Brunt, C.M., Dewdney, P.E., Gray, A.D., Higgs, L.A.,, Kerton, C.R., Knee, L.B.G., Kothes, R., Purton, C.R., Uyanıker, B., Wallace, B.J., Willis, A.G., Martin, P.G., and Durand, D. 2001, AJ submitted
11. Uyanıker, B., Kothes, R., and Brunt, C.M. 2001, ApJ, in press
12. Willis, A.G., A&AS, **136**, 603 (1999)

Counter-intuitive polarization structures

Bülent Uyanıker & Tom Landecker

National Research Council, Herzberg Institute of Astrophysics, Dominion Radio Astrophysical Observatory,
P.O. Box 248, Penticton, B.C., V2A 6K3 Canada

Abstract. We report the detection of two polarization structures in the Cygnus region, using the Canadian Galactic Plane Survey data. One of the structures seems to be associated with a large molecular cloud, east of HB21. The other structure is probably due to an enhancement in electron density but with no total intensity counterpart. The nature of this object is not yet clear. We also discuss the recently discovered rotation measure anomaly towards CTB104A, indicating a magnetic field direction contrary to the "large-scale" Galactic magnetic field.

INTRODUCTION

Structured rotation measure (RM) fluctuations seen in Galactic polarization maps, showing no counterparts in total intensity, are signatures of local magnetic field enhancements or of Faraday screens. Several examples of these objects have already been mapped using single-antenna telescopes [13, 1] and interferometers [3, 16].

The study of linearly polarized Galactic emission through such observations is a new path to understanding the magnetic structure of the Galaxy. Using this tool as a probe of the interstellar medium (ISM) we can now study the details of magnetic field fluctuations in the Galaxy, in contrast to the past where polarization observations were mostly used as a tool for investigating supernova remnants (SNRs). Although considerable effort has been devoted to understanding the magnetic field configuration in the Galaxy, we are still far from putting the pieces together into a more general and coherent picture. We can broaden our understanding by analyzing more of these objects.

In this paper we report recent observations of polarized structures detected in the polarized Galactic "background". We use images of Stokes Q and U made with the DRAO Synthesis Telescope at 1420 MHz as part of the Canadian Galactic Plane Survey [12]. We use the images presented in the companion paper in this volume [6], where the telescope and data reduction are described.

POLARIZATION WITHOUT TOTAL INTENSITY

We have detected an oblate polarization structure towards $\ell = 91°75$ and $b = -2°45$. It has no total-intensity counterpart, but the polarization angle is well-structured. The total intensity and polarized intensity images are shown in Fig. 1 and the polarization angle is displayed in the left panel of Fig. 2.

We have checked known objects in this region for a possible relationship to the smooth structure seen in polarized emission. There is no sign of a coherent structure matching the extent of the polarized emission in our total power maps at 1420 or 408 MHz. A possible SNR was detected at this position by Kassim [5]. However, other surveys, from the 22 MHz DRAO survey [11] to the 2.7 GHz Effelsberg survey [10, 4], do not show any structure resembling an SNR or HII region. An optical nebula, listed in the Lynds Bright Nebula catalog [7] as LBN 416 coincides with the polarization feature. The integrated flux density of LBN 416 is 240 mJy, indicating that the nebula contains enough free electrons to measurably affect the observed polarization. Neither depolarization, nor effects on position angle, can be detected, and we deduce that LBN 416 is behind the polarization feature.

An open cluster, M39 (NGC 7092), whose distance is 265 pc, partly overlaps the polarization structure (Fig. 1). However, the stars in this cluster do not produce enough ionizing flux to form Strömgren spheres of the size required to produce the polarization feature.

CP609, *Astrophysical Polarized Backgrounds*, edited by S. Cecchini et al.

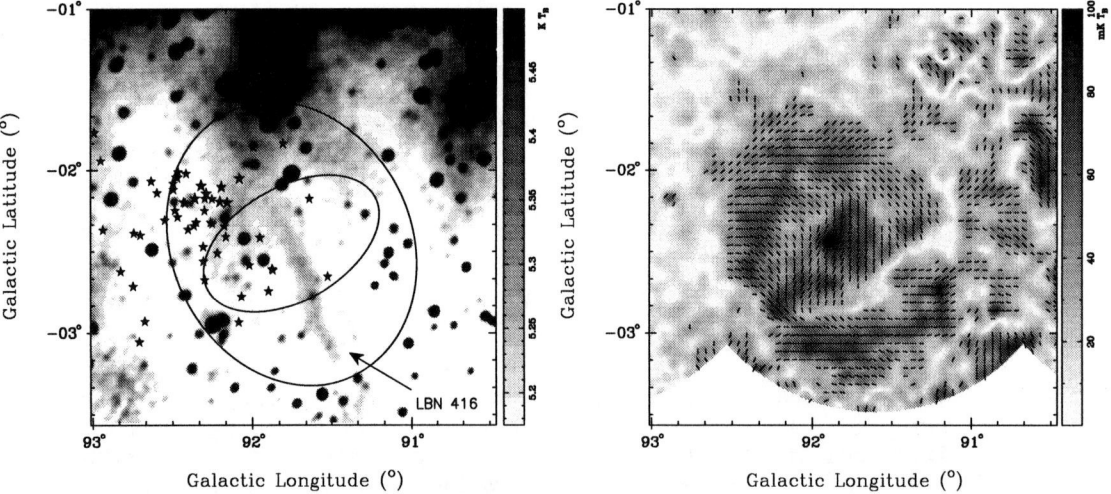

FIGURE 1. Total intensity (left) and polarized intensity images of the observed structure. The members of M39 are marked by star symbols. Electric field vectors are plotted as bars.

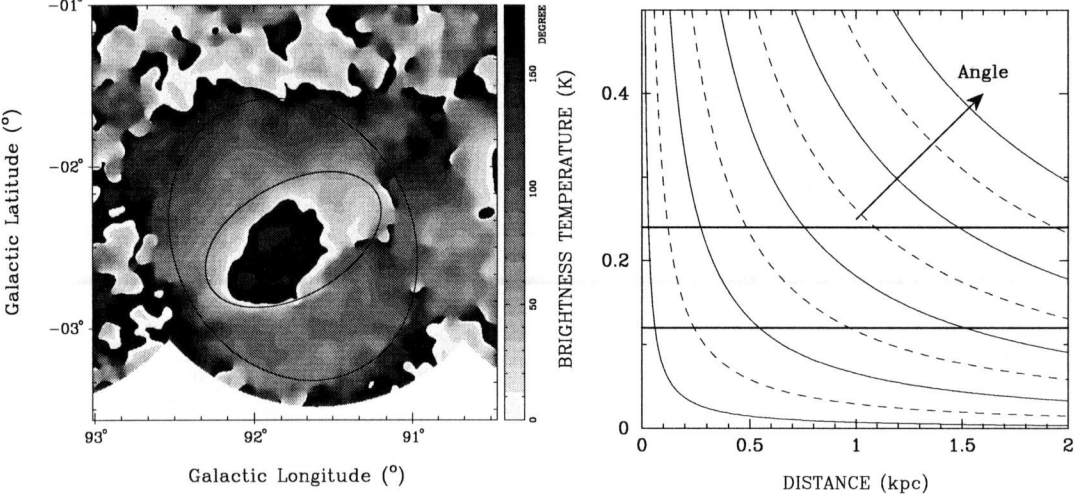

FIGURE 2. Polarization angle image (left) and brightness temperature as a function of Faraday rotation angle and distance. Overlaid ellipses, centered at $\ell = 91°.75$ and $b = -2°.45$, show the approximate size of the structure ($90' \times 108'$). Inner ellipse ($72' \times 42'$) delineates the region where polarization angle jumps from 180° to 0°.

Assuming a magnetic field of 1 μG, we can calculate the amount of ionized material required to produce the observed change in polarization angle (100 degrees). We can constrain the distance to the structure because this ionized material does not produce a detectable total-intensity flux in our telescope (or in other more sensitive total-intensity surveys). The results of this calculation are illustrated in the right-hand panel in Fig. 2. On the basis of these arguments, we conclude that the ionized structure is probably an enhancement in electron density of amount ~ 0.2 cm^{-3} at a distance of about 1 kpc. It is a similar object to the "polarization lens" described in Gray et al. [3].

POLARIZATION TOWARDS AN SNR

Recent DRAO observations of CTB104A [14] reveal a quite remarkable rotation measure gradient towards this SNR, although in this region small scale polarization and rotation measure (RM) structures are turbulent in nature. Fig. 3

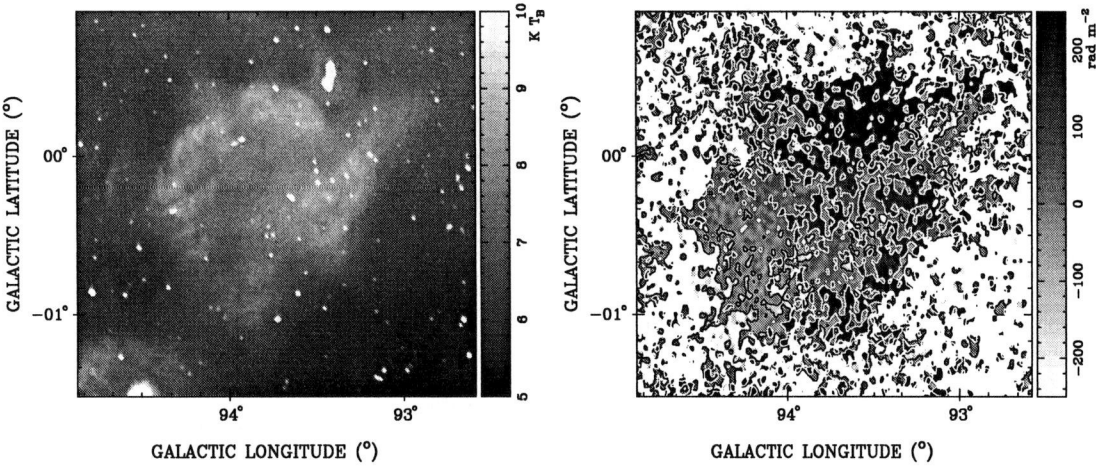

FIGURE 3. Total intensity (left) and rotation measure images of CTB104A. White contour is at -100 rad m^{-2} and black contour is at 100 rad m^{-2}.

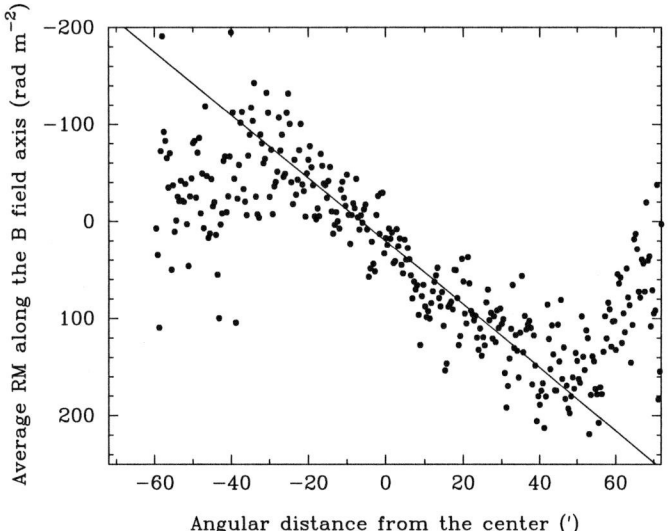

FIGURE 4. Rotation measure distribution as a function of the angular distance from the center of CTB104A, $(\ell, b) = 93°.7, -0°.3$, projected on to the magnetic field axis, which makes an angle of $30°$ with respect to the Galactic plane. The straight line is the fit to the RM gradient.

shows the total intensity and rotation measure images of the remnant. The observed morphology of CTB104A is consistent with expansion in a uniform magnetic field. The direction of the rotation measure gradient defines the axis of the magnetic field around the remnant, lying $30°$ north of west. The rotation measure profile projected on to this axis is plotted in Fig.4. This magnetic field is local to the remnant and generally counter to the global magnetic field of the Galaxy in this area. A fit to the RM gradient gives a magnetic field strength of $B_0 \simeq 2.3$ μG within CTB104A. These observations indicate fluctuations in the Galactic magnetic field.

POLARIZATION TOWARDS MOLECULAR CLOUDS

HB21 (G89.0+4.5) is a large non-circular SNR, and the polarization measurements [9] indicate a complicated magnetic structure. We observe strong polarized emission from this SNR (Fig. 5). Interestingly, polarized emission is not

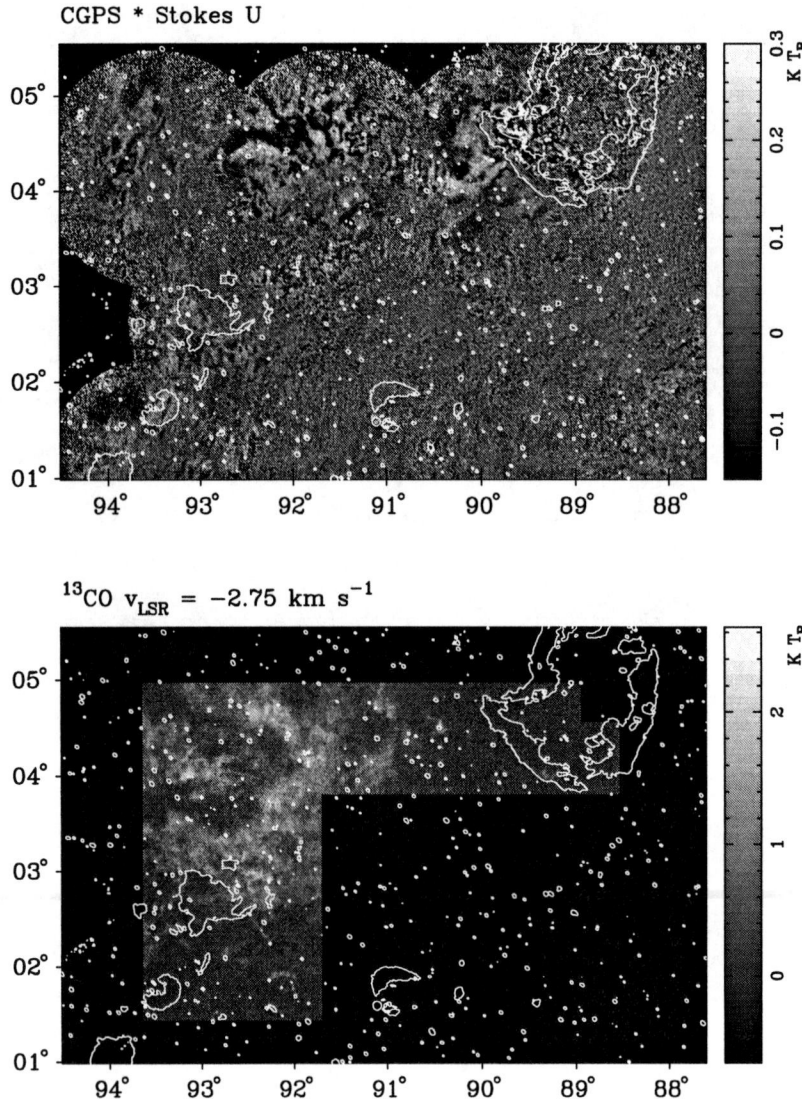

FIGURE 5. Stokes U map (top) showing the region in the vicinity of HB21. Overlaid contours are total intensity at 1420 MHz. The bottom panel shows the same region as observed in the ^{13}CO line with the FCRAO 14m telescope.

confined to the SNR, but is detected at almost the same intensity from adjacent regions, especially to the South-East. There is a large polarized structure to the east of HB21 (at $\ell = 92°, b = 4°.5$) comparable in size to the remnant, without any total-intensity counterpart.

Earlier observations of HB21 indicate that its expansion is strongly influenced by contact with a large molecular cloud lying along the eastern boundary of the SNR [15]. To obtain a more complete understanding of the relationship of the SNR to its environment, we have mapped a large region towards HB21 in ^{13}CO (J= 1 − 0) using the FCRAO 14-m Telescope. Results are shown in the lower panel of Fig. 5. Comparison of the distribution of molecular material with the distribution of polarization shows a striking result. There are filaments in U coinciding with the edges of the molecular material seen at $v_{LSR} = -2.75$ km s^{-1}. The strong polarized feature at $\ell = 92°, b = 4°.5$ has some resemblance in form to the molecular structure.

We know that the SNR and the molecular cloud are at the same distance, but are the two polarization features at the same distance? In Fig. 6 we show Rotation Measure (RM) along a line at $b = 4°.5$ through the SNR and the adjacent polarized region. RM varies smoothly across the boundary between the two objects. This does not conclusively show that both objects are at the same distance. However, if they were at very different distances, we would expect a

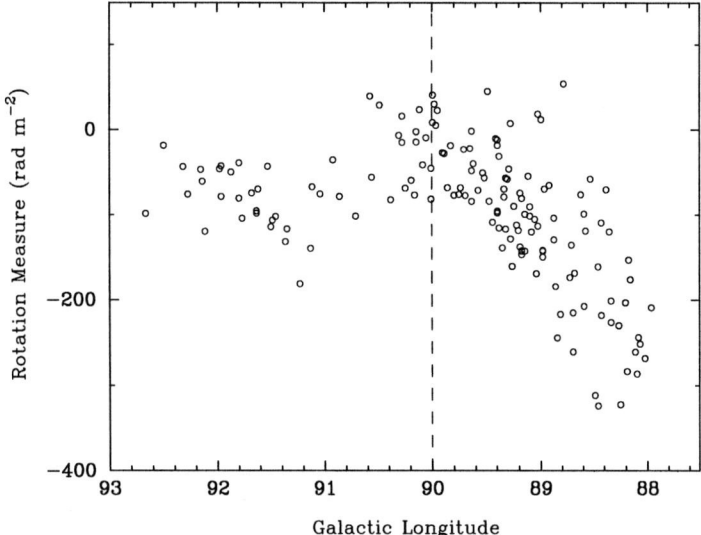

FIGURE 6. Variation of rotation measure with respect to Galactic longitude. The dashed line indicates the boundary between the molecular cloud and HB21

substantial jump in RM at this point.

It is unlikely that the polarization features are produced by Faraday rotation inside the molecular cloud. Although magnetic fields can be high (as much as 10^4 μG, see [2, 8]) the fractional ionization inside such a cloud is extremely low, and we expect an electron density of $< 10^{-4}$. The expected RM from a typical cloud is less than ~ 2 rad m^{-2}, making it unlikely that the molecular cloud itself can form the Faraday screen. We consider it more likely that the polarization features are attributable to Faraday rotation in ionized material on the surface of the molecular cloud. The polarization structures seen in the U image of Fig. 5 very plausibly arise at the outer edge of the molecular cloud. The strong polarized emission patches at $\ell = 90°, b = 4°5$ and $\ell = 92°, b = 4°5$ are probably produced by stars behind the molecular cloud. A region in which a molecular cloud and a SNR are both found is likely to be a region where other stars with considerable ionizing flux will exist.

On the whole, these unconventional objects discussed in this paper are difficult to understand with intuition, but our perception will broaden as more of these structures are discovered and studied.

REFERENCES

1. Duncan A.R., Haynes R.F, Jones K.L. & Stewart R.T., 1997, MNRAS, 291, 279
2. Fiebig D. & Güsten R., 1989, A&A, 214, 333
3. Gray A.D., Landecker T.L, Dewdney P.E & Taylor A.R, Nature, 393, 690
4. Junkes N, Fürst E., Reich W., 1987, A&AS, 69, 451
5. Kassim N.E., 1988, ApJ, 328, 55
6. Landecker T.L, Uyanıker B. & Kothes R., *this volume*
7. Lynds B.T., 1965, ApJS, 12, 163
8. Myers F.C., Goodman A.A, Güsten R. & Heiles C., 1995, ApJ, 442, 177
9. Reich et al. 1983, IAU SYMP. 101, p. 377, Reidel, Dordrecht
10. Reich W., Fürst E., Steffen P., Reif K. & Haslam C.G.T, 1984, A&AS, 58, 197
11. Roger R.S., Costain C.H., Landecker T.L., Swerdlyk C.M., 1999, A&AS, 137, 7
12. Taylor, A.R., Dewdney, P.E., Landecker, T.L., Martin, P.G., Brunt, C., Dougherty, S.M., Durand, D., Gibson, S.J., Gray, A.D., Higgs, L.A., Kerton, C.R., Knee, L.B.G., Kothes, R., Peracaula, M., Purton, C.R., Uyanıker, B., Wallace, B.J., Willis, A.G. 2001, AJ, submitted
13. Uyanıker B., Fürst E., Reich W., Reich P., Wielebinski, R., 1999, A&AS, 138, 31
14. Uyanıker B., Kothes R. & Brunt C.M., ApJ, accepted for publication.
15. Tatematsu K., Fukui Y., Landecker T.L., Roger R.S., 1990, A&A, 237, 189
16. Wieringa M.H., de Bruyn A.G., Jansen D., Brouw W.N. & Katgert P., 1993, A&A, 268 215

19

High Resolution Polarimetry of the Inner Galaxy

Bryan M. Gaensler*, J. M. Dickey†, N. M. McClure-Griffiths**, N. S. Bizunok‡ and
A. J. Green§

*Harvard-Smithsonian Center for Astrophysics, Cambridge MA 02138, USA
†University of Minnesota, Minnesota MN 55455, USA
**Australia Telescope National Facility, Epping NSW 1710, Australia
‡Boston University, Boston MA 02215, USA
§University of Sydney, NSW 2006, Australia

Abstract. We present our results from the Southern Galactic Plane Survey, an effort to map the fourth quadrant of the Milky Way in linear polarization at a frequency of 1.4 GHz and at a resolution of 1–2 arcmin. These data are a powerful probe of both the turbulence and large-scale structure of magneto-ionic gas, and have revealed a variety of new features in the interstellar medium.

INTRODUCTION

The Milky Way was the first celestial radio source discovered, and was subsequently one of the first sources to be detected in linear polarization. There are two sources of this polarization: discrete objects such as supernova remnants (SNRs), and a diffuse polarized background produced by the relativistic component of the interstellar medium (ISM). All of this emission undergoes Faraday rotation as it propagates towards us, either in the source itself or in intervening material. With sufficiently high angular and frequency resolution, we can use the properties of this polarized emission to map out the distribution of ionized gas and magnetic fields in individual sources and in the ambient ISM. Only recently have instruments and techniques advanced to a point where such studies are feasible [1, 2, 3].

Motivated by the spectacular single-dish polarization surveys of Duncan et al [4, 5], we have made polarimetric images of the entire fourth quadrant of the Galaxy with the Australia Telescope Compact Array (ATCA). These data have been taken as part of the Southern Galactic Plane Survey (SGPS; [6]). While the primary focus of the SGPS is to study the Galactic distribution of H I, the ATCA simultaneously receives full polarimetric continuum data, which have allowed us to map out the distribution of linearly polarized emission in the survey region.

While the full survey has now been completed, a detailed analysis has only been carried out on a 28-deg^2 test region, covering the range $325°5 < l < 332°5$, $-0°5 < b < +3°5$. We here summarize the main results of this analysis; this work is described in more detail by Gaensler et al [7].

OBSERVATIONS AND REDUCTION

The ATCA is a 6-element synthesis telescope, located near Narrabri, NSW, Australia. Observations of the test region of the SGPS were carried out in nine observing runs in 1997 and 1998, and comprised 190 separate telescope pointings (see [6] for details). Data were recorded in nine spectral channels spread across 96 MHz of bandwidth and centered at a frequency of 1384 MHz. The two sources MRC B1438–481 and MRC B1613–586 were observed over a wide range in parallactic angle in order to solve for the instrumental polarization characteristics of each antenna [8]. For each spectral channel, images of the field in Stokes I, Q, U and V were deconvolved jointly using the maximum entropy algorithm PMOSMEM [9] and then smoothed to a resolution of $\sim$1 arcmin. The final sensitivity in each image is $\lesssim 0.5$ mJy beam^{-1}.

Images of linearly polarized intensity, $L = (Q^2 + U^2)^{1/2}$, linearly polarized position angle, $\Theta = \frac{1}{2} \tan^{-1}(U/Q)$, and uncertainty in position angle, $\Delta\Theta = \sigma_{Q,U}/2L$, were then formed from each pair of Q and U images. The nine L maps

CP609, *Astrophysical Polarized Backgrounds*, edited by S. Cecchini et al.

(one per spectral channel) were then averaged together to make a final image of L for the entire test region, while the nine Θ and $\Delta\Theta$ maps were used to derive an image of the rotation measure (RM) over the field.

RESULTS

In Figure 1 we show images of I and L from both our 1.4-GHz ATCA observations and from the 2.4-GHz Parkes survey of Duncan et al [4] (resolution 10.'4). The total intensity images show the presence of SNRs and H II regions (see [6] for further discussion). Although the interferometric ATCA observations are not sensitive to the diffuse emission seen by Parkes, it is clear that the same features are present in both data-sets.

At first glance it seems that the L images have very little in common with the Stokes I emission. In particular, the ATCA L image is dominated by diffuse polarization spread all over the field of view, composed of discrete patches separated by narrow "canals" of reduced polarization. While none of this emission is correlated with total intensity, there does seem to be a good match between the brightest polarized regions of the ATCA and Parkes data, despite the differing frequencies and resolutions of these data-sets. Using the images of Θ and $\Delta\Theta$, we can determine the variation of polarization position angle with frequency wherever we detect polarized emission. The resulting RMs are generally small and negative, with a mean RM for the entire field of -12.9 ± 0.1 rad m^{-2}; 50% of the RMs have magnitudes smaller than ± 25 rad m^{-2} and 98% are smaller than ± 100 rad m^{-2}.

The ATCA L image reveals two large voids of reduced polarization, each elliptical and several degrees in extent. One void is centered on $(l,b) = (332°4, +1°4)$ ("void 1") and the other on $(328°2, -0°5)$ ("void 2"); both voids are also seen in the 2.4-GHz Parkes polarization map. The RMs around the edges of these voids range up to ± 400 rad m^{-2}, in distinction to the low RMs seen over the rest of the field.

A careful examination shows one marked correspondence between the ATCA Stokes I and L images: at $(326.3, +0.8)$, the bright H II region RCW 94 shows reduced polarization towards its interior, and is further surrounded by a halo in which no polarization at all is seen. This is shown in more detail in Figure 2.

Finally, of the numerous unresolved sources distributed across the field, 21 of these sources show detectable linear polarization. The RMs for these sources fall in the range -1400 rad m^{-2} to $+200$ rad m^{-2}.

DISCUSSION

Diffuse Emission

We first note that the incomplete $u - v$ coverage of an interferometer affects images of polarization in complicated ways. While it is physically required that $L \leq I$, and we generally expect that structures seen in L might correspond to similar structures in I, neither situation will be generally observed in interferometric data. This is because an interferometer can not detect structures larger than a certain size, corresponding to the closest spacings between its antenna elements (in the case of the ATCA, this maximum scale of $\sim 35'$). A source larger than this maximum scale will not be seen in Stokes I; if it is also a uniformly polarized source, it will not be detected in polarization either. However, magnetic field structure within the source, plus variations in the Faraday rotation along different lines-of-sight, can introduce power in Stokes Q and U on smaller scales, to which the interferometer is sensitive. We thus can often observe complicated structures in polarization which have no counterpart in total intensity [1, 2, 3, 10].

Clearly such an effect is occurring here, and is producing virtually all the linear polarization seen in Figure 1. We can crudely divide up the diffuse polarization we see into two components.

The brightest polarization seen with the ATCA matches well the bright polarized structures seen with Parkes. Since the amount of Faraday-induced polarization is very strongly dependent on both resolution and frequency, the fact that two such disparate data-sets show similar structures implies that these bright polarized structures are intrinsic to the emitting regions. By comparing the RMs observed for this emission to those observed for pulsars in this part of the sky, we can conclude that the distance to this emission is in the range 1.3–4.5 kpc. The depolarizing effects of RCW 94 (discussed further below) imply that the polarized emission is >3 kpc distant, while the lack of depolarization against other H II regions gives an upper limit of 6.5 kpc. Dickey [11] has made H I absorption measurements against this emission to derive a lower limit on its distance of 2 kpc. Taking into account all these constraints, we argue that the mean distance to the source of polarized emission is 3.5 ± 1.0 kpc, corresponding to the Crux spiral arm of our Galaxy.

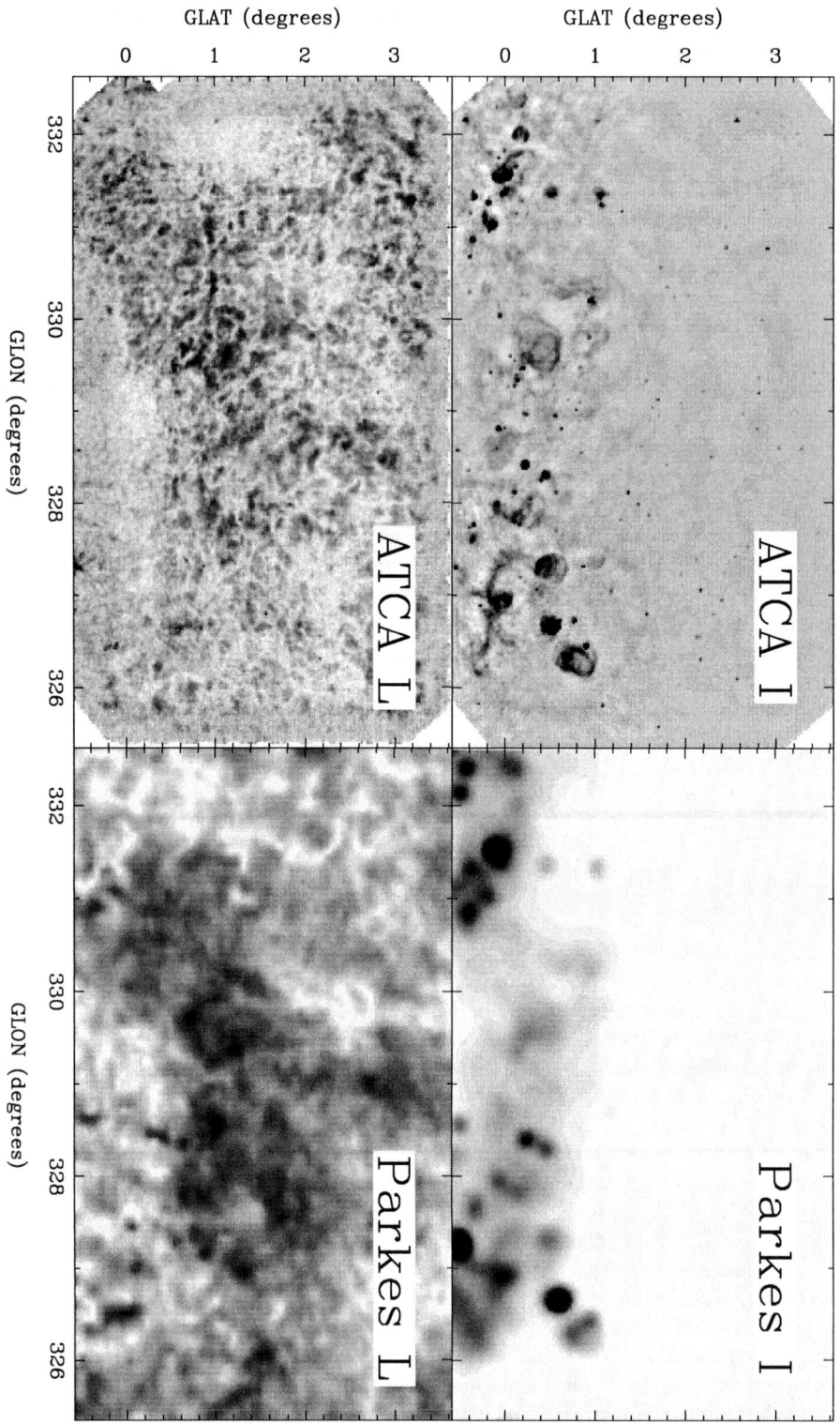

FIGURE 1. Images of the SGPS test region with the ATCA (1.4 GHz, 1 arcmin resolution) and Parkes (2.4 GHz, 10.4 arcmin), in both total and linearly polarized intensity.

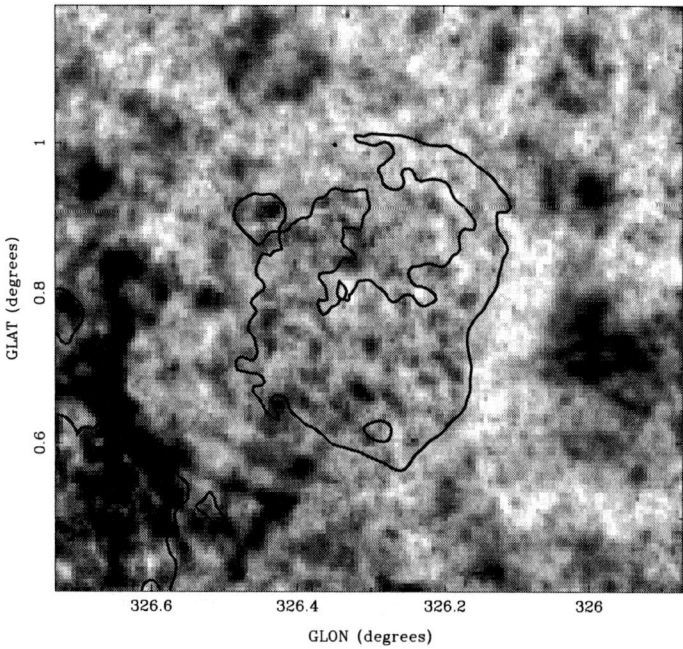

GLAT (degrees)

GLON (degrees)

FIGURE 2. Polarized emission towards the H II region RCW 94. The greyscale represents linearly polarized intensity, while the single contour corresponds to total intensity emission from the same region at the level of 45 mJy beam^{-1}.

The rest of the ATCA field is filled with fainter diffuse polarization, which does not have any counterpart in the Parkes data. This emission is best explained as being due to Faraday rotation in foreground material. The RMs measured for this emission imply that they are caused by foreground clouds of RM ~ 5 rad m^{-2}, consistent with the conclusions made by Wieringa et al [1].

Voids in Polarization

To the best of our knowledge, voids in polarization such as those described here have not been previously reported. There are two possible explanations to account for these structures: either they represent regions where the level of intrinsic polarization is low, or they are the result of propagation through a foreground object, whose properties have depolarized the emission at both 1.4 and 2.4 GHz.

If the voids are intrinsic to the emitting regions, then the distance of 3.5 kpc inferred above implies that they are hundreds of parsecs across — it is hard to see what could produce such uniformly low polarized intensity across such large regions. We thus think it unlikely that the voids are intrinsic to the emitting regions.

We thus favor the possibility that the voids are caused by depolarizing effects in foreground material. We have considered in detail the various ways in which foreground Faraday rotation can produce the observed structure, and can rule out bandwidth and gradient depolarization as possible mechanisms (see [7] for details).

The only remaining possibility is that depolarization in the voids is due to beam depolarization, in which the RM varies randomly on small scales. We have developed a detailed model for "void 1" to confirm this. We consider void 1 to be caused by a sphere of uniform electron density n_e cm^{-3}, centered on (332°5, +1°2) with a radius of 1°4 and at a distance to us of d kpc. Within the sphere, we suppose that there are random and ordered components to the magnetic field, and that these two components have identical amplitudes B μG. The ordered component is uniformly oriented at an angle θ to the line of sight. We assume that the random component is coherent within individual cells of size l pc, but that the orientation from cell to cell is random. Uniformly polarized rays which propagate through a different series of cells will experience differing levels of Faraday rotation, resulting in beam depolarization when averaged over many different paths.

By calculating the properties of the polarized signal which emerges after propagating through this source, we find that we can account for the observed properties of void 1 if $n_e \sim 20$ cm^{-3}, $B \sim 5$ μG, $\theta \gtrsim 80°$, $d \sim 300$ pc and

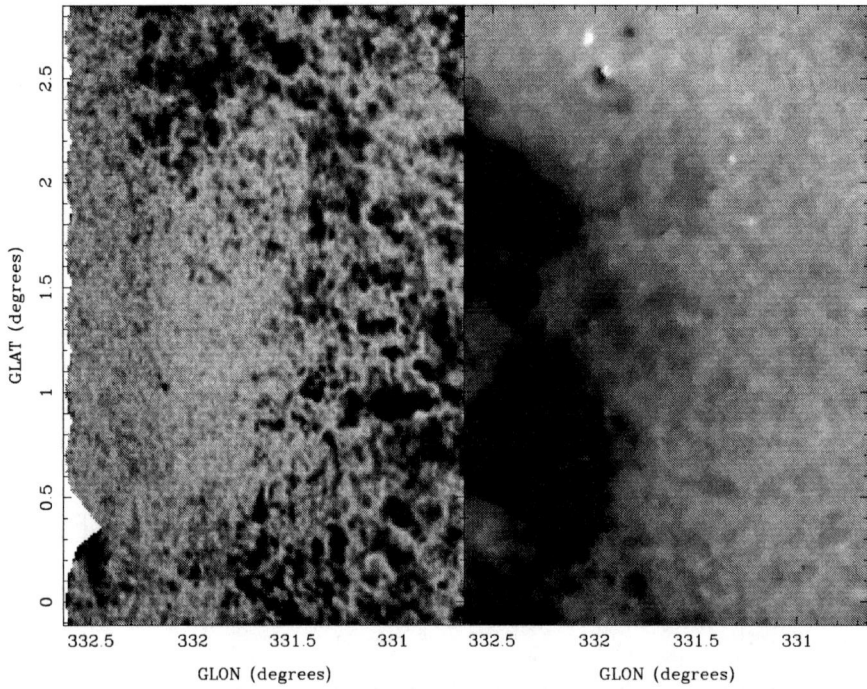

FIGURE 3. Comparison of linear polarization (left, [7]) and Hα emission (right, [12]) towards void 1.

$l \sim 0.2$ pc (see [7] for details). These properties are consistent with those of an H II region of comparatively low emission measure. Indeed Figure 3 demonstrates that Hα emission fills void 1, its morphology and perimeter matching exactly to that of the void. It is interesting to note that the O9V star HD 144695 is very close to the projected center of void 1, and is at a distance of 300 ± 160 pc. The radius of the Strömgren sphere which this star would produce is consistent with the extent of the void. It is thus reasonable to propose that the star is embedded in and powers the surrounding ionized bubble.

Two properties of the voids which our simple model cannot account for are the requirement that the uniform component of the magnetic field be largely oriented in the plane of the sky, but that we generally observe coherent regions of large RM (of the order of a few hundred rad m^{-2}) around the edges of the voids. We suggest that both these results can be explained by the field geometry which arises during the expansion phase of an H II region as it interacts with surrounding material. This produces a magnetic field perpendicular to the line of sight over most of the void, but which is parallel to the line of sight (and can thus potentially produce high RMs) around the perimeter.

Depolarization seen towards RCW 94

The reduced polarization seen coincident with RCW 94 in Figure 2 presumably results from beam depolarization, just as for the H II region argued to produce void 1. However, the effects of beam depolarization are expected to be weakest around the edges of the source, and thus cannot account for the halo of complete depolarization surrounding RCW 94. We rather account for this depolarization halo by requiring the electron density to be approximately constant across RCW 94, but to fall off rapidly beyond the boundaries of the source. This produces a sharp gradient in RM around the edges of the source, resulting in complete depolarization.

The presence of significant CO emission at the same position and systemic velocity as for RCW 94 [13] suggests that the H II region is interacting with a molecular cloud. This possibility is supported by H I observations of the region, which show that RCW 94 is embedded in a shell of H I emission, which is further surrounded by a ring of decreased H I emission [14]. McClure-Griffiths et al [6, 14] argue that this structure in H I confirms that RCW 94 is embedded in a molecular cloud, the shell of emission resulting from H_2 molecules dissociated by the H II region, and the surrounding region of reduced H I corresponding to regions of undisturbed molecular material. Simulations of H II regions evolving within molecular clouds ([15] and references therein) show that for certain forms of the density profile within the parent cloud, the shock driven into the cloud by the embedded expanding H II region can produce a halo of partially ionized material around the latter's perimeter, which would produce the fall-off in n_e required to produce the depolarization halo observed.

Point Sources

With the exception of one source known to be a pulsar, the polarized point sources in our field are presumably extragalactic, and their RMs thus probe the entire line-of-sight through the Galaxy. When combined with information from pulsar RMs, we can use these data to constrain the geometry of the overall Galactic magnetic field. So far we have compared the RMs in our test region to those expected for a bisymmetric spiral configuration, and have found that pitch angles in the lower end of the range allowed by pulsars ($p \sim -4.5°$) are favored [16]. We are in the process of carrying out a more detailed study using the RMs of 163 background sources from the entire SGPS, in which we are comparing these measurements to the distributions expected for a wider variety of geometries and model parameters (Bizunok et al, in preparation).

CONCLUSIONS

The ATCA's sensitivity, spatial resolution and spectral flexibility have allowed us to study linear polarization and Faraday rotation from the inner Galaxy in an unprecedented detail. Even though the test region we have considered covers less than 7% of the full survey, we have been able to identify a variety of distinct polarimetric phenomena, and have used these to map out both global and turbulent structures in the magneto-ionized ISM. We anticipate that our analysis of the full SGPS will result in a comprehensive study of magnetic fields and turbulence in the inner Galaxy.

ACKNOWLEDGMENTS

The Australia Telescope is funded by the Commonwealth of Australia for operation as a National Facility managed by CSIRO. Hα data were taken from the Southern H-Alpha Sky Survey Atlas (SHASSA), which is supported by the National Science Foundation. B.M.G. is supported by a Clay Fellowship awarded by the Harvard-Smithsonian Center for Astrophysics, while J.M.D. acknowledges the support of NSF grant AST-9732695 to the University of Minnesota.

REFERENCES

1. Wieringa, M. H., de Bruyn, A. G., Jansen, D., Brouw, W. N., and Katgert, P., *A&A*, **268**, 215–229 (1993).
2. Gray, A. D., Landecker, T. L., Dewdney, P. E., Taylor, A. R., Willis, A. G., and Normandeau, M., *ApJ*, pp. 221–231 (1999).
3. Haverkorn, M., Katgert, P., and de Bruyn, A. G., *A&A*, **356**, L13–L15 (2000).
4. Duncan, A. R., Haynes, R. F., Jones, K. L., and Stewart, R. T., *MNRAS*, **291**, 279–295 (1997).
5. Duncan, A. R., Reich, P., Reich, W., and Fürst, E., *A&A*, **350**, 447–456 (1999).
6. McClure-Griffiths, N. M., Green, A. J., Dickey, J. M., Gaensler, B. M., Green, A. J., Haynes, R. F., and Wieringa, M. H., *ApJ*, **551**, 394–412 (2001).
7. Gaensler, B. M., Dickey, J. M., McClure-Griffiths, N. M., Green, A. J., Wieringa, M. H., and Haynes, R. F., *ApJ*, **549**, 959–978 (2001).
8. Sault, R. J., Hamaker, J. P., and Bregman, J. D., *A&AS*, **117**, 149–159 (1996).
9. Sault, R. J., Bock, D. C.-J., and Duncan, A. R., *A&AS*, **139**, 387–392 (1999).
10. Gray, A. D., Landecker, T. L., Dewdney, P. E., and Taylor, A. R., *Nature*, **393**, 660–662 (1998).
11. Dickey, J. M., *ApJ*, **488**, 258–262 (1997).
12. Gaustad, J. E., McCullough, P. R., Rosing, W., and Van Buren, D., *Publ. Astr. Soc. Pacific* (2001), in press (astro-ph/0108518).
13. Bronfman, L., Alvarez, H., Cohen, R. S., and Thaddeus, P., *ApJS*, **71**, 481–548 (1989).
14. McClure-Griffiths, N. M., Dickey, J. M., Gaensler, B. M., Green, A. J., Haynes, R. F., and Wieringa, M. H., *Pub. Astr. Soc. Aust.*, **18**, 84–90 (2001).
15. Rodríguez, J. A., Tenorio-Tagle, G., and Franco, J., *ApJ*, **451**, 210–217 (1995).
16. Dickey, J. M., Onken, C. A., McClure-Griffiths, N. M., Gaensler, B. M., Green, A. J., Haynes, R. F., and Wieringa, M. H., "A straw-man model of the Galactic magnetic field", in *Radio Polarization: A New Probe of the Galaxy*, edited by T. L. Landecker, DRAO, Penticton, 2001, pp. 43–54.

Low Frequency Polarization Observations of the Galactic Radio Emission

E.N. Vinyajkin, V.A.Razin

Radiophysical Research Institute (NIRFI), 25 B.Pecherskaya st., 603950, Nizhnij Novgorod, Russia,

Abstract. Multifrequency polarization observations of a number of regions of the Galaxy in the frequency range 200-1250 MHz have been carried out. The area of enhanced linear polarization at meter wavelengths around $\alpha_{1950}=$ 4^h30^m, $\delta_{1950} = 61°$ ($l=146°47'$, $b= 9°03'$) reveals a nonmonotonic, oscillating spectrum of the polarized brightness temperature. On the contrary, the region of the North Celestial Pole ($l=123°$, $b= 27°24'$) has practically monotonic spectrum. The possible reasons for this difference are discussed.

1. INTRODUCTION

Multifrequency polarization observations of the diffuse Galactic radio emission in selected directions of the sky are among basic polarization observations carried out at the Radio Astronomy Observatory "Staraya Pustyn' " of the Radiophysical Research Institute (NIRFI) near Nizhnij Novgorod. Such "frequency surveys" in selected directions, in addition to surveys of extended regions of the sky at particular frequencies, are an effective way to obtain information about the Galactic magnetic field, ionized gas and relativistic electrons in the interstellar medium.

The linear polarization of the diffuse Galactic radio emission was firstly measured [1,2] at wavelengths 1.45 and 3.3 meters. After that many polarization observations of diffuse Galactic radio emission were made in Staraya Pustyn' at a number of frequencies in the meter and decimeter wavebands [3-13]. These observations permitted to get spectra of polarized brightness temperature and position angles for several directions of the sky. The coordinates of these directions are: $\alpha_{1950}= 4^h30^m$, $\delta_{1950} = 61°$ ($l=146°47'$, $b= 9°03'$, the region of enhanced linear polarization at meter wavelengths), $l=123°$, $b= 27°24'$ (the North Celestial Pole), $\alpha_{1950}= 12^h49^m$, $\delta_{1950} = 27°24'$ ($b= 90°$, the North Galactic Pole), $\alpha_{1950}= 3^h48^m$, $\delta_{1950} = 64°$ ($l=141° 09'$, $b= 7°53'$, a well known calibration point) and $\alpha_{1950}= 14^h28^m$, $\delta_{1950} = 14°$ ($l=7° 51'$, $b= 63°20'$, a calibration point in the North Polar Spur). More extended areas in the Loop III and the Loop I were also observed [4, 9, 11]. The linearly polarized component of the Galactic diffuse radio emission at frequencies below ~300 MHz comes from distances not farther than a few hundred parsecs because the Faraday depolarization cancels out the emission from more distant regions of the Galaxy. Thus, it is possible to get information about the local interstellar medium and magnetic field from the low frequency polarization observations. At low frequencies there is a strong influence of the Faraday rotation and the Faraday depolarization in the interstellar medium. The Faraday rotation in the ionosphere is also essential and must be taken into account when measuring the true value of the position angle of the polarization plane.

Let us consider briefly the Faraday depolarization in the interstellar medium. The degree of linear polarization P of a uniform layer synchrotron radio emission for an infinitely narrow band and beam is [1,2]

$$P = \frac{\gamma+1}{\gamma+7/3}\left|\frac{\sin\psi\lambda^2}{\psi\lambda^2}\right|,$$

(1)

where ψ(rad m^{-2})$=0.81B_l$ (μG)N_e(cm^{-3})L(pc) is the Faraday depth of the layer, B_l (μG) is a line-of-sight component of the magnetic field, N_e(cm^{-3}) is the thermal electron density, L(pc) is the layer linear dimension along the line of

CP609, *Astrophysical Polarized Backgrounds*, edited by S. Cecchini et al.

sight, λ(m) is a wavelength, and γ is the exponent of the energy spectrum $N(E)$ of the relativistic electrons ($N(E)\propto E^{-\gamma}$). Taking into account the Faraday depolarization within the received frequency band it follows for rectangular passband Δv [14]

$$P = \frac{\gamma+1}{\gamma+7/3} \frac{1}{2|\psi|\lambda^2} \left[\left(\frac{\sin 2\psi\lambda^2 \frac{\Delta v}{v}}{2\psi\lambda^2 \frac{\Delta v}{v}} \right)^2 - 2\frac{\sin 2\psi\lambda^2 \frac{\Delta v}{v}}{2\psi\lambda^2 \frac{\Delta v}{v}} \cos 2\psi\lambda^2 + 1 \right]^{1/2}, \quad (2)$$

where v is the central frequency of the passband and ($\Delta v/v$)<<1. If ($\Delta v/v$)$\rightarrow$0 formula (2) transforms into (1). The brightness temperature of the linearly polarized component of the synchrotron radio emission T_b^p as a function of the frequency is

$$T_b^p \propto B_\perp^{\frac{\gamma+1}{2}} v^{-\frac{\gamma+3}{2}} P, \quad (3)$$

where $B_\perp$ is the magnetic field component perpendicular to the line of sight. From (3) it is obvious that the nonmonotonic frequency and angular variations of T_b^p are possible for the uniform layer model. In fact the above mentioned region of the Galaxy with coordinates l=146°47′, b= 9°03′ has a nonmonotonic frequency spectrum of the linearly polarized radio emission.

In section 2 we describe the observations. Results and discussion are given in section 3. The conclusion is presented in section 4.

2. OBSERVATIONS

Polarization observations were made using fully steerable alt-azimuth radio telescopes with parabolic reflectors 8, 10, and 14 m in diameter and modulation radio polarimeters. A 14-m radiotelescope was used in the frequency band of 200-448 MHz (HPBW is 7°.5-3°.5), a 10-m radiotelescope was used at 910 MHz (HPBW=2°.5) and a 8-m radiotelescope was used at 1250 MHz (HPBW=2°15′). A coaxial feed with switched orthogonal linear vibrators was mounted in the primary focus of the reflector on a cylindrical tube fixed in the paraboloid vertex without rods. The polarimeters were based on the power-switching principle. A diode switch was mounted with the linear mutually perpendicular vibrators and was driven by a square-wave oscillator at 1 kHz to connect alternately each vibrator to the receiver via a coaxial line and rotatable coaxial connector. During the polarization measurements of the diffuse Galactic radio emission, the feed was continuously rotated by a motor at a rate of one revolution every 5 min. An ideal record of the linearly polarized constant signal in the absence of noise would be a T/2 period (T=5 min) sinusoid. In practice an intensity and position angle of the linearly polarized signal are not constant and there is noise. For each 360° rotation of a feed the Stokes parameters corresponding to the apparent polarization vector were derived from the eight numbers recorded. These numbers correspond to the values 0°, 45°, 90°, 135°, 180°, 225°, 270° and 315° of a feed position angle.

Multifrequency polarization observations in the selected directions were made by the tracking method. Observations of points of the sky with declination larger than the latitude of the observatory (55°39′) consist of a tracking around culmination during an hour angle interval between the point's East and West elongations. Measured values of the Stokes parameters Q_Σ, U_Σ of an incoming radio emission are the sums of the true signals of the Galactic linearly polarized radio emission Q_g and U_g and spurious polarization Q_s, U_s. The main source of spurious polarization is the linearly polarized component of the ground radio emission received by the side lobes of the radio telescope. The contribution from the total ground radio emission converted to the linear polarization via instrumental polarization is much lower than the contribution from spurious polarization. If the value of spurious polarization is nearly constant during the time of a tracking the measured values of the Stokes parameters Q_Σ, U_Σ are approximately lying on a circle in the Q_Σ, U_Σ –plane. The circle radius is the measure of the Galactic linearly polarized radio emission brightness temperature, and the coordinates of the circle center are the Stokes parameters of the spurious signal. Fig. 1 shows an example of such circle for the observation of the point with coordinates α_{1950}= 4^{h}30^m, δ_{1950} = 61° (l=146° 47′, b= 9°03′) at 290 MHz by the 14-m radio telescope (HPBW =5°20′ × 5°15′).

Cassiopeia A was used for calibration. The rotation of the feed and the switching of orthogonal polarizations were stopped and for each of the linear vibrators the calibrations were separately made using this source. Since a relation between the output using one linear vibrator for polarization observation and that using switched orthogonal linear vibrators is known we can get the true scale of polarization intensities. The secular decrease of Cassiopeia A radio emission (e.g. [15,16]) was taken into account. The main beam brightness temperature of the linearly polarized component of the Galactic radio emission is calculated from the relation

$$T_b{}^p = \frac{n_p}{n_{CasA}} \frac{\lambda^2}{8\pi k} D_m S_{\lambda CasA},$$ (4)

where n_p/n_{CasA} is the ratio between the response n_p to the linearly polarized emission and the response n_{CasA} to the radio emission from Cassiopeia A, k is the Boltzmann constant, D_m is the antenna's main beam directivity, and $S_{\lambda CasA}$ is Cas A flux density at wavelength λ. We use here a definition of $T_b{}^p$ according to the following relation between polarization intensity I_p and $T_b{}^P$: $I_p = k T_b{}^P / \lambda^2$. The accuracy of the calibration is 5 –10%.

To check the correctness of the polarization measurements the simultaneous observations of the same point at two or three different frequencies are very useful. Fig.2 shows the correlation diagram between the observed values of equatorial position angles, including variable contribution of the ionosphere, at 408 MHz (10-m radiotelescope) and 290 MHz (14-m radiotelescope) for observations of the North Galactic Pole ($b=90°$). There is a good correlation (coefficient of correlation $\rho = 0.94 \pm 0.03$) and the slope of the line corresponds to the λ^2 dependence of the Faraday rotation.

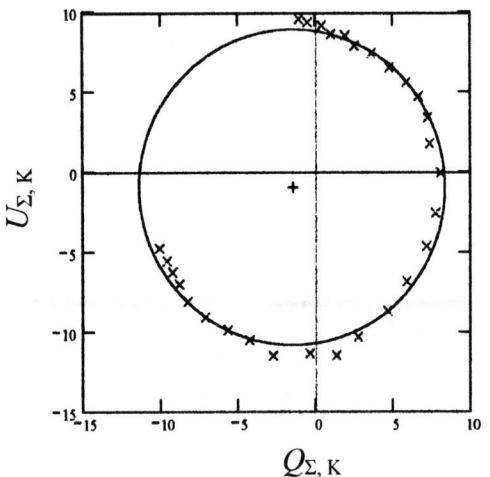

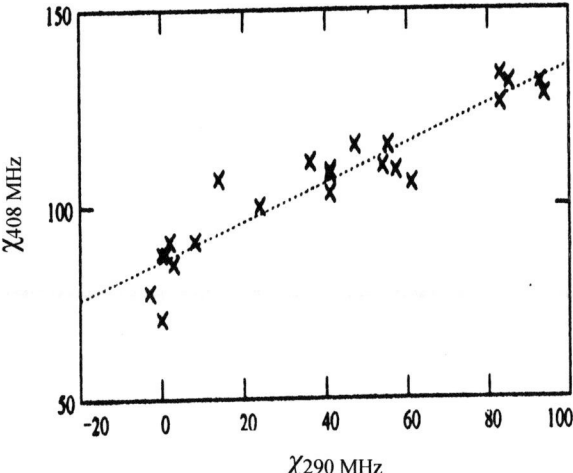

FIGURE 1. This is an example of observations of the region with coordinates $l=146°\,47'$, $b= 9°03'$ at 290 MHz by the 14-m radiotelescope. Each cross "×" corresponds to the feed rotation period T=5m. The best fit circle and its center "+" are also shown.

FIGURE 2. Correlation between observed values of equatorial position angle at 290 MHz by the 14-m radiotele-scope (x-axis) and at 408 MHz by the 10-m radiotelescope (y-axis) for observations of the North Galactic Pole. The cor-relation coefficient is 0.94±0.03.

3. RESULTS AND DISCUSSION

Observations of the area around $l=146°47'$, $b= 9°03'$ at 25 frequencies in the band 195-1250 MHz have revealed its nonmonotonous polarized brightness temperature spectrum [3-10,13]. Fig. 3 shows the values of $T_b{}^p$ as a function of the frequency. There is a maximum near 300 MHz and a deep minimum at 219 MHz. The most simple interpretation of these data involves a homogeneous model that takes into account the Faraday depolarization in the band. Formula (3) with P taken from (2) was used to fit the observational data. The relative value y of the bandwidth was $y= \Delta v/v =0.035$. As a result the following values of the parameters were obtained: the absolute value of the Faraday depth $|\psi|$=1.68 rad m^{-2}, the spectral index of the total emission brightness temperature $\beta =(\gamma+3)/2=2.7$, the amplitude factor $A=T_b{}^P(300 \text{ MHz})\cdot(P_0/P(300 \text{ MHz})) =12.6$ K, where P was taken from (2) and $P_0=(\gamma+1)/(\gamma+7/3)$. The best fit curve (5) is shown in Fig.3. This simple formula describes the observed spectrum only very approxima-

28

$$T_b^P = \frac{A}{2|\psi|}\left(\frac{\nu}{300}\right)^{2-\beta}\left\{\left[\frac{\sin 2\psi\left(\frac{300}{\nu}\right)^2 y}{2\psi\left(\frac{300}{\nu}\right)^2 y}\right]^2 - 2\frac{\sin\left[2\psi\left(\frac{300}{\nu}\right)^2 y\right]}{2\psi\left(\frac{300}{\nu}\right)^2 y}\cos\left[2\psi\left(\frac{300}{\nu}\right)^2\right] + 1\right\}^{1/2}, \quad (5)$$

tely but reflects its main features. It seems the magnetic field is rather regular (quasiuniform) in this region of the Galaxy. The observed value of rotation measure RM for a uniform region is half of the Faraday depth ψ, so $|RM|=0.84$ rad m^{-2}. This value was obtained from the polarized brightness temperature spectrum without using information on position angles. Using the results of our measurements of the position angles in the range 375-910 MHz we obtained $RM=0.6\pm0.15$ rad m^{-2} in a good agreement with the model value (see also [17, 18]). The model predicts the second minimum of T_b^p at 155 MHz which we plan to observe. The spectrum of T_b^p in the direction of $l=146°47'$, $b=9°03'$ is an example of oscillating spectrum. This type of spectrum is probably due to a relatively thin layer with an embedded quasiuniform magnetic field.

Another type of spectrum is revealed by the observations of the North Celestial Pole (NCP) [12]. Fig. 4 shows the measured values of T_b^p at 13 frequencies in the range of 200-1407 MHz. All measurements except that at 1407 MHz (HPBW=2°) [19] were made at Staraya Pustyn'. Values of the NCP polarized brightness temperatures are substantially lower than those in the direction of $l=146°47'$, $b=9°03'$. The spectrum of the NCP looks like an almost monotonic decrease of T_b^p with the frequency. In the frequency range 200-1407 MHz, the spectrum of T_b^p can be formally approximated as

$$T_b^P(K) = (1.95\pm0.05)\left(\frac{\nu(MHz)}{300}\right)^{-1.87\pm0.05}. \quad (6)$$

The value of the spectral index of the NCP polarized brightness temperature we obtain is $\beta_p=1.87\pm0.05$, close to that obtained in [20] (2.06±0.1) for the band of 240 -1415 MHz with HPBW <2° at all frequencies. The analysis showed that the model of the emitting region, which consists of three different homogeneous layers, explains quite satisfactorily this spectrum. It was assumed that $\beta=(\gamma+3)/2=2.6$ [21] and $\Delta\nu/\nu =0.01$. The optimum values of the parameters are as follows: relative contributions to the total intensity $r_1=0.3$, $r_2=0.1$, $r_3=0.6$; Faraday depths $\psi_1=0.003$ rad m^{-2}, $\psi_2=1.51$ rad m^{-2}, $\psi_3=-5.41$ rad m^{-2}; the internal values of the position angles of all three layers are identical; amplitude factor $T_b^P(300\text{ MHz})$ $(P_0/P(300\text{ MHz})) =1.7$ K, where P is the degree of polarization for the three different homogeneous layer model ($\chi^2 =7.4$). It was assumed that the first layer is the outermost. The approximating curve is shown in Figure 4. This curve is weakly oscillating because of the averaging of the contribu-

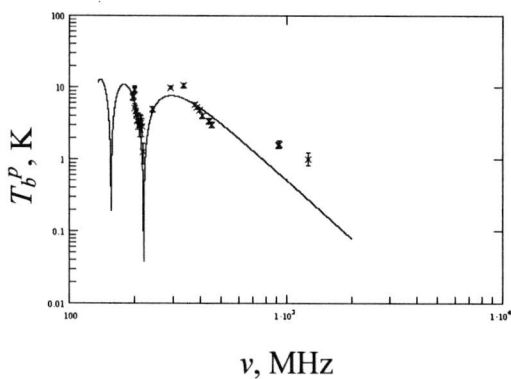

ν, MHz

FIGURE 3. The spectrum of T_b^p (K) and its approximation by (5) for the region with $l=146°47'$, $b=9°03'$.

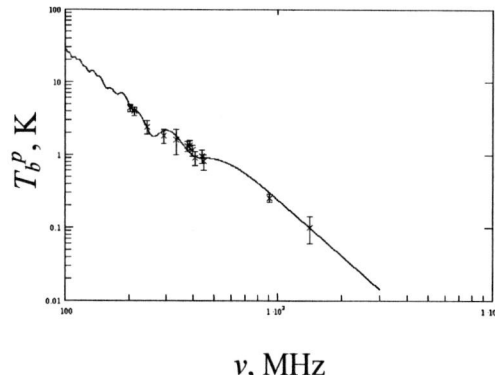

ν, MHz

FIGURE 4. The spectrum of T_b^p (K) and its approximation for the North Celestial Pole.

tions of the separate layers.

Thus, the inhomogeneous and unequal distributions of the emissivity and rotation abilities of the interstellar medium explain the observed almost monotonic spectrum in the direction to the North Celestial Pole. By using the measurements of position angles in the frequency range 408-910 MHz the value of RM= -0.9 rad m^{-2} for the rotation measure was obtained.

Since the values of the rotation measures for both regions are low, a difference in the value of HPBW at different frequencies does not affect substantially the form of their spectra.

Let us estimate now the effective extent along the line of sight, L_{pol}, of the region with coordinates l=146°47′, b= 9°03′, that generates the linearly polarized radio emission we observe at the frequency of 290 MHz. This can be estimated as $L_{pol} \sim T_b^p / 2P\varepsilon_T$, where ε_T is the emissivity. For estimating ε_T we use the results of [22]. Radio emission at 22 MHz generated behind the HII region with coordinates l≈135°, b≈0° placed at the distance 2.2±0.3 kpc is absorbed by it. This gives the possibility to estimate the emissivity at 22 MHz. Using the value β=2.7 for the brightness temperature spectral index we obtain ε_T(290 MHz)=42 K/kpc. The degree of polarization at 290 MHz is equal to 0.42 according to the model, and the polarized brightness temperature T_b^p =(9.7±0.5) K at 290 MHz according to our observations. As a result we obtain that L_{pol} is equal to 275 pc. Knowing the Faraday depth of the region and its extent it is possible to estimate the value of $B_l N_e$. As a result we obtain, that $B_l N_e$ is equal to 7×10^{-3} µG cm^{-3}. The strength of 2.2 µG [23] for the uniform component of the magnetic field in the vicinity of the Solar System leads to the value of the thermal electron density N_e ≈3×10^{-3} cm^{-3}. Thus the region of enhanced linear polarization is characterized by rather a low content of thermal electrons.

4. CONCLUSION

Multifrequency polarization measurements of several regions of the Galaxy were carried out in the frequency range 200-1250 MHz. Two types of spectra of polarized brightness temperature have been revealed. The first type this is a nonmonotonic oscillating spectrum. This spectrum has a region of enhanced polarization with coordinates l=146°47′, 9°03′. Nonmonotonicity of the spectrum testifies that the magnetic field in the region is almost uniform. The region of the North Celestial Pole has practically monotonic spectrum of the polarized brightness temperatures in the same frequency band. This spectrum can be explained by the inhomogeneous and different spatial distributions of the emissivity and the rotation ability of the interstellar medium. Multifrequency polarization measurements permit in principle to recover the distributions of the emissivity and the rotation ability along the line of sight, if additional information on the distance is given.

The linearly polarized component of the Galactic diffuse radio emission at frequencies below 300 MHz comes from distances smaller than a few hundreds of parsecs because of the Faraday depolarization of more distant regions. It is thus possible to get information about the local interstellar medium from the low frequency radio polarization observations.

ACKNOWLEDGMENTS

This work was supported by the Russian Foundation for Basic Research (project no. 00-02-17648), the Federal Scientific Technical Program Astronomy (project no. 1.3.7.2) and the Ministry of Industry and Science (setup no. 06-29). We are thankful to dr. A.M. Paseka and A.I.Teplykh for helpful discussion.

REFERENCES

1. Razin, V.A., Radiotekhnika i Electronica 1, 846-851 (1956), in Russian.
2. Razin,V.A., Soviet Astronomy 2, 216 (1958) (translated from Russian, Astronom. Zh. 35, 241-252 (1958)).
3. Mel'nikov, A.A., Razin, V.A., Khrulev, V.V., Izv. vuzov, Radiofizika 10, 1760-1762 (1967), in Russian.
4. Razin, V.A., Khrulev, V.V., Fedorov, V.T., Volokhov, S.A., Mel'nikov, A.A., Paseka, A.M., Pupysheva, L.V., Radiophysics and Quantum Ellectronics 11, 824-829 (1968) (translated from Russian, Izv. vuzov, Radiofizika, 11, 1461-1472 (1968)).
5. Kapustin, P.A., Petrovskij, A.A., Pupysheva, L.V., Razin, V.A., Izv. vuzov, Radiofizika 16, 1325-1333 (1973), in Russian.
6. Paseka, A.M., Popova, L.V., Razin, V.A., Arkhangel'skij, V.G., Samokhvalov, Yu.E., Izv. vuzov, Radiofizika 18, 926-929 (1975), in Russian.
7. Kuznetsova, I.P., Mel'nikov, A.A., Razin, V.A., Izv. vuzov, Radiofizika 18, 1548-1549 (1975), in Russian.
8. Paseka, A.M., Popova, L.V., Razin, V.A., Soviet Astronomy 20, 162 (1976) (translated from Russian, Astronom. Zh. 53, 286-287, (1976)).
9. Paseka, A.M., Soviet Astronomy 22, 664-666 (1978) (translated from Russian, Astronom. Zh. 55, 1163-1168 (1978)).

10. Paseka, A.M., Astronomy Reports 37, 135-138 (1993) (translated from Russian, Astronom. Zh. 70, 258-264 (1993)).
11. Vinyajkin, E.N., Astronomy Reports 39, 599-610 (1995) (translated from Russian, Astronom. Zh. 72, 674-686 (1995)).
12. Vinyaikin, E.N., Kuznetsova, I.P., Paseka, A.M., Razin, V.A., and Teplykh, A.I., Astronomy Letters 22, 582-588 (1996) (translated from Russian, Pis'ma v Astronom. Zh. 22, 652-659 (1996)).
13. Vinyajkin, E.N., Paseka, A.M., Teplykh, A.I., Izv. vuzov, Radiofizika (in press).
14. Vinyajkin, E.N., Krajnov, I.L., Preprint of Radiophysical Research Inst. (NIRFI), Gorkij, no. 288 (1989).
15. Vinyajkin, E.N., Razin, V.A., Australian J. of Physics 32, 93-94 (1979).
16. Vinyajkin, E.N., Astrophys. and Space Sci. 252, 249-257 (1997).
17. Brouw, W.N., Spoelstra, T.A.Th., Astron. and Astrophys. Suppl. 26, 129-146 (1976).
18. Spoelstra, T.A.Th., Astron. and Astrophys. 135, 238-248 (1984).
19. Bingham, R.G., Mon. Not. R. Astronom. Soc. 134, p.327-345 (1966).
20. Wilkinson, A., and Smith, F.G., Mon. Not. R. Astronom. Soc. 167, 593-611 (1974).
21. Reich, P., Reich, W., Astron. and Astrophys. 196, 211-226 (1988).
22. Roger R.S., Astrophys. J. 155, 831-840 (1969).
23. Heiles, C., In: Roberge, W.R., Whittet, D.C.B. (eds.) Polarimetry of the Interstellar Medium. Conf. Ser. Vol. 97, ASP, San Francisco, 457 (1996).

Polarized Microwave Emission from Dust

A. Lazarian* and S. Prunet[†]

*Department of Astronomy, University of Wisconsin, Madison, WI 53706
[†]Institut d'Astrophysique de Paris, 98bis Bld Arago, 75014 PARIS

Abstract. Polarized emission from dust is an important foreground that can hinder the progress in polarized CMB studies unless carefully accounted for. We discuss potential difficulties associated with the dust foreground, namely, the existence of different grain populations with very different emission/polarization properties and variations of the polarization yield with grain temperature. In this context we appeal for systematic studies of polarized dust emission as the means of dealing with this foreground.

INTRODUCTION

Diffuse Galactic microwave emission carries important information on the fundamental properties of interstellar medium, but it also interferes with the Cosmic Microwave Background (CMB) experiments (see [2, 79]). Polarization of the CMB provides information about the Universe that is not contained in the temperature data. In particular, it offers a unique way to trace specifically the primordial perturbations of tensorial nature (*i.e.* cosmological gravitational waves, see [77, 38]), and allows to break some important degeneracies that remain in the measurement of cosmological parameters with intensity alone ([81, 7, 57, 66]). Therefore, a number of groups around the world (see Table 1 in [78]) work hard to measure the CMB polarization. In view of this work, the issue of determining the degree of Galactic foreground polarization becomes vital.

Among different sources of polarized foregrounds, interstellar dust is probably the most difficult to deal with. We can identify several reason for that. First of all, dust has both a population of tiny grains ([55]), which are frequently called PAH, along with the "classical" power-law distribution of larger grains ([60]). Then the composition of grains changes with their size, which influences both grain temperature and degree of grain alignment. Moreover, both recent experience with microwave emissivity and theoretical studies of expected polarization response ([16]) show that the naive extrapolation of the grain properties from FIR to microwave does not work. If we take into account that the very nature of dust alignment that causes the polarization still remains somewhat mysterious after more than half a century after its observational discovery (see review by [45]), the scope of the problem becomes apparent.

The discovery of the anomalous emission in the range of 10-100 GHz illustrates well the treacherous nature of dust. Until very recently it has been thought that there are three major components of the diffuse Galactic foreground: synchrotron emission, free-free radiation from plasma (thermal bremsstrahlung) and thermal emission from dust. In the microwave range of 10-90 GHz the latter is subdominant, leaving essentially two components. However, it is exactly in this range that an anomalous emission was reported ([40, 41]). In the paper by [11] this emission was nicknamed "Foreground X", which properly reflects its mysterious nature. This component is spatially correlated with 100 μm thermal dust emission, but its intensity is much higher than one can expect by directly extrapolating thermal dust emission spectrum to the microwave range. It is very likely that discoveries of such a nature are expected when the foreground polarimetry is performed.

In this review, we briefly summarize what is known about the grain populations, grain emission and grain alignment. We discuss the origin of the Foreground X and its expected polarization. Earlier reviews of the subject include ([64, 16, 45]).

CP609, *Astrophysical Polarized Backgrounds,* edited by S. Cecchini et al.

OBSERVATIONAL EVIDENCE

Infrared emission: extrapolation to microwave range

Emission spectrum of diffuse interstellar dust was mostly obtained by *InfraRed Astronomy Satellite* (IRAS) and infrared spectrometers on the *COsmic Background Explorer* (COBE) and on the *InfraRed Telescope in Space* (IRTS).

The emission at short wavelength, e.g. $< 50 \, \mu m$, arises from transiently heated very small grains. These grains have so small heat capacity that the absorption of a single 6 eV starlight photon rises their temperature to $T > 200K$. Typically these grains have less than 300 atoms and can be viewed as large molecules rather than dust particles. They are, however, sufficiently numerous to account for $\sim 35\%$ of the total starlight absorption. The contribution of those grains at the microwave frequencies was thought to be negligible.

In terms of CMB studies the most important is the emission from cool classical dust. Far infrared emission in the range from 1 mm (300 GHz) to 100 μm (3000 GHz) is primarily due to dust particles heated by starlight to temperatures around 20 K. Those particles are "classical" grains known from ground-based starlight absorption studies. It is convenient to approximate the far infrared emission with νj_ν peaking at $\lambda_m \sim 130 \, \mu m$ and following the power-law corresponding the absorption cross section $\sim \nu^\beta$, where $\beta \sim 1.7$, and temperature $T_{dust} = \frac{hc}{\lambda_m k(4+\beta)}$ (see discussion in [16]). It was considered natural to extrapolate this fit to frequencies lower than 300 GHz, and no other contribution was expected from large dust particles.

If the extrapolation from infrared to microwave were as simple as it is suggested above, dealing with dust contribution would be trivial. Further research, however, revealed a much more complex picture. Both classical and small grains were shown to be more important microwave emitters than researchers used to assume. For tiny grains a new mechanism of emission was found ([14, 15]) while magnetic properties were shown to be important for microwave emissivity of large grains ([16]). This example should be used to caution against simple minded attempts to extrapolate polarization from infrared to microwave range.

Anomalous microwave emission: unexpected discovery

Until very recently it has been thought that there are three major components of the diffuse Galactic foreground: synchrotron emission, free-free radiation from plasma (thermal bremsstrahlung) and thermal emission from dust. In the microwave range of 10-90 GHz the latter is definitely subdominant, leaving essentially two components. However, it is exactly in this range that an anomalous emission was reported ([40, 41]). In the paper by [11] this emission was nicknamed "Foreground X", which properly reflects its mysterious nature. This component is spatially correlated with 100 μm thermal dust emission, but its intensity is much higher than one can expect by directly extrapolating thermal dust emission spectrum to the microwave range.

Since its discovery the Foreground X has been detected in the data sets from Saskatoon ([8]), OVRO ([56]), the 19 GHz survey ([9]), and Tenerife ([10, 62]). Initially, the anomalous emission was identified as thermal bremsstrahlung from ionized gas correlated with dust ([40]) and presumably produced by photoionized cloud rims ([61]). This idea was subjected to scrutiny in [14] and criticized on energetic grounds. Additional arguments against the free-free hypothesis became available through correlating anomalous emission with ROSAT X-ray C Band ([23]) and Hα with 100 μm emission ([61]). They are summarized in [16]. Recently [11] used Wisconsin H-Alpha Mapper (WHAM) survey data and established that the free-free emission "is about an order of magnitude below Foreground X over the entire range of frequencies and latitudes where is detected". The authors conclude that the Foreground X cannot be explained as free-free emission. Additional evidence supporting this conclusion have come from a study at 5, 8 and 10 GHz by [25].

The spectrum of the Foreground X is not consistent with synchrotron emission, and maps at 408 MHz ([31]) and 1.42 GHz ([71]) do not correlate with the observed 15-100 GHz intensity, so the anomalous emission is evidently not synchrotron radiation from relativistic electrons.

Correlations of the Foreground X with dust induced ([14, 15, 16]) to conjecture that it can be indeed due to dust. It is encouraging that the observational evidence obtained since the theoretical predictions were published has supported the theory.

Polarization from dust: half a century puzzle

Polarization due to interstellar dust alignment was discovered in the middle of the last century ([36, 30]) and was studied initially via starlight extinction and more recently through emission. Correlation of the polarization with the interstellar magnetic field revealed that electric vector of light polarized via starlight extinction tend to be parallel to magnetic field[1]. This corresponds to grains being aligned with their longer axes perpendicular to the local magnetic field. Due to the presence of the stochastic magnetic field, the polarization patterns are pretty involved.

The existing data presents a complex picture. It is generally accepted that the observations indicate that the ability to produce polarized light depends on grain size and grain composition. For instance, a limited UV polarimetry dataset available indicates that graphite grains tend not to be aligned (see [5]), while maximum entropy technique applied to the existing data by [39] show that large $> 6 \times 10^{-6}$ cm grains are responsible for the polarization via extinction.

Moreover, the environment of grains seems to matter a lot ([27, 51]). A study by [1] indicates that grains selectively extinct starlight up to optical depth $A_v < 3$. Recent emission studies ([35, 34]) produced a polarization spectrum for dense clouds that reveal a tight correlation between grain temperature and its ability to emit polarized light. As multicomponent fits invoking grains of different temperature were claimed to provide a better fit for the observed 1 mm-100 μm emission ([24]), this correlation may be very troublesome for the attempts to construct polarization templates.

POLARIZED EMISSION FROM CLASSICAL DUST

Grain alignment: light at the end of the tunnel?

The basic explanation of polarized radiation from dust is straightforward. Aligned dust particles preferentially extinct (i.e. absorb and scatter) the E-component of starlight parallel to their longer axis. Thermal E-component of the emitted radiation, on the contrary, is higher along the longer axis. Thus for aligned grains one must have polarization. What is the cause of alignment?

Grain alignment is an exciting and very rich area of reseach. For example, two new solid state effects have been discovered recently in the process of understanding grain dynamics ([47, 48]). It is known that a number of mechanisms can provide grain alignment (see review by [45] and Table 1 in [51]). Some of them rely on paramagnetic dissipation of rotational energy ([6, 68, 59, 46, 43, 74]) , some appeal to the anisotropic gaseous bombardment when a grain moves supersonically through the ambient gas ([20, 69, 13, 42, 44, 73, 49]). Grains are definitely paramagnetic and sometimes even strongly magnetic. Supersonic grain motions may be due to outflows ([67]), Alfvenic turbulence ([42]) or ambipolar diffusion ([72]).

At present, grain alignment via radiative torques ([18, 19]) looks preferable, although the theory and the understanding of the mechanism are far from being complete. The mechanism appeals to a spin-up of a grain as it differentially scatters left and right polarized photons ([12, 13]). This process acts efficiently if the irregular grain has its size comparable with the photon wavelength. The mechanism can account for the systematic variations of the alignment efficiency with extinction.

However, other mechanisms should also work. For instance, paramagnetic mechanism may preferentially act on small grains ([52], in preparation), while mechanical alignment may act in the regions of outflows ([70]). In general, the variety of astrophysical conditions allows various mechanisms to have their niche.

Note, that in interstellar circumstances grain alignment happens in respect to magnetic field, even if the mechanism of alignment is not of magnetic nature. This is due to the fact that the Larmor precession of grains is so fast compared to the time scales over which either magnetic field changes its direction or the alignment mechanism acts. In general, the alignment may happen both parallel and perpendicular to magnetic field. In most cases, the alignment happens with long grain axes perpendicular to magnetic field, however.

The history of grain alignment research is full of surprises. Initially it looked so ubiquitous that observers were not even interested in the theory of alignment. But then it showed that it may fail within molecular clouds. What will be the next surprise?

---

[1] The polarizations in emission and in extinction are orthogonal if they are produced by the same grains.

Diffuse gas and molecular clouds: different beasts?

Alignment of grains is different in diffuse gas and molecular clouds. [51] showed that in dark clouds without star formation all alignment mechanisms fail. Indeed, grain alignment depends on non-equilibrium processes [2], while interiors of dark clouds are close to thermodynamic equilibrium.

As soon as stars are born within clouds, the conditions in their vicinity become favorable for grain alignment. This explains why far infrared polarimetry detects aligned grains, while near infrared and optical polarimetry fails. The latter point is a subject of controversy. In the recent paper by [63] it is claimed that far infrared polarimetry does not provide us with any new information compared with optical and near-infrared studies. We are worried about this conclusion. Indeed, if anything, radiative torques must be active near the newly born stars and the spectropolarimetric studies of [34] indicate the existence of aligned hot grains. These aligned grains are selectively warmer and should reveal the structure of magnetic field in the star cruddle after the star is born. This information is unlikely to be obtained via short-wave polarimetry. Results by [63] may be relevant, however, to 850 μm polarization observed by [80] from dense pre-stellar cores where radiative torques must be inefficient [3].

We may hope that grain alignment in diffuse clouds is more uniform. Radiation freely penetrates them and therefore the radiative torques must ensure good alignment. This assumption was used in [26] who related the polarization from dust extinction and the polarization from dust emission. Further research in this direction is necessary.

Complications: turbulence, heating etc.

Interstellar medium is very complex and this tells on polarization. As we have discussed earlier, grain alignment traces the direction of the local magnetic field. In the presence of turbulence, this field is very complex. The resulting polarization depends on the telescope resolution at a particular wavelength. A possible way of dealing with this complication is to correct for the field stochasticity. Tensor description of turbulent magnetic field was obtained in [4] and this can be used for the purpose.

Earlier on we mentioned that radiative torques may be responsible for the bulk of grain alignment. As starlight also heats the grains the systematic variations in the alignment efficiency are expected for grains of different temperatures. Moreover, radiative torques depend on grain size and grain composition and so do grain temperatures. These and related issues require a further study and a further work on the radiative torque theory is necessary.

It is unfortunate for the CMB research that we still do not understand many processes related to the polarization arising from classical grains. The good news, however, is that we will have to understand those processes along with the structure of Galactic magnetic field at high latitude if we ever want to understand the CMB polarization well. As the bonus from this research we will get an insight into the operation of Galactic dynamo, high latitude MHD turbulence, turbulent mixing and will make many yet unforeseen discoveries.

POLARIZED EMISSION FROM SPINNING DUST

Can the ultrasmall grains observed via Mid-IR be important at the microwave range? The naive answer to this question is no, as the total mass in those grains is small.

However, [14] appealed to a different mechanism of emission, namely, to the rotational emission [4] that must emerge when a grain with a dipole moment μ rotates with angular velocity ω.

For the model with the most likely set of parameters, [14] obtained a reasonable fit with observations available at that time. It is extremely important that new data points obtained later ([9, 10]) correspond to the already published model. The observed flattening of the spectrum and its turnover around 20 GHz agree well with the spinning dust

---

[2] To avoid confusion we should remind the reader that interstellar grain alignment is very different from the alignment of ferromagnetic particles in the external magnetic field. The latter is the equilibrium process with the align particles corresponding to the lowest energy level of the system. The grain alignment is a *dissipative* process that requires constant driving and vanishes in thermodynamic equilibrium.

[3] Alternatives are discussed in [45].

[4] The very idea of grain rotational emission was first discussed by [21]. More recently, after the discovery of the population of ultrasmall grains, [22] noted that the rotational emission from such grains may be observable, but their treatment assumed Brownian thermal rotation of grains, which is incorrect.

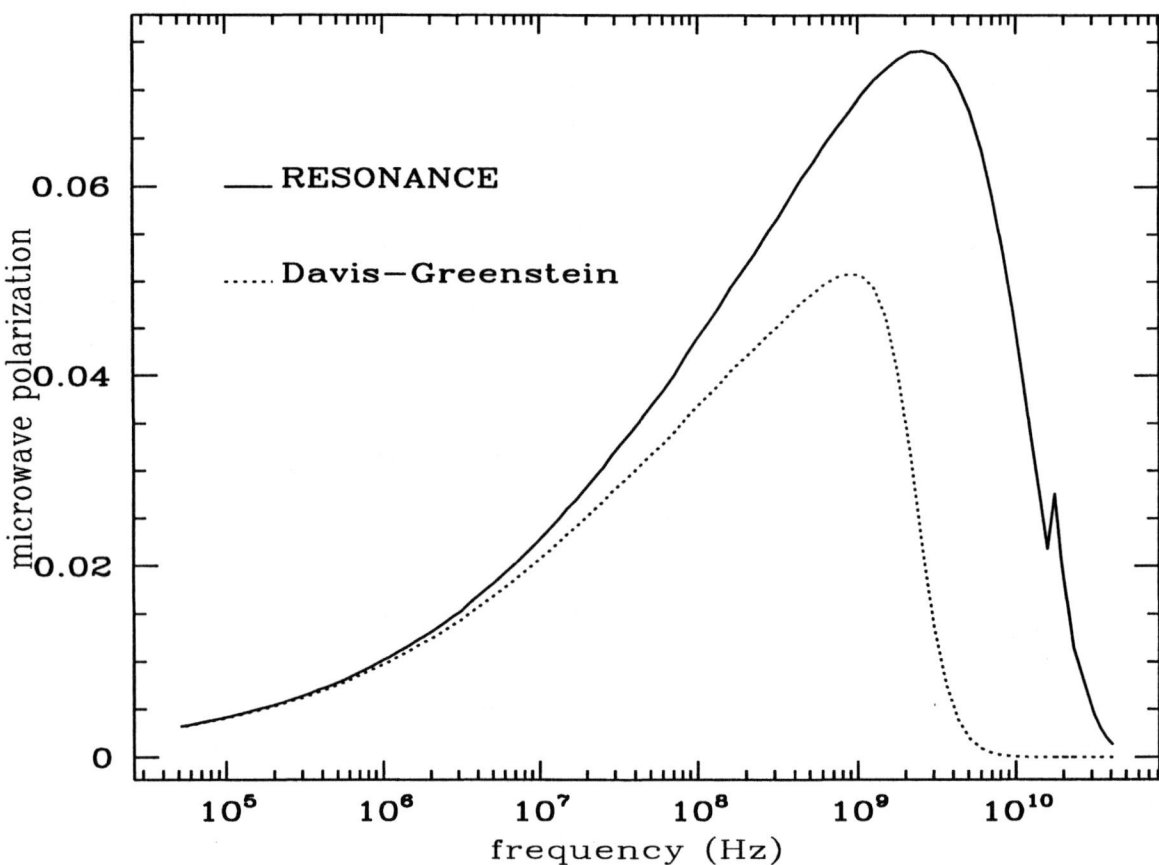

FIGURE 1. Polarization for both resonance relaxation and Davis-Greenstein relaxation for grains in the cold interstellar medium as a function of frequency (from [48]). For resonance relaxation the saturation effects (see eq. (1)) are neglected, which means that the upper curves correspond to the *maximal* values allowed by the paramagnetic mechanism.

predictions.

Microwave emission from spinning grains is expected to be polarized if grains are aligned. Alignment of ultrasmall grains which are essentially large molecules is likely to be different from alignment of large (i.e. $a > 10^{-6}$ cm) grains. One of the mechanisms that might produce alignment of the ultrasmall grains is the paramagnetic dissipation mechanism by [6]. The Davis-Greenstein alignment mechanism is straightforward: for a spinning grain the component of interstellar magnetic field perpendicular to the grain angular velocity varies in grain coordinates, resulting in time-dependent magnetization, associated energy dissipation, and a torque acting on the grain. As a result grains tend to rotate with angular momenta parallel to the interstellar magnetic field.

[48] found that the traditional picture of paramagnetic relaxation is incomplete, since it disregards the so-called "Barnett magnetization" (Landau & Lifshitz 1960). The Barnett effect, the inverse of the Einstein-De Haas effect, consists of the spontaneous magnetization of a paramagnetic body rotating in field-free space. This effect can be understood in terms of the lattice sharing part of its angular momentum with the spin system. Therefore the implicit assumption in [6] that the magnetization within a *rotating grain* in a *static* magnetic field is equivalent to the magnetization within a *stationary grain* in a *rotating* magnetic field – is clearly not exact.

[48] accounted for the "Barnett magnetization" and termed the effect of enhanced relaxation arising from grain magnetization "resonance relaxation". It is clear from Fig. 1 that resonance relaxation persists at the frequencies when the Davis-Greenstein relaxation vanishes. However the polarization is marginal for $\nu > 35$ GHz anyhow. The discontinuity at ~ 20 GHz is due to the assumption that smaller grains are planar, and larger grains are spherical. The microwave emission will be polarized in the plane perpendicular to magnetic field.

Can we check the alignment of ultrasmall grains via infrared polarimetry? The answer to this question is "probably not". Indeed, as discussed earlier, infrared emission from ultrasmall grains, e.g. 12 μm emission, takes place as grains absorb UV photons. These photons raise grain temperature, randomizing grain axes in relation to its angular momentum (see [53]). Taking values for Barnett relaxation from [47], we get the randomization time of the 10^{-7} cm grain to be 2×10^{-6} s, which is less than grain cooling time. As the result, the emanating infrared radiation will be polarized very marginally. If, however, Barnett relaxation is suppressed, the randomization time is determined by inelastic relaxation ([50]) and is ~ 0.1 s, which would entail a partial polarization of infrared emission.

POLARIZED EMISSION FROM MAGNETIC GRAINS

While the spinning grain hypothesis got recognition in the community, the magnetic dipole emission model suggested by [16] was left essentially unnoticed. This is unfortunate, as magnetic dipole emission provides a possible alternative explanation to the Foreground X. Magnetic dipole emission is negligible at optical and infrared frequencies. However, when the frequency of the oscillating magnetic field approaches the precession frequency of electron spin in the field of its neighbors, i.e. 10 GHz, the magneto dipole emissivity becomes substantial.

How likely is that grains are strongly magnetic? Iron is the fifth most abundant element by mass and it is well known that it resides in dust grains (see [75]). If 30% of grain mass is carbonaceous, Fe and Ni contribute approximately 30% of the remaining grain mass. Magnetic inclusions are widely discussed in grain alignment literature ([37, 59, 58, 28]). If a substantial part of this material is ferromagnetic or ferrimagnetic, the magneto-dipole emission can be comparable to that of spinning grains. Indeed, calculations in [16] showed that less than 5% of interstellar Fe in the form of metallic grains or inclusions is necessary to account for the Foreground X at 90 GHz, while magnetite, i.e. Fe_3O_4, can account for a considerable part of the anomalous emissivity over the whole range of frequencies from 10 to 90 GHz. Adjusting the magnetic response of the material, i.e. making it more strongly magnetic than magnetite, but less magnetic than pure metallic Fe, it is possible to get a good fit for the Foreground X ([16]).

How can magneto-dipole emission be distinguished from that from spinning grains? The most straightforward way is to study microwave emission from regions of different density. The population of small grains is depleted in dark clouds ([55]) and this should result in a decrease of contribution from spinning grains. Private communication from Dick Crutcher who attempted such measurements corresponds to this tendency, but the very detection of microwave emissivity is a 3σ result. Obviously the corresponding measurements are highly desirable. As for now, magnetic grains remain a strong candidate process for producing part or even all of Foreground X. In any case, even if magnetic grains provide subdominant contribution, this can be important for particular cases of CMB and interstellar studies. For instance, polarization from magnetic grains may dominate that from spinning grains even if the emission from spinning grains is more of higher level.

The mechanisms of producing polarized magneto-dipole emission is similar to that producing polarization of electro-dipole thermal emission emitted from aligned non-spherical grains (see [32]). There are two significant differences, however. First, strongly magnetic grains can contain just a single magnetic domain. Further magnetization along the axis of this domain is not possible and therefore the magnetic permeability of the grains gets anisotropic: $\mu = 1$ along the domain axis, and $\mu = \mu_\perp$ for a perpendicular direction. Second, even if a grain contains tiny magnetic inclusions and can be characterized by isotropic permeability, polarization that it produces is orthogonal to the electro-dipole radiation emanating through electro-dipole vibrational emission. In case of the electo-dipole emission, the longer grain axis defines the vector of the electric field, while it defines the vector of the magnetic field in case of magneto-dipole emission.

The results of calculations for single domain iron particle (longer axis coincides with the domain axis) and a grain with metallic Fe inclusions are shown in Fig. 2. Grains are approximated by ellipsoids $a_1 < a_2 < a_3$ with $\mathbf{a_1}$ perfectly aligned parallel to the interstellar magnetic field $\mathbf{B}$. The polarization is taken to be positive when the electric vector of emitted radiation is perpendicular to $\mathbf{B}$; the latter is the case for electro-dipole radiation of aligned grains. This is also true (see Fig. 3) for high frequency radiation from single dipole grains. It is easy to see why this happens. For high frequencies $|\mu_\perp - 1|^2 \ll 1$ and grain shape factors are unimportant. The only important thing is that the magnetic fluctuations happen perpendicular to $\mathbf{a_1}$. With $\mathbf{a_1}$ parallel to $\mathbf{B}$, the electric fluctuations tend to be perpendicular to $\mathbf{B}$ which explains the polarization of single domain grain being positive. For lower frequencies magnetic fluctuations tend to happen parallel to the intermediate size axis $\mathbf{a_2}$. As the grain rotates about $\mathbf{a_1} \| \mathbf{B}$, the intensity in a given direction reaches maximum when an observer sees the $\mathbf{a_1a_2}$ grain cross section. Applying earlier arguments it is easy to see that magnetic fluctuations are parallel to $\mathbf{a_2}$ and therefore for sufficiently large a_2/a_1 ratio the polarization is negative. *The variation of the polarization direction with frequency presents the characteristic signature of magneto-dipole emission*

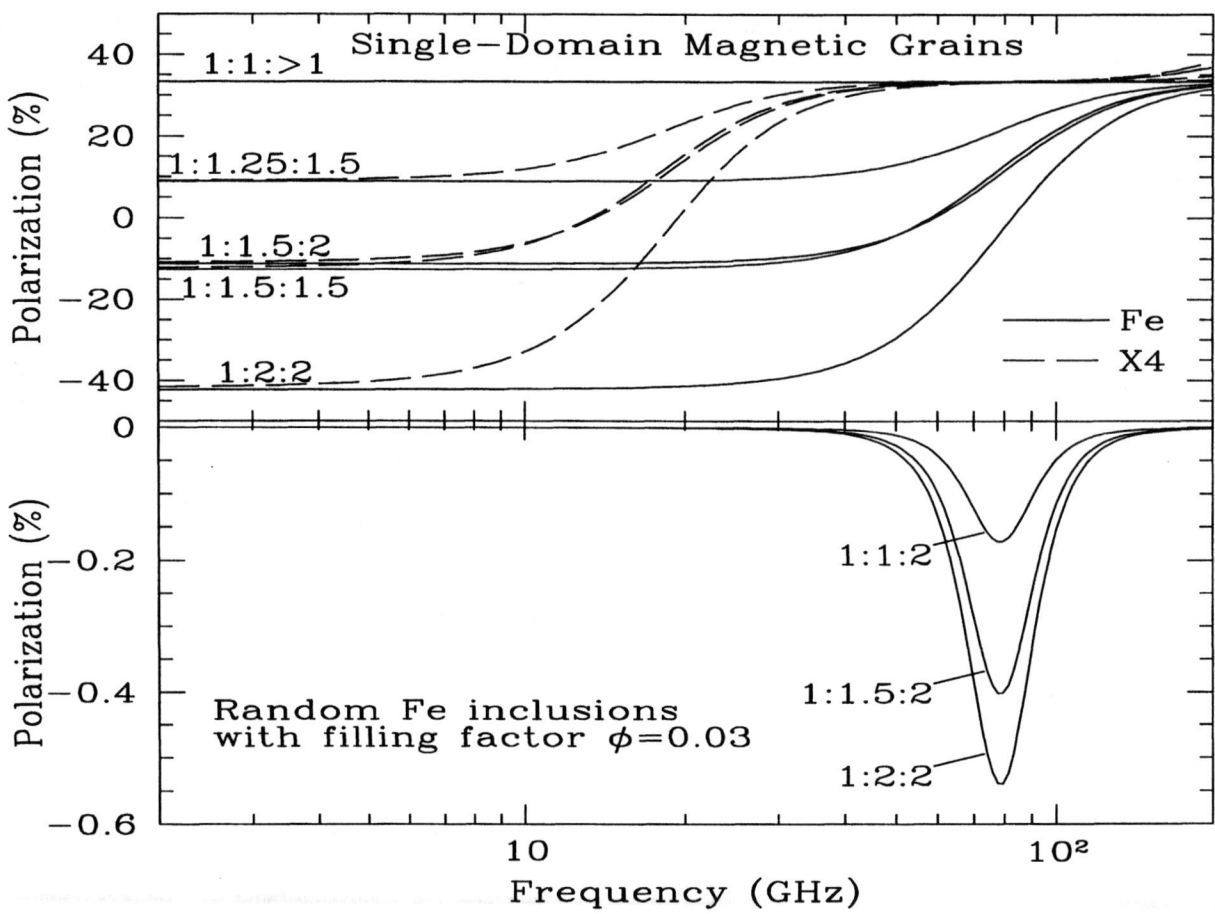

FIGURE 2. Polarization from magnetic grains (from [16]). Upper panel: Polarization of thermal emission from perfectly aligned single domain grains of metallic Fe (solid lines) or hypothetical magnetic material that can account for the Foreground X (broken lines). Lower panel: Polarization from perfectly aligned grains with Fe inclusions (filling factor is 0.03). Grains are ellipsoidal and the result are shown for various axial ratios.

from aligned single-dipole grains and it can be used to separate this component from the CMB signal. Note that the degree of polarization is large, and such grains may substantially interfere with the attempts of CMB polarimetry. Even if the intensity of magneto-dipole emission is subdominant to that from rotating grains, it can still be quite important in terms of polarization. A relatively weak polarization response is expected for grains with magnetic inclusions (see Fig. 2). The resulting emission is negative as magnetic fluctuations are stronger along longer grain axes, while the short axis is aligned with **B**.

Systematic studies of dust foreground polarization should improve our insight into the formation dust, its structure, its composition. For instance, [16] showed that the present-day microwave measurements do not allow more than 5% of Fe to be in the form of metallic iron. More laboratory measurements of microwave properties of candidate materials are also necessary. Some materials, e.g. iron, were studied at microwave range only in the 50's and this sort of data must be checked again using modern equipment.

POLARIZED DUST EMISSION AS A FOREGROUND TO CMB MEASUREMENTS

As we have seen in the previous sections, dust emission at microwave and submillimeter frequencies is likely to be polarized, thus representing a foreground emission to the cosmic microwave background polarized (hereafter CMBP) emission. In order to estimate the impact of this foreground on the CMBP measurements, it is essential to compare not only their relative dominance in different frequency channels, but also their spatial statistical distribution since the CMBP emission is expected to have a very distinctive spatial distribution. It is therefore important to compare foreground and background emission at different wavelengths and *at different scales*. In the following we will concentrate on frequencies where the thermal emission of dust is dominant over the rotational (or magnetic dipole) emission. This corresponds in particular to the frequency range covered by the High Frequency Instrument aboard the Planck satellite, which offers the highest possible sensitivity to the tiny CMBP emission.

Depolarization factors

Given the low optical depth of the diffuse ISM (where the CMBP measurements will be made, i.e. at high Galactic latitude), the polarized emission of dust for a particular line of sight depends on the integrated signal throughout the entire emitting medium. Even if the grains are perfectly aligned with their short axis along the magnetic field lines, the magnetic field lines reversals will act as a depolarization mechanism on the observed dust emission. The statistical distribution of the polarized emission is therefore the result of a complicated interplay between the dust density distribution, possibly its temperature distribution (resulting in varying alignment efficiencies, as previously discussed), and the magnetic field distribution (direction and amplitude a priori). Following [29], we will define a depolarization factor Φ which gives the relation between the intrinsic dichroic polarised cross-section of a grain and the (average) observed polarized cross-section along a typical line of sight. This factor is defined as follows:

$$\Phi = RF\cos^2\gamma \tag{1}$$

$$R = \frac{3}{2}(\langle\cos^2\beta\rangle - \frac{1}{3}) \tag{2}$$

$$F = \frac{3}{2}(\langle\cos^2\theta\rangle - \frac{1}{3}) \tag{3}$$

where β and θ are the angle between the grain major axis (axis of maximal inertia) and the direction of the magnetic field, and the angle between the random (turbulent) component of the magnetic field and the regular component respectively; finally γ is the angle between the regular component of the magnetic field with the plane of the sky. In principle the Rayleigh reduction factor R is decomposed into the partial alignment of the grain angular momentum **J** with its major inertia axis and the partial alignment of **J** with the magnetic field. However the statistical distributions of these two angles are generally not independent (see for instance [74]) and it is precisely the determination of their combined statistical average R that is the goal of grain alignment theory. In this section, we will assume a "worst case scenario" (in terms of CMBP measurement contamination) by assuming that the grain alignment mechanism is perfect (hence $R = 1$). We are then left with the depolarization due to magnetic field line warping, and the value of the polarized cross-section which depends on grain composition, size and axis ratio (we assume grains of spheroidal shape, see e.g. [54]).

Polarized cross-sections

The polarized cross-sections in the infrared wavelengths are primarily a function of the grain composition and shape, and are not very sensitive to the size distributions of the grains provided that the wavelengths considered are much larger than the maximal grain sizes considered. In particular, the spectro-polarimetric properties of grains around absorption features of their component materials can be used to constrain the grain axis ratios in the assumption of spheroidal shapes (see e.g. [54, 33]). For instance, prolate and oblate spheroidal grains will have different locations of the peak polarization in absorption around the $9.7\mu m$ silicate feature. The $3.1\mu m$ ice feature can be used also to distinguish between prolate and oblate grains for a fixed grain core composition. Both [54] and [33] find that oblate grains with axis ratios in the range 1:2 and 2:3 can reproduce the observations reasonably well. For the mixture of silicate and graphite grains needed to reproduce the $9.7\mu m$ and $2.2\mu m$ absorption features, and for a given grain shape one can compute the polarization efficiency in the millimeter range to be of order $\sim 35\%$ for a perfectly aligned grain in the plane of the sky ([33]). The difference between this theoretical upper limit of polarization efficiency and the observed histograms of polarization degree in emission at $100\mu m$ (peak at 2% and maximum around 9%) toward the Orion nebulae would then be explained by the depolarization factors described in the preceding section, i.e. the magnetic field lines entanglement and the partial alignment of grains. To estimate these factors in the diffuse ISM, we will follow a pragmatic approach described in [65].

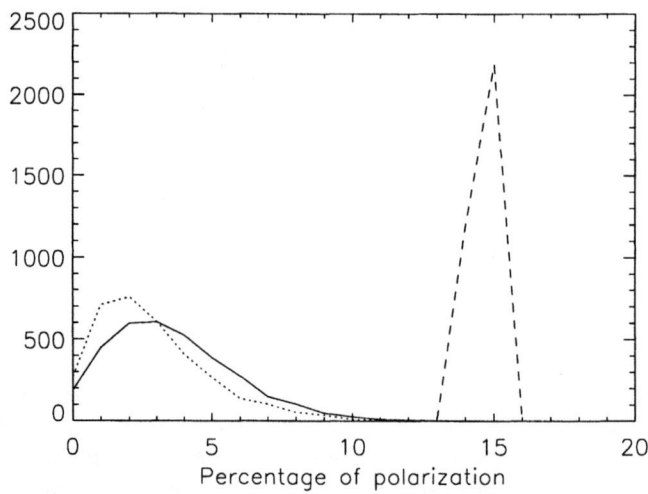

FIGURE 3. Predicted histograms of Galactic dust polarized degree in emission for the three different assumptions for magnetic field lines distribution (see text). The solid, dotted and dashed lines correspond respectively to the three cases described in the text. While two of the assumptions can reproduce the observations reasonably, the constant magnetic field case is clearly rejected.

Estimating the polarization efficiency in emission: a first model

As explained in the beginning of this section, the estimation of the contamination of CMBP measurements by polarized dust emission requires not only the knowledge of the line of sight statistics of the polarization degree of dust emission, but also the spatial distribution of this emission. This in turn requires some knowledge of the 3-dimensional distribution of the dust and magnetic field lines. [65] argued that the Galactic HI maps (with their velocity-space information), together with the observed strong correlation of HI gas and dust distribution for low column densities ([3]) could be used to estimate power spectra of polarized dust emission. It should be noted here that their method relies on very simplistic assumptions to relate the velocity-space structure of the HI gas to the 3d distribution of the gas, as well as simple extreme cases of cross-correlation of the magnetic field distribution with the underlying gas density; so that the inferred power spectra represent at best a guideline to the expected distribution of dust polarized emission. However, the different assumptions concerning the interplay between the magnetic lines orientations and the gas density structures are wide enough that this rough modelling should provide good upper and lower bounds on the slope/normalization of the polarized dust emission power spectra. In order to estimate a "worst case scenario" of CMBP measurement contamination, they assumed a perfect alignment of dust grains on magnetic field lines in the diffuse ISM, so that the problem becomes independent of the magnetic field amplitude, as well as the dust temperature. This is of course not expected to be true for realistic grain alignment mechanisms, be it paramagnetic relaxation of radiative torques. With these assumptions they considered three cases for the relation between magnetic field lines and gas density:

- magnetic field lines parallel to gas filaments, defined as least gradient directions in the HI distribution.
- magnetic field lines randomly distributed in the plane perpendicular to these directions. This could represent the case of helical field configurations.
- constant magnetic field lines orientation.

They applied their method to the HI Dwingaloo survey at intermediate Galactic latitude to compute an average contamination for all-sky CMBP measurements. The first statistics that they derived is the histogram of the polarization degree in emission for the three different assumptions above, shown in figure 3. They used the same method to predict the polarized (spatial) power spectra of dust emission, more specifically the power spectra of the *electric* (E) and *magnetic* (B) modes of polarization, in order to compare them to the theoretical predictions for the CMBP emission (see figure 4). The power spectra for the electric and magnetic modes of polarized dust emission were computed in the flat sky approximation ([76]) from the Fourier coefficients of the Stokes parameters Q and U maps.

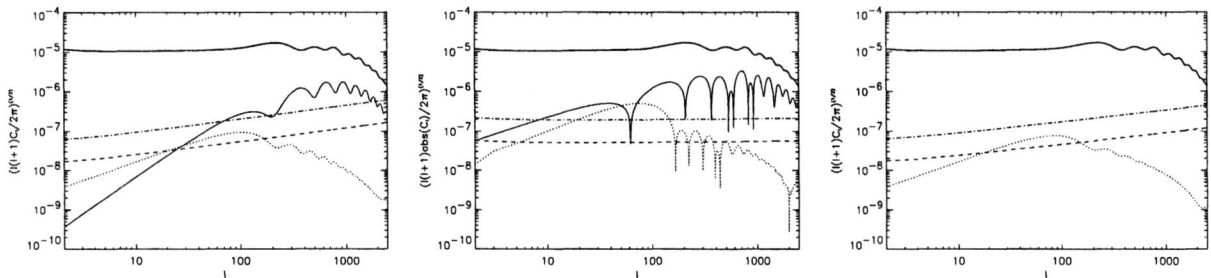

FIGURE 4. Predicted polarized *EE*, *TE* and *BB* power-spectra of Galactic dust, for two Planck HFI channels at 143 and 217 GHz, compared to the predicted spectra of the CMBP for a typical cosmological model. The dashed straight lines correspond to the 143 GHz dust spectra, and the dot-dashed ones to the same spectra at 217 GHz. The thin solid lines (dotted lines) show the different CMBP spectra for the scalar (tensor) perturbations for the chosen cosmological model, and the thick lines gives the level of the temperature *TT* power spectrum for comparison.

They were then fitted by power laws, together with the temperature-polarization cross-correlation spectrum. This decomposition in electric and magnetic modes of polarization is of particular importance for the CMBP emission since scalar primordial perturbations of the metric can only produce electric modes of polarization in the CMBP, thus defining the magnetic modes as a tracer of the primordial tensor (gravitational waves) perturbations (see [77, 38]). The figures show the expected level of contamination by dust polarized emission for the two polarized channels of the Planck High Frequency Instrument that are most sensitive to CMBP (143 and 217 GHz). One can see that for a broad range of scales the measurement of electric modes of polarization in the CMBP should not be too much hindered by Galactic dust emission. This is even more true for the *TE* cross-correlation. The B modes however (shown here for a typical cosmological model with $n_T = -0.1$ and $T/S = -7n_T$) are seriously contaminated by the dust emission, and our ability to measure them will rely heavily on our capacity of using the multi-frequency measurements of the polarized diffuse emissions in the millimeter and sub-millimeter wavelengths to disentangle the cosmological and Galactic signals (see e.g. [2, 79]) using the fact that the intrinsic polarization degree in emission in the submillimeter should be roughly independent of wavelength.

This task could however be complicated by the expected complexity of the dust polarized emission (in particular the possible dependence of polarization efficiency with dust temperature, and more generally on the grains environment). It is therefore of prime importance for cosmology to better understand the polarized emission of Galactic dust and its behaviour, both spatially and spectrally.

SUMMARY

The principal points discussed above are as follows:

- Dust provides the most intricate pattern of polarized radiation. The dependence of polarization of grain temperature, composition, size and environment makes the use of templates difficult.
- If anomalous emission in the range of 10-100 GHz is due to spinning dust particles, the polarization of the emission is marginal for frequencies larger than ~ 35 GHz. If the anomalous emission or part of it is due to magneto-dipole mechanism the polarization may be substantial and may exhibit reversals of direction with frequency.
- To get a better insight into the microwave properties of dust more laboratory studies are necessary. Some of them, e.g. measurements of the magnetic susceptibility of candidate materials at microwave frequencies, are trivial using the modern technology.
- Systematic studies of microwave polarization arising from dust will enable to determine the pattern of the CMB polarization and will shed light on many problems, including those of interstellar magnetic field and dust composition.

ACKNOWLEDGMENTS

AL is thankful to John Mathis for useful exchanges. AL acknowledges the support of NSF Graint AST-0125544.

REFERENCES

1. Arce, H.G. et al. 1998, *ApJ*, 499L, 93
2. Bouchet, F.R., Prunet, S., & Sethi, S.K. 1999, *MNRAS*, 302, 663
3. Boulanger, F., et al. 1996, *A&A*, 312, 256
4. Cho, J., Lazarian, A. & Vishniac, E. 2001, to be published in *ApJ*
5. Clayton et al. 1997, AJ, 114, 1132
6. Davis, L., & Greenstein, J.L. 1951, *ApJ*, 114, 206
7. Davies, R.D., & Wilkinson, A. 1999 in ASP Conf. Ser. Vol. 181, "Microwave Foregrounds", eds. Angelica de Oliveira-Costa and Max Tegmark, (San Francisco: ASP), 77 (henceforth "Microwave Foregrounds")
8. de Oliveira-Costa, *et al.* 1997, *ApJ Lett.*, 482, L17
9. de Oliveira-Costa, *et al.* 1998, *ApJ Lett.*, 509, L9
10. de Oliveira-Costa, *et al.* 1999, *ApJ Lett.*, 527, L9
11. de Oliveira-Costa, Angelica; Tegmark, Max; Devlin, Mark J.; Haffner, L. M.; Herbig, Tom; Miller, Amber D.; Page, Lyman A.; Reynolds, Ron J.; Tufte, S. L. 2000, ApJ, 542, L5 2000
12. Dolginov A.Z. 1972, Asr. and Space Science, 16, 337
13. Dolginov A.Z. & Mytrophanov, I.G. 1976, Asr. and Space Science , 43, 291
14. Draine, B.T., & Lazarian, A. 1998a, *ApJ Lett.*, 494, L19
15. Draine, B.T., & Lazarian, A. 1998b, *ApJ*, 508, 157
16. Draine, B.T., & Lazarian, A. 1999, *ApJ*, 512, 740
17. Draine, B.T., & Lazarian, A. 1999 in "Microwave Foregrounds", 133
18. Draine, B.T., & Weingartner, J.C. 1996, ApJ, 470, 551
19. Draine, B.T., & Weingartner, J.C. 1997, ApJ, 480, 633
20. Gold, T. 1951, Nature, 169, 322
21. Erickson, W.C. 1957, *ApJ*, 126, 480
22. Ferrara, A., & Dettmar, R.-J. 1994, *ApJ*, 427, 155
23. Finkbeiner, D.P., & Schlegel, D.J. 1999 in ASP Conf. Ser. Vol. 181, "Microwave Foregrounds", eds. Angelica de Oliveira-Costa and Max Tegmark, (San Francisco: ASP), 101
24. Finkbeiner, D.P., Davis, M., & Schlegel, D.J. 1999, *ApJ*,524, 867
25. Finkbeiner, D.P., Schlegel, D.J., Curtis, F., & Heiles, C. 2001, *ApJ*, accepted, astro-ph/0109534
26. Fosalba, P., Lazarian, A., Prunet, S. & Tauber, J.A. 2001, ApJ, accepted, astro-ph/0105023
27. Goodman, A.A. 1995, in From Gas to Stars to Dust, ed. J. Davidson, E. Erickson & M. Haas (San Francisco: ASP), APS, vol. 73, 45
28. Goodman, A.A., & Whittet, D.C.B. 1995, *ApJ Lett.*, 455, L181
29. Greenberg, J.M. 1968, in "Stars and Stellar Systems", vol. 7, p221, eds. Middlehurst,R.H., & Aller, L.H., University of Chicago Press
30. Hall, J.S., Mikesell, A.H. 1949, *AJ*, 54, 187
31. Haslam, C.G.T. *et al.* 1982, A&AS, 47, 1
32. Hildebrand, R.H. 1988, QJRAS, 29, 327
33. Hildebrand, R.H., Dragovan, M. 1995, *ApJ*, 450, 663
34. Hildebrand, R.H., Davidson, J.A., Dotson, J.L., Dowell, C.D., Novak, G. & Vaillancourt, J.E. 2001, PASP, 112, 1215
35. Hildebrand, R.H., Dotson, J.L., Dowell, C.D., Schleuning, D.A. & Vaillancourt, J.E. 1999, *ApJ*, 516, 834
36. Hiltner, W.A. 1949, *ApJ*, 109, 471
37. Jones, R.V., & Spitzer, L., Jr. 1967, *ApJ*, 147, 943
38. Kamionkowski, M., Kosowski, A., & Stebbins, A. 1997, *PRL*, 78, 2058
39. Kim, S.H., & Martin, P.G. 1995, *ApJ*, 444, 293
40. Kogut, A., *et al.* 1996a, *ApJ*, 460, 1
41. Kogut, A., *et al.* 1996b, *ApJ Lett.*, 464, L5
42. Lazarian, A. 1994, *MNRAS*, 268, 713
43. Lazarian, A 1997, *MNRAS*, 288, 609
44. Lazarian, A. 1997, *ApJ*, 483, 296
45. Lazarian, A. 2000, in "Cosmic Evolution and Galaxy Formation", ASP v.215, eds. Jose Franco, Elena Terlevich, Omar Lopez-Cruz, Itziar Aretxaga, p. 69-79, astro-ph/0003314
46. Lazarian, A., & Draine, B.T. 1997, *ApJ*, 487, 248
47. Lazarian, A., & Draine, B.T. 1999, *ApJ Lett.*, 520, L67
48. Lazarian, A., & Draine, B.T. 2000, *ApJ Lett.*, 535, L15
49. Lazarian, A., & Efroimsky, M. 1996, *ApJ*, 466, 274

50. Lazarian, A., & Efroimsky, M. 1999, *MNRAS*, 303, 673
51. Lazarian, A., Goodman, A.A., & Myers, P.C. 1997, ApJ, 490, 273
52. Lazarian, A., & Martin, P.G. 2002, in preparation
53. Lazarian, A., & Roberge, W.G. 1997, *ApJ*, 484, 230
54. Lee, H.M., & Draine, B.T., 1985, *ApJ*, 290, 211
55. Léger, A., & Puget, J.L. 1984, *ApJ Lett.*, 278, L19
56. Leitch, E.M, Readhead, A.C.S., Pearson, P.J., & Myers, S.T. 1997, *ApJ Lett.*, 468, 23
57. Lesgourgues, J., Prunet, S., & Polarski, D. 1999, *MNRAS*, 303, 45
58. Martin, P.G. 1995, *ApJ Lett.*, 445, L63
59. Mathis, J.S. 1986, *ApJ*, 308, 281
60. Mathis, J.S., Rumpl, W., & Nordsieck, K.H. 1977, *ApJ*, 217, 425
61. McCullough *et al.* 1999 "Microwave Foregrounds", 253
62. Mukherjee, P. *et al.* 2000, astro-ph/0002305
63. Padoan, P., Goodman, A., Draine, B.T., Juvela, M., Norlund, A. & Rognvaldsson, O.E. 2001, ApJ, 559, 1005
64. Prunet, S., & Lazarian, A. 1999 "Microwave Foregrounds", 113
65. Prunet, S., Sethi, S.K., Bouchet, F.R., & Miville-Deschênes, M.-A. 1998, *A&A*, 339, 187
66. Prunet, S., Sethi, S.K., Bouchet, F.R. 2000, *MNRAS*, 314, 348
67. Purcell, E.M. 1969, On the Alignment of Interstellar Dust, Physica, 41, 100
68. Purcell, E.M 1979, *ApJ*, 231, 404
69. Purcell, E.M., & Spitzer, L., Jr 1971, *ApJ*, 167, 31
70. Rao, R, Crutcher, R.M., Plambeck, R.L., Wright, M.C.H. 1998, *ApJ*, 502, L75
71. Reich, P., & Reich, W. 1988, *A&A Supp.*, 74, 7
72. Roberge, W.G., & Hanany, S. 1990, B.A.A.S., 22, 862
73. Roberge, W.G., Hanany, S. & Messinger, D.W. 1995, *ApJ*, 453, 238
74. Roberge, W.G., & Lazarian, A. 1999, *MNRAS*, 305, 615
75. Savage, B.D., & Sembach, K.R. 1996, *ARAA*, 34, 279
76. Seljak, U. 1996, astro-ph/9608131
77. Seljak, U., & Zaldarriaga, M. 1997, *PRL*, 78, 2054
78. Staggs, S.T., Gundersen, J.O., & Church, S.E. 1999 "Microwave Foregrounds", 299
79. Tegmark et al. 2000, *ApJ*, 530, 133 in "Microwave Foregrounds", 3
80. Ward-Thompson, D., Kirk, J.M., Crutcher, R.M., Greaves, J.S., Holland, W.S., & Andre, P. 2000, *ApJ*, 537L, 135
81. Zaldarriaga, M., Spergel, D.N., & Seljak, U. 1997, *ApJ*, 488, 1

Dust Polarization From Starlight Data

Pablo Fosalba*, Alex Lazarian[†], Simon Prunet** and Jan A. Tauber[‡]

*Institut d'Astrophysique de Paris, France
[†]Department of Astronomy, University of Wisconsin, Madison, USA
**Canadian Institute for Theoretical Astrophysics, Toronto, Canada
[‡]Astrophysics Division, ESA-ESTEC, Noordwijk, The Netherlands

Abstract. We present a statistical analysis of the interstellar medium (ISM) polarization from the largest compilation available of starlight data, which comprises ~ 5500 stars. The measured correlation between the mean polarization degree and extinction indicates that ISM dust grains are not fully aligned with the uniform component of the large-scale Galactic magnetic field. Moreover, we estimate the ratio of the uniform to the random plane-of-the-sky components of the magnetic field to be $\mathbf{B_u}/\mathbf{B_r} \approx 0.8$. From the analysis of starlight polarization degree and position angle we find that the magnetic field broadly follows Galactic structures on large-scales. On the other hand, the angular power spectrum C_ℓ of the polarization degree for Galactic plane data is found to be consistent with a power-law, $C_\ell \propto \ell^{-1.5}$ (where $\ell \approx 180°/\theta$ is the multipole order), for angular scales $\theta > 10'$. We argue that this data set can be used to estimate diffuse polarized emission at microwave frequencies.

INTRODUCTION

The Milky Way Galaxy emits polarized radiation at radio, mm-wave, far-infrared and optical wavelengths (see e.g, [4] for a recent review). The different mechanisms which cause the emission to be polarized at each of these wavelengths are all related to the Galactic magnetic field. Therefore the measurement of the polarized Galactic emission should yield valuable information on our Galaxy's magnetic field (see e.g, [32, 19, 14]).

Observed starlight polarization is believed to be caused by selective absorption by magnetically aligned interstellar dust grains along the line of sight. Since these measurements are limited by dust extinction, they provide us with a picture of the magnetic field only in the vicinity of the sun. Despite this limitation, recent analyses of such measurements (see [13] and references therein) suggest that they do contain information about the uniform and random components of the magnetic field on large scales. In particular, starlight polarization vectors trace the plane-of-the-sky projection of the Galactic magnetic field [32] and measurements of polarization for stars of different distances reveals the 3D distribution of magnetic field orientations averaged along the line of sight.

The Milky Way magnetic field has also been studied from measurements at far-infrared and mm wavelengths [18, 26]; see also [19, 14] for recent reviews. However, the regions surveyed correspond to few very small regions, largely dense dark clouds, mostly in the Galactic plane and they reflect rather local distortions of the large-scale magnetic field, while a global view has only been obtained for external spiral galaxies similar to our own [32].

At optical wavelengths many polarization measurements do exist, which offer an alternative view to our Galaxy. We shall present below the most complete compilation to date of starlight polarization observations. This analysis will allow us to extract basic information on the large scale statistical properties of the polarization field in the visible. We do this by studying the correlations between stellar parameters and computing the angular power spectrum of the optical polarization degree from Milky Way stars. A more detailed discussion of the analysis and results presented here is given in [7].

CP609, *Astrophysical Polarized Backgrounds*, edited by S. Cecchini et al.

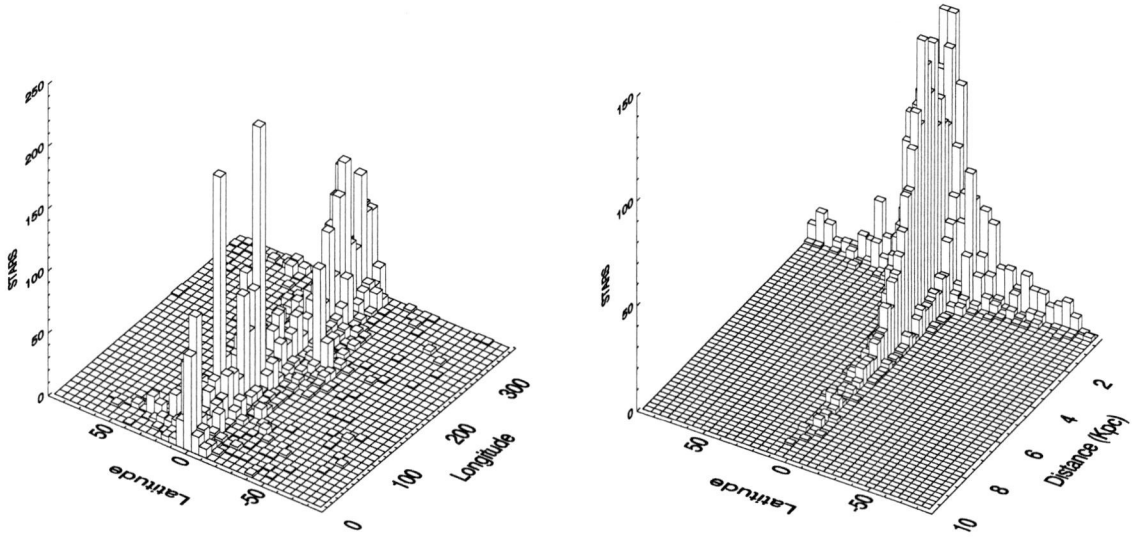

FIGURE 1. Distribution of starlight polarization data in Galactic longitude and latitude bins (*Left*) and distance and latitude bins (*Right*) for the subsample of 5513 stars analyzed.

DATA

The starlight polarization data used in this analysis is taken from the compilation by Heiles (see [13] for details and references to the original catalogues; see also [7]). This compilation includes data from 9286 sources taken from a dozen of catalogs combining multiple observations, providing accurate positions and reliable estimates for extinction and distance of stars. From this catalog, we have selected a subsample of 5513 stars (60% of the data) based on the following criteria: (1) the degree and angle of polarization are given, (2) small absolute error in the polarization degree ($< 0.25\%$) and (3) a (positive) extinction is given. All the stars in the Heiles compilation fulfilling the above requirements also have quoted distance with an estimated 20 % error for most of the sources [13].

Table 1. Mean Stellar Parameters. High latitude (low latitude) means $|b| > 10°$ ($|b| < 10°$) and nearby (distant) denotes d < 1 Kpc (d > 1 Kpc). The quantities between brackets denote amount % of all stars in the sample.

Latitude	Distance	Stars (%)	P(%)	E(B-V)
	Total	4114(75)	1.69	0.49
Low Latitude	Nearby	1451(26)	0.94	0.29
	Distant	2663(48)	2.09	0.60
	Total	1399(25)	0.45	0.15
High Latitude	Nearby	1315(24)	0.42	0.14
	Distant	84(1)	0.89	0.26

Fig 1 shows the distribution of sources in our subsample for data binned in Galactic coordinates (left panel) as well as in distance and latitude (right panel). As shown in the latter, practically all high latitude ($|b| > 10°$) stars are nearby (d < 1 Kpc). Within the Galactic plane one can find relatively distant stars, though the vast majority are within 2 Kpc. Therefore, this is a rather local sample. This is also clearly displayed in the starlight polarization map of the subsample of 5513 stars analyzed (see Fig 2).

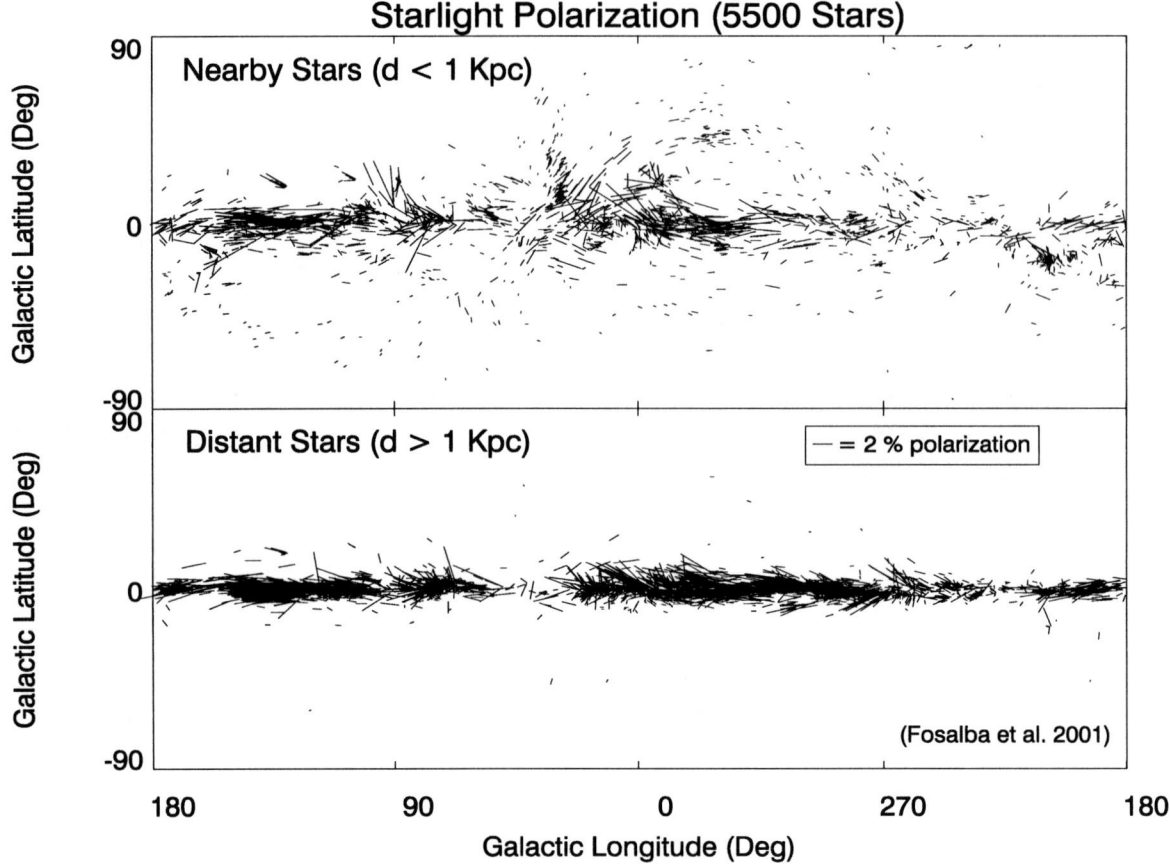

FIGURE 2. Starlight polarization vectors in Galactic coordinates for a sample of 5513 stars. The upper panel shows polarization vectors in local clouds, while the lower panel displays polarization averaged over many clouds in the Galactic plane. The length of the vectors is proportional to the polarization degree and the scale used is shown in the lower panel.

A more quantitative account of this fact is summarized in Table 1, where we give the mean stellar parameters (i.e, polarization degree P(%) and extinction as measured by the color excess E(B-V)) in the subsample as a function of latitude and distance. It is seen that low-latitude stars have large values of the polarization degree P(%) ≈ 1.7, and extinction E(B-V) ≈ 0.5, while high-latitude sources exhibit significantly lower values, P(%) ≈ 0.5, E(B-V) ≈ 0.15. Polarization vectors (defined with respect to Galactic coordinates) are typically oriented along the Galactic plane ($\theta_p \approx 90°$) although a more detailed analysis reveals a rich spatial distribution (see Fig 2 and discussion below).

STARLIGHT POLARIZATION AND GALACTIC MAGNETIC FIELD

It is generally accepted that grains in diffuse interstellar gas tend to be aligned with their major axes perpendicular to the magnetic field. [1] According to this picture, the electric field of radiation transmitted by an interstellar dust grain is less absorbed along the grain minor axis and therefore polarized in that direction which is parallel to the external magnetic field orientation. Thus, polarized starlight radiation vectors are oriented parallel to the Galactic magnetic field [32]. This polarization mechanism is usually referred to as *differential absorption*.

Since starlight polarization vectors are only seen as projected in the plane of the sky, they just give us direct information on the plane-of-the-sky projection of the Galactic magnetic field orientation. As shown in Fig 2, there

---

[1] There is no consensus as yet in relation to what alignment mechanism is the dominant in the interstellar environment [23], [25]. Currently, the radiative torque mechanism seems the most promising [5, 6].

is a strong net alignment of starlight polarization vectors averaged over many clouds with the Galactic plane structures (see lower panel) as well as a clear alignment with the spherical shell of Loop 1 as seen from the polarization vectors in local clouds (see upper panel). This is in remarkable qualitative agreement with previous studies, namely [31] (based on the catalog of [1]) and [32]. In fact, for an homogeneous distribution of intervening dust, the larger the path-length starlight travels to reach the observer, the larger the polarization degree and extinction are expected to be. According to this simple picture, regions with measured low starlight polarization degree (and extinction) correspond to the *local* ISM, while highly polarized ISM regions are observed in the line-of-sight to distant stars. This is actually observed in the sample of starlight data we have analyzed (see Fig 2). Most of the nearby stars (d < 1 Kpc) are found at high Galactic latitudes, while distant sources (d > 1 Kpc) lie mainly in the Galactic plane.

On the other hand, the spatial distribution of the polarization degree and position angle are expected to be highly correlated and this is also observed in the starlight polarization map (Fig 2), where highly polarized regions exhibit position angles aligned with the Galactic plane. There is a (sinusoidal) *modulation* of this correlation with Galactic longitude due to a projection effect: one observes the plane-of-the-sky projection of the polarization vectors that are aligned with the various Galactic spiral arms (see right panel in Fig 3).

In summary, we find evidence that *there is a net alignment of the magnetic field (as seen from its plane-of-the-sky projection) with Galactic structures on large-scales*. However, we stress that the full reconstruction of the 3D magnetic field orientations (and strength) requires additional complementary data from radio (synchrotron), sub-mm/IR (dust) observations and rotational measures from distant pulsars (see [32, 10, 11] and references therein).

POLARIZATION DEGREE AND EFFICIENCY OF MAGNETIC FIELD ALIGNMENT

As discussed in the previous section, it follows from simple arguments that the starlight polarization degree and extinction should be correlated. We do find such correlation for individual sources in our sample. However, *on the mean*, the measured correlation, $P(\%) = 0.39 \, E(B-V)^{0.8}$ (see top panel in Fig 3) has a lower amplitude than what is expected from complete dust-grain alignment from homogeneous magnetic fields [20], $P(\%) = 9 \, E(B-V)$ (see lower panel in Fig 3). The observed roughly linear correlation for individual sources is in agreement with measurements at $2.2 \, \mu m$ [20] & $100 \, \mu m$ [17]. The fact that starlight data exhibits a lower polarization degree as a function of extinction than the theoretical upper limit, suggests that *either the grain alignment is not optimal or the Galactic magnetic field has a significant random component*.

In general, we can decompose the total magnetic field as the addition of a *uniform* (coherent), $\mathbf{B_u}$, and a *random* (incoherent) $\mathbf{B_r}$ component. A random component of the magnetic field smears to some degree the correlation introduced by the uniform component [20]. This smearing effect is likely to affect the observed stars as supported by the high degree of incoherence observed for the starlight position angle. This is especially evident from nearby sources (top panel in Fig 2). Assuming that the depolarization is mainly caused by the randomness of the magnetic field, one can relate the observed polarization degree to the ratio of uniform to random plane-of-the-sky components of the underlying magnetic field (see e.g, [12]). In particular, assuming Burn's model [3], one finds for the starlight sample, $\mathbf{B_u}/\mathbf{B_r} \approx 0.80$, for $E(B-V) \approx 1$, corresponding to distant stars as an unbiased estimate of the ratio. This value is roughly consistent with previous estimates from starlight data (see [12] for a review and references therein): $\mathbf{B_u}/\mathbf{B_r} \approx 0.68$, and is typically larger than estimates from synchrotron polarization or rotational measures of distant pulsars. The discrepancy can be explained as every data set basically samples a different component of the interstellar medium: pulsars mainly trace the warm ionized medium, starlight data samples primarily the neutral media while synchrotron data seems to sample all components [12].

LARGE-SCALE PATTERN OF THE POLARIZED ISM: A POWER SPECTRUM ANALYSIS

The large-scale statistical properties of the ISM polarization from *absorption* of starlight by dust grains might give direct statistical information on the polarized diffuse *emission* by dust: if the grains that extinct starlight and emit constitute the same grain population, the power spectrum of starlight polarization degree is directly related to the power spectrum of polarized emission from dust. Starlight polarization is caused by aligned grains with sizes $10^{-4} > a > 10^{-5}$ cm [21], which generate polarized emission in diffuse media [27], [25]. Therefore if aligned grains

47

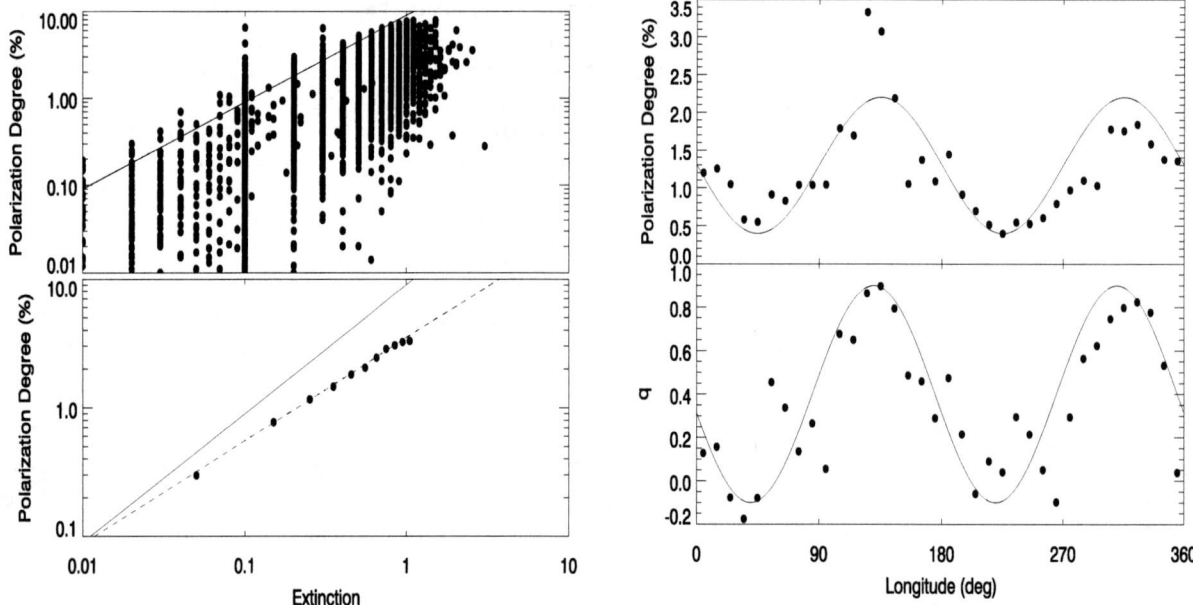

FIGURE 3. (*Left*) Correlation between polarization degree P(%), and extinction E(B-V). Upper panel shows all individual sources while lower panel displays data averaged in extinction bins. Solid line shows the theoretical upper limit, P(%) = 9 E(B-V), for completely aligned grains by external (regular) magnetic fields. Dashed line in lower panel shows P(%) = 0.39 E(B-V)$^{0.8}$, which is a good fit to the data up to E(B-V) $\approx$ 1. (*Right*) Starlight Polarization Degree (top panel) and the parameter $q = \cos 2(\theta_p - 90°)$ (bottom panel) for data averaged in 10° longitude bins. The solid line shows a best fit to a sinusoidal dependence.

in diffuse medium have the same temperature the power spectrum of the starlight polarization should be identical to the spectrum of the polarized continuum from dust in the FIR range (e.g, 100μm).

In a fully-sampled map, the two-point correlation function $\xi(\theta)$ of the scalar field S is simply related to the angular power spectrum (APS):

$$\xi(\theta) = <S(\mathbf{q_1})S(\mathbf{q_2})> = \sum_\ell \frac{\ell + 1/2}{\ell(\ell+1)} C_\ell P_\ell(\cos\theta) \tag{1}$$

where the APS, C_ℓ, estimates autocorrelations of the field at an angular scale $\theta \approx 180°/\ell$, ℓ being the so-called multipole order.

We have focused on Galactic plane data ($|b| < 10°$) as it concentrates most of the sources in the sample and therefore makes the statistical analysis more reliable. To compute the APS, we use a *hybrid* approach, that we shall call the *improved correlation function analysis*, that combines the advantages of real (or pixel) and harmonic (or multipole) space approaches [28, 29]. For this purpose we first compute the two-point correlation function, $\xi(\theta)$, of the polarization degree data using a quadratic estimator where shot-noise and edge effects intrinsic to the starlight data set are adequately corrected in pixel space. [2]

Our analysis shows that the starlight polarization degree S is well fitted by a power-law behavior, $C_\ell \propto \ell^{-1.5}$ for $\ell < 1000$ (where the multipole order $\ell \approx 180°/\theta$) which translates into angular scales $\theta > 10'$ (see left panel in Fig 4). This is approximately the pixel resolution scale used, 3.5'. We have assessed how the above results are affected by the *clustering* or non-Gaussianity in the distribution of sources by simulating a *mock starlight map* for the polarization degree (see right panel in Fig 4). We found that the efficiency with which one measures the power spectrum of the underlying densely-sampled signal is not significantly altered by the clustering of the sources, although for the actual data, that is non-Gaussian distributed, shot-noise dominates at smaller scales (i.e, in the direct harmonic approach, $C_\ell = Constant$ for a larger multipole in the actual data than in the simulation; see solid thin lines in Fig 4) and the estimated power spectrum is noisier than the simulated random-Gaussian case (see solid think lines in Fig 4).

---

[2] This method uses the *anafast* program of the HEALpix package [9] for the fast computation of the APS.
See http://www.eso.org/science/healpix/

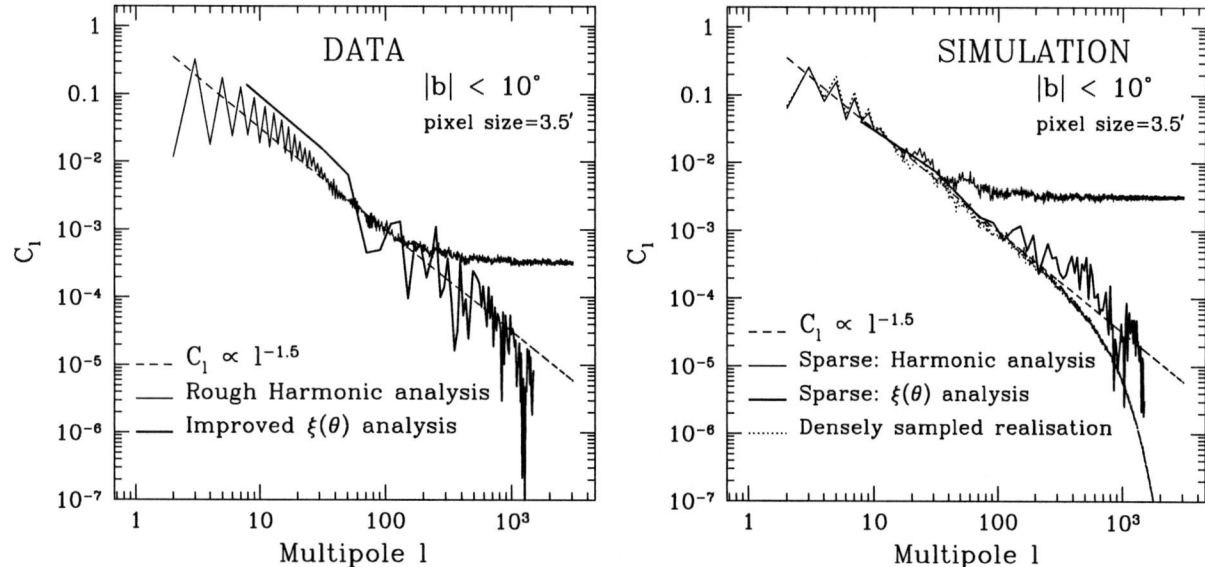

FIGURE 4. Angular power spectrum of the starlight polarization degree map in the Galactic plane, $|b| < 10°$, for the real data (*Left*) and the simulated mock catalog (*Right*). *Rough harmonic analysis* denotes a direct approach in harmonic space, while *improved* $\xi(\theta)$ *analysis* is a hybrid approach (uses techniques both in pixel and harmonic space) that corrects for shot-noise and edge effects. As a reference, the dotted line shows a densely sample realization of $C_\ell \propto \ell^{-1.5}$.

The above results provide evidence that the use of the polarization degree of the ISM as sparsely-sampled from lines-of-sight to several thousand (Galactic-plane) stars allows a clean reconstruction of the APS of an underlying *homogeneously sampled* (continuum) polarization degree of the ISM. In particular, we find that the ISM polarization degree in the continuum has the same APS slope than that measured from sparsely-sampled data, $C_\ell \propto \ell^{-1.5}$. It is interesting to note that this slope is consistent with that estimated from surveys of polarized Galactic synchrotron emission [30, 2], and the possible underlying common cause certainly deserves investigation. In a future work [8] we shall discuss how to use the starlight data set to estimate diffuse polarized emission at microwave frequencies by relating polarization by differential absorption in the optical with polarized emission in the sub-mm/FIR using dust grain alignment models [15, 16]. An accurate knowledge of such polarized emission is a critical issue in the process of component separation in cosmic microwave background experiments [27, 25].

ACKNOWLEDGMENTS

We acknowledge the use of the starlight data compilation by C. Heiles who has kindly made it publicly available. PF is supported by a CMBNET fellowship of the European Comission. AL would like to acknowledge the NSF grant AST-0125544.

REFERENCES

1. Axon D.J., Elis R.S., 1976, MNRAS, 177, 499
2. Baccigalupi C., Burigana C., Perrotta F., De Zotti G., La Porta L., Maino D., Maris M., Paladini R., 2001, A&A, 372, 8
3. Burn B.J., MNRAS, 1966, 133, 67
4. de Oliveira-Costa A., Tegmark M., (eds) 1999, "Microwave Foregrounds", ASP, V181
5. Draine B.T., Weingartner J.C., 1996, ApJ, 470, 551
6. Draine B.T., Weingartner J.C., 1997 ApJ, 480, 633
7. Fosalba P., Lazarian A., Prunet S., Tauber J.A., 2001, ApJ, accepted, [astro-ph/0105023]
8. Fosalba P., et al. 2002, in preparation

9. Gorski K.M., Hivon E., Wandelt B.D., in "Evolution of Large-Scale Structure: From Recombination to Garching " eds. A.J. Banday, R.K. Sheth and L. Da Costa [astro-ph/9812350]

10. Han J.L., 2001, Ap&SS, 277, in press [astro-ph/0010537]

11. Han J.L., 2001, to appear in proc. of the AIP conf. "Astrophysical Polarized Backgrounds", eds. S. Cecchini, S. Cortiglioni, R. Sault and C. Sbarra [astro-ph/0110319]

12. Heiles C., 1996, ASP, in "Polarimetry of the Interstellar Medium" eds. W.G. Roberge and D.C.B. Whittet, V97, 457

13. Heiles C., 2000, AJ, 119, 923
    The data is publicly available by anonymous ftp at vermi.berkeley.edu. See pol15.out (compilation of data), and pol1.ps (description and references for the compiled data) in directory pub/polcat . It has also been recently made available through the web-sites:
    `http://vizier.u-strasbg.fr/viz-bin/VizieR?-source=II/226`
    `http://adc.gsfc.nasa.gov/viz-bin/VizieR?-source=II/226`
    However, note that in these http addresses, no starlight extinction information is provided.

14. Heitsch F., Zweibel E.G., Mac Low M-M., Li, P.S., Norman M.L., 2001, ApJ, in Press [astro-ph/0103286]

15. Hildebrand R.H., 1988, QJRAS, 29, 327

16. Hildebrand R.H., Dragovan M., 1995, ApJ, 450, 663

17. Hildebrand R.H., Dotson J.L., Dowell C.D., Platt S.R., Schleuning D., Davidson J.A., Novak G., 1995, in "Airborne Astronomy Symposium on the Galactic Ecosystem: From Gas to Stars to Dust", ASP, V73, 97

18. Hildebrand R.H., Dotson J.L., Dowell C.D., Schleuning D.A., Vaillancourt J.E., 1999, ApJ, , 516, 834

19. Hildebrand R.H., Davidson J.A., Dotson J.L., Dowell C.D., Novak G., Vaillancourt J. E., 2000, PASP, 112, 1215

20. Jones T.J., 1989, ApJ, 346, 728

21. Kim S.-H., Martin P.G., 1995, ApJ, , 444, 293

22. Lazarian A., Goodman A., Myers P., 1997, ApJ, 490, 273

23. Lazarian A., 2000, in "Cosmic Evolution and Galaxy Formation", ASP, eds. J. Franco, E. Terlevich, O. Lopez-Cruz, I. Aretxaga, V215, 69, [astro-ph/0003314]

24. Lazarian A., 2001, astro-ph/0101001

25. Lazarian A., Prunet S., 2001, to appear in proc. of the AIP conf. "Astrophysical Polarized Backgrounds", eds. S. Cecchini, S. Cortiglioni, R. Sault and C. Sbarra [astro-ph/0111214]

26. Novak G., Dotson J.L., Dowell C.D., Hildebrand R.H., Renbarger T., Schleuning D.A., 2000, ApJ, 529, 241

27. Prunet S., Lazarian A., 1999, in "Microwave Foregrounds" eds. A. de Oliveira-Costa and M. Tegmark, ASP, V181, 113

28. Szapudi I., Szalay A.S., 1998, ApJ, 494, L41

29. Szapudi I., Prunet S., Pogosyan D., Szalay A.S. Bond J.R., 2001, ApJ, 548, L115

30. Tucci M., Carretti E., Cecchini S., Fabri R., Orsini M., Pierpaoli E., 2000, NA, 5, 181

31. Whittet D.C.B., 1992, "Dust in the Galactic Environment", IOP Publishing, London.

32. Zweibel E.G., Heiles C., 1997, Nature, 235, 131

Small Scale I, U, Q Galaxy Noise at cm. Waves

Yu.N. Parijskij*, A.B.Berlin*, A.Bogdantsov*, N.N.Bursov*, N.A.Nizhelskij*,
P.Tsibulev*, I.D.Novikov† and P.D.Naselskij†

*369167 Nizhnij Arkhyz, Karachay-Cherkessia Republic, RUSSIA
†Astronomical Observatory of the Copenhagen University, Julian Maries Vej 30, DK2100, Copenhagen,
DENMARK

Abstract. It is shown that the "spinning dust" component does not prevent observation of the CMB anysotropy at high ℓ values, as well as observation of the CMB polarization at scales which are the most interesting ($\ell = 1000$). These results were obtained with the 600-meter reflector RATAN-600, equipped with a multi-frequency feed array with an angular resolution up to 0.1x1 arcmin and with a temperature resolution below 100 μK.

The CMB polarization anisotropy experiments need direct observations of the Galaxy noise in intensity and polarization at wavelengths and scales as close to the optimum ones as possible. RATAN-600 is the world biggest reflector with a high brightness temperature sensitivity at wavelengths close to those of the SPOrt, MAP and PLANCK experiments. We present here our new data collected with the help of a multi-frequency focal receiver array (25 channels in 0.6 GHz - 22 GHz range, see [1] for details).

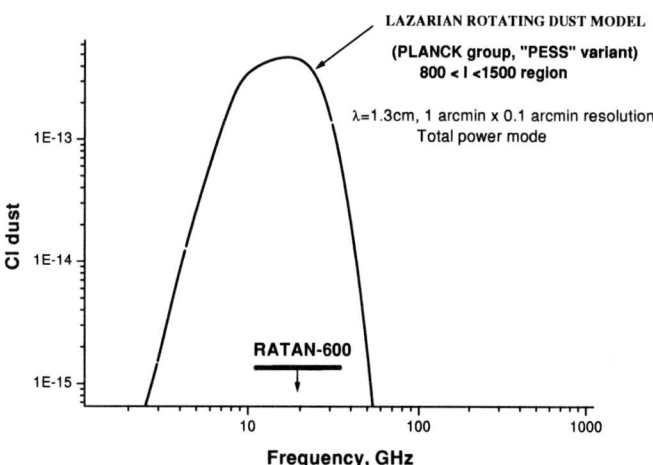

FIGURE 1. Comparison of the expected intensity from "spinning dust" in the pessimistic case ([5]) with new RATAN-600 data at the predicted frequency of the greatest effect for $\ell = 1000$.

New RATAN-600 data on the synchrotron and free-free emission can be found in [2]. Here we discuss new deep RATAN-600 observations of the spinning dust component at frequencies predicted by the Princeton group (20-30 GHz for intensity and 5-10 GHz for polarization [3, 4]) at the scales most important for CMB polarization ($\ell = 1000$, scalar mode).

Observations were performed in 1998-2001, and about 200 $0h < R.A. < 24h$ daily scans were accumulated at the declination of 3C84 at all frequencies with resolution up to 0.1x1 arcmin at the shortest wavelength (1.38 cm). At

CP609, *Astrophysical Polarized Backgrounds,* edited by S. Cecchini et al.
© 2002 American Institute of Physics 0-7354-0055-5/02/$19.00

all cm. waves we used cryogenically cooled broad band HEMT receivers (see [1]) with a sensitivity of about a few mK/$sec^{1/2}$, or better than 1 mK pixel sensitivity at $\ell = 1000$ in the single day record. In practice, at 1.38 cm we have reached about 60 μK pixel sensitivity, and sum and difference between the two big groups of observations were very close, with NET result of about 10 μK at $\ell = 1000$. At 3.9 cm we had about 100 μK pixel sensitivity, and NET result by comparing sum and difference between two big sets was below 20 μK.

The results were compared with the predictions, ("pessimistic" case, [5]), see Figure 1 (1.38 cm, spinning dust intensity limit at $\ell = 1000$ scale), and Figure 2, (3.9 cm in U Stokes parameter, at $500 < \ell < 1500$ scales). We are strongly below the theory predictions for that "pessimistic" case.

It is interesting to compare our polarization data with the best one at longer wavelengths [6]. We used the recommended extrapolation in scale and frequency, see Figure 1. Our polarization noise turned out to be smaller than predicted, and some model parameters should be changed: see possible variants in Figure 3.

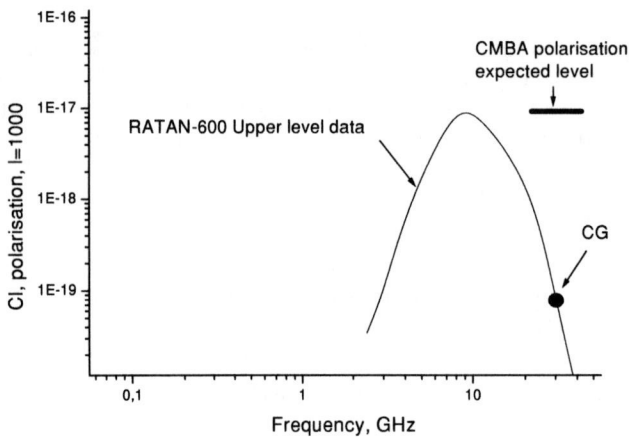

FIGURE 2. Comparison of the expected polarized signal from the dust ([4]) with new RATAN-600 data at the frequency where the polarization effect is maximum.

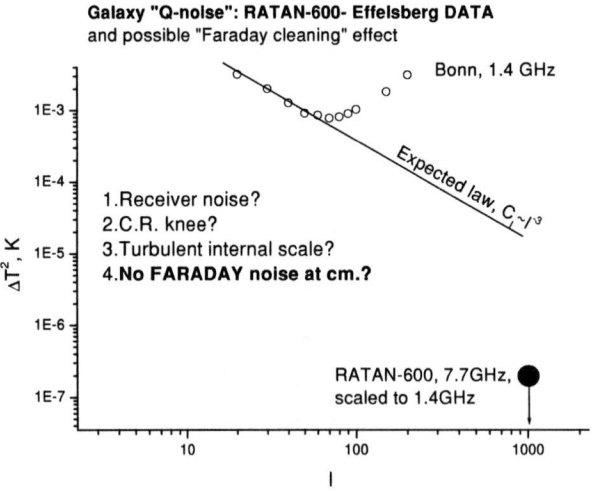

FIGURE 3. Comparison of the RATAN-600 polarization data at 7.7GHz, $500 < \ell < 1500$, with observations at longer wavelength, [6], if the extrapolation to high frequencies and small scale as suggested by [6] is applicable. We write in the figure several possible explanations for the observed behaviour.

"Faraday cleaning" at short waves seems to be a very probable reason for the dissappearing of synchrotron noise visible at longer waves. In any case, Galaxy screen does not prevent polarization CMB experiments at $\ell = 1000$ scale not only at the PLANCK main frequencies, but also at the MAP and Russian "Cosmological Gene" project frequencies (http://www.sao.ru/science/projects/; http://brown.nord.nw.ru/CG), about 1 cm.

CONCLUSIONS

We are sure that the "pessimistic" variant [5] overestimates strongly all components of the Galaxy noise and μK level sensitivity may be achieved by the proper receiving systems even in the low frequency range of the Galaxy window.

ACKNOWLEDGMENTS

This contribution was supported in part by INTAS 97-01192, RFFI 99-2-17114, ASTRONOMY and INTEGRATION projects.

REFERENCES

1. A.B. Berlin, G.M. Timofeeva, N.A. Nizhelsky, A.V. Bogdantsov, A.M. Pillipenko, V.M. Chmil, Yu. N. Meshkov *A&ApTr*, **19**, 558-565, (2000).
2. Y. Parijskij, "The Cosmic Microwave Background" in *Current Topics in Astrofundamental Physics*, 2001, p. 219
3. Drane B.T. and Lazarian A., *ApJ* **L19**, 494 (1998)
4. Lazarian A., astro-ph/0101001 30 Dec 2000, Lzarian A., Draine B.T. astro-ph/0003312, v2, 25 Apr 2000, Lazarian A., 1999; Astro-ph/9902356
5. M. Tegmark, D. Eisestein, W. Hu, A. de Oliveira-Costa, astro-ph/9905257 20 May 1999
6. C. Baccigalupi, C. Burigana, F. Perrotta, G. De Zotti, L. La Porta, D. Maino, M. Maris, R. Paladino A&A in press, Astro-ph/0009135, v2, 23 Mar 2001

A Multifrequency Analysis of the Polarized Diffuse Galactic Radio Emission at Degree Scales

C. Burigana* and L. La Porta*

Istituto TeSRE/CNR, via Gobetti 101, I-40129 Bologna, Italy

Abstract. The polarized diffuse Galactic radio emission, mainly synchrotron emission, is expected to be one of the most relevant source of astrophysical contamination at low and moderate multipoles in cosmic microwave background polarization anisotropy experiments at frequencies $\nu \leq 50 \div 100$ GHz. We present here preliminary results based on a recent analysis of the Leiden surveys covering about 50% of the sky at low as well as at middle and high Galactic latitudes. By implementing specific interpolation methods to deal with these data, which show a large variation of the sampling across the sky, we produce maps of the polarized diffuse Galactic synchrotron component at frequencies between 408 and 1411 MHz with pixel sizes larger or equal to $\simeq 0.92°$. We derive the angular power spectrum of this component for the whole covered region and for three patches in the sky significantly oversampled with respect to the average and at different Galactic latitudes. We find multipole spectral indices typically ranging between ~ -3 and $\sim -1 \div -1.5$, according to the considered frequency and sky region. At $\nu \geq 610$ MHz, the frequency spectral indices observed in the considered sky regions are about -3.5, compatible with an intrinsic frequency spectral index of about -5.8 and a depolarization due to Faraday rotation with a rotation measure RM of about 15 rad/m^2. This implies that the observed angular power spectrum of the polarized signal is about 85% or 20% of the intrinsic one at 1411 MHz or 820 MHz respectively.

INTRODUCTION

The polarized diffuse Galactic synchrotron emission, whose study provides important insights into the properties of the Galactic magnetic field and the interstellar ionized matter, is expected to be one of the most relevant source of astrophysical contamination at low and moderate multipoles in cosmic microwave background (CMB) polarization anisotropy experiments at frequencies $\nu \leq 50 \div 100$ GHz. At frequencies about 1 GHz, the Galactic synchrotron emission dominates over the Bremsstrahlung emission which, although expected to be weakly polarized, significantly increases with the frequency, ν, in comparison with the synchrotron one.

While ongoing and future experiments with high sensitivity and resolution are expected to cover large sky areas at different Galactic latitudes and frequencies (see, e.g., these Proceedings), the Leiden surveys [1], covering about 50% of the sky, can be used to derive the angular power spectrum of the polarized diffuse Galactic radio emission at frequencies between 408 and 1411 MHz. We have implemented specific interpolation methods to project surveys with large variations of the sampling across the sky into maps; we work here with pixel sizes larger or equal to $\simeq 0.92°$, appropriate to the case of the Leiden surveys. Some sky areas, both at low and middle/high Galactic latitudes, show a much better sky sampling than the average and are particularly suitable for an analysis in terms of angular power spectrum. Their extents, few tens of degrees both in Galactic latitude b and longitude l, together with the beamwidths (HPBW from 2.3° to 0.6° for ν from 408 to 1411 MHz) and the measure sensitivities (of about 100 mK at 610 GHz and about 1.5 (3) times better (worst) at the highest (lowest) frequency) imply that, in principle, we can study the angular power spectrum, C_ℓ^{ant} (here in terms of antenna temperature), in a multipole range from $\ell \sim$ few tens, where boundary effects begin to be neglibible, to about $\ell \sim 100$, where, even for the better sampled regions, the noise power is close to the signal power. In addition, the survey full coverage analysis allows to recover the C_ℓ^{ant}s at ℓ between ~ 5 and few tens.

As a by-product of this work, these maps of polarized diffuse Galactic emission, once properly rescaled in frequency, can be used as inputs for simulation activities in current and future microwave polarization anisotropy experiments.

CP609, *Astrophysical Polarized Backgrounds*, edited by S. Cecchini et al.
© 2002 American Institute of Physics 0-7354-0055-5/02/$19.00

MAP PRODUCTION AND CONSISTENCY TESTS

The main problem in the analysis of the Leiden surveys is due to their poor sampling across the sky; it is necessary to project them into maps with pixel size of about $3.7°$ (i.e. $n_{side} = 16$; we use here the HEALPix scheme [2] in which the number of pixel in the sky is $12n_{side}^2$), to find about one observation into each pixel of the whole observed sky region. For the same reason, a smoothing of the whole data with a window function with comparable size can not be properly applied. On the other hand, for some sky regions the sampling of the surveys is significantly much better, by a factor $\simeq 4$. We identified three regions (see [3]): patch 1 [$(110° \leq l \leq 160°, 0° \leq b \leq 20°)$]; patch 2 [$(5° \leq l \leq 80°, b \geq 50°)$ together with $(0° \leq l \leq 5°, b \geq 60°)$ and $(335° \leq l \leq 360°, b \geq 60°)$]; patch 3 [$(10° \leq l \leq 80°, b \geq 70°)$].

Therefore, we chose to derive the angular power spectrum at ℓ between $\simeq 5$ and $\simeq 50$, as representative of the "whole" sky, by working with the survey full coverage, and at ℓ between $\simeq 30$ and $\simeq 100$ by working with the above patches, which allow to analyze both low and middle/high Galactic latitudes. We anticipate that in these patches the polarized signal is typically higher than the average and it also shows relevant intensity variations on scales of some degrees. We then expect to find in the patches an angular power spectrum relatively higher than the spectrum obtained from the survey full coverage analysis in the common multipole range.

By simply averaging the survey observations in a given pixel, we can produce maps of polarized signal P (and also of Stokes parameters Q and U) that can be analyzed in terms of C_ℓ^{ant}s of the polarized signal (and E and B modes). On the other hand, we verified that the results produced in this way, although in rather good agreement with those derived from the improved method described below both in terms of maps and of C_ℓ^{ant}s, are affected by discontinuity effects on scales of the order of the pixel size. This tends to add power in the recovered C_ℓ^{ant}s, particularly at multipoles close to that corresponding to the pixel size [$\ell \sim 180/(\theta_{pix}/\deg)$], because of its "euristic" similarity with point source confusion noise.

To overcome this problem, we implemented a specific "interpolation" method and decided to produce maps at resolution of about $0.92°$ (in order to smooth the discontinuities discussed above on scales of about $2°$), which, of course, provide reliable information only on scales larger than $\simeq 2°$, i.e. only up to $\ell \simeq 100$. We assign to each sky pixel the average of the signals falling close to the pixel centre, properly weighted according to a certain power of the distance from the pixel. Different powers have been tested (from ~ 0.5 to ~ 2): clearly, the higher the power the higher the map contrast. On the other hand, we verified that the results depend very weakly on the adopted power. The algorithm searches, pixel by pixel, for a suitable number of observations to use in the weighted average, according to the following basic recipes: (i) to have enough observations (typically more than 3, if compatible with the following criteria); (ii) to use only observations quite close to the considered pixel centre (typically, less than few degrees, if compatible with the other criteria); (iii) to obtain the convergency of the result (i.e. minimize its fractional variation) with the variation of the number of observations and of the circle around the pixel centre that contain them. By applying our algorithm with different resolutions (we consider here n_{side} from 16 to 64), we can test its dependence on the details of its practical application[1]. Clearly, each interpolation method may alter the power at small scales; in principle, it operates as a kind of filter or regularization of the map and may decrease the power at small scales. Then, a reasonable agreement with the angular power spectrum derived, whenever possible, from other data with a better sampling across the sky (at the same ν and in same sky region) is crucial to probe the validity of the code recipes: in particular, for the reason discussed above, we require that it does not produce an underestimation of the power spectrum.

We obtained maps for P, Q, and U at all frequencies and with different resolutions ($16 \leq n_{side} \leq 64$) by using the different methods discussed above[2].

From each map, by using the HEALPix package (properly the "anafast" code) we derive the angular power spectrum of the polarized signal for the survey full coverage and for the three considered patches (we renormalized the C_ℓ^{ant}s to the case of whole sky coverage). Preliminary results at 1411 MHz as well as tests of the applicability of this method to derive the C_ℓ^{ant}s on relatively small patches have been reported in [3] (in our case the limited patch extent is a minor problem, because of the larger dimension of the considered sky areas).

In the next section we present some of our results and show how both the maps and the angular power spectra derived from them pass all the consistency tests discussed above.

---

[1] We tested also a mix of a simple average of the signals in the pixel - whenever possible - and of this "interpolation" scheme. We found preferable to apply everywhere the "interpolation" scheme, in order to reduce better the discontinuity effects.

[2] We verified also, pixel by pixel in the maps, that $P \simeq (Q^2 + U^2)^{0.5}$ with an accuracy significantly better than the rms noise in each pixel.

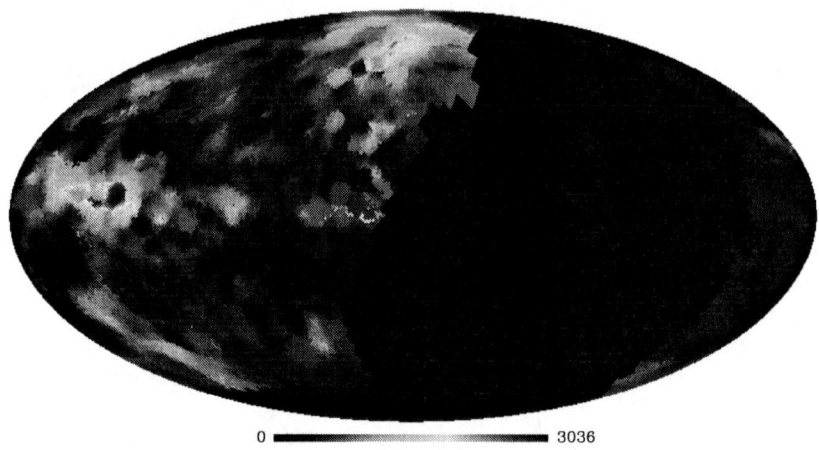

0 ▬▬▬▬▬▬▬ 3036

FIGURE 1. Map of polarized signal obtained at $n_{side} = 64$ ($\simeq 0.92°$ of pixel size).

RESULTS

In Fig. 1 we show one of the maps of polarized signal produced with the code described in the previous section[3].

In Fig. 2 we compare the angular power spectra of the patch 1 at 610 MHz obtained by working at different resolutions: note the good agreement in the common multipole range. The same holds at all frequencies for the different sky regions considered here and also by extending the comparison to the lowest resolution maps ($n_{side} = 16$), a crucial test in the case of the survey full coverage analysis.

The angular power spectrum derived from each map is, of course, mainly given by the sum of the astrophysical diffuse component relevant here, the synchrotron, and of the instrumental noise; in particular, the latter dominates at large multipoles, as it is evident also from the flattening we find there. We fit the recovered C_ℓ^{ant}s as sum of two components, represented by a set of parameters: (i) the synchrotron component, smoothed with the beam (assumed to be perfectly symmetric and Gaussian), is approximated as $k\ell^\alpha\exp[-(\sigma_b\ell)^2]$, σ_b being the 1σ beamwidth (of course, the window function is not crucial in this context at the highest frequencies); (ii) the noise contribution is approximated as a flat, white noise, component, c_{wn}. All the parameters of the fit depend on ν and on the sky region. In each case, we find the best fit parameters at ℓ between ~ 30 and $\sim 100 \div 200$; we separately repeat the fit at ℓ between ~ 5 and ~ 30 in the case of the survey full coverage analysis.

Of course, it is extremely important to check that the level of the noise power derived from the fit is similar to that derived on the basis of the rms noise per pixel in the map. Given rms noise maps, we generate simulated maps of white noise and extract their C_ℓ^{ant}s. This is shown in Fig. 3. We quote the rms noise per pixel by using three different methods: (i) the rms noise calculated by using the rms noise of each observation and propagating the error according to the mathematical rules applied to produce the maps (solid lines); (ii) the same as in the case (i), except for pixels in which survey observations fall, in which case we apply the standard weighted error on the standard weigthed average (dot-dashed line in the left panel); (iii) the standard weighted error on the standard weigthed average for pixels in which survey observations fall, and the average of these errors for the other pixels (three dots-dashes). Of course, for the patches, and particularly for the patch 3, the evaluations (ii) and (iii) provide practically the same results, given the good sampling across the sky; thus, we only report the evaluation (iii). Note that, for the survey full coverage analysis, the noise spectrum evaluated according to (iii) is of the order of that derived from the signal map at $\ell \sim 50$: therefore, at ℓ larger than ~ 50 no reliable information can be derived. For the patches, on the contrary, the noise spectrum is below that derived from the signal map, independently of the different noise evaluations, up to $\ell \sim 100$. This confirms that, not only from the point of view of the sampling across the sky, but also from the point of view of the sensitivity,

---

[3] A color figure may be requested via e-mail to the authors.

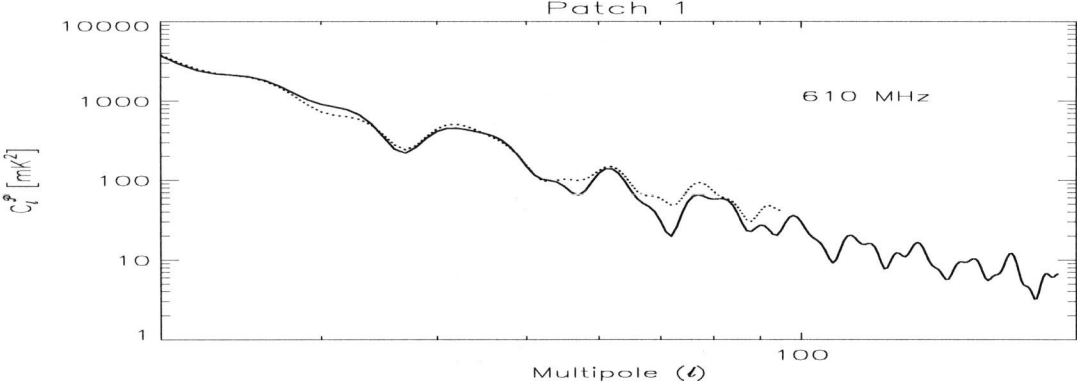

FIGURE 2. Comparison between the angular power spectra obtained by projecting the Leiden surveys into maps at different resolution, $n_{side} = 32$ (i.e., pixel size $\simeq 1.8°$, dotted line) and 64 (i.e., pixel size $\simeq 0.92°$, solid line).

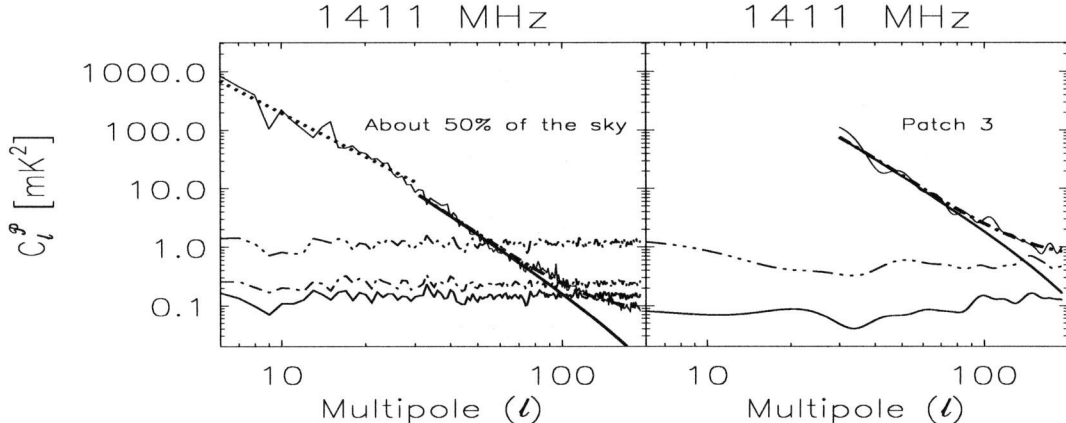

FIGURE 3. Angular power spectra derived from the map (thin solid lines) and synchrotron (thick solid lines) and synchrotron plus noise (thick dot-dashed lines) best fit spectra at multipoles larger than ~ 30 compared with the those derived from simulated maps of pure white noise with different estimates of the rms noise per pixel (roughly horizontal lines). In the case of the survey full coverage analysis we report also the C_ℓ^{ant}s obtained at lower multipoles (again, thin solid line) together with the best fit spectrum (dotted line) dominated by the synchrotron component (see also the text).

the patches can be used to extract the synchrotron C_ℓ^{ant}s up to $\ell \sim 100$. We note that this is not valid, in principle, for possible analyses of patches in other sky regions less sampled and of lower polarized signal intensity, where we expect a signal to noise ratio close to that derived here for the survey full coverage analysis, i.e. a reliable information can be obtained only up to $\ell \sim 50$.

From the survey full coverage analysis, we derive reliable information also at $5 \leq \ell \leq 30$ (of course, only nearly full sky observation can provide reliable information at very low multipoles). In this case, both the noise and possible effects introduced by the poor sampling or by the adopted "interpolation" technique are clearly negligible. We find a flattening of the spectrum with respect to that at higher multipoles; the slope (dotted line of left panel of Fig. 3) at $\nu = 1411$ MHz, crucial for the extrapolation to higher frequencies, is close to -2.5. This implies an essentially flat (slope ~ -0.25) behaviour of $\delta T_\ell = \sqrt{\ell(2\ell+1)C_\ell/4\pi}$ at $5 \leq \ell \leq 30$ (δT_ℓ is a quantity particularly relevant in the comparison with the CMB temperature and polarization anisotropy observations).

We focus now on the angular power spectrum of the synchtrotron component at ℓ between $\simeq 30$ and $\simeq 100 \div 200$. Our results are summarized in Fig. 4. We find slopes varying from about -3 to about $-1 \div -1.5$, mainly depending on the frequency, although, as expected, different sky regions show different slopes even at the same frequency.

As already recognized (see [3]), we find a particularly impressive agreement between the result obtained for the patch 1 and the angular power spectrum (long dashes in Fig. 4) obtained at 1411 MHz in a smaller region (the region 2 in eq. (20) of [3]) inside the patch 1 by exploiting the data from [4] (which have better sampling across the sky,

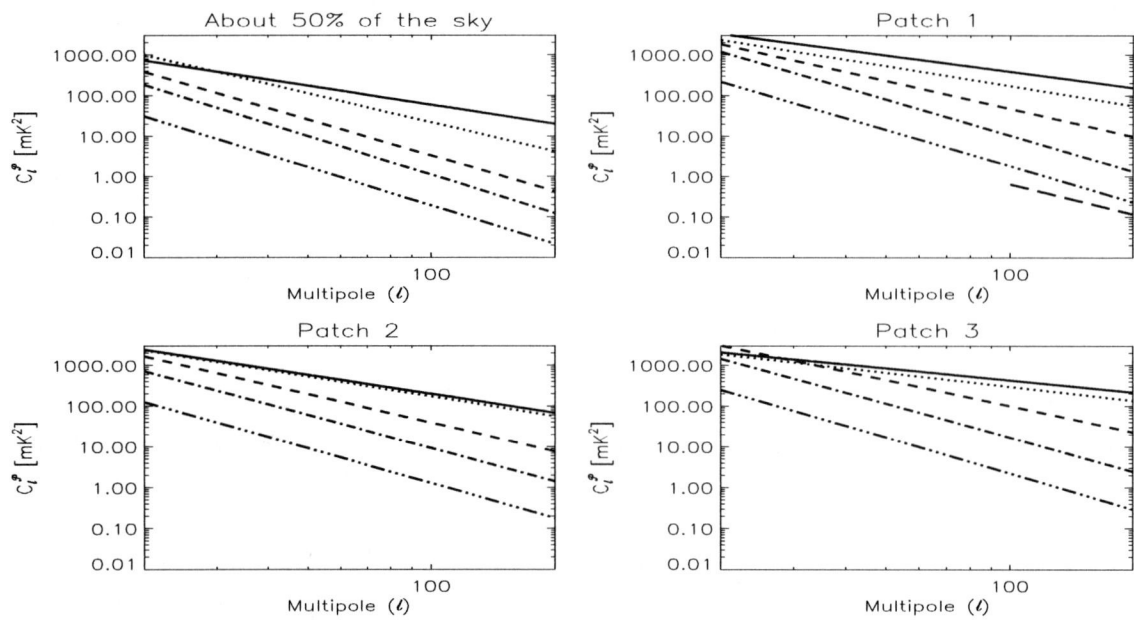

FIGURE 4. Synchrotron component of the angular power spectrum derived for the entire map and the three patches at the different frequencies: 408 (solid lines), 465 (dotted lines), 610 (dashed lines), 820 (dot-dashed lines) and 1411 MHz (three dots-dashes). In the case of the patch 1, note the good agreement (within a factor ~ 2) with the spectrum (long dashes) obtained from a smaller patch inside the patch 1 by exploiting the survey [4] at 1411 MHz (see also the text).

sensitivity and resolution and in which only the absolute calibration takes advantage of the Leiden surveys). This strongly supports the validity of our code to project the Leiden surveys into HEALPix maps or, at least, indicates that it does not introduce significant errors in the evaluations of the C_ℓ^{ant}s even at $\ell \sim 100$ and in the worst case of $\nu = 1411$ MHz (here the ratio between the beamwidth and the typical angular distance among adjacent survey pointings is minimal and the survey sky coverage is not the best – it occurs at 610 MHz).

DISCUSSION AND CONCLUSIONS

We presented here preliminary results based on a recent analysis of the Leiden surveys for about 50% of the sky and three patches at low as well as at middle/high Galactic latitudes significantly better sampled than the average. By implementing specific interpolation methods, we produce maps of the polarized diffuse Galactic synchrotron component at frequencies between 408 and 1411 MHz with pixel sizes larger or equal to $\simeq 0.92°$.

We derive the angular power spectrum of the polarized diffuse Galactic synchrotron emission for the whole covered region and for the three patches. The multipole spectral indices typically range between ~ -3 and $\sim -1 \div -1.5$, depending on the frequency and on the sky region.

Together with the large sky coverage, the multifrequency Leiden surveys offer the opportunity to study the dependence of the Galactic synchrotron polarized signal on ν. Clearly, depolarization effects and possible transitions from essentially optically thin regimes to significantly self-absorbed ones may "obscure" the intrinsic synchrotron spectral behaviour. We find large variations of the spectral index observed in the different sky pixels, probably due to real variations, but also to the limited sensitivity and, perhaps, to differential depolarization effects related to the ν-dependent beamwidth or, finally, to possible systematic and not well understood effects in the data. We try to circumvent, at least in part, these difficulties by exploiting the "statistical" information contained in the C_ℓ^{ant}s.

In Fig. 5 we show the angular power spectrum as a function of ν at the multipole $\ell = 50$ (a middle, representative value considered in this study); we find very similar results also at $\ell = 30$ and 100. The observed frequency spectral indices at $\ell = 50$ are ~ -3.2 (survey full coverage analysis), -3.3 (patch 1), -3.6 (patch 2) and -3.9 (patch 3). At the highest frequencies (820, 1411 MHz), that are crucial for the extrapolation at frequencies of $20 \div 100$ GHz relevant for CMB anisotropy polarization radiometric experiments and where beamwidth depolarization effects are expected

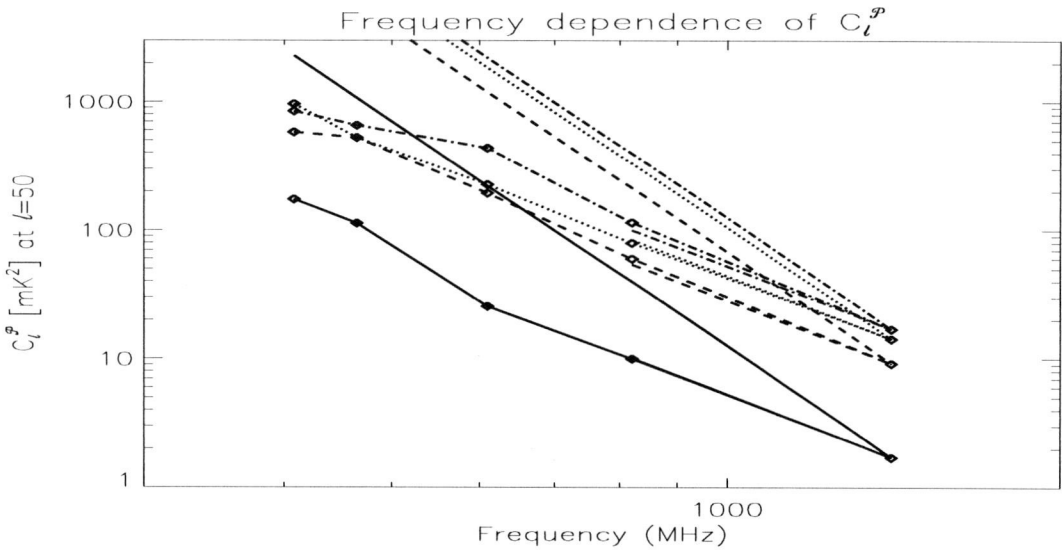

FIGURE 5. Angular power spectrum at multipole $\ell = 50$ as function of ν for the different sky regions: the survey full coverage (solid lines), the patch 1 (dotted lines), the patch 2 (dashed lines) and the patch 3 (dot-dashed lines). The straight longer lines represent a simple power law dependence of the C_ℓ^{ant}s (at $\ell = 50$) on ν corresponding to an "intrinsic" slope of -5.8, rescaled to the angular power spectra derived here at 1411 MHz. The straight shorter lines represent the dependence of the C_ℓ^{ant}s on ν between 820 and 1411 MHz for a Faraday rotatation depolarization with RM = 15 rad / m² applied to the above "intrinsic" slope (in the case of the survey full coverage analysis, note the practically perfect agreement with the observed spectrum between 820 and 1411 MHz). Note also the further flattening of the spectrum at the lowest frequencies in the case of the two patches at relatively high Galactic latitudes (see also the text).

to be quite similar (HPBW varying only of $\sim 50\%$ between 820 and 1411 MHz), the observed slope can be easily explained by assuming an intrinsic slope of -5.8 (as in the case of a slope -0.9 in terms of flux) and a depolarization due to Faraday rotation with a rotation measure RM of about 15 rad/m². This results is in quite good agreement with that previously obtained from the analysis of a sky region partially overlapped with the patch 1, as well as clearly compatible with the upper limits on RM found in the Leiden surveys (see [5]). This implies that the observed angular power spectrum of the polarized signal is about 85% or 20% of the intrinsic one (strictly, in absence of Faraday rotation depolarization) at 1411 MHz or 820 MHz respectively.

ACKNOWLEDGMENTS

It is a pleasure to thank C. Baccigalupi, G. De Zotti and R. Fanti for many constructive comments. We acknowledge also D. Maino, M. Maris, R. Paladini and F. Perrotta for the fruitful collaboration on this topic. We are grateful to T.A.T. Spoelstra for his kind clearifications and to L. Chiappetti and P. Platania for useful explanations on the database of the Leiden surveys, provided us in a more user-friendly form. The HEALPix package is acknowledged.

REFERENCES

1. Brouw W.N., Spoelstra T.A.T., A&AS, **26**, 129 (1976)
2. Gòrski K.M., Hivon E., Wandelt B.D. 1998, "Analysis Issues for Large CMB Data Sets", in *MPA/ESO Conference on Evolution of Large-Scale Structure: from Recombination to Garching*, edited by A.J. Banday, R.K. Sheth, L. Da Costa, 37, astro-ph/9812350
3. Baccigalupi C., Burigana C., Perrotta F., et al., A&A, **372**, 8 (2001)
4. Uyaniker B., Furst E., Reich W., Reich P., Wielebinski R., A&AS, **138**, 31 (1999)
5. Spoelstra T.A.T., A&A, **135**, 238 (1984)

Angular Spectra of Galactic Polarized Synchrotron Emission at Arcmin Scales

M. Tucci[*], E. Carretti[†], S. Cecchini[†], L. Nicastro[**], R. Fabbri[‡], B.M. Gaensler[§], J.M. Dickey[¶] and N.M. McClure-Griffiths[‖]

[*]Dipartimento di Fisica, Università di Milano – Bicocca, Piazza della Scienza 3, I-20126 Milano, Italy
[†]I.Te.S.R.E./CNR, Via P. Gobetti 101, I-40129 Bologna, Italy
[**]I.F.C.A.I./C.N.R., Via Ugo la Malfa 153, I-90146 Palermo, Italy
[‡]Dipartimento di Fisica, Università di Firenze, Via Sansone 1, I-50019 Sesto Fiorentino, Italy
[§] Harvard-Smithsonian Center for Astrophysics, Cambridge MA 02138, USA
[¶]Department of Astronomy, University of Minnesota, Minneapolis, MN 55455, USA
[‖] Australia Telescope National Facility, CSIRO, Epping, NSW 1710, Australia.

Abstract. We study the polarization angular spectra of the Galactic synchrotron emission from the 28–deg^2 Test Region of the Southern Galactic Plane Survey at 1.4 GHz. These data were obtained by the Australia Telescope Compact Array and allow us to investigate angular power spectra down to arcminute scales. We find that, at this frequency, the polarization spectra for $E-$ and $B-$modes are strongly affected by Faraday rotation produced in compact foreground screens. A different behavior is shown by the angular spectrum of the polarized intensity $PI = \sqrt{Q^2 + U^2}$. This is well fitted by a power law ($C_{PI\ell} \propto \ell^{-\alpha_{PI}}$) with slope ~ 1.7, which agrees with higher frequency results and can probably be more confidently extrapolated to the cosmological window.

INTRODUCTION

Large efforts have been recently performed to compute the angular power spectra (APS) of the polarized component of the Galactic synchrotron emission in order to evaluate its contribution at high frequencies and see if it can be an important contamination for the forthcoming measurements of the CMB polarization [15, 16, 3, 1, 7]. These works investigated large angular scales making use of Brouw & Spoelstra data [2], or data with sub–degrees scales thanks to low–latitude data provided by Duncan et al. [5, 6], and Uyaniker et al. [17]. In the present contribution, we extend such analyses to arcminute scales ($\ell = 10^3 \div 10^4$), i.e. nearly one order of magnitude smaller than in previous works. We make use of high–resolution polarization data taken from the Test Region of the Southern Galactic Plane Survey [13, 8, 9]. The observations were carried out with the Australia Telescope Compact Array (ATCA) at 1.4 GHz.

The study of the synchrotron contribution on arcminute scales is relevant for CMB observations. In fact, the angular scales $300 \lesssim \ell < 2000$ are expected to be those where CMB should exhibit the highest level of polarized signal, and a useful strategy to detect CMB polarization could be to concentrate observations on small sky regions with a resolution of few arcminutes. Moreover, at $\ell > 3000$ non-linear effects on CMB become important, producing polarized signal stronger than the primary spectrum.

However, the computation of polarization APS of Galactic synchrotron at such scales is important by itself. It can provide, in fact, a large number of information on spatial properties of the interstellar medium (ISM): the small scale structures we observe in the diffuse polarized emission are generated both in the emitting regions by non–uniform magnetic fields and along the line of sight, due to Faraday effects produced by small inhomogeneities in the electron–density distribution.

We compute the APS for the $E-$ and $B-$modes and for the polarized intensity, $PI = \sqrt{Q^2 + U^2}$. The results show strong differences between $C_{E,B\ell}$ and $C_{PI\ell}$. The formers appear to be much steeper than both $C_{PI\ell}$ and results from previous works on larger scales. These peculiar behaviours are supposed to be due to effects of Faraday rotation along the line of sight, which, instead, should not affect the polarized intensity spectrum. The latter is, in fact, well fitted by a

CP609, Astrophysical Polarized Backgrounds, edited by S. Cecchini et al.

power law with $\alpha_{PI} = 1.7 \div 1.8$ on the whole ℓ–range, and in a good agreement with higher frequency results. Finally, no evidence is found for a contribution of extragalactic point sources.

In particular, we stress that the sum $C_{P\ell} = C_{E\ell} + C_{B\ell}$, which is the quantity usually considered in the CMB polarization analyses, and $C_{PI\ell}$ are not equivalent and provide different information. Such important difference has not yet been outlined in the CMB analyses and in previous works on synchrotron APS [1, 7].

COMPUTATION OF POLARIZATION APS

The Test region of the Southern Galactic Plane Survey (SGPS)

The data we use here were obtained with the ATCA interferometer. Observations were made in the ATCA's multichannel continuum mode, resulting in 9 spectral channels spread across a total bandwidth of 96 MHz, with a center frequency of 1384 MHz (for more details on observations see [13, 8]). In the analysis we have only utilized images from the single channel centered at 1404 MHz, in order to avoid Faraday effects across the 96 MHz bandwidth. These data cover a small portion of the sky, with Galactic coordinates $325^\circ\!.5 < l < 332^\circ\!.5$, $-0^\circ\!.5 < b < 3^\circ\!.5$. This area was a pilot survey for the recently completed Southern Galactic Plane Survey (SGPS), and the analysis of which is in progress. Maps of the Stokes parameters Q and U were derived from the interferometric visibility data and then smoothed with a gaussian beam of FWHM $87'' \times 67''$. The sensitivity of the images was ~ 1.5 mJy beam^{-1}, except in a strip of width $0^\circ\!.5$ around the edges of the field for which the sensitivity was ~ 3 mJy beam^{-1} (we disregard this latter area in the analysis).

Interferometric observations are sensitive only to a limited range of spatial scales. The largest one depends on the shortest antenna spacing; in the case of ATCA this spacing is 31 m, which corresponds to an angular scale of $\sim 23'$ ($\ell \gtrsim 500$). Even though mosaicing processes allow information on scales up to $\sim 35'$, we will consider the APS to be reliable only from $\ell > 500$.

The same area of the sky was observed by the Parkes telescope at 2.4 GHz and at the resolution of $10'\!.4$ (for a direct compare of the two images see [8]). In general, a good correspondence between the two data-sets is found for the high–polarized structures, while areas of fainter polarized emission appear poorly correlated. In these regions, effects of foreground Faraday rotation, which are strongly dependent on both frequency and angular resolution, are relevant.

Differences among polarization APS estimators

To estimate the APS from ATCA observations, we use the Q and U data resulting from a single channel at the frequency of 1404 MHz. The analysis is performed in square patches of the survey of different dimension; because of the limited sky area considered, the Fourier approach is applied (see, e.g., [15]). In the small–scale limit the relation between E, B modes and the Q, U Stokes parameters assumes a very simple expression, consisting of a rotation in the ℓ–space:

$$
\begin{aligned}
E(\vec{\ell}) &= Q(\vec{\ell})\cos(2\phi_{\vec{\ell}}) + U(\vec{\ell})\sin(2\phi_{\vec{\ell}}) \\
B(\vec{\ell}) &= -Q(\vec{\ell})\sin(2\phi_{\vec{\ell}}) + U(\vec{\ell})\cos(2\phi_{\vec{\ell}}),
\end{aligned}
\tag{1}
$$

with $\phi_{\vec{\ell}}$ the direction angle of $\vec{\ell}$.

Then, after computing the Fourier components of $Q(\vec{\ell})$ and $U(\vec{\ell})$, equations (1) are used to compute the estimator of the angular power spectrum for E and B modes by means of

$$
C_{X\ell} = \left\{ \frac{\Omega}{N_\ell} \sum_{\vec{\ell}} X(\vec{\ell}) X^*(\vec{\ell}) - w^{-1} \right\} b_{\vec{\ell}}^{-2}
\tag{2}
$$

where $X(\vec{\ell})$ is the discrete Fourier transform of $X = E, B$. The sums are performed over the N_ℓ independent modes with wavevector magnitude around ℓ and $w^{-1} = \Omega\sigma^2/N$ is the pixel–independent measure of noise (σ is the rms noise amplitude). In the present case the term $b_{\vec{\ell}}$ takes into account that ATCA images were smoothed with an elliptical Gaussian window function.

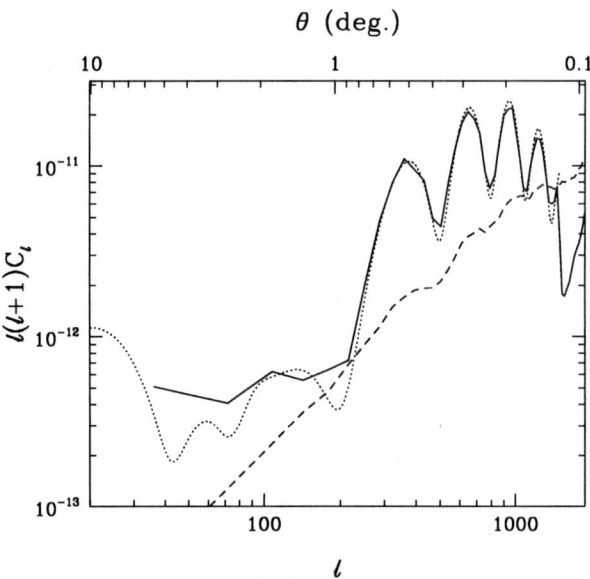

FIGURE 1. The CMB spectra for the E–mode (dotted line is the input spectra, the solid line is the resulting spectra from Fourier analysis) and for the PI component (dashed line). A CDM model with a secondary ionization optical depth $\tau_{ion} = 0.2$ has been adopted.

Together with the electric and magnetic modes, in the following analysis we consider the APS also for the polarized intensity, $PI = \sqrt{Q^2 + U^2}$. This is a scalar quantity and can be expanded on the whole sky in ordinary spherical harmonics. In the small scale limit, an expression equivalent to equation (2) can be used to compute $C_{PI\ell}$. Some caution must be taken for the value of w^{-1} and to the $b_{\vec{\ell}}$ shape. If Q and U are two Gaussian quantities, PI will have a Ricean distribution and its variance will be $\sigma_{PI} = 0.65\sigma_{Q,U}$. In expression (2), therefore, we consider the contribution of the noise of polarized intensity (w_{PI}^{-1}) lower than E and B one by a factor ~ 0.42. The function $b_{\vec{\ell}}$ is not exactly the antenna beam used for Q and U, however this can be considered a good approximation for $b_{\vec{\ell}}$. The differences with the actual smoothing function become meaningful only at angular scales very close to the telescope resolution.

The spectrum $C_{PI\ell}$ must not be confused with the polarization power spectrum $C_{P\ell}$ defined by Seljak [14] where $C_{P\ell} = C_{Q\ell} + C_{U\ell} = C_{E\ell} + C_{B\ell}$ and clearly has a different physical meaning with respect to $C_{PI\ell}$. The latter, unlike the E– and B–modes, does not provide a complete description of the polarization field because it is related only to the intensity of the polarization without giving any information on its direction on the sky. Only if the polarization angle is uniform inside all the survey area, the equality $C_{PI\ell} = C_{P\ell}$ is warranted, while in general we expect $C_{PI\ell}$ and $C_{P\ell}$ to present a different behavior. These differences will be bigger when the direction of polarization changes very rapidly on the map. This is surely the case for CMB polarization, as shown in Fig. 1. There we present the E–mode (solid line) and PI (dashed line) spectra resulting from a Fourier analysis on simulated maps for CMB (we chose a cosmological model with only scalar perturbations, where therefore $C_{E\ell} = C_{P\ell}$). This is not unexpected if we consider the geometry of the polarization angle in the E–mode spots.

In the case of synchrotron emission, the polarization direction for the diffuse component is quite smooth on large scales following the Galactic magnetic field. From low–resolution surveys the estimates of $C_{E,B\ell}$ and $C_{PI\ell}$ give only moderate differences in the spectral shape (see [8]). However, it is shown there that the polarized intensity APS decreases systematically faster with ℓ than do the E and B spectra. This is not surprising because we expect that the E and B modes vary more rapidly than PI, due to changes both in the intensity and in the angle of polarization. Anyway, as we will highlight in the present analysis, when small scales are considered, fluctuations in magnetic fields, discrete sources and also Faraday effects contribute to amplify the variations in the polarization spectra.

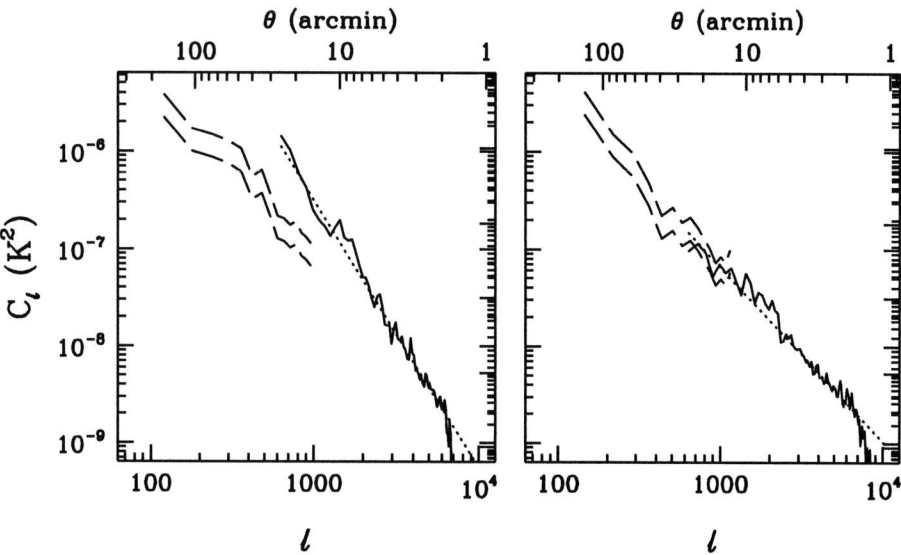

FIGURE 2. **Left panel:** E–mode spectrum from the $4° \times 4°$ region (solid lines), compared to the result from Parkes data (dashed lines) in a $5° \times 5°$ area centered on the same position. The dotted line is the best–fit power law in the ℓ–range between 600 and 6000 (see Table 1). **Right panel:** as in right panel, but for polarization intensity spectra.

RESULTS AND DISCUSSION

We compute the polarization APS in a $4° \times 4°$ box centered in $(l, b) = (329°, 1°.5)$, which is the largest square area that can be extracted from the survey and in two $3° \times 3°$ boxes covering nearly all the high–sensitivity part of survey. Then, a standard least–squared method has been used to fit each curve to a power law, $C_{X\ell} = A_X \ell^{-\alpha_X}$, in the ℓ–range [600, 6000].

In Fig. 2 we report $C_{E\ell}$ (left panel) and $C_{PI\ell}$ (right panel) obtained from the $4° \times 4°$ box; the results for B–mode are similar to those for E–mode. We find that, in the range $600 \leq \ell \leq 6000$, $C_{E,B\ell}$ are well approximated by a power law with $\alpha_E \simeq \alpha_B \simeq 2.7 \div 2.8$ (dotted line; see Table 1). A quite different behaviour is found for the spectrum of PI, which shows a flatter slope with $\alpha_{PI} = 1.80 \pm 0.06$.

In Fig. 2, we compare such spectra to those resulting from the observations in the same sky area by the Parkes telescope [5]. In this plot the Parkes spectra have been scaled to the ATCA frequency, assuming a spectral index -2.5 and -3. The best fit for E–mode is a power law index of $\alpha_E = 1.78 \pm 0.18$, which is significantly lower than α_E from the ATCA survey. Moreover, as it is clearly shown in Fig. 2, the amplitude of the Parkes and ATCA spectra presents a difference of nearly one order of magnitude at scales $600 \lesssim \ell \lesssim 1000$. Nevertheless, in this range, corresponding to angular dimensions between $20'$ and $10'$, both Parkes and ATCA telescope are able to detect polarization signals. This gap, therefore, must be related to the different frequency at which the observations were made. A possible cause is Faraday rotation, whose effects are strongly dependent on frequency and angular resolution. On the contrary, if we consider the PI spectra, the result from the two surveys are in a good agreement and the ATCA spectrum looks like the extension at high ℓ of the 2.4 GHz curves.

All the features found in polarization APS can be interpreted as due to the effects of Faraday rotation along the line of sight and variations in Faraday rotation in the plane of the sky. ATCA data and other polarimetric observations [18, 10, 11, 12] have shown a wealth of small–scale structures in the diffuse polarized emission. These structures are attributed both to a non–uniform distribution of magnetic fields in the emitting regions and to inhomogeneities in ISM in the form of clouds or filaments. In particular, the latter can produce, by Faraday rotation, small–scale modulations of a relatively uniform polarized background, which cross them. If these ISM inhomogeneities have angular dimensions to which the interferometer is sensitive, the Faraday rotation will induce detectable variations in Q and U. This has a relevant effect on APS: angular scales, corresponding to the dimension of these compact ISM structures, will gain extra–power, reducing, instead, the power on larger scales. All of these features are clearly observed in the E– and

TABLE 1. Best–fit parameters for angular power spectra

Survey	Box (deg)	Centre (l,b)	A_E (K^2)	α_E	α_B	A_{PI} (K^2)	α_{PI}	ℓ–range ($\times 10^3$)
[5] (2.4 GHz)	5×5	(329,1.5)	9.1×10^{-4}	1.78 ± 0.18	1.62 ± 0.19	3.6×10^{-4}	1.68 ± 0.30	(0.1,0.8)
[8] (1.4 GHz)	4×4	(329,1.5)	60.2	2.76 ± 0.05	2.71 ± 0.05	1.6×10^{-2}	1.80 ± 0.06	(0.6,6)
	3×3	(330.5,1.5)	108	2.82 ± 0.06	2.64 ± 0.06	9.6×10^{-3}	1.72 ± 0.05	(0.6,6)
	3×3	(327.5,1.5)	6.3	2.51 ± 0.08	2.50 ± 0.08	3.8×10^{-3}	1.65 ± 0.05	(0.6,6)

B–mode spectra.

The situation is quite different for the polarized intensity spectrum, $C_{PI\ell}$. In the absence of depolarization, a Faraday rotation of the polarization angle does not affect the intensity of polarization, PI. Apart from small peculiar regions, the RM values measured in the ATCA survey are quite low (see [8]), suggesting that Faraday depolarisation along the line of sight should not affect our estimates of the polarized intensity spectrum. We can conclude that $C_{PI\ell}$ describes the intrinsic spatial distribution of the polarized emission fairly well.

Our interpretation explains the increase of the $C_{E,B\ell}$ slope and the gap in the amplitude in 1.4 and 2.4–GHz spectra, as well as the good agreement of the PI spectra at the two different frequencies.

Similar results come from the analysis of the two $3° \times 3°$ regions that sample all the low–noise part of the survey. Table 1 reports the best–fit parameters of the curves.

Although at the ATCA resolution the contribution of extragalactic point sources could be relevant, no peculiar features are observed in the polarization spectra at high ℓ. This is consistent with our estimates of the point–sources contribution at 1.4 GHz. We consider, in fact, as a conservative upper limit, the intensity spectra derived by Tucci et al. ([16]; see also [8]) from Uyaniker et al. survey [17]. These data are obtained at 1.4 GHz and, in some patches, exhibit very flat spectra in the total intensity, with $\alpha_I \simeq 0$; assuming that such spectra are dominated by point sources, we get the limit $C_{I\ell}^{PS} < 5 \times 10^{-7}$ K^2. From this number, adopting a radio-source polarization degree of 5% (in agreement with [4]) we get $C_{X\ell}^{PS} < 1.3 \times 10^{-9}$ K^2 for $X = P, PI$, and we conclude that the contribution of point sources should be negligible in the whole range $\ell \lesssim 6000$.

An extrapolation of the results for $C_{E\ell}$ and $C_{B\ell}$ to higher frequencies should not be regarded as a trivial matter, since Faraday effects are substantial at 1.4 GHz while they become negligible at a few tens of GHz. The electric and magnetic spectra are strongly affected by Faraday rotation along the line of sight, and the results found for the slope of E and B spectra probably overestimate the actual values. Different conclusions can be drawn for polarized intensity. We argue, in fact, that the polarized intensity spectrum can be more reliably extrapolated to the "cosmological" frequencies, because, as discussed above, it is not affected by Faraday depolarisation. The power index α_{PI} is less than 2 independent of the region analysed, with a value of 1.80 ± 0.06 in the $4° \times 4°$ box. This result is in good agreement with spectra found by [3]: their best value for α_{PI} is $1.7 \div 1.8$ in the full range $10 \leq \ell \leq 800$, for any Galactic latitudes and for frequencies between 1.4 and 2.7 GHz.

ACKNOWLEDGMENTS

This work was performed within the SPOrt Collaboration, and is supported by Agenzia Spaziale Italiana (ASI). The Australia Telescope is funded by the Commonwealth of Australia for operation as a National Facility managed by CSIRO. B.M.G. acknowledges the support of NASA through Hubble Fellowship grant HST-HF-01107.01-A awarded by the Space Telescope Science Institute, which is operated by the Association of Universities for Research in Astronomy, Inc., for NASA under contract NAS 5–26555. N.M.Mc-G. and J.M.D. acknowledge the support of NSF grant AST-9732695 to the University of Minnesota. N.M.Mc-G. is supported by NASA Graduate Student Researchers Program (GSRP) Fellowship NGT 5-50250.

REFERENCES

1. Baccigalupi, C., Burigana, C., Perrotta, F., De Zotti, G., La Porta, L., Maino, D., Maris, M., & Paladini, R., 2001, A&A, 372, 8.
2. Brouw, W.N., & Spoelstra, T.A., 1976, A&AS, 26, 129.

3. Bruscoli, M., Tucci, M., Natale, V., Carretti, E., Fabbri, R., & Sbarra, C., 2001, this volume.
4. De Zotti, G., et al., 1999, NewA, 4, 481.
5. Duncan, A.R., Haynes, R.F., Jones, K.L., & Stewart, R.T., 1997, MNRAS, 291, 279.
6. Duncan, A.R., Reich, P., Reich, W., & Fürst, E., 1999, A&A, 350, 447.
7. Giardino, G., et al., 2000, astro–ph/0011084.
8. Gaensler, B.M., Dickey, J.M., McClure–Griffiths, N.M., Green, A.J., Wieringa, M.H., & Haynes, R.F., 2001a, ApJ, 549, 959.
9. Gaensler, B.M., Dickey, J.M., McClure–Griffiths, N.M., Bizunok, N.S., & Green, A.J., 2001b, this volume; astro-ph/0111114.
10. Gray, A.D., et al., 1999, ApJ, 221.
11. Haverkorn, M., Katgert, P., & de Bruyn, A.G., 2000, A& A, 356, L13.
12. Haverkorn, M., Katgert, P., & de Bruyn, A.G., 2001, this volume; astro–ph/0111381.
13. McClure–Griffiths, N.M., Green, A.J., Dickey, J.M., Gaensler, B.M., Haynes, R.F., & Wieringa, M.H., 2001, ApJ, 551, 394.
14. Seljak, U., 1997, ApJ, 482, 6.
15. Tucci, M., Carretti, E., Cecchini, S., Fabbri, R., Orsini, M., & Pierpaoli, E., 2000, New Astronomy, 5, 181.
16. Tucci, M., Carretti, E., Cecchini, S., Cortiglioni, S., Fabbri, R., & Pierpaoli, E., 2001, Proc. 20th Texas Symposium on Relativistic Astrophysics, 184.
17. Uyaniker, B., et al., 1999, A&AS, 138, 31.
18. Wieringa, M.H., de Bruyn, A.G., Jansen, D., Brouw, W.N., & Katgert, P., 1993, A&A, 268, 215.

Galactic Synchrotron Foreground and the CMB Polarization Measurements

M.V.Sazhin

Sternberg Astronomical Institute, Universitetsky pr.13, 119899, Moscow, Russia

Abstract. The polarization of the CMBR represents a powerful test for modern cosmology. It allows to break the degeneracy of fundamental cosmological parameters, and also to observe the contribution of gravitational wave background to the CMBR anisotropy. To observe the CMBR polarization several experiments are either in progress or planned and SPOrt is one of the most promising planned by ASI/ESA [1]. At the same time the observation of the CMBR polarization is a difficult task and one of the reasons is the presence of polarized foreground emission. For instance, galactic polarized synchrotron emission (according to some estimates) can completely mimic the polarization of the CMBR. Nevertheless, one can use the mathematical properties of the spherical harmonics of the distribution of radiation over the sky to separate different contributions. In this paper the mathematical properties of the polarized synchrotron foreground and the physical mechanism that produces it are discussed. The separation of synchrotron polarization from the polarization generated by density cosmological perturbations is discussed as well. A new method to separate the polarization of the galactic synchrotron emission from the polarization of the CMBR without the need of multifrequency analysis is presented.

INTRODUCTION

Next year it will be a decade since the first detection of the Cosmic Microwave Background Radiation (CMBR) anisotropy [2, 3]. Afterwards the CMBR anisotropy has been observed in a wide angular range including the intermediate angular scales. The first and second Doppler peaks as well as the anisotropy angular spectrum have been observed [4, 5, 6]. These observations allow cosmologists to determine the spectrum of primordial cosmological perturbations, prove the inflation theory and allow us to extend our knowledge to the very early Universe. At the same time new problems arise requesting new approaches. The most natural way to solve them is the measurement of the CMBR polarization.

The polarization of the CMBR was generated during the recombination epoch of our Universe at $z \approx 1000$. M.Rees was the first to recognize that polarization promised to be a new tool of cosmological investigation. Afterwards the polarization of the CMBR was considered by many authors (see, for instance [7, 8, 9]).

The polarization of the CMBR, indeed, can provide very important information about the early Universe either proving or disproving gravitational wave background originated in the very early stages of our Universe. Moreover it can resolve the degeneracy of fundamental cosmological parameters: density of matter, density of dark energy etc. At the same time the observation of the CMBR polarization is a very difficult task because it is expected to be at least 10 times smaller than the CMBR anisotropy and it is mimicked by the inhomogeneous foreground polarization.

This paper is devoted to the discussion of both the inhomogeneous foreground polarization and the CMBR polarization generated by density cosmological perturbations. The problem of separation of these component is discussed too.

CP609, *Astrophysical Polarized Backgrounds*, edited by S. Cecchini et al.

The microwave foreground radiation of our Galaxy consists of three main components:

- the synchrotron radiation (strong polarized component),
- the free-free radiation (polarization is supposed to be negligible),
- dust radiation (the situation is uncertain, but presumably it is negligible [10]).

As far as the second component is unpolarized and the polarization of dust emission is uncertain, this preliminary consideration is devoted to the contribution of galactic synchrotron to the CMBR Stokes parameters, comparing the Stokes parameters of the inhomogeneous synchrotoron radiation with those generated by cosmological density perturbations.

SYNCHROTRON RADIATION AND ITS POLARIZATION

The angular distribution of the synchrotron radioemission polarized component is very inhomogeneous (see, for instance, [11]). It can affect our ability to observe the polarization of the CMBR.

This section discusses main principles of the synchrotron radiation, resulting from the spiral motion of an electron around magnetic lines (see, for instance [12, 13, 14]).

The galactic synchrotron radiation in fact is produced by relativistic electrons which move into interstellar magnetic fields. As far as they are relativistic, their energy is much higher than the rest energy of the electron $E \gg mc^2$.

The frequency of the circular motion is determined by the value of the magnetic field H. In case of relativistic particles the frequency is determined by the ratio of the strengh of the magnetic field to the energy of the particle:

$$\omega_H = \frac{eH}{mc} \frac{mc^2}{E}$$

This is the frequency of the circular motion of the electron, but the frequency of the radioemission is much higher. The maximum in the spectrum of one emitting electron corresponds to the frequency

$$\omega_m = \frac{eH_p}{mc} \left(\frac{E}{mc^2} \right)^2 \qquad (1)$$

Here H_p is the component of the magnetic field perpendicular to the direction of the particle velocity vector. The additional multiplication factor $\frac{E}{mc^2}$ is a consequence of the Doppler shift. Therefore, the maximal frequency is much higher than the circular motion frequency of the electron.

The galactic magnetic field determines both the intensity and the frequency of the radiation. The distribution of the magnetic field in our Galaxy is random and inhomogeneous. Therefore, both the intensity and the frequency of the synchrotron radiation reflect the angular structure of the distribution of the magnetic field [15]. The polarization of the synchrotron radiation results to be anisotropic.

First of all I would like to discuss the synchrotron radiation of an electron and its Stokes parameters. One can plot perpendicular axes in an observer's plane perpendicular to the direction of the propagation of an electromagnetic wave. I designate them as $\vec{l}$ and $\vec{r}$. The intensity of any harmonic vibration of the electromagnetic field can be projected into these vectors. The intensity along $\vec{l}$ will be identified by I_l, along $\vec{r}$ by I_r. The component describing the correlation of the intensity between $\vec{l}$ and $\vec{r}$ will be designated I_u. The Stokes parameters are: $I = I_l + I_r$, $Q = I_l - I_r$, and $U = I_u$. One additional Stokes parameter V defines the circular polarization. The parameters I_l, I_r, and I_u are used in the theory of polarization of the CMBR ($V = 0$). The Stokes parameters, I_l and I_r are also used in the theory of synchrotron radiation, but instead of I_u the angle χ, defined as

$$\tan 2\chi = \frac{U}{Q}$$

in the interval $0 < \chi < \pi$, is more convenient in the synchrotron theory. It is the angle between the vector $\vec{l}$ and the main axis of the polarization ellipse.

The Stokes parameters of the synchrotron radiation of a particle can be written as a function of the amplitude of the magnetic field, of the angle between the magnetic field direction and the line of sight, and of the dimensionless

67

frequency of the radiation $\frac{v}{v_c}$, where $v_c = 1.5\frac{\omega_m}{2\pi}$ is determined by (1) and consequently depends of the magnetic field too [12].

The electron moving around magnetic lines produces all Stokes parameters: I (intensity), Q and U (describing the linear polarization), and V (describing the circular polarization). However, relativistic electron produce mainly linear polarization. More exactly the degree of circular polarization over linear polarization is of the order of

$$\sim O(\frac{mc^2}{E}).$$

The projection of the magnetic vectors on the observer's plane defines the polarization ellipse, being the minor axis of the ellipse along the projection. The angle between the vector $\vec{l}$ and the major axis of the polarization ellipse on the observer's plane is χ. The distribution of the angle χ is random inside the interval $0 < \chi < \pi$ and it is independent of the direction of observation.

The observed synchrotron radioemission is produced by an ensemble of relativistic electrons, and as far as the Stokes parameters are additive, the contributions of separate particles are summed. Although the intensity of the radiation produced by an ensemble of particles having distribution $N(E,\vec{R},\vec{k})$ can be found in many books devoted to the subject, finding the expressions for Q and U is less usual. They can be rewritten as:

$$Q = w_0 \int dEdlN(E,l,\theta,\varphi,\vec{k})H(l,\theta,\varphi)\sin\mu(\theta,\varphi)\cos 2\chi \frac{v}{v_c}K_{2/3}\left(\frac{v}{v_c}\right) \qquad (2)$$

$$U = w_0 \int dEdlN(E,l,\theta,\varphi,\vec{k})H(l,\theta,\varphi)\sin\mu(\theta,\varphi)\sin 2\chi \frac{v}{v_c}K_{2/3}\left(\frac{v}{v_c}\right) \qquad (3)$$

Here $w_0 = \frac{\sqrt{3}e^2}{mc^2}$ is a constant, $H(\vec{r}) = H(l,\theta,\varphi)$ is the spatial distribution of the magnetic field, μ is the angle between the line of sight and the magnetic vector ($H_p = H\sin\mu$), $K_{2/3}$ is the modified Bessel function and v_c was defined above.

The only difference between equation (2) and (3) is that the term $\cos 2\chi$ is substituted by $\sin 2\chi$. This angle is random over the sky.

Under reasonable assumptions one can get the Stokes parameters as functions of frequency, magnetic field and angle χ:

$$Q = c_1 \left(H(l,\theta,\varphi)\sin\mu(\theta,\varphi)\right)^{\gamma+1}v^{-\gamma}\cos 2\chi$$

$$U = c_1 \left(H(l,\theta,\varphi)\sin\mu(\theta,\varphi)\right)^{\gamma+1}v^{-\gamma}\sin 2\chi$$

Here γ and c_1 are constants related to the energy spectrum of relativistic particles which is assumed to follow a power law.

What is very important for our considerations is that both Q and U are associated to the synchrotron radiation of our Galaxy and they can be rewritten as a function $F(\theta,\varphi)$ which depends on the magnetic field, the distribution of relativistic particles etc. The only difference is that $Q = F\cos 2\chi$ and $U = F\sin 2\chi$, so that

$$\langle Q^2 \rangle = \langle U^2 \rangle$$

THE CMB AND ITS STOKES PARAMETERS

The polarizarion tensor of the CMBR can be obtained as a solution of the Boltzman equations, which describes the transfer of radiation in nonstationary plasma and in presence of variable and inhomogeneous gravitational field [16, 17, 18, 7]. The linear polarization of the CMBR is produced mainly at the recombination epoch by Thomson scattering on free electrons in the primordial cosmological plasma.

The gravitational field in the Universe can be separated into the background gravitational field, that is described by the homogeneous and isotropic FRW metric, and inhomogeneous and variable waves of three types: density perturbations, vector fluctuations, and gravitational waves. In a homogeneous and isotropic expanding Universe only one parameter of the CMBR is changed: the temperature, which decreases adiabatically. This expansion does not produces neither anisotropy nor polarization. Therefore the intensity $I = I_l + I_r$ decreases adiabatically during the expansion, being this valid both for I_l and I_r separately. As a consequence $Q = 0$ and $U = I_u = 0$.

On the contrary, the inhomogeneous and variable perturbations of the gravitational field produce both anisotropy and polarization of the CMBR. In this case one can introduce small variations of I_l, I_r and I_u (δ_l, δ_r, δ_u) describing both anisotropy and polarization of the CMBR.

Here only the contribution of density perturbations into polarization will be considered.

The equations and details of their solutions can be found in [7, 19, 20] and in reference therein.

One can introduce auxiliary functions α and β:

$$\delta_l + \delta_r = (\mu^2 - \tfrac{1}{3})\alpha,$$
$$\delta_l - \delta_r = (1 - \mu^2)\beta,$$
$$\delta_u = 0,$$

so that in case of a density perturbation taken as a single plane wave and in a defined reference frame the Boltzman equations can be re-written in terms of these parameters:

$$\frac{d\alpha}{d\eta} = F - \frac{9}{10}\sigma_T n_e a(\eta)\alpha - \frac{6}{10}\sigma_T n_e a(\eta)\beta$$
$$\frac{d\beta}{d\eta} = -\frac{1}{10}\sigma_T n_e a(\eta)\alpha - \frac{4}{10}\sigma_T n_e a(\eta)\beta$$

Here F is the gravitational force which drives both anisotropy and polarization, σ_T is the Thomson cross-section, n_e is the density of free electrons, $a(\eta)$ is a scale factor and μ is the angle between the wave vector of the perturbation and the line of sight.

The solution of these equations, which produce only the Stokes parameter Q (in this case $U = 0$), is:

$$Q = \frac{1}{7}(1 - \mu^2) \int F(\eta) \left(e^{-\tau} - e^{-\frac{3}{10}\tau}\right) d\eta$$

where $\tau(\eta)$ is the variable optical depth.

The distribution of Q over the sky can be obtained by adding the contribution from all plane waves.

MATHEMATICAL PROPERTIES OF STOKES PARAMETERS

The Stokes parameters of the CMBR are rank 2 tensor on the sphere. The rotationally invariant values are I (intensity of radiation), $Q + iU$ and $Q - iU$. The intensity I is decomposed into usual (scalar) spherical harmonics $Y_{lm}(\theta, \varphi)$.

$$I = \sum_{l,m} a_{lm} Y_{lm}(\theta, \varphi)$$

Moreover, the two values $Q \pm iU$ can be decomposed into ± 2 spin harmonics [17, 19, 20, 21] $Y_{lm}^{\pm 2}(\theta, \varphi)$:

$$Q \pm iU = \sum_{l,m} a_{lm}^{\pm 2} Y_{lm}^{\pm 2}(\theta, \varphi)$$

That form a complete orthonormal system (see, for instance, [22, 23, 24, 25]). They can be re-written in term of the generalized Jacobi polynomials [20], [26]:

$$Y_{lm}^{2}(\theta, \varphi) = N_{lm}^{2} P_{lm}^{2}(\theta) e^{im\varphi}$$

$$Y_{lm}^{-2}(\theta, \varphi) = N_{lm}^{-2} P_{lm}^{-2}(\theta) e^{im\varphi}$$

which, as a function of φ, are similar to scalar spherical harmonics. The difference of tensor harmonics lies in P, which can be expressed in terms of Jacobi polynomials:

$$P_{lm}^{s}(x) = (1 - x)^{\frac{(m+s)}{2}} (1 + x)^{\frac{(s-m)}{2}} P_{lm}^{(m+s, s-m)}(x) \tag{4}$$

Alternatively, the equivalent polynomials derived in [23] can be used.

In equation (4) $s = \pm 2$ and the normalization factor is:

$$N_{lm}^s = \frac{1}{2^s} \sqrt{\frac{2l+1}{4\pi}} \sqrt{\frac{(l-s)!(l+s)!}{(l-m)!(l+m)!}}$$

The harmonic amplitudes $a_{lm}^{\pm 2}$ correspond to the Fourier spectrum of the angular decomposition of rotationally invariant combinations of the Stokes parameters.

According to [21] one can introduce the E (electric) and B (magnetic) modes of these harmonics

$$a_{lm}^E = \frac{1}{2} \left(a_{lm}^{+2} + a_{lm}^{-2} \right)$$

$$a_{lm}^B = \frac{i}{2} \left(a_{lm}^{+2} - a_{lm}^{-2} \right)$$

that have different parity. In order to clarify this mathematical statement let us consider the following situation. Two observers define the main axes in a different way: while the $\vec{l}$ axes are directed in the same direction, the $\vec{r}$ axes have different directions. Let both observers measure the Stokes parameters: the Q parameters look similar to both observers, while U parameters have different sign.

It means that by transforming the coordinate system $Oxyz$ into the new coordinate system $\tilde{O}\tilde{x}\tilde{y}\tilde{z}$ the vectors $\vec{l}$ and $\vec{r}$ are transformed according to:

$$\tilde{\vec{l}} = \vec{l} \qquad \text{and} \qquad \tilde{\vec{r}} = -\vec{r}$$

Similarly, the E and B modes are transformed as the vectors $\vec{l}$ and $\vec{r}$:

$$\tilde{a}^E = a^E \qquad \text{and} \qquad \tilde{a}^B = -a^B$$

It is necessary to mention that a^E and a^B are not correlated.

The density perturbations generate polarization in such a way that $Q \neq 0$ and $U = 0$. This is equivalent to electric mode excitation and vanishing of magnetic modes in the CMBR polarization [21]:

$$a_d^E \neq 0 \qquad\qquad a_d^B = 0 \tag{5}$$

Since the synchrotron radiation produces both Q and U, both electric and magnetic modes exist:

$$a_s^E \neq 0, \qquad\qquad a_s^B \neq 0$$

because the synchrotron radiation is connected with the axial vector. The electric and the magnetic components of the synchrotron emission obey to the equation:

$$\langle (a_s^E)^2 \rangle = \langle (a_s^B)^2 \rangle \tag{6}$$

SEPARATION OF SYNCHROTRON RADIATION AND THE CMB RADIATION

To separate the synchrotron polarization from the CMBR polarization generated by density perturbations one has to separate the polarized components obeying

$$a^E \neq 0 \qquad\qquad\qquad a^B = 0$$

E and B modes of synchrotron foreground do not correlate each other and they do not correlate with polarization of the CMBR.

To estimate the contribution of E modes generated by density perturbations one can choose the estimator:

$$D = \langle (a^E)^2 \rangle - \langle (a^B)^2 \rangle$$

Let us consider this estimator in more detail: the a^E component is the sum of two components, $a^E = a_s^E + a_d^E$, and the indexes s and d represent the synchrotron component and the CMBR component, respectively. The same equation is valid for the B component: $a^B = a_s^B + a_d^B$. The mean square of the electric component is: $\langle (a^E)^2 \rangle = \langle (a_s^E)^2 \rangle + \langle (a_d^E)^2 \rangle + 2 \langle a_s^E a_d^E \rangle$ Since density fluctuations and synchrotron fluctuations do not correlate the last term vanishes and the equation can be written as $\langle (a^E)^2 \rangle = \langle (a_s^E)^2 \rangle + \langle (a_d^E)^2 \rangle$ The same equation is valid for B components. As far as the squares of both the E and B components of the synchrotron radiation cancel out (6) and the B component of density fluctuations is equal to zero, the estimator D is equal to the square of the E mode of density fluctuations:

$$D = \langle (a_d^E)^2 \rangle$$

Thus the D estimator can be used to separate the polarization of galactic synchrotron emission from the polarization of the CMBR generated by density perturbation.

CONCLUSION

A method to separate the polarization of the galactic synchrotron emission from the polarization of the CMBR without the need of multifrequency analysis has been presented. The method is based on the mathematical properties of the spin ± 2 spherical harmonics.

ACKNOWLEDGMENTS

The author is indebted to Drs. S.Cortiglioni, E.Caretti, and E.Vinjakin for many helpfull discussions and suggestions. Also, he would like to acknowledge the Te.S.R.E. CNR for hospitality and Italian Space Agency for financial support during the preparation of this paper.

REFERENCES

1. S.Cortiglioni et al., in: *3K Cosmology. EC -TMR Conference*, Ed., L.Maiani, F.Melchiorri, N.Vittorio, AIP Conference Proc. 476, 1999, p.186.
2. Strukov, I.A., et al., Mon. Not. R. Astron. Soc., 1992, v. 258, p. 37P.
3. Smoot, G., et al., Astrophys. J., 1992, v.396, p.L1.
4. de Bernardis, P.,et al., Nature, 2000, v.404, p.955.
5. de Bernardis, P.,et al., astro-ph/0105296, 17 May 2001.
6. Netterfield, C.B., et al., astro-ph/0104460, 17 May 2001.
7. Sazhin M.V., Benitez N., Astrophys. Lett. Commun., 1995, v. 32, p.105;
8. Ng K.L., Ng K.W., Astrophys.J., 1996, v. 456, p.413;
9. Melchiorri A., Vittorio N., in: *The Cosmic Microwave Background*, 1997, p.419;
10. Sethi, S,K, Prunet, S., Bouchet, F.R., 1998, astro-ph/9803158.
11. Reich, W., et al., 2001, this volume.
12. Ginzburg V.L., Syrovatskii S.I., *Ann. Rev. Astron. Astrophys.*, The Annual Review Publ., 1965, v.3, pp.297 - 350
13. Ginzburg V.L., Syrovatskii S.I., *Ann. Rev. Astron. Astrophys.*, The Annual Review Publ., 1969, v.7, pp.375 - 420
14. Westfold K.S., Astrophys.J., Chicago Univ.Press, Chicago, 1959, v.130, 241 - 258.
15. Wielebinski, R., 2001, this volume.
16. Basko, M.M., Polnarev, A.G., Mon.Not.R.Astron.Soc., 1980, v. 191, p.L47.
17. Sazhin, M.V., *Modern Theoretical and Experimental Problems of General Relativity and Gravitation*, Moscow Pedagogical Inst. Publ., Moscow, 1984, p. 88.
18. Harrari, D., Zaldarriaga, M., Phys.Lett., 1993, v. B 319, p.96.
19. Sazhin M.V., and Shulga V.V., Vestnik MSU, Moscow State University Publ., Moscow, 1996, ser.3, N 3, p. 69.
20. Sazhin M.V., and Shulga V.V., Vestnik MSU, Moscow State University Publ., Moscow, 1996, ser.3, N 4, p. 87.
21. Seljak U., Zaldarriaga M., Phys.Rev.Lett., 1997, v.78, p.2054.
22. Gelfand, I.M., Minlos, R.A., Shapiro, Z, Ya., *The Representation of the rotation group and the Lorenz group*. FizMatGiz Publ., Moscow, 1958.
23. Goldberg, J.N., et al., J. Math. Phys., 1967, v.8, p.2155.
24. Zerilli, F. J., J. Math. Phys., 1970, v. 11, p.2203.
25. Thorn, K.S., Rev. Mod. Phys., 1980, v.52, p.299.
26. Sazhin, M.V., Sironi, G., New Astronomy, 1999, v. 4, p. 215.

Parsec-scale Structure in the Warm ISM from Polarized Galactic Radio Background Observations

M. Haverkorn[*], P. Katgert[*] and A. G. de Bruyn[†]

[*]*Leiden Observatory, P.O. Box 9513, 2300 RA Leiden, the Netherlands*
[†]*ASTRON, P.O. Box 2, 7990 AA Dwingeloo, the Netherlands*

Abstract. We present multi-frequency polarization observations of the diffuse radio synchrotron background modulated by Faraday rotation, in two directions of positive latitude. No extended total intensity I is observed, which implies that total intensity has no structure on scales smaller than approximately a degree. Polarized intensity and polarization angle, however, show abundant small-scale structure on scales from arcminutes to degrees. Rotation Measure (RM) maps show coherent structure over many synthesized beams, but also abrupt large changes over one beam. RM's from polarized extragalactic point sources are correlated over the field in each of the two fields, indicating a galactic component to the RM, but show no correlation with the RM map of the diffuse radiation. The upper limit in structure in I puts constraints on the random and regular components of the magnetic field in the galactic interstellar medium and halo. The emission is partly depolarized so that the observed polarization mostly originates from a nearby part of the medium. This explains the lack of correlation between RM from diffuse emission and from extragalactic point sources as the latter is built up over the entire path length through the medium.

INTRODUCTION

Synchrotron radiation emitted in our galaxy provides a diffuse radio background, which is altered by Faraday rotation and depolarization when it propagates through the galactic halo and interstellar medium. Multi-frequency observations of the linearly polarized component of this radio background allow determination of the Rotation Measure (RM) of the medium along many contiguous lines of sight, from which the structure of the small-scale galactic magnetic field, weighted with electron density, can be probed. After the discovery paper ([1]), many other high resolution radio background observations have shown intriguing polarization structure, mostly in the galactic plane ([2], [3], [4], [5]). Here, we focus on regions of positive galactic latitude to evade complexities such as HII regions, supernova remnants etc. Furthermore, these are sensitive and multi-frequency observations so that accurate RM values can be determined.

OBSERVATIONS

With the Westerbork Synthesis Radio Telescope (WSRT) we mapped the polarized radio background in two fields of over 50 square degrees each in the second galactic quadrant at positive latitudes. All four Stokes parameters I, Q, U, and V were imaged in five frequency bands centered at 341, 349, 355, 360, and 375 MHz simultaneously (bandwidth 5 MHz), at a resolution of $\sim 4'$. The first field, in the constellation Auriga, is $7° \times 9°$ in size and centered on $(l,b) = (161,16)°$; the second field, in the constellation Horologium, is $8° \times 8°$ wide and centered on $(l,b) = (137,7)°$. No total intensity I was detected (besides point sources) in either field down to ~ 0.7 K, which is $< 1.5\%$ of the expected sky brightness in these regions, indicating that I does not vary on scales detectable to the interferometer, i.e. below about a degree. However, linearly polarized intensity P and polarization angle φ show abundant small-scale structure. Other fields observed with the WSRT at a single frequency around 350 MHz also show small-scale structure in polarized intensity and polarization angle, but of very different topologies ([6]).

CP609, *Astrophysical Polarized Backgrounds*, edited by S. Cecchini et al.

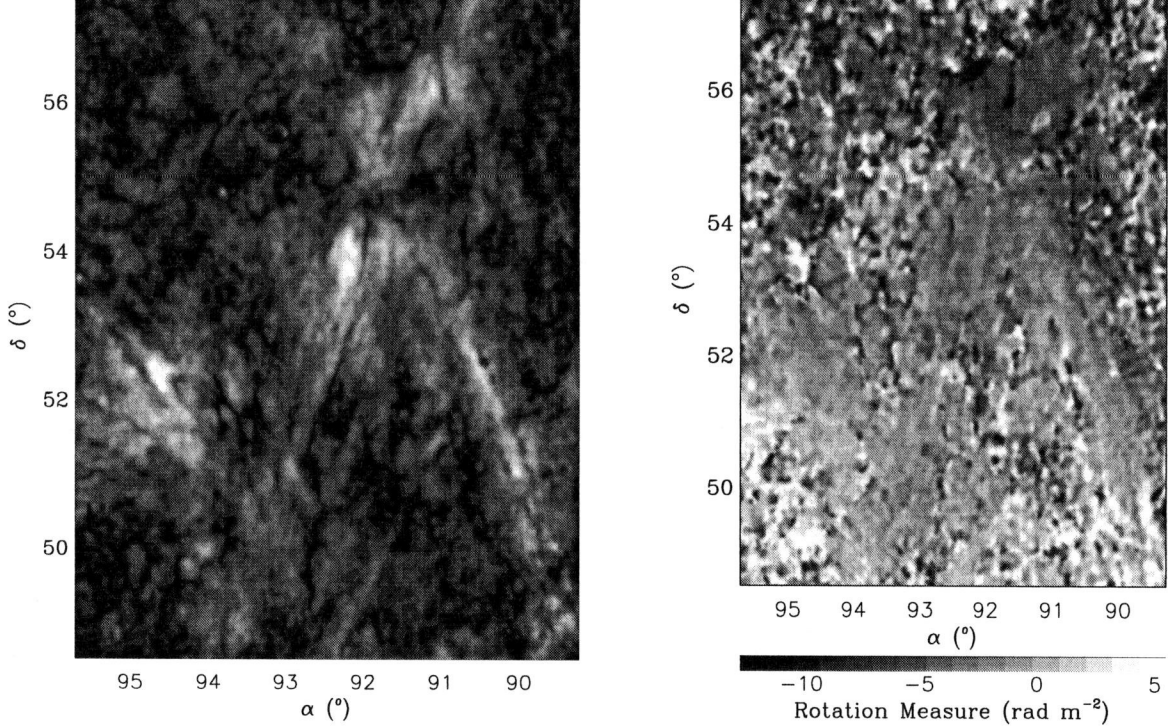

FIGURE 1. *Left:* polarized intensity P at 349 MHz in the Auriga field at $4'$ resolution. White denotes a maximum $T_{b,pol} \approx 18$ K. *Right:* RM in the Auriga field. Very high or low RM values ($|\text{RM}| \approx 30 - 60$ rad m^{-2}) in the field have been removed from the maps (see text).

ANALYSIS OF THE AURIGA FIELD

The structure in polarized intensity P in the Auriga field shows a wide variety in topology on several scales, as shown in the left plot of Fig. 1, where P at 349 MHz is mapped at $4'$ resolution. The typical polarization brightness temperature is $T_{b,pol} \approx 6 - 8$ K, with a maximum of ~ 18 K. From the Haslam continuum survey at 408 MHz ([7]), the I-background at 408 MHz in this region of the sky is ~ 33 K. Extrapolating this to our frequencies with a temperature spectral index of -2.5 between 341 MHz and 408 MHz ([8]), the total intensity background is $\sim 41 - 52$ K. So the maximum degree of polarization $p_{max} \approx 35\%$, with an average p of 15%.

In addition, a pattern of black narrow wiggly canals is visible (see e.g. the canal around $(\alpha, \delta) = (92.7, 49 - 51)°$). These canals are all one synthesized beam wide and have been shown to separate regions of fairly constant polarization angle φ where the difference in φ is approximately $90°$ ($\pm n 180°$, $n = 1, 2, 3 \ldots$), which causes beam depolarization ([9]). The angle changes are due to abrupt changes in RM. Hence, the canals reflect specific features in the angle distribution. Other angle (and RM) changes within the beam cause less or no depolarization, so that they do not leave easily visible traces in the polarized intensity distribution.

The RM of the Faraday-rotating material can be derived from $\varphi(\lambda^2) \propto \text{RM} \lambda^2$ (see [10] for details and pitfalls). The right plot in Fig. 1 gives a $4'$ resolution RM map of the Auriga field. The average RM ≈ -3.4 rad m^{-2}, and in general $|\text{RM}| < 15$ rad m^{-2}. Very high or low RM values ($|\text{RM}| \approx 30 - 60$ rad m^{-2}) in the field occur only at positions where polarized intensity P is very low, so noise errors in polarization angle are very large. Therefore the RM's at these positions are not reliable and have been removed from the maps. RM's show structure on scales of many beams (up to degree scales), but also abrupt changes from one beam to another. This is illustrated in the left map of Fig. 2, where a small part of the Auriga field is shown. Here, the lines are graphs of polarization angle against wavelength squared, so that the slope of the line is RM. Each graph is an independent synthesized beam. The greyscale denotes polarized intensity P at 349 MHz, five times oversampled. Large sudden RM changes occur: e.g. at $(\alpha, \delta) = (94.68, 53.15)°$, RM changes from -9 rad m^{-2} to 7 rad m^{-2}, and at $(\alpha, \delta) = (94.60, 53.00)°$ from 3 rad m^{-2} to -11 rad m^{-2}. A change in *sign* of RM indicates in general a change in *direction* of the galactic magnetic field along the line of sight, although

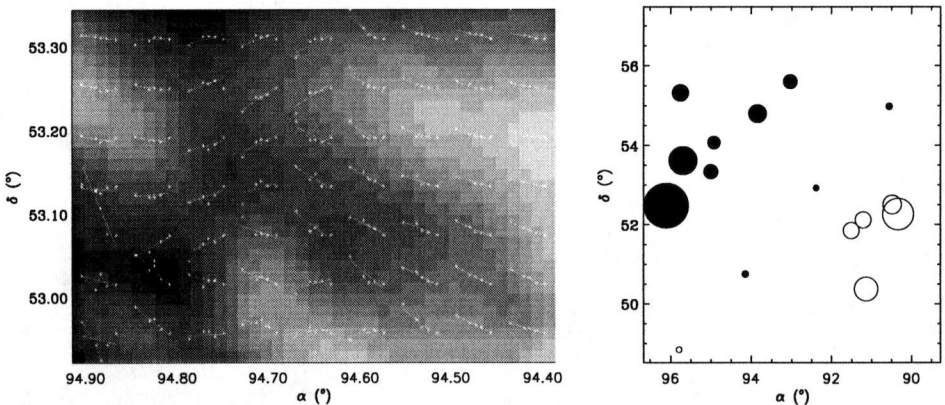

FIGURE 2. *Left:* Graphs of polarization angle against the square of the wavelength, so that the slope is RM. Each graph denotes an independent beam. The greyscale denotes polarized intensity at 349 MHz, five times oversampled. RM's range from ~ -10 to 10 rad m^{-2}, ignoring the one anomalously large negative RM on the left side. Abrupt changes in Rotation Measure over one beam (4′) are coherent along several beams. *Right:* RM's of observed polarized extragalactic point sources in the Auriga field. The radii of the circles are scaled with magnitude of RM, where filled circles are positive RM's. Maximum (minimum) RM is 19.5 (-13.6) rad m^{-2}.

numerical models of propagation of polarized radiation through a mixed (synchrotron emitting and Faraday-rotating) medium show a change of sign of RM also without a reversal in the magnetic field ([11]).

We detected seventeen polarized extragalactic sources in the Auriga field at a higher resolution ($\sim 1'$), with RM's from -13.6 to 19.5 rad m^{-2}. The right plot in Fig. 2 shows the RM's and positions of the sources, where the sizes of the circles are proportional to RM, and open (filled) circles denote negative (positive) RM's. The RM's of the extragalactic sources exhibit a clear gradient across the field of ~ 5 rad m^{-2} per degree roughly in the direction of galactic latitude, indicating a galactic component to the RM's of the sources. The change of sign over the field means a (local) reversal in the magnetic field parallel to the line of sight. We estimate a RM component intrinsic to the source of < 5 rad m^{-2}, consistent with earlier estimates ([12]). The RM structure of the diffuse galactic radiation is independent from the observed RM of the extragalactic sources (see below).

ANALYSIS OF THE HOROLOGIUM FIELD

The left map in Fig. 3 shows polarized intensity at 349 MHz in the Horologium field. The average polarization brightness temperature is ~ 5 K, and $T_{b,pol}$ in the ring-like structure is ~ 11-15 K. The average degree of polarization is 5% and the maximum 25%, again derived using the Haslam survey ([7]). The ring with diameter $\sim 2.7°$ is visible in all frequency bands, although the ring becomes more diffuse and smeared out towards higher frequencies and the left side is clearer than the right side. Beam-wide depolarized canals are again caused by beam depolarization. In the lower right, the depolarized canals are aligned along constant latitude. Caution is required in interpreting the P map, as the Horologium field is imbedded in a region of very high constant polarization ([13]). Due to missing small spacings, we cannot detect this large-scale component in Q and/or U, which causes a distorted image in P as well as RM. Possible solutions to this problem will be given in a forthcoming paper. A ring in P was also detected by Verschuur ([14]) at 40′ resolution with a single dish, although not as an enhancement but as a deficiency in P.

In the RM map of the Horologium field (the right hand map of Fig. 3), a circular structure is clearly visible, which is slightly bigger than the ring in polarized intensity. Inside this RM disk, RM's decrease from the edge of the ring to the center, from ~ 0 to -10 rad m^{-2}. Outside the disk, RM's vary around zero without a clear gradient, with a maximum of ~ 7 rad m^{-2}. Note that if P is low outside the disk, the influence of undetected large-scale polarization becomes larger and RM's may not be well-determined. The left and center plots of Fig. 4 give horizontal cross sections through the center of the RM disk and through the P ring at 349 MHz at the same position. The decrease in RM within the RM disk can be modeled with a homogeneous sphere of constant electron density and magnetic field. For values of electron density $n_e = 0.03$ cm^{-3} and parallel magnetic field $B_{\parallel} = 2\mu$G, the distance to the sphere is about 300 pc.

In the Horologium field, we detected 18 polarized extragalactic point sources as shown in the right plot of Fig. 4.

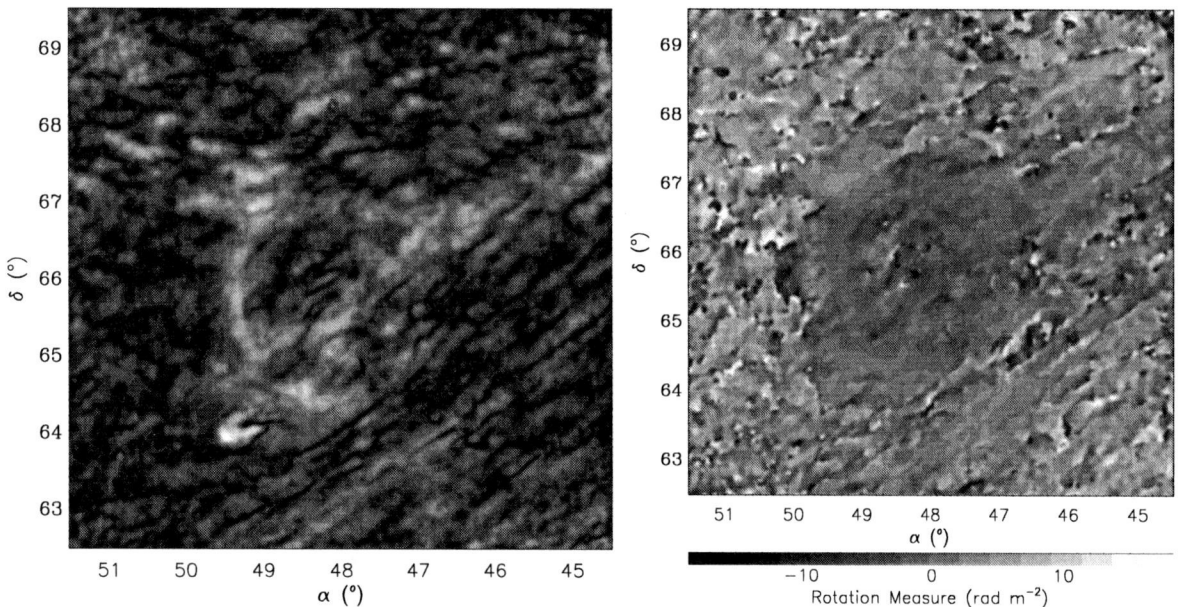

FIGURE 3. *Left:* polarized intensity P at 349 MHz in the Horologium field at $4'$ resolution. White denotes a maximum $T_{b,pol} \approx 15$ K. *Right:* RM in the Horologium field. Very high or low RM values ($|RM| \approx 30-60$ rad m^{-2}) in the field have been removed from the maps (see text).

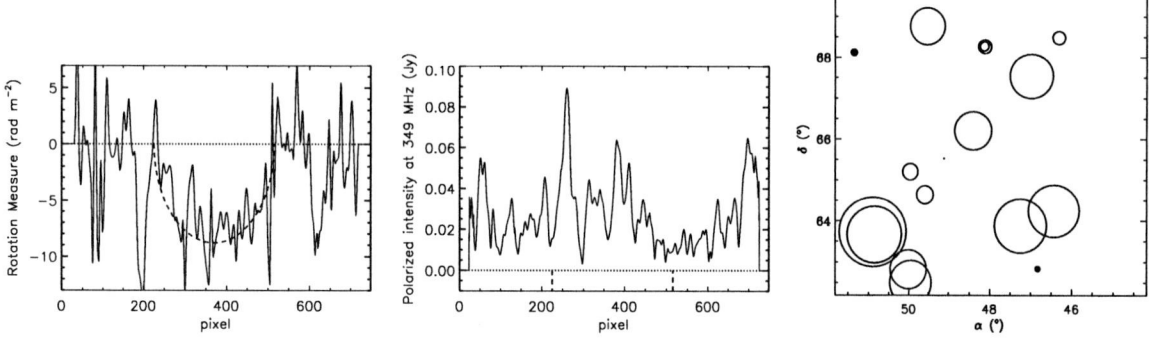

FIGURE 4. *Left and center:* horizontal cross sections through the RM map and the P map of the Horologium field in Fig. 3 (at $4'$ resolution) through the center of the RM disk. The dashed line in the RM plot is a fit to the decrease in RM, calculated for a sphere of constant thermal electron density and line-of-sight magnetic field. In the P plot, the same fit is indicated at the bottom. *Right:* RM's of observed polarized extragalactic point sources in the Horologium field. The radii of the circles are again scaled with magnitude of RM, but with radii twice as small as in Fig. 2. Maximum (minimum) RM is 7.4 (-67.9) rad m^{-2}.

RM's of these sources are in the range of -67.9 to 7.4 rad m^{-2}. These RM values are not correlated with the RM's from the diffuse radiation, similar as in the Auriga field. Extragalactic source RM's are higher than in the Auriga field, which is most likely due to the lower latitude of the Horologium field.

INTERPRETATION OF THE OBSERVATIONS

The galactic synchrotron emission is thought to originate from two separate domains centered on the galactic plane: a thin and a thick disk, with scale heights of 180 pc and 1.8 kpc respectively ([16]). The thick disk emits 90% of the synchrotron radiation, and the thin disk coincides approximately with the stellar disk and the HI disk. Pulsar RM observations have shown that the random magnetic field in the thin disk $B_{ran} \approx 5\mu G$ is of the same order or larger than

TABLE 1. Three schematic domains with different characteristics in the galactic ISM and halo.

domain	components			scale height (pc)	relevant constituents	B_{ran}/B_{reg}
III			Thick disk	1800	n_{rel}, B	≤ 0.1
II		Reynolds layer		1000	n_{rel}, B, n_{th}	≤ 0.1
I	Thin disk			180	n_{rel}, B, n_{th}	≥ 1

the regular magnetic field ([17],[18]).

The thermal electrons that cause the Faraday rotation of the synchrotron background are contained in the Reynolds layer, with a scale height of about a kpc ([19]). The thermal electrons are more confined to the galaxy than the relativistic electrons, so there is a "halo" above the Reynolds layer that only contains relativistic but no thermal electrons.

These two mediums thus define three domains in the galactic ISM and halo with different characteristics, as sketched in Table 1. The thin synchrotron disk (domain I) extends to a few hundred parsec, and is mixed with the lower parts of the Reynolds layer and the thick disk. The Local Bubble is not taken into account. The upper part of the Reynolds layer is also mixed with the thick synchrotron disk (domain II), whereas the highest part of the thick synchrotron disk is so high above the galactic plane that it doesn't contain a significant amount of thermal electrons anymore (domain III). Our observations, in particular the very low upper limit on small-scale structure in total intensity I, put strict constraints on the characteristics of these three domains.

The first constraint is a small scale height of the thin disk. Due to the large random magnetic field component in the thin disk, the emissivity I has a fluctuating component. If the observed integrated emissivity of this fluctuating component is only a few Kelvin, the I-structure can be averaged out if it is on small enough scales, i.e. if there are enough "turbulent cells" along the line of sight. Therefore only a thin layer is allowed where the B_{ran} component is large, with structure on small enough scales to average out the small-scale structure in I that is created in this layer. Typically, in a layer with a scale height of 200 pc, B_{ran}-structure on scales of 10 to 20 pc can smooth I down to the observed limits.

The second constraint is a constraint on the magnetic field in the layers above the thin synchrotron disk. Small-scale structure in I emitted in these layers has to be negligible. Assuming equipartition in energy between the relativistic electrons and magnetic field, the synchrotron emissivity $\varepsilon \propto B^2$. As the fluctuation in I requires $\Delta\varepsilon < 1\%$, this implies that $B_{ran}/B_{reg} < 0.1$. Therefore, the absence of small-scale structure in I dictates a regular magnetic field dominating over the random magnetic field component by more than a factor ten in the halo of the galaxy, i.e. above a scale height of a few hundred parsecs, at least for the component of the magnetic field perpendicular to the line of sight.

The observed structure in polarized intensity is made by depolarization on small scales in the layers containing thermal electrons, by three mechanisms. The first mechanism is beam depolarization, as discussed above. Second, linearly polarized radiation emitted at different depths is Faraday-rotated by different amounts (differential Faraday rotation [20]) and third, small-scale structure in thermal electron density and/or magnetic field causes spatial structure in Faraday rotation (internal Faraday dispersion [21]). These depolarization mechanisms define a wavelength dependent Faraday depth for the polarized radiation, which indicates that most of the observed polarized intensity comes from the nearer part of the medium. This explains the difference between RM structure from the diffuse emission and from extragalactic point sources, as the latter is built up over the whole path length through the Faraday-rotating medium (domains I and II). A more quantitative discussion can be found in Haverkorn et al. ([10])

CONCLUSIONS

We observed two fields of over 50 square degrees, both located in the second galactic quadrant at positive latitude, in the constellations Auriga and Horologium. All four Stokes parameters were derived from observations done at five frequencies 341, 349, 355, 360, and 375 MHz simultaneously. The total intensity I emission is featureless on scales smaller than approximately a degree, while linear polarizations Q and U show abundant structure on arcminute to degree scales. Polarized intensity has a maximum $T_{b,pol} \approx 15 - 18$ K.

The observed structure in polarized intensity P is 'cloudy' on scales from arcminutes up to a degree. Long canals of one synthesized beam wide where no polarization is detected are caused by beam depolarization in a beam which separates two regions where the polarization angle changes abruptly by 90° (or 270°, 540° etc).

Values of RM in the two fields are in general small: $|RM| < 15$ rad m^{-2}. RM maps show coherent structure in RM over several independent beams up to a degree, but also sudden large RM changes across one beam of more than 100%. Not only the magnitude of the RM changes but regularly also the sign, which is most easily explained by a change in direction of $B_{\parallel}$. In the Horologium field, a ring-like structure in polarized intensity coincides with a disk of radially increasing RM, although the RM disk is slightly larger than the ring in P.

The lack of observed small-scale structure in total intensity I puts constraints on the medium (the galactic ISM and halo). In the thin disk of synchrotron radiation, comparable to the stellar and HI disk, the random magnetic field component B_{ran} is comparable to or larger than the regular field component B_{reg}. As B_{ran} causes fluctuating I, the layer cannot be thicker than a few hundred parsecs, and the structure has to be on small enough scales to average out the created I fluctuations. Furthermore, in the thick synchrotron disk above the thin disk, the random magnetic field has to be very small: $B_{ran} \leq 0.1 B_{reg}$. Because of depolarization in the synchrotron emitting and Faraday-rotating medium most of the polarized emission we observe originates from the nearby medium. As the RM of an extragalactic point source is built up along the entire path length through the medium, the RM structure of the diffuse emission can be uncorrelated with the structure in RM values of extragalactic point sources, as is observed.

ACKNOWLEDGMENTS

The Westerbork Synthesis Radio Telescope is operated by the Netherlands Foundation for Research in Astronomy (ASTRON) with financial support from the Netherlands Organization for Scientific Research (NWO). This work is supported by NWO grant 614-21-006.

REFERENCES

1. Wieringa, M. H., de Bruyn, A. G., Jansen, D., Brouw, W. N., and Katgert, P., A&A, 268, 215 (1993).
2. Uyanıker, B., Fürst, E., Reich, W., Reich, P., and Wielebinski, R., A&AS, 138, 31 (1999).
3. Gray, A. D., Landecker, T. L., Dewdney, P. E., Taylor, A. R., Willis, A. G., and Normandeau, M., ApJ, 514, 221 (1999).
4. Gaensler, B. M., Dickey, J. M., McClure-Griffiths, N. M., Green, A. J., Wieringa, M. H., and Haynes, R. F., ApJ, 549, 959 (2001).
5. Duncan, A. R., Haynes, R. F., Jones, K. L., and Stewart, R. T., MNRAS, 291, 279 (1997).
6. Katgert, P., and de Bruyn, A. G., "Small-scale structure in the diffuse polarized radio background: WSRT observations at $\lambda \approx$ 90 cm", in New Perspectives on the Interstellar Medium, edited by A. R. Taylor, T. L. Landecker, and G. Joncas, Astronomical Society of the Pacific, 1999, p. 411.
7. Haslam, C. G. T., Stoffel, H., Salter, C. J., and Wilson, W. E., A&AS, 47, 1 (1982).
8. Roger, R. S., Costain, C. H., Landecker, T. L., and Swerdlyk, C. M., A&AS, 137, 7 (1999).
9. Haverkorn, M., Katgert, P., and de Bruyn, A. G., A&A, 4, L245 (2000).
10. Haverkorn, M., Katgert, P., and de Bruyn, A. G., A&A, (2002), submitted.
11. Sokoloff, D. D., Bykov, A. A., Shukurov, A., Berkhuijsen, E. M., Beck, R., and Poezd, A. D., MNRAS, 299, 189 (1998).
12. Leahy, J. P., MNRAS, 226, 433 (1987).
13. Brouw, W. N., and Spoelstra, T. A. T., A&AS, 26, 129 (1976).
14. Verschuur, G. L., Obs, 88, 15 (1968).
15. Simard-Normandin, M., and Kronberg, P. P., ApJ, 242, 74 (1980).
16. Beuermann, K., Kanbach, G., and Berkhuijsen, E. M., A&A, 153, 17 (1985).
17. Rand, R. J., and Kulkarni, S., ApJ, 343, 760 (1989).
18. Ohno, H., and Shibata, S., MNRAS, 262, 953 (1993).
19. Reynolds, R. J., ApJ, 339, 29 (1989).
20. Gardner, F. F., and Whiteoak, J. B., ARA&A, 4, 245 (1966).
21. Burn, B. J., MNRAS, 133, 67 (1966).

CMB Polarization: Scientific Case and Data Analysis Issues

A. Balbi*[†], P. Cabella[†], G. de Gasperis[†], P. Natoli*[†] and N. Vittorio*[†]

*INFN, Sezione di Roma II
[†]Dipartimento di Fisica, Università Tor Vergata, Roma I-00133, Italy

Abstract. We review the science case for studying CMB polarization. We then discuss the main issues related to the analysis of forth-coming polarized CMB data, such as those expected from balloon-borne (e.g. BOOMERanG) and satellite (e.g. Planck) experiments.

INTRODUCTION

Strong theoretical arguments suggest the presence of fluctuations in the polarized component of the cosmic microwave background (CMB) at a level of 5-10% of the temperature anisotropy. A wealth of scientific information is expected to be encoded in this polarized signal. However, while the existence of anisotropies in the temperature of the Cosmic Microwave Background (CMB) has now been firmly established by several experiments [1, 2, 3], only upper limits are currently available for fluctuations in the polarization of the CMB radiation.

The prospect of detecting CMB polarization anisotropy at small angular scales is now more promising than in the past. In the next few years, a number of experiments (e.g. BOOMERanG, Planck) will have the right sensitivity, as well as the necessary control on systematic effects, to make the measurement of polarization an achievable goal.

In this contribution we will first quickly review the major features of CMB polarization, then we will address some of the issues that will have to be faced in order to analyze the data collected by the forthcoming experiments.

CMB POLARIZATION: THEORETICAL FRAMEWORK

There are at least three features of CMB polarization that make it an appealing target for observation. First, CMB polarization anisotropies are generated at last scattering, so they are not affected by effects taking place after recombination, as opposed to temperature anisotropies. Second, distinctive polarization patterns are produced by different kind of density perturbations (e.g. scalar, vector, tensor). Finally polarization provides information complementary to temperature, helping in clarifying issues such as cosmological parameter degeneracies in the temperature power spectrum.

The rest of this chapter covers basic theoretical aspects of CMB polarization. Excellent reviews on the subject are [4, 5].

Formalism

A useful way to characterize the polarization properties of the CMB is to use the Stokes parameters formalism [6]. For a nearly monochromatic plane electromagnetic wave propagating in the z direction,

$$E_x = a_x(t)\cos\left[\omega_0 t - \theta_x(t)\right], \qquad E_y = a_y(t)\cos\left[\omega_0 t - \theta_y(t)\right], \tag{1}$$

the Stokes parameters are defined by:

$$I \equiv \langle a_x^2 \rangle + \langle a_y^2 \rangle, \qquad Q \equiv \langle a_x^2 \rangle - \langle a_y^2 \rangle, \qquad U \equiv \langle 2a_x a_y \cos(\theta_x - \theta_y) \rangle, \qquad V \equiv \langle 2a_x a_y \sin(\theta_x - \theta_y) \rangle, \tag{2}$$

CP609, *Astrophysical Polarized Backgrounds,* edited by S. Cecchini et al.

where the brackets $\langle\rangle$ represent time averages. The parameter I is simply the average intensity of the radiation. The polarization properties are described by the remaining parameters: Q and U describe linear polarization, while V describes circular polarization. Unpolarized radiation (or natural light) is characterized by having $Q = U = V = 0$. CMB polarization is produced through Thomson scattering (see below) which, by symmetry, cannot generate circular polarization. Then, $V = 0$ always for CMB polarization. The Stokes parameters Q and U are not scalar quantities. If we rotate the reference frame of an angle ϕ around the direction of observation, Q and U transform as:

$$Q' = Q\cos(2\phi) + U\sin(2\phi), \qquad U' = -Q\sin(2\phi) + U\cos(2\phi). \tag{3}$$

We can define a *polarization vector* **P** having:

$$|\mathbf{P}| = \left(Q^2 + U^2\right)^{1/2}, \qquad \alpha = \frac{1}{2}\tan^{-1}\left(\frac{U}{Q}\right). \tag{4}$$

Although **P** is a good way to visualize polarization, it is not properly a vector, since it remains identical after a rotation of π around z, thus defining an orientation but not a direction. Mathematically, Q and U can be thought as the components of the second-rank symmetric trace-free tensor:

$$\mathbf{P}_{ab} = \frac{1}{2}\begin{pmatrix} Q & -U\sin\theta \\ -U\sin\theta & -Q\sin^2\theta \end{pmatrix}, \tag{5}$$

where the trigonometric functions come from having adopted a spherical coordinate system.

Physical Mechanisms

The CMB photons interact before recombination with the free electrons of the primeval plasma through Thomson scattering. The dependence of the Thomson scattering cross-section on polarization is given by:

$$\frac{d\sigma}{d\Omega} = \frac{3\sigma_T}{8\pi}\left|\hat{\varepsilon}\cdot\hat{\varepsilon}'\right|^2, \tag{6}$$

where $\hat{\varepsilon}$ and $\hat{\varepsilon}'$ are incident and scattered polarization directions. After scattering, initially unpolarized light has:

$$I = \frac{3\sigma_T}{16\pi}I'\left(1 + \cos^2\theta\right), \qquad Q = \frac{3\sigma_T}{16\pi}I'\sin^2\theta, \qquad U = 0. \tag{7}$$

Integrating over all incoming directions gives:

$$I = \frac{3\sigma_T}{16\pi}\left[\frac{8}{3}\sqrt{\pi}a_{00} + \frac{4}{3}\sqrt{\frac{\pi}{5}}a_{20}\right], \qquad Q - iU = \frac{3\sigma_T}{4\pi}\sqrt{\frac{2\pi}{15}}a_{22}. \tag{8}$$

Then, polarization is only generated when a quadrupolar anisotropy in the incident light at last scattering is present. This has two important consequences. Because it is generated by a causal process, CMB polarization peaks at scales smaller than the horizon at last scattering. Moreover, the degree of polarization depends on the thickness of last scattering surface. As a result, the polarized signal for standard models at angular scales of tens of arcminutes is about 10% of the total intensity (even less at larger scales). Typically, this means a polarized signal of a few μK.

Statistics

Since polarization is not a scalar, it cannot be expanded over the sky using spherical harmonics, as it is done with temperature. We can however expand the polarization tensor using the *tensor spherical harmonics* basis [7], as:

$$\mathbf{P}_{ab} = T_0\sum_{l=2}^{\infty}\sum_{m=-l}^{l}\left[a_{(lm)}^{\mathrm{E}}Y_{(lm)ab}^{\mathrm{E}} + a_{(lm)}^{\mathrm{B}}Y_{(lm)ab}^{\mathrm{B}}\right] \tag{9}$$

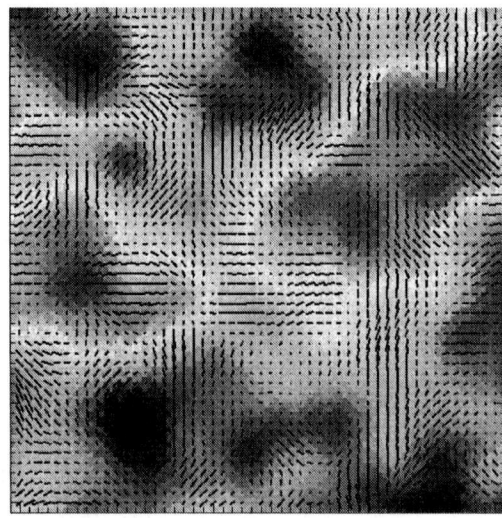

FIGURE 1. A simulated CMB temperature map and the corresponding polarized component represented by the polarization vector $|\mathbf{P}|$, for a standard cosmological model where only scalar density perturbations are present. The field is $6° \times 6°$, the resolution is $10'$ FWHM.

where the E and B labels refer to the scalar and pseudo-scalar components of the polarization tensor. The statistical properties of the CMB anisotropies polarization are then characterized by six power spectra: C_l^T for the temperature, C_l^E for the E-type polarization, C_l^B for the B-type polarization, $C_l^{TE}, C_l^{TB}, C_l^{EB}$ for the cross correlations. For the CMB, $C_l^{TB} = C_l^{EB} = 0$. Furthermore, since B relates to the component of the polarization field which possesses a handedness, one has $C_l^B = 0$ for scalar density perturbations. The detection of a non zero B component would point to the existence of a tensor contribution to density perturbations.

The relation relating (E,B) to (Q,U) has a non-local nature. In the limit of small angles it can be written as:

$$E(\theta) = -\int \mathbf{d^2\theta'}\, \omega(\tilde{\theta})\, \mathbf{Q_r}(\theta'), \qquad \mathbf{B}(\theta) = -\int \mathbf{d^2\theta'}\, \omega(\tilde{\theta})\, \mathbf{U_r}(\theta'), \tag{10}$$

where the 2D angle θ defines a direction of observation in the coordinate system perpendicular to z,

$$Q_r(\theta) = Q(\theta')\cos(2\tilde{\phi}) - U(\theta')\sin(2\tilde{\phi}), \qquad U_r(\theta) = U(\theta')\cos(2\tilde{\phi}) + Q(\theta')\sin(2\tilde{\phi}). \tag{11}$$

and $\omega(\tilde{\theta})$ is a generic window function.

Theoretical Predictions

The theoretical study of the CMB, for what concerns both its polarized and unpolarized components, is in a fully mature stage. We can produce high-precision predictions of the expected statistical CMB pattern for any given cosmological model. Figure 1 shows a simulated map of the CMB temperature anisotropy, and the corresponding polarization field, represented by the polarization vector $|\mathbf{P}|$. This kind of simulation can be very helpful in investigating optimal observational strategy for future experiments. In particular, now that high-resolution CMB temperature maps are available for certain areas of the sky, one can use this information to predict the statistical properties of the expected polarized signal in those regions, and tailor the polarization observations to enhance the likelihood of a detection. Figure 2 shows an example of how a polarization measurement could complement the information from temperature. Two models that would be undistinguishable from their temperature power spectra (because of the degeneracy between the effect of reionization and of tensor modes) can be discerned by their signature in the polarization power spectrum: only the model having a tensor contribution produces B-type polarization. Furthermore, reionization produces a bump at low l's in the polarization spectrum.

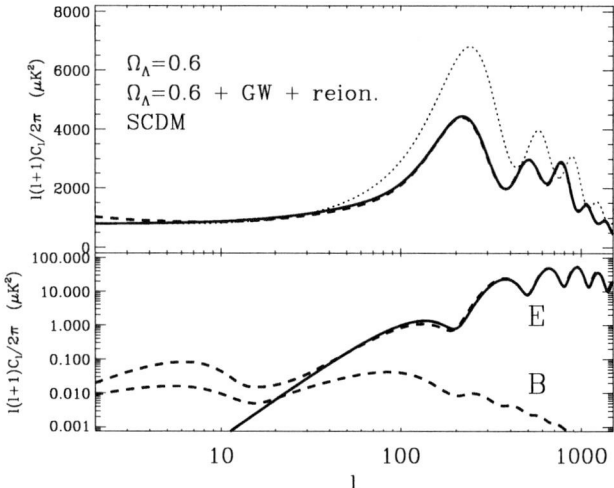

FIGURE 2. *Top panel* – Temperature power spectra for a standard CDM model (solid line), and the same CDM model with a fraction of the critical density coming from a cosmological constant (dotted line). If we add to the latter a contribution from tensor perturbations (gravitational waves background) and reionization (dashed line) we can make it indistinguishable from the standard CDM model. *Bottom panel* – The same models and their polarization power spectra. The CDM model can be identified by its polarized signal, because it does not generate a B-type component

CMB POLARIZATION: DATA ANALYSIS

In order to extract the cosmological information encoded in the CMB polarization, one has to face the challenge of a complicated data analysis stage after the observations are performed. CMB data analysis was successfully performed for recent CMB temperature experiments. While analyzing polarized data is in principle not different from unpolarized data, further complications have to be addressed. In the following we show how the problem of producing polarized maps from time-ordered observations can be addressed using the same kind of algorithms and formalism developed for the temperature case [8].

Map-making

We can write the total signal measured from a generic noiseless polarimeter as:

$$\mathcal{M} = \frac{1}{2}\left(I + Q\cos 2\phi + U\sin 2\phi\right). \tag{12}$$

Consider the experimental set-up shown in Figure 3, where the four polarimeters are assumed to observe the same

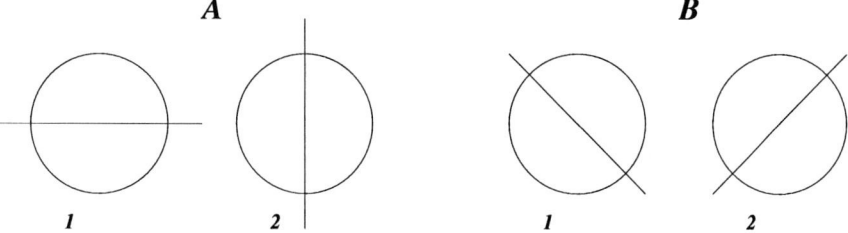

FIGURE 3. Experimental set-up for two radiometers, each measuring two polarization states at 90° orientation. Furthermore, the radiometer B polarimeters are at 45° with respect to radiometer A

point on the sky. Then, one can write:

$$Q = \left(\mathcal{M}_{A1} - \mathcal{M}_{A2}\right)\cos 2\phi - \left(\mathcal{M}_{B1} - \mathcal{M}_{B2}\right)\sin 2\phi \tag{13}$$

$$U = (\mathcal{M}_{A1} - \mathcal{M}_{A2})\sin 2\phi + (\mathcal{M}_{B1} - \mathcal{M}_{B2})\cos 2\phi. \tag{14}$$

More generally, the time-ordered data stream for a noisy polarimeter (i) is:

$$d_t^{(i)} = \frac{1}{2}A_{tp}^{(i)}\left[I + Q_p\cos 2\phi_t^{(i)} + U_p\sin 2\phi_t^{(i)}\right] + n_t^{(i)}. \tag{15}$$

For the previous set-up:

$$\widetilde{d}_t^{\,A} = A_{tp}^A\left[Q_p\cos 2\psi_t + U_p\sin 2\psi_t\right] + \widetilde{n}_t^{\,A} \tag{16}$$

$$\widetilde{d}_t^{\,B} = A_{tp}^B\left[-Q_p\sin 2\psi_t + U_p\cos 2\psi_t\right] + \widetilde{n}_t^{\,B}, \tag{17}$$

$$\tag{18}$$

where:

$$\widetilde{d}_t^{\,A} \equiv d_t^{A1} - d_t^{A2}, \quad \widetilde{d}_t^{\,B} \equiv d_t^{B1} - d_t^{B2}, \quad \widetilde{n}_t^{\,A} \equiv n_t^{A1} - n_t^{A2}, \quad \widetilde{n}_t^{\,B} \equiv n_t^{B1} - n_t^{B2}. \tag{19}$$

We can recast everything in a matrix formalism:

$$\mathbf{D}_t = \mathbf{A}_{tp}\mathbf{S}_p + \mathbf{n}, \tag{20}$$

where:

$$\mathbf{D}_t \equiv \begin{pmatrix} \widetilde{d}_t^{\,A} \\ \widetilde{d}_t^{\,B} \end{pmatrix}, \quad \mathbf{A}_{tp} \equiv \begin{pmatrix} \cos 2\psi_t A_{tp}^A & \sin 2\psi_t A_{tp}^A \\ -\sin 2\psi_t A_{tp}^B & \cos 2\psi_t A_{tp}^B \end{pmatrix}, \quad \mathbf{S}_p \equiv \begin{pmatrix} Q_p \\ U_p \end{pmatrix}, \quad \mathbf{n}_t \equiv \begin{pmatrix} n_t^A \\ n_t^B \end{pmatrix}. \tag{21}$$

We then obtain the standard map-making solution:

$$\widetilde{\mathbf{S}}_\mathbf{p} = \left(\mathbf{A}^t\mathbf{N}^{-1}\mathbf{A}\right)\mathbf{A}^t\mathbf{N}^{-1}\mathbf{D}, \tag{22}$$

where:

$$\mathbf{N} \equiv \langle \mathbf{n}_t\mathbf{n}_{t'}\rangle = \begin{pmatrix} \langle n_t^A n_{t'}^A\rangle & \langle n_t^A n_{t'}^B\rangle \\ \langle n_t^B n_{t'}^A\rangle & \langle n_t^B n_{t'}^B\rangle \end{pmatrix}, \tag{23}$$

which can be further simplified if:

$$\langle n_t^A n_{t'}^B\rangle = \langle n_t^B n_{t'}^A\rangle = \mathbf{0}. \tag{24}$$

An example of the application of this procedure is shown in Figure 4, for the 100 GHz channel of Planck Low Frequency Instrument.

CONCLUSIONS

Temperature anisotropy measurements have just started to have the accuracy required for high precision cosmology. Polarization has enormous scientific potential, but is still a big challenge, both experimentally (low signal, fine-scale structure, systematics, etc.) and for data analysis (which must be both accurate and efficient). The next few years will most likely bring us definitive high-resolution high-sensitivity maps of the CMB temperature anisotropy by satellites (MAP, Planck). The new frontier of cosmological exploration will then shift towards observations of CMB polarization, which will certainly provide us with new insights about the physics of the early universe.

ACKNOWLEDGMENTS

It is a pleasure to thank the organizers of this interesting workshop for the invitation and for the stimulating environment. We acknowledge use of HEALPix (http://www.eso.org/science/healpix/) and CMBFAST.

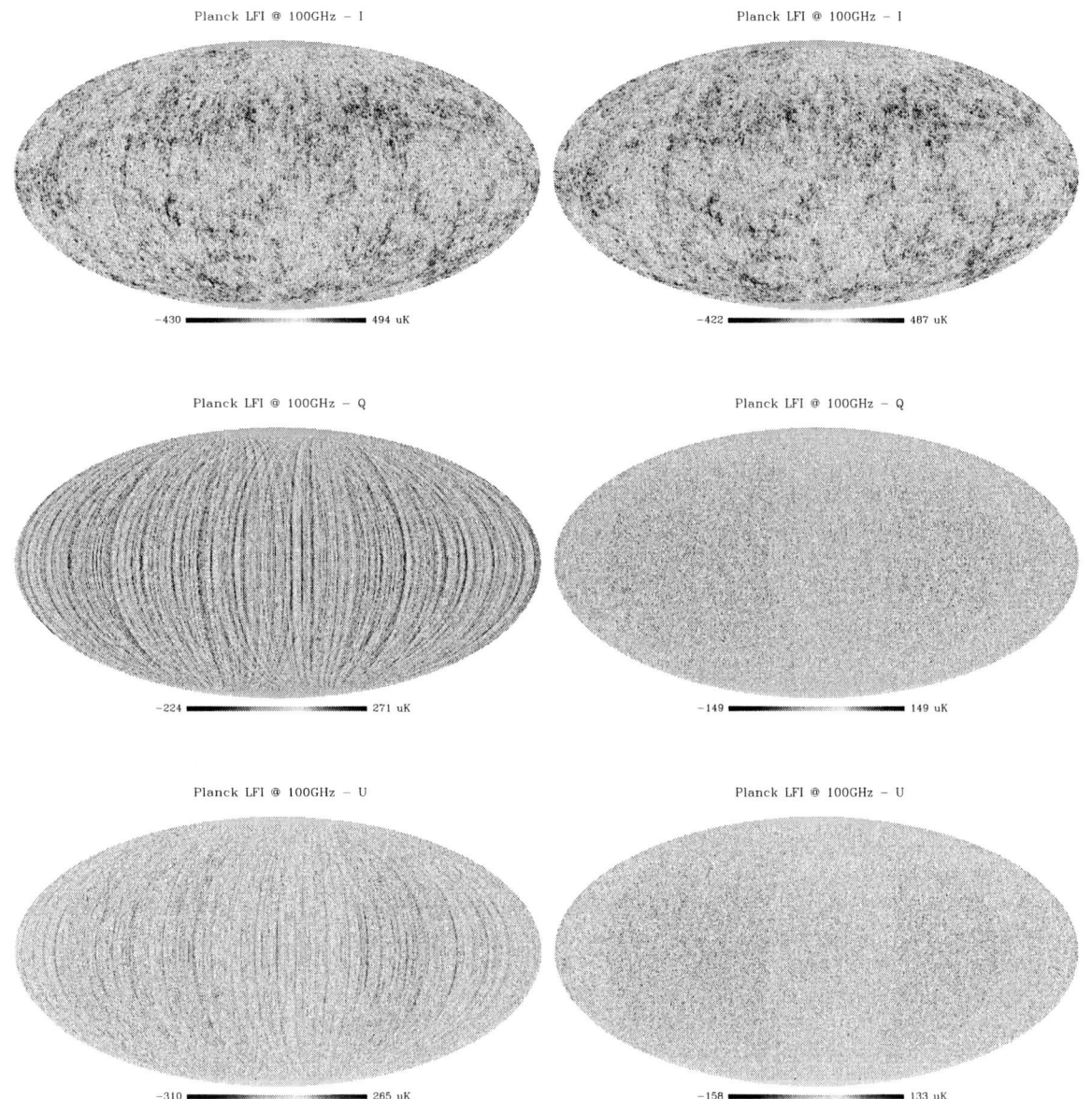

FIGURE 4. Simulated maps for the 100 GHz channels (32 radiometers) of Planck/LFI. Shown from top to bottom are the I, Q and U components, obtained by a naive coadding of observations in each pixel (left) and using the map-making procedure described in the text (right).

REFERENCES

1. De Bernardis, P., et al., *Nature* 404 (2000) 955-959
2. Hanany, S., et al., *Astrophys.J.* 545 (2000) L5
3. Halverson, N.W., et al., *Astrophys.J.* in press (2001)
4. Kosowsky, A., *Annals Phys.* 246 (1996) 49-85
5. Hu, W. & White, M., *New Astron.* 2 (1997) 323
6. Chandrasekhar, S., *Radiative Transfer* (1960) Dover, New York
7. Varshalovich, D.A., Moskalev, A.N. and Khersonskii, V.K., *Quantum Theory of Angular Momentum* (1988) World Scientific, Singapore
8. Natoli, P., et al., *A&A* 372 (2001) 346-356

Polarized Foregrounds Power Spectra vs CMB

Carlo Baccigalupi*, Gianfranco De Zotti†, Carlo Burigana** and Francesca Perrotta†

*SISSA/ISAS, Via Beirut 2-4, 34014 Trieste, Italy
†Osservatorio Astronomico di Padova, Vicolo dell'Osservatorio 5, 35122 Padova, Italy
**ITeSRE-CNR, Via Gobetti, 101, I-40129 Bologna, Italy

Abstract. We briefly review our work about the polarized foreground contamination of the Cosmic Microwave Background maps. We start by summarizing the main properties of the polarized cosmological signal, resulting in "electric" (E) and "magnetic" (B) components of the polarization tensor field on the sky. Then we describe our present understanding of sub-degree anisotropies from Galactic synchrotron and from extra-Galactic point sources. We discuss their contamination of the cosmological E and B modes.

INTRODUCTION

Several ongoing or planned experiments are designed to reach the sensitivities required to measure the expected linear polarization of the Cosmic Microwave Background (CMB), see e.g. [1]. The forthcoming space missions MAP and PLANCK aim at obtaining full sky high resolution maps of Cosmic Microwave Background (CMB) anisotropies, up to several arcminute resolution (see e.g.[2, 3]; MAP webpage: http://map.gsfc.nasa.gov/; PLANCK webpage: http://astro.estec.esa.nl/SA-general/Projects/Planck/).

They will also probe the polarization of the CMB radiation. MAP has polarization sensitivity in all channels. The current design of instruments for the PLANCK mission provides good sensitivity to polarization at all LFI (Low Frequency Instrument) frequencies (30–100 GHz) as well as at three HFI (High Frequency Instrument) frequencies (143, 217 and 545 GHz).

While there is a very strong scientific case for CMB polarization measurements (cf., e.g., [4] and references therein), they are very challenging both because of the weakness of the signal and because of the contamination by foregrounds that may be more polarized than the CMB.

In this paper, we begin by giving a description of the key features and meaning of the polarized CMB component. The latter is usually described in terms of the angular power spectra of two components of the CMB polarization signal, namely electric (E) and magnetic (B) modes (see [5] for an extensive treatment). Then we summarize the main results of our recent work [6] on the diffuse Galactic synchrotron polarized emission, focusing on sub-degree anisotropies. These results have been obtained by analyzing the existing high resolution data in the radio band: the Parkes and Effelberg surveys at 2.4 and 2.7 GHz [9] along the Galactic plane, and the medium latitude data at 1.4 GHz [10]. Moreover we present some preliminary results on the power spectrum of polarized emission from extragalactic radio sources obtained exploiting data from the NVSS survey ([7]; http://www.cv.nrao.edu/jcondon/nvss.html). In the last Section we give some concluding remarks.

COSMOLOGICAL POLARISATION MODES: ELECTRIC AND MAGNETIC TYPE

We give here only the basic features of the current description of the cosmological polarized signal. A detailed treatment can be found, e.g., in [5].

Given two orthogonal axes i, j in the plane perpendicular to the photon propagation direction $\hat{n}$, the 2×2 linear polarization tensor I_{ij} is represented by the Stokes parameters Q and U, with $Q = (I_{11} - I_{22})/4$, $U = I_{12}/2$. It is convenient to define the complex quantities $Q \pm iU$, which transform like a definite spin state under rotation by an

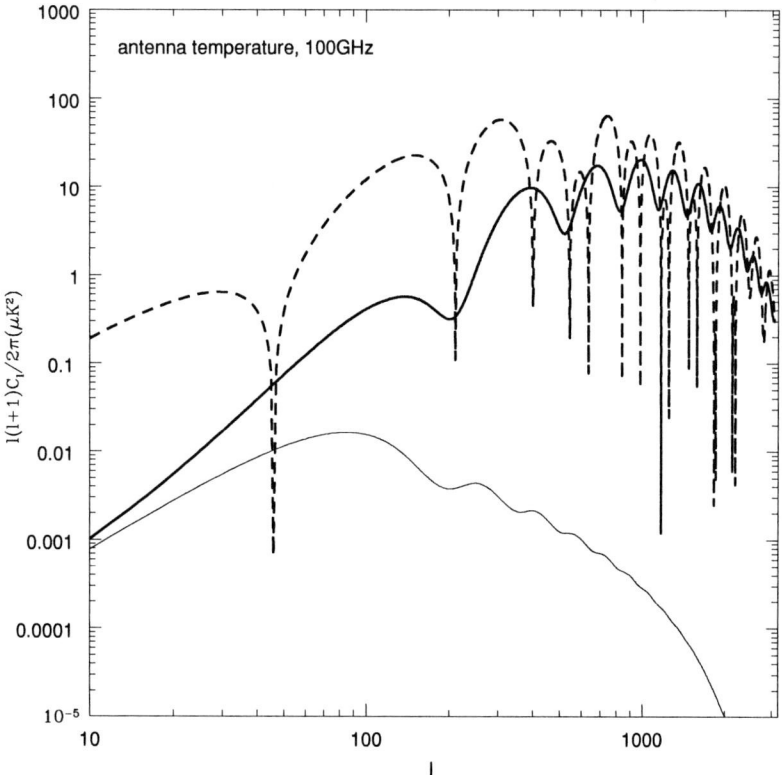

FIGURE 1. Angular power spectrum of CMB polarization for the cosmological model described in the text. Dashed lines refer to the TE correlation, light solid lines to B and heavy solid lines to E modes.

angle ψ around $\hat{n}$:

$$(Q \pm iU) \rightarrow e^{\mp 2i\psi}(Q \pm iU) \; . \tag{1}$$

These quantities can be expanded in the tensor spherical harmonics $\pm 2 Y_l^m$ as

$$(Q \pm iU)(\hat{n}) = \sum_{lm} a_{\pm 2,lm} \pm 2 Y_{lm}(\hat{n}) \; . \tag{2}$$

The expansion coefficients for E and B modes can then be defined as:

$$a_{E,lm} = -(a_{+2,lm} + a_{-2,lm})/2 \; , \; a_{B,lm} = i(a_{+2,lm} - a_{-2,lm})/2 \; . \tag{3}$$

The electric and magnetic analogy comes from the properties of E and B modes under parity transformation $\hat{n} \rightarrow -\hat{n}$: while the $a_{E,lm}$ remain unchanged, the $a_{B,lm}$ change sign [5]. The power spectra associated with E and B modes, as well as their relation with the power spectrum of Q and U can be easily evaluated as:

$$C_l^E = \frac{1}{2l+1} \sum_m |a_{E,lm}|^2 \; , \; C_l^B = \frac{1}{2l+1} \sum_m |a_{B,lm}|^2 \; , \; C_l^E + C_l^B = C_l^Q + C_l^U \; . \tag{4}$$

Due to the opposite parity properties, no correlation exists between E and B. It is also useful to recall that E modes are correlated with the total intensity fluctuations, giving rise to a C_l^{TE} power spectrum. The latter can be stronger than that of E and B CMB spectra since it receives contributions from total intensity fluctuations that are expected to be 10 times or so larger than the polarization ones. The description of the polarization field in terms of E and B modes is more convenient than the classical one in terms of local Q and U Stokes parameter because while E receives contribution from all the types of cosmological perturbations, B is non-zero only if vector or tensor fields are present

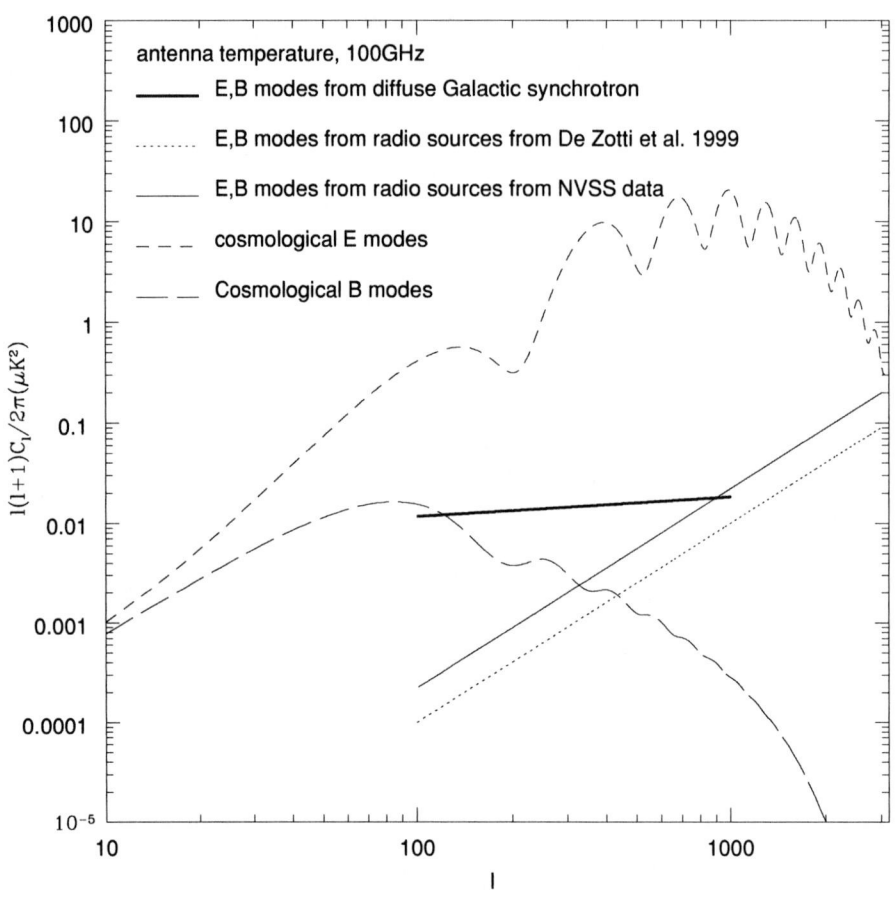

FIGURE 2. Comparison of cosmological E and B power spectra (short- and long-dashed lines, respectively) with those from diffuse synchrotron at low and medium Galactic latitudes (solid line at $100 \leq l \leq 1000$), and from extragalactic radio sources (solid line at $100 \leq l \leq 3000$) at 100 GHz.

in the cosmological perturbations [5]. Of particular interest are tensor perturbations associated to gravitational waves because their amplitude is directly related to the vacuum energy density during inflation.

To give a worked example, consider the constraints on cosmological parameters set by the recent data from BOOMERanG [8]. According to these data, the present cosmological energy density is consistent with the critical one, being made by a 70% of vacuum energy ($\Omega_\Lambda = 70\%$), a 25% of dark matter and a 5% of baryons with an Hubble constant of 70 km/sec/Mpc ($\Omega_b h^2 = 0.022$). In Fig. 1 the TE, E and B power spectra are shown for this cosmological model, further assuming a contribution to the temperature quadrupole of tensor perturbations equal to 30% of the contribution of scalar perturbations. The main power resides in TE and E since these modes receive inputs from acoustic oscillations occurring inside the horizon at decoupling, corresponding to $l \geq 200$ in the figure. The B component is subdominant since it is excited by gravitational waves which decay rapidly inside the horizon.

This example gives an idea of the importance of measuring E and B modes of the CMB polarization fluctuations. It is therefore extremely important to study the power of the foregrounds as contaminants to this signal. In the next two Sections we give our present guess of the contamination coming from the low frequency Galactic and extragalactic emissions, taking as reference model the one presented in Fig. 1.

POLARIZED GALACTIC SYNCHROTRON EMISSION

We summarize here the main results of our recent paper [6] in which we analyzed data from low and medium Galactic latitudes at 1.4, 2.4, 2.7 GHz [9, 10], having resolution of several arcminutes, and from high Galactic latitudes on large angular scales [11]. We focus here on the results concerning the power spectrum on sub-degree angular scales, that we recast in terms of E and B modes.

By comparing total with polarized emissions we were able to observe the following facts. The polarized emission does not show any significant decrease with increasing Galactic latitude, up to the highest latitudes considered ($|b| \simeq 20°$), while the total intensity decreases by a large factor. Correspondingly, the polarization degree increases from typical values of a few percent on the Galactic plane to about 30% at latitudes $10° \leq b \leq 20°$.

We found that the low polarization degree on the Galactic plane can be largely explained by the contribution to the observed total intensity from known intrinsically unpolarized HII regions, catalogued by [12], which are concentrated on the plane. We verified that, after removal of the contributions from HII regions, the polarization degree drops to values consistent with those found at medium latitudes. Of course, HII regions themselves also contribute to Faraday depolarization of synchrotron emission coming from outer Galactic regions.

Regions were identified where rotation measures towards pulsars and extragalactic sources, the high polarization degree and, in some cases, data on the distribution of polarization vectors and on the Galactic magnetic field, consistently indicate low Faraday depolarization. The mean Galactic synchrotron power spectrum was estimated as the average power spectrum of several such regions. In terms of the E and B modes, and assuming a spectrum of the form $S_\nu \propto \nu^{-0.9}$, i.e. antenna temperature $T_A \propto \nu^{-2.9}$, we have, on degree and subdegree angular scales ($100 \leq l \leq 1000$):

$$C_l^E \simeq C_l^B = (1.2 \pm 0.8) \cdot 10^{-9} \cdot \left(\frac{l}{450}\right)^{-1.8 \pm 0.3} \cdot \left(\frac{\nu}{2.4\,\text{GHz}}\right)^{-5.8} \text{K}^2 \,. \tag{5}$$

The power is almost equally distributed among E and B modes, as is expected since the alignment is preferentially determined by magnetic fields, which do not have the characteristic parity properties of scalar density perturbations (see [13]).

In Fig. 2 we plot this results (solid line at $100 \leq l \leq 1000$), scaled to 100 GHz, against the different components of the CMB spectrum shown in Fig. 1. This, albeit preliminary, estimate, suggests that contamination from diffuse synchrotron is not a serious hindrance for measuring the the CMB E-mode polarization, but poses a serious challenge for measurements of the B-mode power spectrum.

EXTRAGALACTIC RADIO SOURCES

The confusion fluctuations due to a Poisson distribution of extragalactic sources in the case of a polarimetric survey have been discussed by [14] and [15]. Briefly, in the case of a population with uniform evolution properties and constant (time-independent) polarization degree Π, the polarization fluctuations σ_P^2 for cells of solid angle ω are simply given by

$$\sigma_P^2 = \sigma_I^2 \langle \Pi^2 \rangle \,, \tag{6}$$

where σ_I^2 is the amplitude of intensity fluctuations for the given cell size (see, e.g. [16]) and

$$\langle \Pi^2 \rangle = \int_0^1 \Pi^2 p(\Pi) d\Pi \,, \tag{7}$$

$p(\Pi)$ being the distribution function of the polarization degree. Clearly, an uncorrelated source distribution give equal contributions to the E- and B-mode power spectra.

The estimates by [15] exploited the models by [16] to estimate σ_I^2. To estimate the mean polarization degree, they defined a complete sub-sample of BL-Lacs for which polarization measurements at cm wavelengths are available. The mean polarization degree at $\lambda = 2\,\text{cm}$ was found to be 5%. The available data at shorter wavelengths suggest that the polarization degree remains constant down to $\lambda \simeq$ few mm. The E-mode (or B-mode) power spectrum of polarization fluctuations due to radio sources, assuming $\Pi = 0.5\%$ for all populations contributing to the 100 GHz counts, is shown in Fig. 2.

A new analysis, currently underway by [17], exploits the NRAO VLA Sky Survey (NVSS) [7] which has provided I, Q, and U data at 1.4 GHz for almost 2×10^6 discrete sources brighter than $s \simeq 2.5\,\text{mJy}$ over about 10.3 sr of sky (about 82% of the celestial sphere). Whenever possible, spectral indices of sources have been determined combining the 1.4 GHz flux densities with those given by the GB6 [18] and PMN [19] catalogues at $\simeq 5\,\text{GHz}$. Extrapolations of polarized fluxes to higher frequencies have been made assuming that the polarization degree is frequency independent. A very preliminary estimate of the derived polarization power spectrum is shown in Fig. 2.

Advantages of this latter approach are that it automatically takes into account the real space distribution of point sources, including clustering effects, as well as the actual distribution of their polarization properties. The large extrapolations in frequency introduce, however, substantial uncertainties. On one side, the polarization degree may be higher at higher frequencies both because the Faraday depolarization becomes negligible and because additional polarized components become optically thin. A hint in this direction is provided by the fact that the mean polarization degree of NVSS sources turns out to be $\simeq 1.4\%$, to be compared with the 5% mean polarization at 15 GHz found by [15]. On the other side, it is known that many sources with flat or inverted spectrum up to $\sim 5\,\text{GHz}$ show spectral breaks at higher frequencies. Thus, the assumption of a constant spectral index up to 100 GHz leads to an overestimate of polarization fluctuations. The two effects go in opposite directions and therefore tend to counterbalance each other. To the extent that the hypothesis of constant spectral indices holds, the effective spectral index is found to be $\alpha_{\text{eff}} \simeq 0.1$ ($S_\nu \propto \nu^{-\alpha}$, which becomes $T_A \propto \nu^{-2-\alpha}$ in antenna temperature). In order to estimate the polarization fluctuation power spectrum we need to specify the maximum flux of contributing sources (i.e. the minimum flux of sources that can be individually detected and subtracted out). Assuming that all sources with total flux larger than 5 times the global rms fluctuations (including contributions of noise, CMB and Galactic foregrounds), as estimated by [16], can be removed, the E- and B-mode power spectrum of polarization fluctuations due to extragalactic sources is described by:

$$C_l^E \simeq C_l^B = 1.4^{+0.7}_{-0.4} \cdot 10^{-7} \mu K^2 \cdot \left(\frac{\nu}{100\,\text{GHz}} \right)^{-4.2}, \qquad (8)$$

This result is represented by the solid line extending up to $l = 3000$ in Fig. 2. We must caution that the assumption about the flux limit for source subtraction may be somewhat optimistic, so that the amplitude of fluctuations may be somewhat underestimated. On the other hand it is reassuring that the two totally independent estimates mentioned above give quite similar results. It is also interesting to note that the power spectrum derived from NVSS data is fully consistent with a Poisson distribution of sources: clustering effects turn out to be essentially negligible, as argued by [16].

CONCLUSIONS

There is growing interest and excitement about CMB polarization studies. Measurements are extremely challenging because of the extreme weakness of the signal to be detected. So, advances in experimental techniques will be crucial, particularly to measure the B-mode power spectrum, induced by gravitational waves. On the other hand, it is not yet clear whether our ability to measure the CMB polarization power spectrum will be limited by detector sensitivity or by foregrounds. In fact, polarized foregrounds are currently very poorly understood.

On the other hand, new surveys are providing important pieces of information, on which we can found preliminary but quantitative estimates of the effect of foregrounds. We have focussed here on polarized synchrotron emission from our own Galaxy and on extragalactic radio sources. As for synchrotron emission, recent high resolution and high sensitivity polarization maps at frequencies in the range 1.4–2.7 GHz ([9, 10]), although covering rather limited regions of the sky, have allowed to estimate the power spectrum at sub-degree angular scales.

We have also presented and briefly discussed polarization fluctuations due to extragalactic radio sources, based on two approaches. On one side there are estimates based on counts as a function of total flux, complemented with estimates of the mean polarization degree. On the other side, the polarization measurements provided by the NVSS were used together with estimates of the spectral index of individual sources derived by combining NVSS data with higher frequency catalogues (GB6 and PMN). The two approaches yield results very close to each other.

Although the analysis is admittedly preliminary and does not consider yet other potential polarized foregrounds at cm/mm wavelengths (e.g. magnetic or spinning dust grains, see [21]), some indications are already emerging. Polarized foregrounds do not seem to be a serious hindrance for measurements of the CMB E-mode power spectrum on degree and sub-degree angular scales, particularly in the frequency range 60–100 GHz (see [20] for a discussion of polarized foregrounds at higher frequencies). However, foregrounds appear to be a potentially serious limiting factor for experiments aimed at detecting B-mode CMB polarization. More data and more detailed analyzes will therefore be essential for designing future experiments.

ACKNOWLEDGMENTS

We thank Dino Mesa for communicating results on extragalactic radio sources in advance of publication. Work supported in part by ASI and MIUR.

REFERENCES

1. Staggs S.T. et al., "CMB Polarization Experiments", in *Microwave Foregrounds*, edited by A. De Oliveira Costa & M. Tegmark, ASP Conf. Ser. 181 ISBN 1-58381-006-4, ASP San Francisco 1999, pp. 299-310.
2. Wright E.L. et al., *New Astr.Rev.* **43**, 257-262, (1999).
3. Mandolesi N. et al., PLANCK Low Frequency Instument, a proposal submitted to ESA (1998); Puget J.L. et al., High Frequency Instrument for the PLANCK mission, a proposal submitted to ESA (1998).
4. Zaldarriaga M., *ApJ* **503**, 1-15 (1998).
5. Hu W., Seljak U., White M., Zaldarriaga M., *Phys.Rev.D* **57**, 3290-3301 (2001).
6. Baccigalupi C. et al., *A&A* **372**, 8-21 (2001).
7. Condon, J. J. et al., *AJ* **115**, 1693-1716 (1998).
8. De Bernardis P. et al., submitted to *ApJ*, preprint astro-ph/0105296, (2001).
9. Duncan A.R. et al., *MNRAS* **291**, 279-295 (1997); Duncan et al., *A& A* **350**, 447-456 (1999).
10. Uyaniker B., et al., A& AS **138**, 31-45 (1999).
11. Brouw W.N., Spoelstra T.A.T., *A& AS* **26**, 129-146 (1976).
12. Paladini R. et al., in preparation, (2002).
13. Seljak U., *ApJ* **482**, 6-16 (1997).
14. Sazhin M.V. & Korolëv V.A., *Sov. Astr. Lett.* **11**, 204-206 (1985).
15. De Zotti G. et al. *New Astr.* **4**, 481-488 (1999).
16. Toffolatti L. et al. *MNRAS* **297**, 117-127 (1998).
17. Mesa D., Baccigalupi C., De Zotti G., in preparation (2002).
18. Gregory, P. C. et al. *ApJS* **103**, 427-432 (1996).
19. Griffith M.R., Wright A.E., *AJ* **105**, 1666-1679 (1993).
20. Prunet S., Lazarian A., "Thermal Dust Emission as a Polarized Foreground", in *Microwave Foregrounds*, edited by A. De Oliveira Costa & M. Tegmark, ASP Conf. Ser. 181 ISBN 1-58381-006-4, ASP San Francisco 1999, pp. 113-134.
21. Draine B.T., Lazarian A., "Microwave Emission from Galactic Dust Grains", in *Microwave Foregrounds*, edited by A. De Oliveira Costa & M. Tegmark, ASP Conf. Ser. 181 ISBN 1-58381-006-4, ASP San Francisco 1999, pp. 133-150.

Cosmic Magnetic Fields:
What we know, what we must discover

R. Wielebinski

Max-Planck-Institut für Radioastronomie, Auf dem Hügel 69, 53121 Bonn, Germany

Abstract. Cosmic magnetic fields have now been studied for over 100 years. Although the initial detection was done at optical wavelengths, it was radio astronomy that gave us the bulk of information on cosmic magnetic fields in the recent years. In this review the observational results obtained by radio astronomy methods will be discussed. In particular maps of Galactic magnetic fields as well as magnetic fields in nearby galaxies will be presented. A discussion of possible developments in the near future will conclude this review.

THE HISTORY

Remote sensing of magnetic fields goes back to the experiments of Piet Zeeman who in 1886 showed that a strong magnetic field would split a spectral line. The Zeeman effect was used by Hale in 1908 to demonstrate the presence of magnetic fields in solar sunspots. In 1946 Babcock observed the Zeeman effect in 'magnetic' stars. In 1949 Hiltner & Hall independently observed the optical polarization of stars. At first these observations were interpreted to be due to scattering only. An extensive theoretical work by Davis & Greenstein in 1951 proposed an alternative interpretation, namely, that the optical polarization was due to dust particles aligned in magnetic fields. There was a controversy surrounding this proposal for many years. Present knowledge, based on radio astronomy observations, confirmed that the 'magnetic' interpretation was correct.

The next important development of or knowledge about magnetic fields came from studies of the Crab nebula. Several observers like Dombrovsky [1], and Woltjer [2] followed up the suggestion of Shklovsky [3] that the optical light of Crab nebula was due to synchrotron mechanism and hence linearly polarized. The successful optical polarization observations were followed by the first positive radio detection of the polarization of the Crab nebula by Mayer et al. [4]. The next breakthrough came with the detection of linear polarization of Galactic radio waves by Westerhout et al. [5] and Wielebinski et al. [6] implying the presence of magnetic fields in a large volume of the Milky Way. In the same year the first linear polarization detection in radio source was made by Mayer et al. [7]. The effects of Faraday rotation by the ionosphere were observed by Wielebinski & Shakeshaft [8] and of the interstellar medium by Muller et al. [9]. The scene was now set for studies of magnetic fields. It was radio astronomy, in particular the use of the linear polarization of the radio continuum emission, that gave us the gross of data on cosmic magnetic fields in a variety of sources. The observations of linear polarization can give information about $B_\perp$, the study of Faraday rotation gives us $B_\parallel$. Combining the two sets of data should allow us to perform a 'tomography' of the magnetic field. The Zeeman effect, a more basic method of measurement, has given so far only sparse results on molecular clouds and Galactic maser sources.

WHAT WE KNOW

We know that magnetic fields are everywhere. The Earth, the planets, Sun, stars, galaxies, radio galaxies, clusters of galaxies are permeated with magnetic fields. We have no observational evidence for a magnetic field of primordial origin. There have been repeated reports of a primordial magnetic field, that have turned out to be false and we now have an upper limit of $B < 10^{-9}$ Gauss. Observations of clusters of galaxies suggest the presence of intergalactic magnetic fields with $B \sim 1 - 2 \cdot 10^{-6}$ Gauss. In nearby galaxies magnetic fields in the range of $2 \times 10^{-6} < B <$

5×10^{-5} Gauss. Magnetic fields in supernova remnants (SNRs) have been estimated to be up to $B \sim 10^{-4}$ Gauss. Magnetic fields of the Earth is well known with $B \sim 0.5$ Gauss. Magnetic fields of planets, measured by in-situ space probe observations, are known to exist with $0.1 < B \lesssim 20$ Gauss. Here a comment must be made, namely, that the rotation rate of a planet (with a warm inner core) seems to be proportional to the magnetic field intensity. On the other end of the scale of magnetic field intensities we find pulsars with $10^8 < B < 10^{12}$ Gauss and magnetars with $B \sim 10^{15}$ Gauss. An intensity range of over 20 orders of magnitude exists in cosmic magnetic fields. In this review I will concentrate on cosmic magnetic fields that are in our foreground, in the Milky Way and nearby galaxies.

The magnetic field of the Galaxy was the first to be investigated. The observations of star polarization, that were collected by Mathewson & Ford [10], gave us the first indication of the large-scale organization of the magnetic field of a galaxy. Most of the 1800 stars measured were within 1000 pc and hence the local magnetic field could be delineated by this method. The orientation of the B vectors was very well organized along the Galactic plane suggesting the existence of an azimuthal magnetic field. Some local anomalies, with B vectors perpendicular to the plane of the Galaxy, were also observed. The early radio observations showed the presence of magnetic fields almost everywhere with one region towards $\ell = 140°$, $b = 8°$ being highly organized. This highly polarized region, studied in great detail by Brouw & Spoelstra [11], shows a magnetic field reversal (positive and negative Rotation Measures) and no associated radio source in total intensity. Also here the magnetic field orientation, deduced from radio observations, was parallel to the plane of the Galaxy. Polarization observations of the southern sky, with considerable angular resolution, were conducted by Mathewson & Milne [12]. From all this work we could say that magnetic fields (polarization) are present in many regions of our Galaxy.

A Galactic plane survey by Junkes et al. [13] at $\lambda 11$ cm found polarized areas that could be attributed to distant ($\sim$ kpc) emission regions. The mapping of the southern Galactic plane at $\lambda 13$ cm by Duncan et al. [14] and the $\lambda 11$ cm northern plane analysis by Duncan et al. [15] showed depolarization along the galactic plane that could be attributed to Faraday effects. Wieringa et al. [16] made observations at 327 MHz with the Westerbork synthesis radio telescope showing unusual filamentary structures on scales of $\sim$ arcmin, well above the plane of the Galaxy, that were definitely due to Faraday depolarization. The Canadian Galactic Plane Survey (CGPS), made at $\lambda 21$ cm in radio continuum, led to delineation of depolarization structures by Gray et al. [17] on arcmin scales. Similar depolarization structures, seen as 'snakes' or 'rings' were found in the Effelsberg $\lambda 21$ cm data of Uyaniker et al. [18] (e.g. Fig. 1). Low-frequency studies by Haverkorn et al. [19] found significant RM variations that can be attributed to the very nearby medium. Observations by Gaensler et al. [20] showed RM variations on 5 arcmin scales with values of $\pm 150 \, \mathrm{rad \, m^{-2}}$ in the direction near $\ell = 327°$, $b = 1°$. From all these data we can say that Faraday depolarization is serious on scales of arcmins at $\lambda 21$ cm. Studies at very low frequencies ($f \sim 327$ MHz) sample the very local Magnetic Interstellar Medium (MISM) while observations at $\lambda 11$ cm sample deep into the Galactic plane.

The polarized emission towards $\ell = 140°$, $b = 8°$ was already noted in the first paragraphs to have no apparent associated radio source. The mapping of the medium latitudes of the Galactic plane by Uyaniker et al. [18] has revealed further such areas of considerable polarization with no apparent total intensity sources. One such region is seen towards $\ell = 74°$, $b = -12°.5$ near the Cygnus loop. This region of $\sim 4°$ extent seems to be well organized with vectors performing a fan-like orientation change. A second region is seen towards $\ell = 42°$, $b = -4°$ in the $\lambda 11$ cm survey of Duncan et al. [15]. All these regions are at present not understood and must be followed up at several additional radio frequencies. Limited studies of the second region (see Reich et al., this conference) shows considerable Faraday rotation, suggesting a deep MISM.

Towards the Galactic Center rotation measures of $\sim 1000 \, \mathrm{rad \, m^{-2}}$ have been observed (e.g. [21]). Possibly these are the highest values for any Faraday depth in the Galaxy. Since these RM's occur in a relatively short line of sight the product of path length, magnetic field intensity and thermal electron density must be very high.

There are two further important methods to study the Galactic magnetic field. The information contained in the Faraday rotation of extragalactic radio sources (e.g. Simard-Normandin & Kronberg [22]) can be used to delineate the structure of the local magnetic field. The surveys of RM's of sources contain only few sources so far in the Galactic plane, within $b = \pm 10°$. The RM's of extragalactic sources are good indicators of the $B_\parallel$ magnetic fields in the thick disk of the local spiral arm. More important are the observations of the pulsar RM and dispersion measure (DM). With this method the average magnetic field along the line of sight can be derived (see e.g. Smith [23], Han et al. [24]). However magnetic field reversals as well as variations of the thermal electron content can have decisive effects on these derivations. The pulsars are observed through the local MISM and this has to be taken account in these studies. A comment in this place must be made that the Galaxy is very different when viewed from the northern or the southern hemisphere. The northern hemisphere emission is dominated by the local (Perseus) spiral arm. In the southern hemisphere the distribution of the emission is more 'correct' and should be more important in the determination of any models of the Galactic magnetic field.

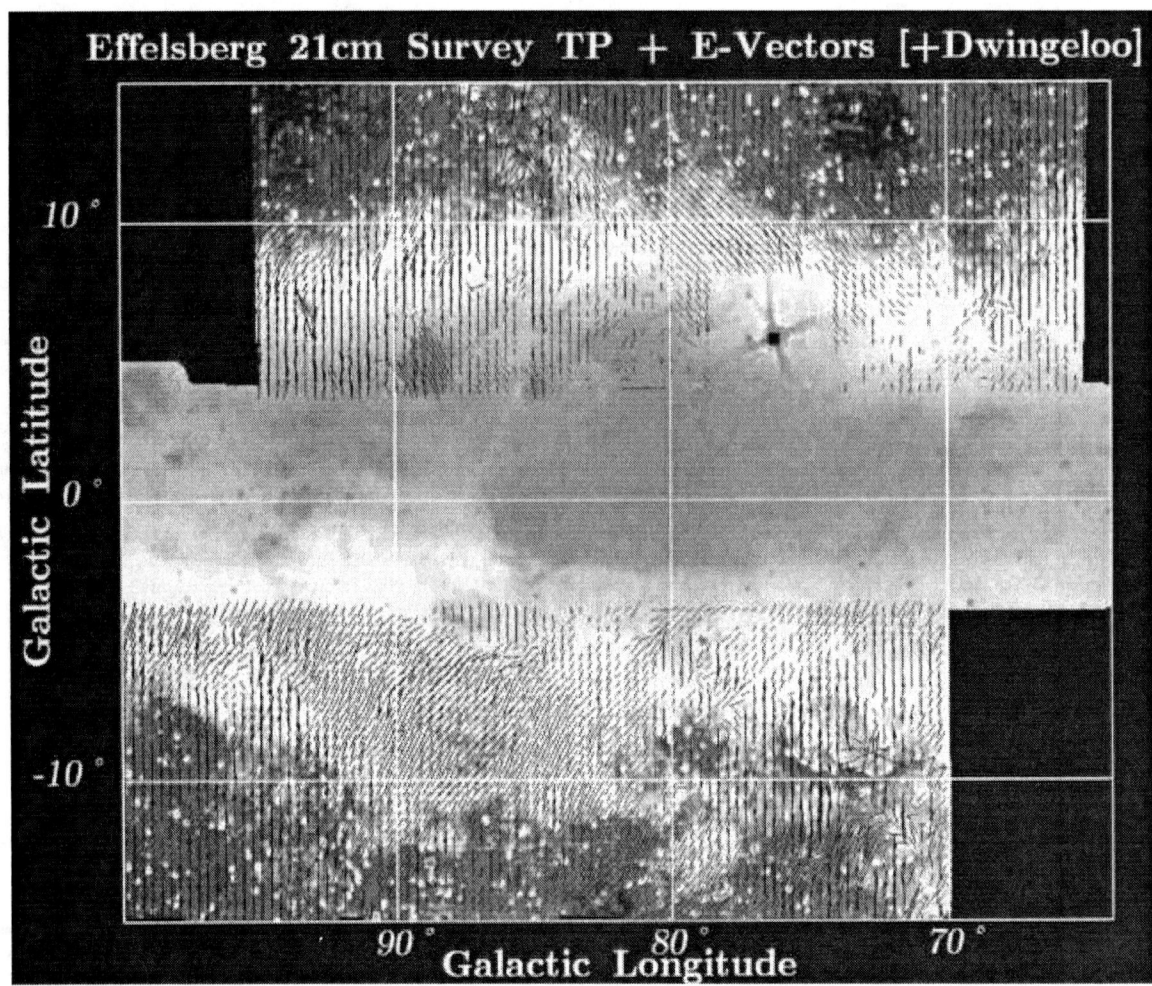

FIGURE 1. A section of the Effelsberg 21 cm medium-latitude survey in the Cygnus A area (courtesy B. Uyaniker)

The linear polarization of the nearby galaxy M51 was first observed by Mathewson et al. [25] with the Westerbork synthesis radio telescope. This early result suggested that the magnetic fields align with the spiral arms and that they are in agreement with earlier optical polarization observations. The study of magnetic fields in galaxies became a domain of research for the 100-m Effelsberg radio telescope of the MPIfR. Polarization of virtually all nearby galaxies was measured at several radio frequencies allowing the delineation to the magnetic field morphology. Some typical results of these observations have been presented by Beck [26] or Wielebinski [27]. Observations of the Magellanic clouds were performed with the Parkes telescope by Klein et al. [28]. Southern galaxies have been studied with the Australia Telescope Compact Array (e.g. [29]). To improve the angular resolution of the maps of magnetic fields Very Large Array data has been combined with Effelsberg data in polarization. An atlas of the magnetic fields in galaxies can be looked up in: http://www.mpifr-bonn.mpg.de (Research/Atlas of Magnetic Fields). The general result is that the magnetic fields are surprisingly homogenous on the large scale, aligned with the spiral arms. In some galaxies (e.g. NGC 6946, NGC 2997) anomalous 'magnetic arms' are observed that follow the optical inter-arm region. In some galaxies (e.g. NGC 4631, M82, Milky Way) vertical magnetic fields are observed in the nuclear areas. In general the Faraday rotation observed in nearby galaxies is low: up to $100\,\mathrm{rad\,m^{-2}}$ maximum. The improvement of angular resolution seems to change our picture. High angular observations of NGC 4631 (M. Krause, priv. comm.) and of NGC 6946 (R. Beck, priv. comm.) seem to indicate that the Faraday depolarization effects of the sort that were observed in our Galaxy are also present in nearby galaxies.

Radio galaxies have been easier to observe because of their higher radio intensities. These high intensity emissions were also clearly magnetic effects, since the radio galaxy emission in both the jets and the lobes is observed to be highly polarized. A beautiful example of this is seen in the mapping of Centaurus A by Junkes et al. [30]. The lobes are very highly polarized with homogenous magnetic fields on scales of 100's of kpc. The inner jet is also highly polarized. The jet is surrounded by a rotating molecular ring, suggesting rotation has an effect on the formation of the aligned magnetic field. A large volume of data on polarization of radio galaxies can be found in the literature.

Another type of extragalactic objects that has been found to have magnetic fields are clusters of galaxies. The large halo of diffuse radio continuum emission of Coma A has been studied by Deiss et al. [31]. These radio continuum observations did not show up in polarization but the presence of a diffuse total intensity implies the existence of magnetic fields. Observations of the Faraday rotation of radio sources, seen through the cluster (Kim et al. [32]) support this 'magnetic' interpretation of the observed halo emission. Similar halos have been found in further clusters of galaxies (e.g. M. Thierbach, priv. comm.) indicating the presence of intergalactic magnetic field on the Mpc scales with intensities of $B \sim 10^{-6}$ Gauss.

WHAT WE MUST DISCOVER

We must proceed towards mapping of the sky polarization at several well-spaced frequencies with sufficient angular resolution. What is particularly missing is a determined effort to map the sky at say $\lambda = 11$ cm, 6 cm and 3 cm from the ground. The present conference showed the efforts of the space community to map polarization at f = 20, 40, 60 and 80 GHz from the International Space Station and with balloons. Obviously these efforts are complimentary. The time needed to perform such observations is however enormous. For ground-based observations smaller telescopes (Diameter $\sim$ 15–30 m) are needed but at sites with good atmospheric conditions. The study of all known pulsars and of thousands of extragalactic sources, in particular in the region of the Galactic plane, would also help to unravel the structure of the Galactic plane. The combination of the data at many frequencies may allow us to perform a 'tomography" of the MISM and hence to determine the 3D structure of the magnetic fields. In addition this work will allow the microwave background investigations to be on a firm ground, enabling a high degree of certainty that measured effects are real, not due to the foreground.

We must invest in instrumental developments to allow the instantaneous determination of the Rotation Measure. This is a basic parameter too little known at the present. This can be achieved with multi-filter devices with polarimeter banks or by digital methods. This instrumental development will also help in the suppression of interference. Some slow beginnings have been made but in future all data should collect RM information.

We must learn more about the coupling of magnetic fields to the ISM. So far the usual argument has been to say that magnetic fields were 'frozen' in the ISM. The magnetic permeability must play a role in these situations. Magnetic permeability does depend on the temperature of the ISM. Recently warm CO gas has been detected in the Milky Way and in nearby galaxies. Since linear polarization has also been observed in the regions of warm gas the E-field and B-field must surely lead to a velocity component. This would suggest that magnetic fields do have an effect on the dynamics of galaxies (e.g. [33] and Fig. 2).

The final topic that we must develop further is the question of the origin of the magnetic fields. Some authors have suggested that primordial fields are compressed sufficiently to give us the $\sim \mu$Gauss fields that are observed. An effect described as the Biermann battery has been considered as the originator of seed fields but its strength can be at the most $B \sim 10^{-20}$ Gauss and hence insignificant. The most viable effect to generate magnetic fields is the Dynamo effect that has been developed in considerable detail (e.g. [34], [35]). Some doubt about the applications of the dynamo effect to galactic systems has been made [36], but for the present it remains so far the only feasible explanation. Possibly a combination of various effect contributes to the generation of magnetic fields that we observe through our polarization measurements.

REFERENCES

1. Dombrovsky, N., *Dokl. Akad. Nauk USSR*, **94**, 1021 (1954).
2. Woltjer, L., *Bull. Astron. Inst. Netherlands*, **13**, 301 (1957).
3. Shklovsky, I. S., *Dokl. Akad. Nauk USSR*, **90**, 983 (1953).
4. Mayer, C. H., McCullough, T. P., and Sloanaker, R. M., *Astrophys. J.*, **126**, 468 (1957).
5. Westerhout, G., Seeger, Ch. L., Brouw, W. N., and Tinbergen, J., *Bull. Astron. Inst. Netherlands*, **16**, 187 (1962).
6. Wielebinski, R., Shakeshaft, J. R., and Pauliny-Toth, I. I. K., *The Observatory*, **82**, 158 (1962).
7. Mayer, C. H., McCullough, T. P., and Sloanaker, R. M., *Astrophys. J.*, **135**, 656 (1962).

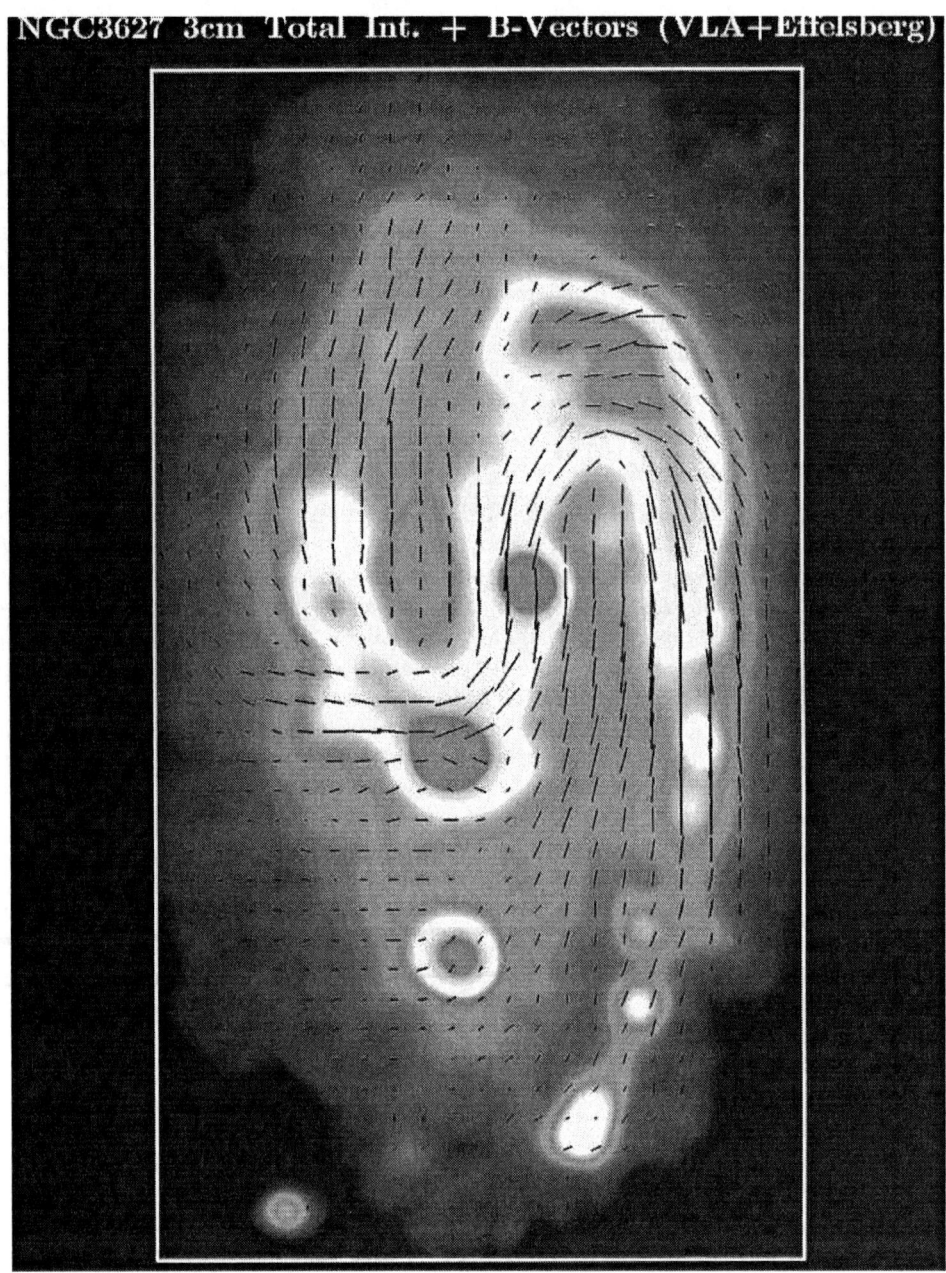

FIGURE 2. The barred galaxy NGC 3627 (from Soida et al. [37])

8. Wielebinski, R., and Shakeshaft, J. R., *Nature*, **195**, 982 (1962).
9. Muller, C. A., Berkhuijsen, E. M., Brouw, W. N., and Tinbergen, J., *Nature*, **200**, 155 (1963).
10. Mathewson, D. S., and Ford, V. L., *Mem. Roy. Astron. Soc.*, **74**, 139 (1970).
11. Brouw, W. N., and Spoelstra, T. A. Th., *Astron. Astrophys.*, **26**, 129 (1976).
12. Mathewson, D. S., and Milne, D. K., *Aust. J. Phys.*, **18**, 635 (1965).
13. Junkes, N., Fürst, E., and Reich, W., *Astron. Astrophys. Suppl.*, **69**, 451 (1987).
14. Duncan, A. R., Haynes, R. F., Jones, K. L., and Stewart, R. T., *Mon. Not. Roy. Astron. Soc.*, **291**, 279 (1997).
15. Duncan, A. R., Reich, P., Reich, W., and Fürst, E., *Astron. Astrophys.*, **350**, 447 (1999).
16. Wieringa, M. H., de Bruyn, A. G., Jansen, D., Brouw, W. N., and Katgert, P., *Astron. Astrophys.*, **268**, 215 (1993).
17. Gray, A. D., Landecker, T. L., Dewdney, P. E., and Taylor, A. R., *Nature*, **393**, 660 (1998).
18. Uyanıker, B., Fürst, E., Reich, W., Reich, P., and Wielebinski, R., *Astron. Astrophys. Suppl.*, **138**, 31 (1999).
19. Haverkorn, M., Katgert, P., and de Bruyn, A.G., *Astron. Astrophys.*, **356**, 13 (2000).
20. Gaensler B. M., Dickey, J. M., McClure-Griffiths, N. M., Green, A. J., Wieringa, M. H., and Haynes, R. F., *Astrophys. J.*, **549**, 959 (2001).
21. Lesch, H., and Reich, W., *Astron. Astrophys.*, **264**, 493 (1992).

22. Simard-Normandin, M., and Kronberg, P. P., *Astrophys. J.*, **242**, 74 (1980).

23. Smith, F. G., *Nature*, **218**, 891 (1968).

24. Han, J. L., Manchester, R. N., and Qiao, G. J., 1999, *Mon. Not. Roy. Astron. Soc.*, **306**, 371 (1999).

25. Mathewson, D. S., van der Kruit, P. C., and Brouw, W. N., 1972, *Astron. Astrophys.*, **17**, 468 (1972).

26. Beck, R., *Phil. Trans. R. Soc. Lond. A*, **385**, 777 (2000).

27. Wielebinski, R., in *Encyclopedia of Astronomy and Astrophysics*, Vol. 3, edited by P. Murdin, Institute of Physics Publishing, Bristol, 2001, pp. 1947–1955.

28. Klein, U., Haynes, R. F., Wielebinski, R., and Meinert, D., *Astron. Astrophys.*, **271**, 402 (1993).

29. Han, J. L., Beck, R., Ehle, M., Haynes, R. F., and Wielebinski, R., *Astron. Astrophys.*, **348**, 405 (1999).

30. Junkes, N., Haynes, R. F., Harnett, J. I., and Jauncey, D. L., *Astron. Astrophys.*, **269**, 29 (1993).

31. Deiss, B. M, Reich, W., Lesch, H., and Wielebinski, R., *Astron. Astrophys.*, **321**, 55 (1997).

32. Kim, K.-T., Kronberg, P. P., Dewdey, P. E., and Landecker, T.L., *Astrophys. J.*, **355**, 29 (1990).

33. Beck, R., Ehle, M., Shoutenkov, V., Shukurov, A., and Sokoloff, D., *Nature*, **397**, 324 (1999).

34. Wielebinski, R., and Krause, F., *Astron. Astrophys. Rev.*, **4**, 449 (1993).

35. Beck, R., Brandenburg, A., Moss, D., Shukurov, A., and Sokoloff, D., *Ann. Rev. Astron. Astrophys.*, **34**, 155 (1996).

36. Kulsrud, R. M., *Ann. Rev. Astron. Astrophys.*, **37**, 37 (1999).

37. Soida, M., Urbanik, M., Beck, R., Wielebinski, R., and Balkowski, C., *Astron. Astrophys.*, **378**, 40 (2001).

Magnetic fields in our Galaxy: How much do we know? (II) Halo fields and the global field structure

JinLin Han

The partner group of MPIfR
National Astronomical Observatories, Chinese Academy of Sciences
Jia-20 DaTun Road, ChaoYang District, Beijing 100012, China
Email: hjl@bao.ac.cn

Abstract. I review the large scale global magnetic field structure of our Galaxy, using all information available for disk fields, halo fields and magnetic fields near the Galactic center (GC). In the local disk of our Galaxy, RM and dispersion measure (DM) data of nearby pulsars yield the strength of regular field as 1.8μG, with a pitch angle of about $8°$, and a bisymmetric spiral structure. There are at least four, maybe five, field reversals from the Norma arm to the outskirts of our Galaxy. The regular fields are probably stronger in the interarm regions. The directions of regular magnetic fields are coherent along spiral arms over more than 10kpc. The regular fields get stronger with decreasing radius from the GC. In the thick disk or Galactic halo, large scale toroidal magnetic fields, with opposite field directions in the Southern and Northern Galaxy, have been revealed by the antisymmetric RM sky towards the inner Galaxy. This signature of the A0 dynamo-mode field structure is strengthened by the indication of a poloidal field of dipole form, that is the transition of the RM signs probably shifted from $l \sim 0°$ to $l \sim +10°$. The local vertical field is probably a part of this dipole field. The field structure of the A0 dynamo-mode strikingly continues towards the region near the GC. We predict that the direction of the dipole fields near the GC should point towards the Southern pole. In short, the magnetic fields in the Galactic disk have a bisymmetric spiral structure of primordial nature, while in the halo and near the GC the fields consist of toroidal fields with opposite directions below and above the Galactic plane and poloidal fields of dipole form.

INTRODUCTION

The magnetic field structure in the Milky Way Galaxy is not yet fully known. It probably will never be observed completely, analogous to the situation of mapping the large scale structure of the universe. However, our Galaxy is a very special case for study, partly because we can see more "trees" though not the "forest". One can observe many more details of the magnetic fields, and study their roles in star formation regions [37]. The details of local field structure and strength are very fundamental to understanding cosmic rays [7], especially those of high energy. The diverse polarized features observed in the diffuse radio emission [13, 14, 42, 10] are closely related to the small- and large-scale of magnetic fields. The fields are also important to the hydrostatic balance [6] and stability. While in the Galactic disk, the spiral structure itself is not yet clear, measuring the regular magnetic field structure could be an approach to mapping the Galaxy.

The major advantages of studying the magnetic field of our Galaxy are the fact that the Galaxy fills the sky and that a very number of pulsars and polarized extragalactic radio sources can be used as probes of the three dimensional magnetic field structure. These are unique to our Galaxy and extremely powerful for the halo field study. For the second largest galaxy in the sky, M31, there are only 21 bright polarized background radio sources[19].

I review the new progress obtained mainly in the last decade, and point to several earlier reviews (e.g. Sect.3 of [44], [40]) which clearly show the situation about 10 years ago. This is a companion paper for the review of observational facts of disk fields within a few kpc from the Sun [17]. This paper will focus on the field structure in the halo and near the Galactic center, with brief discussion of the disk fields so that the global field structure can be delineated.

CP609, *Astrophysical Polarized Backgrounds,* edited by S. Cecchini et al.

MAGNETIC FIELDS IN THE GALACTIC DISK

Significant progress has been made in the last decade on the magnetic fields in the Galactic disk, mainly because of many pulsars newly discovered in the nearby half of the whole Galactic disk [26, 28, 27] and extensive observations of pulsar RMs [16, 36, 22]. Analyses of pulsar RM data [18, 36, 15, 22, 23] yield most definite key parameters of the regular magnetic field, namely the pitch angle of local fields, the field strength, and the field reversals.

Determining the local magnetic field was the main objective of measuring the polarization of star light. Using the largest data set of >7500 stars [3, 25, 12], local magnetic fields were found to be concentrated in the spiral arms and directed along the axes of the arms, with a pitch angle of $p = -7.2° \pm 4.1°$. About a decade ago, a few dozen pulsar RMs were not enough to determine the field pitch angle more accurately than 10°. After Hamilton & Lyne obtained the RMs of 185 pulsars[16], Han & Qiao obtained the pitch angle $p = -8.2° \pm 0.5°$ from model-fitting to the data of carefully selected pulsars within 3 kpc[18]. The result was confirmed later using more pulsar RM data [15, 22], and is consistent with the value from optical data. Nowadays, there is no longer any doubt on the spiral feature of the local regular field, which has a pitch angle, $p = -8°$, with a maximum uncertainty of 2° from different tracers.

Talking about the field strength, we have to be aware of the difference between the total (rms) field strength, the strength of the regular field, the average field strength over a line of sight, and the maximum strength of the reversed field model. Total fields obviously are stronger in the arm regions, mainly contributed by random fields. That is about 5 μG in the vicinity of the Sun. However, regular fields are stronger in the interarm regions [18, 15]. The average field strengths *directly* determined from pulsar DM and RM are mostly in a range of $1 \sim 2\mu$G, with a maximum about 5μG (see Fig.1), suggesting a regular field of $1 \sim 2\mu$G and a random field about 5μG. These are exactly the values obtained by sophisticated analyses [34, 35, 18]. Note that the magnetic field energy stored in the random component is 3.7 times than the regular field [18], indicating that the random field always dominates. Our new measurements of magnetic field in the Norma arm[23] show that the regular fields themselves could be as strong as 5μG, indicating that the field strength probably increases smoothly towards the Galactic center.

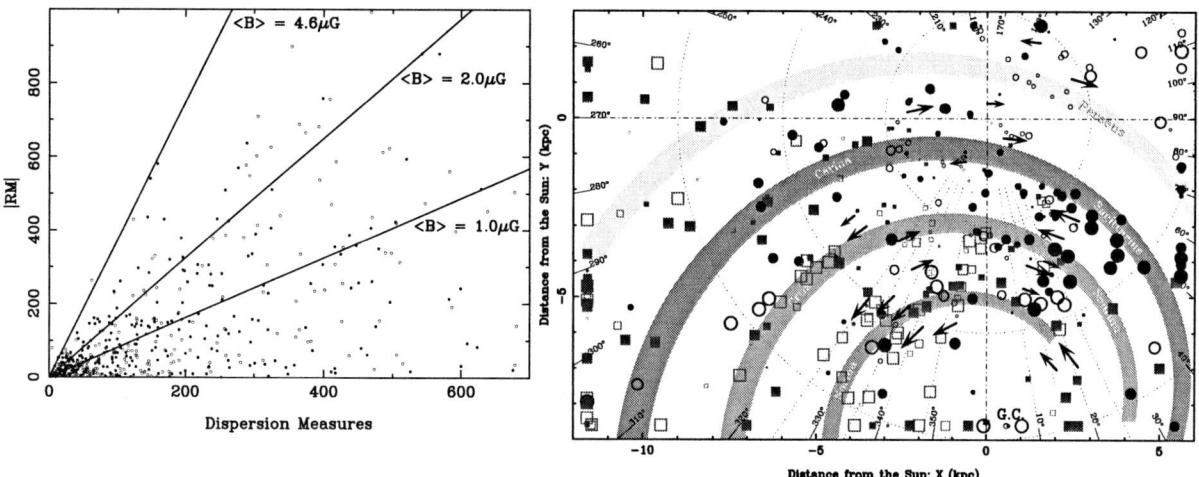

FIGURE 1. *Left:* Pulsar rotation measure values plotted against pulsar dispersion measure values, directly show the field strength averaged over the paths from pulsars to us. *Right:* The distribution of pulsar rotation measures projected onto the Galactic plane. Square symbols represent new measurements (Han, Manchester, Lyne & Qiao, unpublished). Open symbols represent negative RM values. The field directions along the four spiral arms are indicated by arrows.

Pulsar rotation measures can tell the *directions* of the averaged field, in contrast to the *orientation* obtained from polarization of starlight or mapping at radio bands. The local regular magnetic field is pointing towards $l \sim 82°$. There are at least four, maybe five, field reversals (see Fig.1) from the Norma arm to the outskirts of our Galaxy[23, 22, 15, 36, 18]. By comparison of the RM values of distant pulsars with those of extragalactic radio sources, at least one, probably two field reversals have been identified near and beyond the Perseus arm[29, 9, 22], while Canadian groups do not agree on this[43, 8]. In the inner Galaxy, the fields first reverse their direction at about 0.2 kpc inside the solar ring, near the Carina-Sagittarius arm. The fields are reversed back again near or interior to the Crux-Scutum arm, as first shown by pulsar RM data [36]. New RM observations of more distant pulsars suggested that the fields near the Carina-Sagittarius arm and the Crux-Scutum arm are coherent in direction over more than 10 kpc along the spiral arms. From the most recent data we have identified the coherent counterclockwise fields near the

Norma arm, indicating a third field reversal [23, 22].

Compared to that of external galaxies, the magnetic fields in the disk of our Galaxy is exotic because of these field reversals. Similar phenomena have not been observed in external galaxies [4]. A proper model to available data is the only way when astronomers have only the parts but like to understand the entire story. Three models were proposed for the global structure of magnetic fields in the disc of our Galaxy: the concentric ring model [35, 36], the axisymmetric spiral (ASS) model [43] and a bisymmetric spiral (BSS) [38, 39, 18, 15].

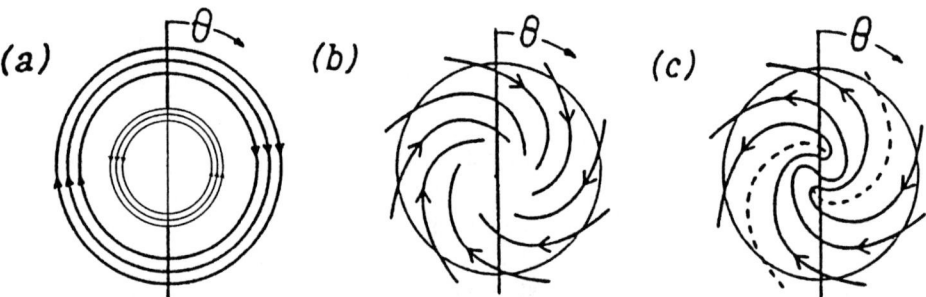

FIGURE 2. A sketch of three models for Galactic magnetic fields, namely, (a) the concentric ring model, (b) the axi-symmetric spiral model, and (c) the bi-symmetric spiral model.

Only the concentric-ring model [35, 36] has a pitch angle of zero but the observed pitch angle of fields (see above) argues for a spiral form of the fields. Nevertheless, this model allows the fields to be reversed in different ranges of Galacto-radius, so this model explain the RM data for the zero-order of approximation. In the ASS model [43], the reversed field occurs only in the range of Galacto radius from 5 to 8 kpc. No field reversals are allowed beyond 8 kpc from the GC center or within 5 kpc of the GC. Field reversals beyond the solar circle or interior to the Crux-Scutum arm are difficult to reconcile with this model. The spiral nature of the regular field and the field reversals strongly support the BSS model for the disk fields in our Galaxy. Model fitting to available pulsar RM data [18, 15] have confirmed this conclusion.

Although only the BSS model has survived confrontation with RM data and star polarization data, it may be worth reminding that, though the model fits the available data, it could fail to fit data from these part of the disk as yet unexplored. There are two possibilities for farther tests of the model. One is to check closely the coherence of field directions; the other is to determine the fields in more distant regions. The essential step for both avenues is to obtain more pulsar RM data in the distant parts of the galactic disk.

MAGNETIC FIELDS IN THE GALACTIC HALO

As was shown by previous radio background observations, our Galaxy consists of two components[5], a thin disk and a thick disk. The magnetic field in the thin disk is more dominated by the spiral structure, as we discussed above. The thick disk has a scale height of 1.5 kpc near the solar circle [38, 18]. We refer to the thick disk, together with the more extended component, as the Galactic halo. The magnetic fields in the thin disk diffuse into the halo, as shown by the spurs or plumes emerging from the Galactic plane [24, 11] and as needed by dynamo actions.

Our Galaxy is the largest visual galaxy in the sky, and the RMs of extragalactic radio sources (EGRS) and pulsars are the best probes for the halo field. Our Galaxy provides a unique opportunity to study halo field structure, on either small or large scales. However, the RM sky has many features [38] related to local high-latitude phenomena, such as, the North Polar Spur region (Loop I), Region A (loop II), both of which are probably superbubbles or supernova remnants. Faint large-scale H_α filaments sometimes also affect the RM distribution [22]. Halo fields have not been well studied well, mainly because of these dominant local features.

After carefully filtering out deviant points, we found the antisymmetry of the rotation measure sky in the *inner* Galaxy[21]. Such a highly symmetric pattern could not be produced by just the coincidence of many local perturbations, instead, it is of large, probably galactic, scale. Such an antisymmetric sky suggests a magnetic field in the halo with opposite directions in the northern and southern Galactic hemispheres. We noticed later that Andreasyan & Makarov independently made a similar suggestion[2] but for only local fields. A very important fact is that such an antisymmetry only occurs at high latitudes for the inner part of the Galaxy, indicating that it is related to the field structure mainly in the Galactic halo. The field structure so revealed is amazingly consistent with that produced by an

A0 dynamo. Together with the vertical field near the Galactic center (see below), we believe that an A0-mode dynamo is operating in the Galactic halo. This is the first time that a dynamo mode has been identified on a galactic scale.

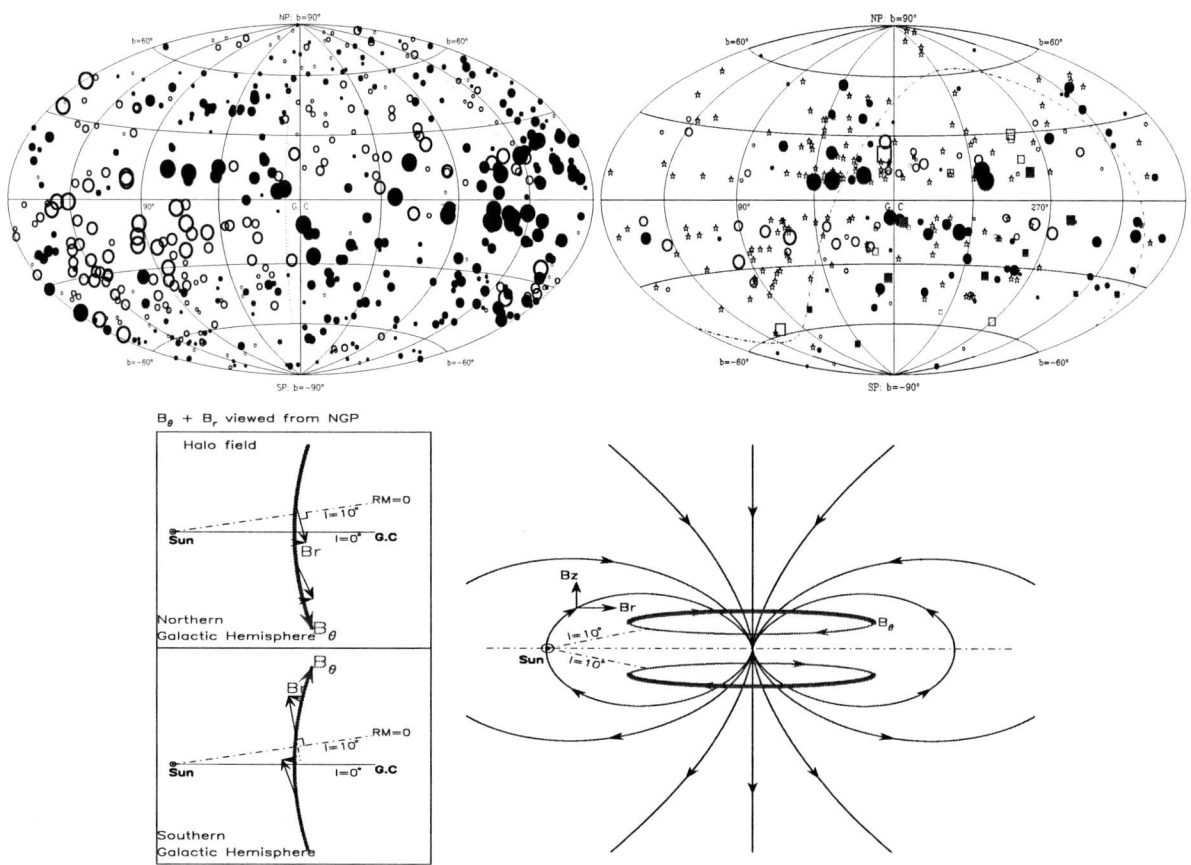

FIGURE 3. The antisymmetric rotation measure sky of extragalactic radio sources (*upper left*) and pulsars (*upper right*) directly indicates the reversed direction of the azimuthal magnetic fields in the southern and northern hemisphere of our Galaxy. The filled symbols represent positive RMs, indicating the averaged field towards us. The effect of the dipole field on the longitude transition of RM signs (*lower left*) and the magnetic field configuration of an A0-mode dynamo (*lower right*) are shown for comparison.

Two facts further support the argument for the dynamo origin of the antisymmetric rotation measures. One is the local vertical fields. After all local perturbations were discounted, we found that the vertical fields have a strength of $0.2{\sim}0.3\,\mu G$ and point from the South Galactic pole to the North [18, 22]. If this is a part of the dipole field expected from the A0 mode dynamo, then the longitude transition of RM signs should be slightly shifted from $l \sim 0°$ to $l \sim$ a few degrees, depending to the strengths of the dipole fields and the toroidal fields. The halo toroidal fields have a strength of about $1\,\mu G$ [22]. The transition shift, though marginally shown by the available RM data of extragalactic radio sources and pulsars, was first noticed by Han et al. in 1999 [22]. We understand that this second fact needs more data to confirm, which we are currently working on.

In short, the RM sky demonstrates the azimuthal fields with opposite directions in the southern and northern galactic hemispheres and also dipole fields. All pieces of observational evidence are nicely consistent for these field components of the A0-mode dynamo field structure. Computer simulation of the toroidal and dipole fields, together with the electron distribution model[41] have confirmed all observational features.

MAGNETIC FIELDS NEAR THE GALACTIC CENTER

Two aspects of the magnetic fields near the Galactic center should be discussed – first how the disk fields continue towards the center, and second, the possible presence of poloidal fields and toroidal fields there and any possible connections with the fields in the Galactic halo. On the continuation of disk fields, there is a knowledge gap to fill. We

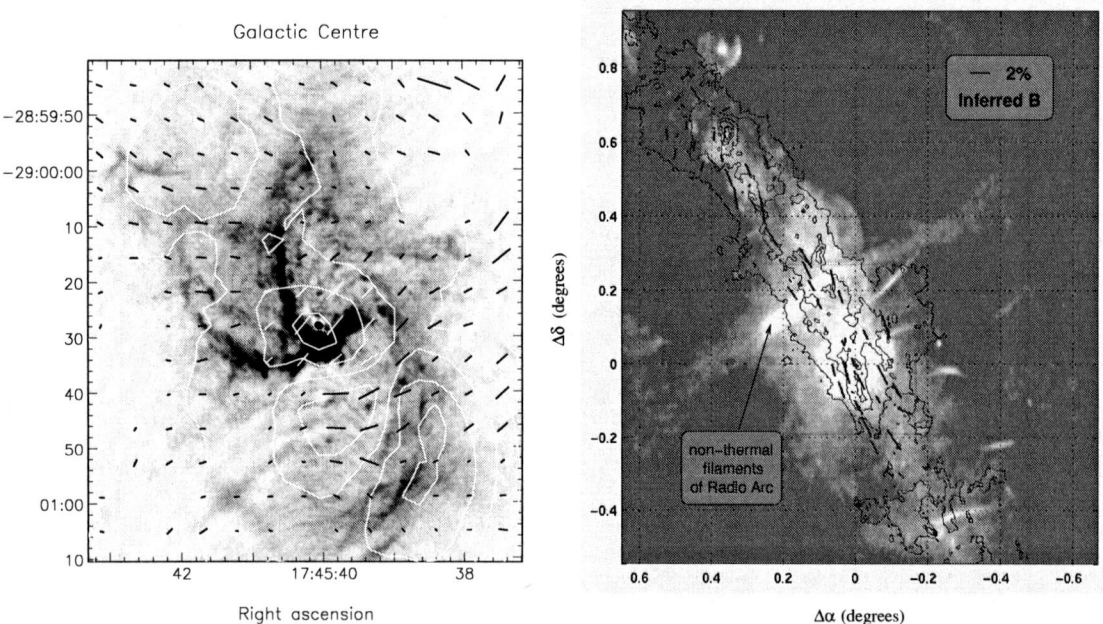

FIGURE 4. The diverse structure of radio emission near the Galactic center at 3.6cm (*left*) and 90cm (*right*) overlaid by the polarization observations at sub-millimeter [1, 33]. The poloidal fields are revealed by the filaments perpendicular to Galactic plane, while toroidal fields parallel to the plane are clearly detected in the central molecular zone of 170 pc × 30 pc (*right*) [33].

have only recently determined the field structure near the Norma arm and have no idea at all about the fields interior to this arm. It is not clear whether the strong magnetic fields observed in the central molecular zone [33, 32] are continued and connected to the large scale fields in the disk. It is the case for the nearby galaxy NGC 2997[20].

As a matter of fact, recognition of the dipole field in the Galactic halo is fundamentally important, since that field is naturally the extension of the dipole field in the Galactic center. The central dipole field then must be very strong and should show many features (see Fig. 4) [31]. The filaments[46], plumes [11] and even threads [30] near the Galactic center all exactly reflect these strong dipole fields. They are the locally illuminated flux tubes in large-scale pervasive fields! Their locations, high linear polarization of 50% to 70% [30], and their curvatures, as well as their perpendicularity to the Galactic plane, all suggest the dipole nature of the poloidal field structure. The strength of the fields should be of the order of mG [45]. As hinted by the direction of the local vertical fields, I predict that the dipole fields in the Galactic center then should be directed from the Northern pole to the Southern pole (Fig. 3).

As well as the poloidal fields, Zeeman splitting [47] and polarization observations at sub-millimeter wavelength [32, 33] have revealed strong toroidal magnetic fields of 2 – 4 mG in the central molecular zone (see Fig. 4). Nova et al. found some evidence for the directions of these toroidal fields [33], exactly the same field configuration as we obtained for the halo field (see Fig. 3). This has led us to believe that the A0 dynamo probably is working from the GC to the halo.

GLOBAL STRUCTURE OF GALACTIC MAGNETIC FIELDS

In the last decade, knowledge of the magnetic field structure of our Galaxy has improved in many aspects through efforts in determining pulsar RMs and mapping near the Galactic center. However, the story is far from complete. Using the presently available information, we can conclude that our Galaxy has such an odd symmetry of the toroidal and poloidal fields in the Galactic halo and near the Galactic center, showing that the dynamo is probably working through the inductive effects of fluid motions in the interstellar medium. However, bi-symmetric spiral magnetic fields in the Galactic disk suggest that the disk keeps some kind of memory of the field reversals from seed fields or the primordial fields. For any modeling to understand the origin of cosmic rays, both the large scale field in the disk and the halo should be considered. Polarization observations of background radiation can directly show the connections of large-scale disk fields to halo fields.

ACKNOWLEDGMENTS

I am very grateful to many colleagues, especially, Prof. R.N. Manchester and Prof. G.J. Qiao for working together with me to improve the knowledge on the magnetic fields of our Galaxy, some of which was presented here. Mr. Xu Dong is acknowledged for assistance on the field model simulations and Dr. Tom Landecker for language editing. I thank the partner group frame between Max-Planck Society and Chinese Academy of Sciences for a long-term cooperation between the MPIfR and NAOC. My research in China is supported by the National Natural Science Foundation of China (10025313 and 19903003) and the National Key Basic Research Science Foundation of China (G19990754).

REFERENCES

1. Aitken, D. K., Greaves, J., Chrysostomou, A., et al., *ApJ*, **534**, L173–L176 (2000).
2. Andreasyan, R. R., and Makarov, A. N., *Astrophysics*, **28**, 247 (1988).
3. Andreasyan, R. R., and Makarov, A. N., *Afz*, **31**, 247 (1989).
4. Beck, R., *Space Science Review*, in press (2001).
5. Beuermann, K., Kanbach, G., and Berkhuijsen, E. M., *A&A*, **153**, 17–34 (1985).
6. Boulares, A., and Cox, D. P., *ApJ*, **365**, 544–558 (1990).
7. Brunetti, M. T., and Codino, A., *ApJ*, **528**, 789–798 (2000).
8. Brown, J. C., and Taylor, A. R., *ApJ*, **563**, L31–L34 (2001).
9. Clegg, A. W., Cordes, J. M., Simonetti, J. M., and Kulkarni, S. R., *ApJ*, **386**, 143–157 (1992).
10. Duncan, A. R., Haynes, R. F., Jones, K. L., and Stewart, R. T., *MNRAS*, **291**, 279–295 (1997).
11. Duncan, A. R., Haynes, R. F., Reich, W., Reich, P., and Gray, A. D., *MNRAS*, **299**, 942–954 (1998).
12. Fosalba, P., Lazarian, A., Prunet, S., Tauber, J. A., *ApJ*, **564**, 762–772 (2002).
13. Gaensler, B. M., Dickey, J. M., McClure-Griffiths, N. M., Green, A. J., and et al., *ApJ*, **549**, 959–978 (2001).
14. Gray, A. D., Landecker, T. L., Dewdney, P. E., Taylor, A. R., Willis, A. G., and Normandeau, M., *ApJ*, **514**, 221–231 (1999).
15. Indrani, C., and Deshpande, A. A., *New Astronomy*, **4**, 33–40 (1998).
16. Hamilton, P. A., and Lyne, A. G., *MNRAS*, **224**, 1073–1081 (1987).
17. Han, J. L., *Ap&SS*, **278**, 181–184. astro–ph/0010537 (2001).
18. Han, J. L., and Qiao, G. J., *A&A*, **288**, 759–772 (1994).
19. Han, J. L., Beck, R., and Berkhuijsen, E. M., *A&A*, **335**, 1117–1123 (1998).
20. Han, J. L., Beck, R., Ehle, M., Haynes, R. F., and Wielebinski, R., *A&A*, **348**, 405–417 (1999).
21. Han, J. L., Manchester, R. N., Berkhuijsen, E. M., and Beck, R., *A&A*, **322**, 98–102 (1997).
22. Han, J. L., Manchester, R. N., and Qiao, G. J., *MNRAS*, **306**, 371–380 (1999).
23. Han, J. L., Manchester, R. N., Lyne, A. G., and Qiao, G. J., *ApJ Letter*, submitted (2002).
24. Haslam, C. G. T., Stoffel, H., Salter, C. J., and Wilson, W. E., *A&AS*, **47**, 1 (1982).
25. Heiles, C., *ApJ*, **462**, 316 (1996).
26. Manchester, R. N., Lyne, A. G., D'Amico, N., and et al., *MNRAS*, **279**, 1235–1250 (1996).
27. Manchester, R. N., Lyne, A. G., Camilo, F., Bell, J. F., and et al., *MNRAS*, **328**, 17–35 (2001).
28. Lyne, A. G., Manchester, R. N., Lorimer, D. R., and et al., *MNRAS*, **295**, 743–755 (1998).
29. Lyne, A. G., and Smith, F. G., *MNRAS*, **237**, 533–541 (1989).
30. Lang, C. C., Morris, M., and Echevarria, L., *ApJ*, **526**, 727–743 (1999).
31. LaRosa, T. N., Kassim, N. E., Lazio, T. J. W., and Hyman, S. D., *AJ*, **119**, 207–240 (2000).
32. Novak, G., Dotson, J. L., Dowell, C. D., Hildebrand, R. H., Renbarger, T., and Schleuning, D. A., *ApJ*, **529**, 241–250 (2000).
33. Novak, G., Chuss, D. T., Renbarger, T., and et al., *ApJ Letter*, submitted. astro–ph/0109074 (2001).
34. Ohno, H., and Shibata, S., *MNRAS*, **262**, 953–962 (1993).
35. Rand, R. J., and Kulkarni, S. R., *ApJ*, **343**, 760–772 (1989).
36. Rand, R. J., and Lyne, A. G., *MNRAS*, **268**, 497–505 (1994).
37. Rees, M. J., *QJRAS*, **28**, 197–206 (1987).
38. Simard-Normandin, M., and Kronberg, P. P., *ApJ*, **242**, 74–94 (1980).
39. Sofue, Y., and Fujimoto, M., *ApJ*, **265**, 722–729 (1983).
40. Sofue, Y., Fujimoto, M., and Wielebinski, R., *ARA&A*, **24**, 459–497 (1986).
41. Taylor, J. H., and Cordes, J. M., *ApJ*, **411**, 674–684 (1993).
42. Uyaniker, B., Fuerst, E., Reich, W., Reich, P., and Wielebinski, R., *A&A Supl.S.*, **132**, 401–411 (1998).
43. Vallee, J. P., *A&A*, **308**, 433–440 (1996).
44. Wielebinski, R., and Krause, F., *A&AR*, **4**, 449–485 (1993).
45. Yusef-Zadeh, F., and Morris, M., *ApJ*, **320**, 545–561 (1987).
46. Yusef-Zadeh, F., Morris, M., and Chance, D., *Nature*, **310**, 557–561 (1984).
47. Yusef-Zadeh, F., Roberts, D. A., Goss, W. M., Frail, D. A., and Green, A. J., *ApJ*, **466**, L25 (1996).

Whistlers in Interstellar Gas Dynamics

Sergey Bastrukov[†‡*], Jongmann Yang[†‡], Dmitry Podgainy[*]

[†]*Department of Physics, Ewha Womans University, Seoul 120-750, Korea*
[‡]*Center for High Energy Physics, Kyungpook National University, Daegu 702-701, Korea*
[*]*Laboratory of Informational Technologies, Joint Institute for Nuclear Research, 141980 Dubna, Russia*

Abstract. Based on equation of electromagnetic induction for the magnetic flux density driven by Hall and Ohmic components of electric field generated by flows of thermal electrons an analytic model of helicoidal magneto-electron waves, whistlers, is developed. We found that the whistlers can propagate in the interstellar medium without noticeable attenuation. The presented numerical estimates for the group velocity of whistlers in the dark molecular clouds suggest that this kind of spiral circularly polarized magneto-electron waves can be responsible for the broadening of molecular lines detected from dark interstellar clouds.

INTRODUCTION

Understanding gasdynamical processes governing the structure and the evolution of dense star-forming clouds is one of the outstanding challenges in current research on the astrophysics of interstellar medium (ISM). While the central role of magnetic fields in such processes was recognized many years ago, major uncertainties regarding the character of the motions in interstellar clouds followed from inadequate knowledge of the material composition of the intercloud medium. Over the years, convincing evidence has been obtained that the composition of dark molecular clouds is dominated by molecular hydrogen with some admixture of OH and CO molecules whose linewidths are found to exhibit the supersonic character of intercloud motions [1]. That the intercloud motions cannot be accounted for the isothermal sound waves has stimulated search for alternative models of interstellar gasdynamics and has led to hypothesis of the presence in dark molecular clouds of a sizable fraction of charged particles whose collective flows coupled with the intercloud magnetic field can support propagation of magnetohydrodynamical (MHD) waves [2]. Since then the MHD model is extensively exploited in interpreting supersonic broadening of molecular lines in terms of hydromagnetic waves of the Alfvén type (e.g.[3, 4, 5]). On average, the model provides a fairly reasonable account of data in CO regions of clouds where the temperature and the ionization factor are pretty high.

In the meantime, recent Zeeman measurements of magnetic fields in dense cores of molecular clouds, highly obscured from ionizing ultraviolet radiation, have revealed a predominately sub-Alfvénic character for intercloud motions [6]. The latter circumstance can be regarded as an indication that the composition and the patterns of the motions in the cores of molecular clouds might be quite different from those which are implied by using the single-component MHD model of interstellar gasdynamics [3]. The observed filamentary structure in dark molecular clouds, like Ophiuchus [7], suggests that the filaments could be regarded as a manifestation of a superparamagnetic state of a gas-dust ISM considered long ago by Jones-Spitzer in paper [10] which is often mentioned in the context of the starlight polarization problem [11]. This Jones-Spitzer ISM can be thought of as highly polarized, but poorly conducting gas-based ferrocolloid consisting of tiny ferromagnetic grains suspended in the dense gas of molecular hydrogen. The most striking feature is its capability of sustaining long-ranged magnetization in form of magnetic chains of ferrograins extended along the intercloud magnetic fields. In [8, 9] it is shown that gas-dynamics of Jones-Spitzer ISM can be properly described by equations of magneto-elastodynamics borrowed from theory of non-conducting magneto-elastic materials; similar equations are used in the theory of magnetic ferrofluids. Adhering to this attitude we have found that the gas-based ferrocolloidal ISM can transmit perturbations by transverse magneto-elastic waves with sub-Alfvénic group velocity, in a line with data on the molecular linewidths. Physically, the magnato-elastic waves in the non-conducting magnetically polarized gas-dust ISM have many features in common with transverse hydromagnetic waves in the perfectly conducting non-magnetic ISM [8, 9].

CP609, *Astrophysical Polarized Backgrounds*, edited by S. Cecchini et al.
© 2002 American Institute of Physics 0-7354-0055-5/02/$19.00

In this communication we discuss one more possible wave process in magnetically supported star-forming molecular clouds which is determined by flows of thermal electrons. The electrons, as is known, are the most abundant and mobile charged components of ISM, and their small mass provides the most strong coupling with intercloud magnetic fields. Therefore it is natural to expect that the collective behavior of electrons may essentially affect the interstellar gasdynamics.

ELECTRON TRANSPORT OF MAGNETIC FLUX DENSITY IN DARK MOLECULAR CLOUDS

The model under consideration implies that intercloud velocity gradients are dominated by thermal electrons whereas neutral molecules and ions frozen in the background magnetic field threading the cloud are regarded as being immobilized. This special case was first discussed by Mouschovias [12] in the context of the magnetic flux transport throughout the cloud core. In Ref. [12], the transport of the magnetic flux density $\mathbf{B}$ in the cores of molecular clouds with their internal magnetic fields frozen in the ions is governed by the equation

$$\frac{\partial \mathbf{B}}{\partial t} - \nabla \times [\mathbf{v}_i \times \mathbf{B}] = -\frac{1}{n_e e} \nabla \times [\mathbf{B} \times \mathbf{j}] - c\nabla \times \left[\frac{\mathbf{j}}{\sigma_c}\right]$$
$$-\frac{c}{n_e^2}[\nabla n_e \times \nabla P_e] - \frac{c m_{en}}{e} \nabla \times [\nu_{en}(\mathbf{v}_n - \mathbf{v}_i)], \tag{1}$$

where the electrical conductivity σ_c and the electron-neutral mass m_{en} are given by

$$\sigma_c = \frac{n_e e^2}{m_e}(\nu_{ei}^{-1} + \nu_{en}^{-1}), \quad m_{en} = \frac{m_e m_n}{m_e + m_n}, \tag{2}$$

with ν_{ei} and ν_{en} being the frequencies of electron-ion scattering and scattering of electrons by neutrals, respectively. $P_e = n_e kT$ is the electron pressure; n_e stands for the electron density, $\mathbf{v}_i$ and $\mathbf{v}_n$ for the velocities of ions and neutrals, respectively, and T for the temperature; $\mathbf{j} = (c/4\pi)[\nabla \times \mathbf{B}]$ is the density of the Ampére current. The first term on the right hand side of Eq. (1) describes advection of the field by Hall's drift, the second term is due to Ohmic diffusion, the third term is associated with thermo-electron diffusion, and the last term belongs to the ambipolar diffusion. Understandably, such a highly involved combination of effects makes the problem of magnetic flux transport too complicated to handle. Therefore we confine our consideration to a highly idealized model of an isothermal ($T = \text{constant}$) interstellar cloud with homogeneous density electrons ($n_e = \text{constant}$), and ions and neutrals are considered to be immobilized, i.e. $\mathbf{v}_i = \mathbf{v}_n = 0$. We further simplify the situations by assuming that the Ohmic conductivity of the electron gas σ_c is much larger than the Hall conductivity, $\sigma_H = n_e ec/B$ [13], so that the second term on the right hand side of Eq. (1) can be omitted. As a result, Eq. (1) takes the form

$$\frac{\partial \mathbf{B}}{\partial t} = -\frac{c}{4\pi e n_e} \nabla \times [\mathbf{B} \times [\nabla \times \mathbf{B}]], \quad \nabla \cdot \mathbf{B} = 0. \tag{3}$$

Our goal is to point out that in the regions of the ISM where magnetic flux density undergoes a Hall drift induced by flows of thermal electrons, the perturbations can be propagated by circularly polarized waves in which the electron current and the magnetic field undergo coupled oscillations. It is to be noted that this kind of wave motions has first been discovered in the solid-state plasmas for which the name helicons was coined. Similar waves are observed in planetary magnetospheres. In particular, propagation of these waves in the Earth's ionosphere causes the whistling audio noise in radio, therefore are often called whistlers. The possibility of propagation of whistlers in interstellar medium comes from the following consideration.

INTERSTELLAR WHISTLERS

Let us consider a perturbation producing fluctuations of lines of magnetic force, $\delta \mathbf{B}$, superimposed on the permanent magnetic field $\mathbf{B}$:

$$\mathbf{B}(\mathbf{r}, t) \to \mathbf{B} + \delta \mathbf{B}(\mathbf{r}, t), \quad \mathbf{B} = \text{constant}. \tag{4}$$

By linearizing Eq. (3), with help of Eq. (4), we obtain

$$\frac{\partial \delta \mathbf{B}}{\partial t} = \frac{c}{4\pi n_e e} (\mathbf{B} \cdot \nabla) [\nabla \times \delta \mathbf{B}].$$ (5)

By applying the plane-wave representation of a fluctuating field, $\delta \mathbf{B} = \mathbf{b} \exp i(\mathbf{k} \cdot \mathbf{r} - \omega t)$ in right hand side of (3) one has

$$\frac{\partial \delta \mathbf{B}}{\partial t} = c \frac{(\mathbf{k} \cdot \mathbf{B})}{4\pi n_e e} [\mathbf{k} \times \delta \mathbf{B}].$$ (6)

To see that the wave motions in question bear a circularly polarized character, let us consider a perturbation that does not affect the intensity of the magnetic field in the z direction, but only the components along x an y directions. Then, the cartesian components of Eq. (6) read

$$\delta \dot{B}_x = +\frac{(\mathbf{k} \cdot \mathbf{B})}{4\pi n_e e} c \, k_z \delta B_y, \qquad \delta \dot{B}_y = -\frac{(\mathbf{k} \cdot \mathbf{B})}{4\pi n_e e} c \, k_z \delta B_x.$$ (7)

One sees that these equations amounts to the cartesian components of the vector equation $\delta \dot{\mathbf{B}} = [\mathbf{\Omega} \times \delta \mathbf{B}]$ for rotation about z-axis

$$\delta \dot{B}_x = -\Omega \delta B_y, \quad \delta \dot{B}_y = \Omega \delta B_x, \quad \Omega = -\frac{(\mathbf{k} \cdot \mathbf{B})}{4\pi n_e e} c k_z.$$ (8)

To describe the transverse wave in which two components of the vector field are coupled by a rotation, it is convenient to introduce a complex field

$$\delta B_+ = \delta B_x + i \delta B_y = B_\perp \exp(-i\omega t).$$ (9)

After some algebra couple of Eqs. (7) are reduced to a single equation:

$$\frac{\partial}{\partial t} \delta B_+ = -i \frac{(\mathbf{k} \cdot \mathbf{B})}{4\pi n_e e} c k_z \delta B_+.$$ (10)

Making use of Eq. (9) to eliminate the time derivative, we arrive at the dispersion equation of a right-hand circularly polarized whistler

$$\omega = \frac{(\mathbf{k} \cdot \mathbf{B})}{4\pi n_e e} c k_z.$$ (11)

Notice that for the left-hand polarized wave, $\delta B_- = \delta B_x - i \delta B_y = B_\perp \exp(i\omega t)$, holds the same dispersion relation. From Eq. (11) it follows that this wave most efficiently propagates when $\mathbf{k}$ and $\mathbf{B}$ point in the same direction. In terms of cyclotron and plasma frequencies

$$\omega_c = \frac{eB}{m_e c}, \qquad \omega_p^2 = \frac{4\pi e^2 n_e}{m_e},$$

the dispersion equation (11) reads

$$\omega = \frac{\omega_c}{\omega_p^2} c^2 k^2 = 4.97 \times 10^{18} \frac{B}{n_e} k^2.$$ (12)

The corresponding group velocity is given by

$$v = \frac{2c^2 \omega_c}{\omega_p^2} k \approx 9.58 \times 10^{18} \frac{B}{n_e} k \quad \text{cm/sec}.$$ (13)

Making use of Eq. (12), we can represent the latter formula in terms of ω as follows:

$$v = \frac{2c\sqrt{\omega_c}}{\omega_p} \sqrt{\omega} \approx 4.46 \times 10^9 \sqrt{\frac{B}{n_e}} \sqrt{\omega} \quad \text{cm/sec}.$$ (14)

The whistlers represent a low-frequency branch of electromagnetic excitations in a non-compensated electron-dominated magnetoplasma; the frequency of oscillation of the electrons in this wave is less than the cyclotron frequency. In the electron magneto-hydrodynamics , the whistlers play the same role as the transverse Alfvén waves for single-component interstellar magnetohydrodynamics. The essential kinematic difference between them is that the group velocity of the whistlers depends on the frequency whereas the Alfvén wave is characterized by a dispersion-free law of propagation.

DISCUSSION

Let us briefly discuss an inference that could be made from the propagation of whistlers in the ISM. The average density of electrons and the magnetic field in the Galactic disc are estimated to be $n_e \approx 0.03\,\mathrm{cm}^{-3}$ and $B \approx 10^{-5} - 10^{-6}$ Gauss, respectively. With above parameters the group velocity $v \approx 10^7 - 10^8$ cm/s, and the wavelength $\lambda = (2\pi/k) \approx 1000$ km. It is easy to check that the criterion of dissipation-free propagation of whistler, $\Gamma = \nu_c/\omega_c << 1$, is fulfilled. The cyclotron frequency ω_c, setting the upper limit for the frequency of the dissipative-free propagation of whistlers, falls into the interval $10 < \omega_c < 100\,\mathrm{s}^{-1}$, whereas the frequency of elastic collisions $\nu_c \approx 10^{-2} - 10^{-3}\,\mathrm{s}^{-1}$. That means that whistlers can fairly freely propagate throughout the ISM with the above parameters. They might be relevant to the interstellar scintillations of the pulsar signals. The latter effect is customarily attributed to the scattering of radio waves on fluctuations in electron density. These fluctuations exhibit features typical of turbulent motions whose spectrum shows very much similar dispersion, $\omega \sim k^{5/3}$. It seems not implausible that whistlers might provide notable contribution to the effect of interstellar scincillations of the pulsar radio signal.

In the ISM of dark molecular clouds, $B \approx 10\,\mu G$, $n_e \approx 10^{-3}$ cm^{-3}, and $T \approx 10$ K [12]. By taking the group velocity equal to the velocity dispersion typical of the widths of molecular lines, $v \sim 0.3 - 5.0$ km/s, one finds $\lambda \sim 10^{12} - 10^{13}$ cm. This space scale is much less than the linear sizes of clouds, $L \sim 10^{17}$ cm. For the same velocity, the period of oscillations of electron flow in the spiral magneto-electron wave falls into the interval $P \sim 0.1 - 10$ years. If one takes into account the effect of the Ohmic decay of the magnetic flux density (second term in the right hand side of Eq. (1)), one can show that the dispersion relation (12) takes the form

$$\omega = \frac{\omega_c}{\omega_p^2} c^2 k^2 \left(1 + i\frac{\sigma_H}{\sigma_c}\right). \tag{15}$$

One sees that the question about a long-term stability of magneto-electron oscillations demands careful analysis of the ISM conductivity. This problem is carefully discussed in above mentioned work [13]. For a typical dark molecular cloud, $\omega_c \approx 10^2$ s^{-1} and $\nu_c \approx 10^{-3} - 10^{-1}$ s^{-1} [12] so that the criterion of dissipation-free propagation of whistlers $\Gamma = \sigma_H/\sigma_c = \nu_c/\omega_c << 1$ is well justified, and its validity remains quite robust to the changes in ν_c up to $\nu_c = 100$ s^{-1}. These estimates suggest that the whistlers can freely propagate in the dark molecular clouds and we conjuncture that they can be responsible for the broadening of molecular line along with weakly damped Alfvén waves [2] and magnetomechanical waves [8].

ACKNOWLEDGMENTS

This work is partly supported by the MOST through the National R & D program for women's universities.

REFERENCES

1. Zuckerman B., and Evans N.J., Astrophys. J. **192**, L152 (1974).
2. Arons J., and Max C.E., Astrophys. J. **196**, L77 (1975).
3. McKee C.F., and Zweibel E.G., Astrophys. J. **440**, 686 (1995).
4. Gammie C.F., and Ostriker E.C., Astrophys. J. **446**, 814 (1996).
5. Padoan P., and Nordlund Å., Astrophys. J. **526**, 526 (1999).
6. Crutcher R.M., Astrophys. J. **520**, 706 (1999).
7. Goodman A.A., and Heiles C., Astrophys. J. **424**, 208 (1994).
8. Yang J., and Bastrukov S.I., JETP Lett. 71, **395** (2000).
9. Bastrukov S.I., and Yang J., Astrofizika **43**, 405, Astrophysics **43**, 295 (2000).

10. Jones R.V., and Spitzer L., Astrophys. J. **146**, 943 (1967).
11. Lazarian A., Goodman A.A., P. C. Myers and A. Goodman, Astrophys. J. **490**, 273 (1997).
12. Mouschovias T.Ch., in: Physical Processes in Interstellar Clouds, Morfill, E. & Scholer, M., eds., (Dordrecht, Reidel, 1986) p.491.
13. Wardle M., and Ng C., Mon. Not. R. Astron. Soc. **303**, 239 (1999).

POLARIMETRY EXPERIMENTS AND INSTRUMENTATION

The SPOrt Experiment

E. Carretti*, M. Baralis†, G. Bernardi*, G. Boella**, S. Bonometto**, M. Bruscoli‡,
S. Cecchini*, S. Cortiglioni*, R. Fabbri§, M. Gervasi**, C. Macculi*, J. Monari¶,
K.W. Ng‖, L. Nicastro††, A. Orfei¶, O. Peverini†, S. Poppi¶, V.A. Razin‡‡,
M.V. Sazhin§§, C. Sbarra*, G. Sironi**, I.A.Strukov¶¶, R. Tascone†, M. Tucci**,
E.N. Vinyajkin‡‡ and M. Zannoni**

*Istituto Te.S.R.E./C.N.R., via P. Gobetti 101, I–40129 Bologna, Italy
†I.R.I.T.I./C.N.R., c.so Duca degli Abruzzi 24, I–10129 Torino, Italy
**Dip. di Fisica, Univ. di Milano - Bicocca, P.za della Scienza 3, I–20126 Milano, Italy
‡Dip. di Astronomia, Univ. di Firenze, Largo E. Fermi 5, I–50125 Firenze, Italy
§Dip. di Fisica, Univ. di Firenze, Via Sansone 1, I–50019 Sesto Fiorentino (FI), Italy
¶I.R.A./C.N.R., via P. Gobetti 101, I–40129 Bologna, Italy
‖Academia Sinica, 11529 Taipei, Taiwan
††I.F.C.A.I./C.N.R., via U. La Malfa 153, I–90146 Palermo, Italy
‡‡NIRFI, 25 B.Pecherskaya st., Nizhnij Novgorod 603600/GSP-51, Russia
§§Schternberg Astronomical Institute, Moscow State University, Moscow 119899, Russia
¶¶I.K.I., Profsojuznaja ul. 84/32, Moscow 117810, Russia

Abstract. SPOrt is a space experiment aimed at studying the polarization of the CMB and of the diffused Galactic Background in the microwave range (22-90 GHz). Here we present the project as well as its main scientific goals.

INTRODUCTION

The Cosmic Microwave Background (CMB) is a powerful tool to understand origin and evolution of the Universe. The CMB looks like a Black Body at 2.725 K [1] almost isotropic and unpolarized: Any detection of deviations from its ideal behaviour allows the estimate of cosmological parameters [2, 3, 4, 5]. Very small temperature anisotropies have been detected at both large [6, 7] and small [8, 9, 10] angular scales, but only upper limits on the CMB polarization (CMBP) have been set up to now. The information contained in the CMB polarization can solve the degeneracies among cosmological parameters that CMB anisotropy alone is not able to remove [4]. Figure 1 presents a comparison between temperature and polarization (E-mode) power spectra for two cosmological models, which differ only in the optical depth τ of the re-ionized medium in the dark ages. It is clear that the E-mode spectrum is much more sensitive to τ than the temperature one and that these new information are found at large angular scales ($l < 20$, i.e. $\theta > 10°$). In addition, the CMBP brings important information also at subdegree angular scales. For instance, the *coherent* primordial fluctuations foreseen by inflation produce a well defined Doppler peak pattern: The peaks in the T and E power spectra are alternate. Thus, the detection of the CMBP at subdegree scales allows a test of the inflationary model [11]. Finally, the CMBP allows the separation between the scalar and the tensorial components of the primordial fluctuations providing a way to disentangle among different inflationary models [12]. Figure 1 also shows that the polarized emission peaks at sub-degree scales, suggesting where a first detection should be easier.

In spite of its importance, the CMBP predicted level is very low (few μK on sub-degree scales and less than 1 μK at large scales). Current experimental upper limits are still one order of magnitude higher than the expected level [13, 14, 15, 16, 17, 18, 19, 20, 21, 22, 23].

Surveys of the diffuse polarized emission in the microwave range are crucial also for the study of the Galactic contribution. In fact, our Galaxy is featured by a smoothed linearly polarized background emission, carrying information on the Galactic structure. Moreover, besides its intrinsic interest, the Galaxy acts as a foreground for CMB experiments

CP609, *Astrophysical Polarized Backgrounds*, edited by S. Cecchini et al.

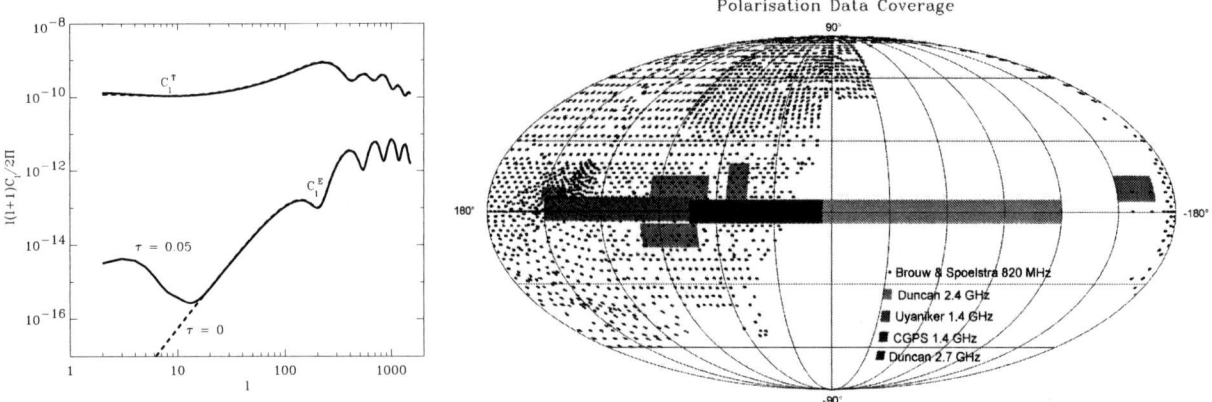

FIGURE 1. **Left:** Temperature anisotropy and E-mode power spectra. Two $\Omega_\Lambda = 0.7$ cosmological models which differ only in the re-ionization optical depth τ are shown. **Right:** Sky coverage of existing continuous polarization surveys in Galactic coordinates (The Galactic centre is in the middle).

and only its accurate knowledge will allow measurements of CMB features ([24, 25] and references therein).

So far, observations have been carried out only at frequencies up to 2.7 GHz [26, 27, 28, 29, 30], where the Galactic emission appears to be dominated by synchrotron. Such observations either are widely undersampled [26] or cover narrow stripes around the Galactic Plane [27, 28, 29, 30] (see Figure 1). For these reasons further observations are necessary.

THE DESIGN

The Sky Polarization Observatory (SPOrt), an Italian Space Agency (ASI) funded experiment, is aimed at filling the current gap in measurements of the diffuse polarized emission in the 22-90 GHz range. It has been selected by ESA to be flown on board the International Space Station (ISS) for a minimum lifetime of 18 months. The flight is scheduled for early 2005.

SPOrt is the first space mission devoted to Q & U Stokes parameters measurements in the microwave domain, and has been designed to be as much as possible insensitive to instrumental polarization.

Moreover, being the new information carried by CMBP mainly contained in large angular scales (multipoles $l < 10$, corresponding to $\theta > 20°$ angular scales), a very simple optics has been adopted (corrugated feed horns), which allows a resolution down to $7°$.

Large scale polarization can be detected only by all-sky surveys which, due to the low level of the signal, require a very stable environment. Such investigations can be carried on by space missions only.

Ground-based and ballon-borne experiments, limited to small sky patches, can concentrate their efforts to sub-degree scales attempting a first CMBP detection. However, Carretti et al. [31] have shown that ground-based instruments can be limited by atmospheric emission: Instrumental polarization can correlate the unpolarized atmospheric signal, whose fluctuations can then degrade the expected long term sensitivity.

The main features of SPOrt are the following (see also Table 1):

- four frequency channels at 22, 32, 60 and 90 GHz to match the best band for CMBP observation (90 GHz), while checking the Galactic contributions (22-90 GHz) and mapping the Galactic synchrotron emission (22 and 32 GHz).

- $7°$ (FWHM) angular resolution to obtain the new information on cosmological parameters contained in CMBP on large scales, while minimizing optics systematic effects.

- a nearly all-sky survey ($\sim 80\%$ sky coverage);

Due to the low level of the expected signal, great care has been taken to optimize the instrument design with respect to systematics generation, long term stability and observing time efficiency. The following major points were thus taken into account for the SPOrt design:

- correlation polarimeters to improve the stability (see Figure 2);
- correlation of the two circularly polarized components E_L and E_R as to measure directly and simultaneously both Q and U (100% observing time efficiency).
- on axis and simple optics (corrugated feed horns) in order to minimize the spurious polarization induced by both the f pattern (a combination of co-polar and cross-polar pattern, see [31] for details) and the CMB temperature anisotropy at the beam scale: With such a configuration ~ -35 dB of cross-polarization translates into a contamination $< 0.2\ \mu$K;
- high OMT isolation (> 60 dB) and low cross-talk (< -60 dB) [32], since these parameters are among the major responsibles for Q & U offset generation in correlation polarimeters [31].
- very small difference (< -30 dB) between the attenuations of the two polarizations in the polarizer, which is the other main responsible for offset generation [31].

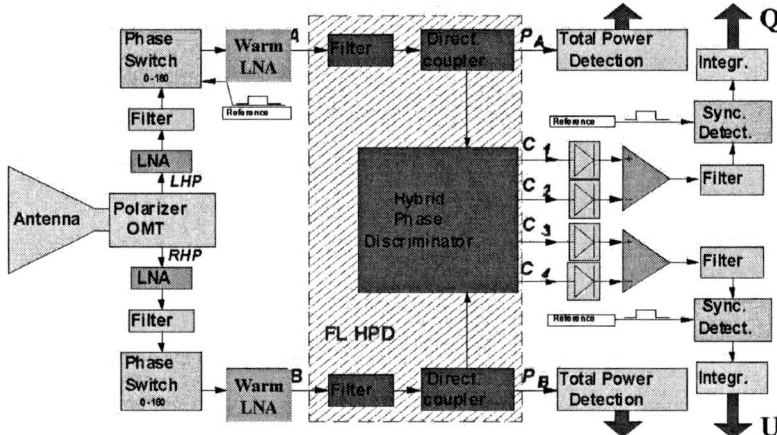

FIGURE 2. Schematic block diagram of the SPOrt radiometers. The antenna collects the signal and the combination of polarizer and OMT extracts the two circularly polarized components *LHP* & *RHP*. After amplification, the signals are correlated by the correlation unit (based on a Hybrid Phase Discriminator, see [33] for details) providing directly both Q & U.

The spurious polarization generated by the optics is due to the anisotropy distribution of the unpolarized radiation (see [31]) modulated by the f pattern:

$$
T^{\mathrm{horn}} = \frac{1}{\Omega_A} \int_0^\pi \sin\theta\, d\theta \int_0^{\pi/2} d\phi\, [\Delta T_b(\theta,\phi) - \Delta T_b(\theta,\phi+\pi/2)+
$$
$$
\Delta T_b(\theta,\phi+\pi) - \Delta T_b(\theta,\phi+3/2\pi)]\cdot f(\theta,\phi), \tag{1}
$$
$$
f(\theta,\phi) = -P(\theta,\phi)\chi^*(\theta,\phi+\pi/2) + \chi(\theta,\phi)P^*(\theta,\phi+\pi/2), \tag{2}
$$

where P and χ are the co-polar and cross-polar patterns, respectively, and Ω_A is the antenna beam. In the case of the SPOrt feed horns, the contribution of the f pattern is ~ -24 dB and the rms contamination from the 30 μK of the CMB anisotropy is lower than 0.2 μK. Due to its intrinsic asimmetry, off-axis optics with the same cross-polar pattern level would imply a spurious contribution 8-10 dB higher.

The instability of a radiometer can be measured in terms of the knee frequency (f_{knee}). This provides the time scale at which the $1/f$ component of the noise power spectrum prevails on the white noise. Destriping techniques can remove most of the effects of the $1/f$ noise, provided the knee frequency is lower than the signal modulation frequency. For SPOrt this corresponds to the orbit frequency $f_{\mathrm{orbit}} = 1.8 \times 10^{-4}$ Hz.

Currently available InP LNA amplifiers have rather high knee frequencies ($f_{\mathrm{knee}}^{\mathrm{lna}} \sim 100$-$1000$ Hz) making correlation architectures more convenient. In fact, the knee frequency of a correlation receiver is related to that of its amplifiers by the formula

$$
f_{\mathrm{knee}} = \left(\frac{T_{\mathrm{offset}}}{T_{\mathrm{sys}}}\right)^2 f_{\mathrm{knee}}^{\mathrm{lna}} \tag{3}
$$

111

where T_{offset} is the radiometric offset and T_{sys} is the system temperature.

The SPOrt team put big efforts in analizing and identifying the offset sources [31], and developed components good enough to satify the stability criteria. The main offset sources were identified with the OMT cross-talk and with the difference between the attenuations of the two polarizations of the polarizer. SPOrt needs were quantified in -60 dB and -30 dB, respectively. The hardware development performed by the team allowed us to achieve such goals. The final result is an offset value as low as $T_{\text{offset}} \sim 50$ mK which, combined with a $T_{\text{sys}} \sim 100$ K, gives the knee frequency

$$f_{\text{knee}} \sim 2.5 \times 10^{-7} f_{\text{knee}}^{\text{lna}} \tag{4}$$

matching the condition for a succesful destriping ($f_{\text{knee}} < f_{\text{orbit}}$)

The scanning strategy of SPOrt is bounded by the ISS motion and consists of great circles $51.6°$ tilted with respect to the celestial equator. The precession of the ISS orbit ensures $\sim$80% sky coverage every 70 days. A satisfactory destriping technique has been presented by Sbarra et al. [34] at this meeting.

In conclusion, driven by the long modulation time, a big effort has been spent by the SPOrt team to keep the receiver stability at proper level.

TABLE 1. SPOrt main features: σ_{1s} is the istantaneous sensitivity (1 second), σ_{PX} and $\sigma(P_{\text{rms}})$ are the final sensitivity per pixel and for the P_{rms}, respectively, considering a 18 month mission lifetime and 50% observing efficiency.

Frequencies [GHz]	BW	FWHM	Orbit Time [s]	Coverage	N_{PX}	$\sigma_{1s}[\text{mKs}^{1/2}]$	$\sigma_{PX}[\mu\text{K}]$	$\sigma(P_{\text{rms}})[\mu\text{K}]$
22, 32, 60, 90	10%	7°	5400	80%	660	1.0	5.2	0.3

SCIENTIFIC GOALS

The goal of SPOrt is to measure the diffuse polarized emission in the 22-90 GHz microwave range on large angular scales as to obtain:

- maps of the Galactic synchrotron emission at the lowest frequencies (22-32 GHz);
- tentative detection of (or more stringent limits on) CMBP on large angular scales.

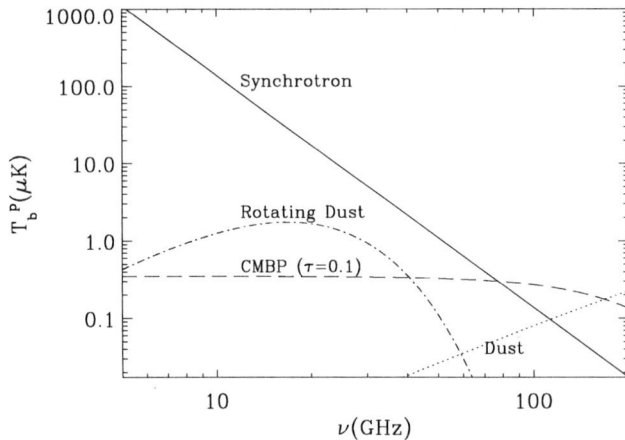

FIGURE 3. Expected polarized brightness temperature for the relevant polarized Galactic foregrounds. The levels are evaluated on 7° scale. See text for the normalization of the synchrotron emission. The partameters of the other components are from [25]. The CMBP behaviour for a ΛCDM model ($\Omega_\Lambda = 0.7$) with optical depth $\tau = 0.1$ is also shown.

As already mentioned, there are no data about the diffuse polarized emission at frequencies greater than 2.7 GHz. However, extrapolating down to 30 GHz the Galactic background emission from the Duncan et al. data [27] a level of $T_{\text{syn}}(30\,\text{GHz}) \sim 5\,\mu\text{K}$ on 7° scales can be evaluated. This estimate has an independent confirmation from unpolarized data: Assuming a 30% polarization, the COBE-DMR data provide a similar value.

Figure 3 shows the expected frequency behaviour of synchrotron emission as well as other relevant foregrounds. Taking into account the SPOrt sensitivities reported in Table 1 we should be able to provide maps of the Galaxy at both 22 and 32 GHz.

The pixel sensitivity does not allow the building of CMBP maps, but we can use full-sky statistical analyses to estimate the mean polarized signal $P_{rms} = \sqrt{\langle Q^2 + U^2 \rangle}$. By applying the flat spectrum analysis developed by Zaldarriaga [35] and already used with the PIQUE and POLAR data [22, 23], simulations performed with the SPOrt specifications show that a detection with an error $\sigma(P_{rms}) = 0.3\,\mu K$ (1σ C.L.) will be possible, including the sensitivity degradation due to the foreground subtraction.

The detection of the P_{rms} on large angular scales is very important because it permits a clean measurement of the optical depth τ of the re-ionized medium during the dark ages. This is shown in Figure 4: The P_{rms} behaviour with respect to τ is mostly independent of other cosmological parameters. The capability of SPOrt to determine τ has been evaluated by a Fisher matrix analysis that gives a sensitivity of $\Delta\tau = 0.13$ for a model with $\tau = 0.2$.

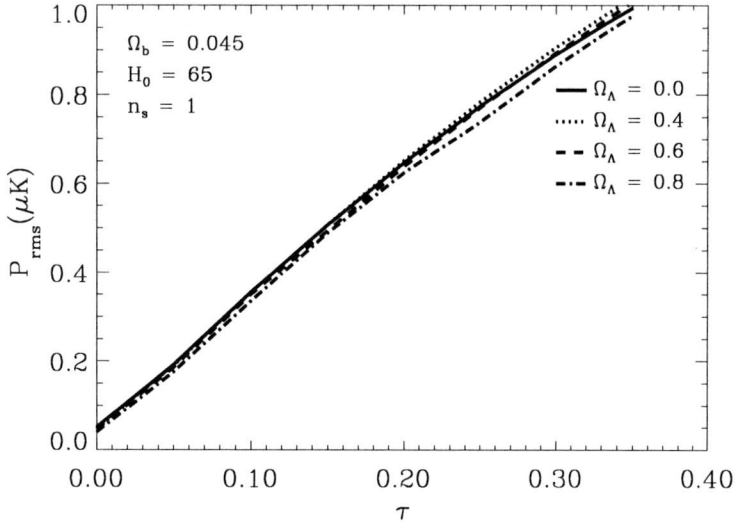

FIGURE 4. Mean polarized signal P_{rms} on 7° scale versus the re-ionized mean optical depth τ for different cosmological models.

ACKNOWLEDGMENTS

We thank V. Natale for useful discussions. We acknowledge use of CMBFAST and HEALPix packages for performing our analysis. SPOrt is an ASI funded project.

REFERENCES

1. Mather, J. C., Fixsen, D. J., Shafer, R. A., Mosier, C., Wilkinson, D. T., *ApJ*, 512, 511-520, 1999.
2. Sazhin, M. V., Benitez, N., *Astrophys. Lett. Commun.*, 32, 105, 1995.
3. Jungman, G., Kamionkowski, M., Kosowsky, A., Spergel, D. N., *PRD*, 54, 1332-1344, 1996.
4. Zaldarriaga, M., Spergel, D. N., Seljak, U., *ApJ*, 488, 1, 1997
5. Efstathiou, G., Bond, J. R., *MNRAS*, 304, 75, 1999.
6. Smoot, G.F., et al., *ApJ*, 396, L1, 1991.
7. Bennet, C.L., et al., *ApJ*, 464, L1, 1996.
8. De Bernardis, P., et al., *Nature*, 404, 955, 2000.
9. Hanany, S., et al., *ApJ*, 545, L5, 2000.
10. Miller, A. D., et al., *ApJ*, 524, L1, 1999.
11. Kosowsky, A., *NewAR*, 43, 157-168, 1999.
12. Kamionkowski, M., Kosowsky, A., *PRD*, 57, 685, 1998.

13. Penzias, A. A., and Wilson, R. W., *ApJ*, 142, 419-221, 1965.
14. Caderni, N., et al., *Phys. Rev. D*, 17, 1901-1907, 1978.
15. Nanos, G. N., *ApJ*, 232, 341-347, 1979.
16. Lubin, P. M., and Smoot, G. F., *ApJ*, 245, 1-17, 1981.
17. Partridge, R. B., et al., *Nature*, 331, 146-147, 1988.
18. Wollack, E. J., et al., *ApJ*, 419, L49-L52, 1993.
19. Netterfield, C. B., et al., *ApJ*, 445, L69-L72, 1995.
20. Sironi, G., Boella, G., Bonelli, G., Brunetti, L., Cavaliere, F., Gervasi, M., Giardino, G., Passerini, A., *NewA*, 3, 1-13, 1998.
21. Subrahmanyan R. et al., *MNRAS*, 315, 808-822, 2000.
22. Hedman M. M. et al., *ApJ*, 548, L111-L114, 2001.
23. Keating, B. G., et al., *ApJ*, 560, L1, 2001.
24. Tucci, M., Carretti, E., Cecchini, S., Fabbri, R., Orsini, M., Pierpaoli, E., *NewA*, 5, 181, 2000.
25. Tegmark, M., Eisenstein, D. J., Hu, W., de Oliveira–Costa, A., *ApJ*, 530, 133, 2000.
26. Brouw, W. N., Spoelstra, T. A. Th., *A&AS*, 26, 129, 1976.
27. Duncan, A. R., Haynes, R. F., Jones, K. L., Stewart, R. T., *MNRAS*, 291, 279, 1997.
28. Duncan, A. R., Reich, P., Reich, W., Fürst, E., 1999, *A&A*, 350, 447, 1999.
29. Uyaniker, B., Fürst, E., Reich, W., Reich, P., Wielebinski, R., *A&AS*, 138, 31, 1999.
30. Gaensler, B. M., Dickey, J. M., McClure–Griffiths, N. M., Green, A. J., Wieringa, M. H., Haynes, R. F., *ApJ*, 549, 959, 2001.
31. Carretti, E., Tascone, R., Cortiglioni, S., Monari, J., Orsini, M., *NewA*, 6, 173-187, 2001 (*astro-ph*/0103318).
32. Tascone, R., et al., in Experimental Cosmology at millimetre wavelengths, AIP Conf. Proc., in press, 2001.
33. Tascone, R., et al., *this volume*.
34. Sbarra, C., et al., *this volume*.
35. Zaldarriaga, M., *ApJ*, 503, 1, 1998.

The BaR-SPOrt Experiment

M. Zannoni*, M. Baralis†, G. Bernardi**, G. Boella*, S. Bonometto*, A. Boscaleri‡,
E. Carretti**, S. Cecchini**, S. Cortiglioni**, R. Fabbri§, M. Gervasi*, C. Macculi**,
J. Monari¶, E. Morelli**, V. Natale‖, R. Nesti††, L. Nicastro‡‡, E. Pascale‡, O.
Peverini†, S. Poppi¶, C. Sbarra**, G. Sironi*, R. Tascone†, M. Tucci* and G.
Ventura**

*Dip. di Fisica, Univ. di Milano - Bicocca, P.zza della Scienza 3, I–20126 Milano
†I.R.I.T.I./C.N.R., c.so Duca degli Abruzzi 24, I–10129 Torino
**I.Te.S.R.E./C.N.R., via P. Gobetti 101, I–40129 Bologna
‡I.R.O.E./C.N.R., Via Panciatichi 64, I–50127 Firenze
§Dip. di Fisica, Univ. di Firenze, Via Sansone 1, I–50019 Sesto Fiorentino, Firenze
¶I.R.A./C.N.R., via P. Gobetti 101, I–40129 Bologna
‖C.A.I.S.M.I./C.N.R., Largo E. Fermi 5, I–50125 Firenze
††Osservatorio Astrofisico di Arcetri, Largo E. Fermi 5, I–50125 Firenze
‡‡I.F.C.A.I./C.N.R., via U. La Malfa 153, I–90146 Palermo

Abstract. BaR-SPOrt (Balloon-borne Radiometer for Sky Polarisation Observations) is an experiment to measure the linearly polarized emission of sky patches aboard a long duration stratospheric balloon. It consists of high sensitivity correlation polarimeters operating in the millimeter wavelength region and coupled to a telescope to obtain a sub-degree angular resolution for direct measurements of the Q and U Stokes parameters. This project shares most of the know-how and sophisticated technology developed for the SPOrt experiment aboard the International Space Station. The instrument design, the various solutions to reduce the systematics and the observing strategy are here described.

INTRODUCTION

The fine characteristics of Cosmic Microwave Background (CMB), spectral distortions, anisotropies and polarization, are the most sensitive tools to investigate the high red-shift Universe [1, 2, 3]. Since its discovery in 1965 [4], many efforts have been concentrated in a precise determination of the spectrum [5, 6, 7] and spatial distribution [8, 9, 10]. Current and near future space missions like MAP[1] and Planck[2] are mainly devoted to the all-sky mapping of CMB small-scale anisotropies for which they will reach the highest sensitivities. On the polarization side, against its unique capability to solve the degeneracy among cosmological parameters that anisotropy alone is not able to remove [2], only upper limits are available so far (see table 1), since either the foreseen polarized component of the CMB is definitely lower than the instrumental sensitivities or the contribution of the systematics dominates the final error budget. For these reasons dedicated experiments must be designed to attempt the CMB polarization detection. One of them will be the space mission SPOrt[3] which is described by Carretti et al. in this volume [11]. Direct spin-off of SPOrt is BaR-SPOrt, a balloon experiment funded by ASI (Italian Space Agency), which shares most of the know-how and technological development of the SPOrt program. Differently from SPOrt, the scientific goal of BaR-SPOrt is the mapping of the sky polarization of some low foreground regions with sub-degree angular resolution. In the following sections such a goal together with a review of the radiometer design and the observing strategies will be discussed in

---

[1] http://map.gsfc.nasa.gov/
[2] http://astro.estec.esa.nl/Planck/
[3] http://sport.tesre.bo.cnr.it/

CP609, *Astrophysical Polarized Backgrounds,* edited by S. Cecchini et al.
© 2002 American Institute of Physics 0-7354-0055-5/02/$19.00

more detail.

TABLE 1. Existing upper limits for the CMB linear polarization.

Resolution (deg)	Frequency (GHz)	Sky Coverage	Upper Limit	Reference
15	4	Scattered	300 mK	[4]
$1.5 - 40$	$100 - 600$	GC	3-03 mK	[12]
15	9.3	$\delta = +40°$	1.8 mK	[13]
15	33	$+37° \leq \delta \leq +63°$	$180\,\mu K$	[14]
$18'' - 160''$	5	$\delta = +80°$	4.2 mK - $120\,\mu K$	[15]
1.2	$26 - 36$	NCP	$30\,\mu K$	[16]
1.4	$26 - 36$	NCP	$18\mu K$	[17]
7	33	SCP	$267\,\mu K$	[18]
$6'$	8.7	$\delta = -50°$	$16\,\mu K$	[19]
0.24	90	NCP	$13\,\mu K$	[20]
7	$26 - 36$	$\delta = +43°0$	$10\,\mu K$	[21]

THE SCIENTIFIC MOTIVATION AND POSSIBLE TARGETS

Large scale CMB polarization (CMBP) measurements, due to the low level of the signal expected (see figure 1), need a very stable environment to be performed. For this reason all-sky surveys can be only carried out in space where very quiet and stable conditions exist. Ground-based and balloon-borne experiments can be devoted to small sky patches where they can reach sensitivities comparable with the expected level of polarization. As pointed out in [22], ground-based instruments operating in the millimeter domain are plagued by atmospheric emission which is the main source of spurious polarizations even in the best observing site like the Antarctic plateau. The instrumental polarization can correlate the unpolarized atmospheric signal, whose fluctuations can then degrade the expected sensitivity. The obvious consequence is to perform the observations from stratospheric altitude where the residual atmosphere and its contribution are negligible. Due to the limited observing time (a long duration flight can last about ten days - cfr [9]), typical targets are small ($\sim 20°x20°$) patches. On such portions of sky it is possible to study the sub-degree scales where the expected polarization peaks (see figure 1).

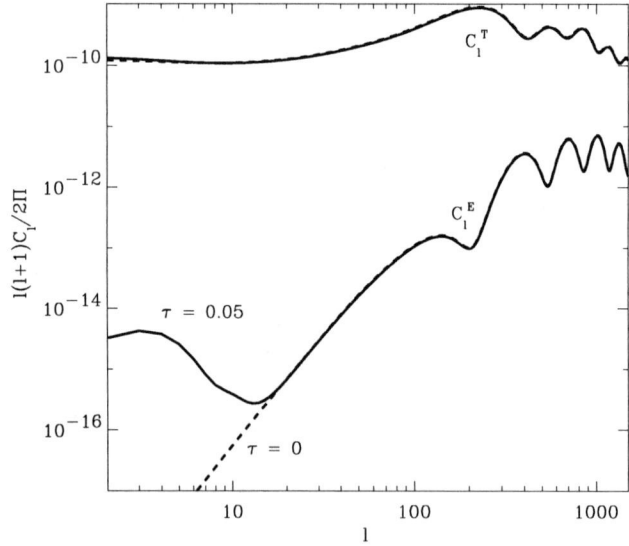

FIGURE 1. Anisotropy and E-mode power spectra. The expected polarized component of the CMB peaks in the sub-degree (high *l*) scales while the larger (degree) angular scales are more sensitive to the reionization scenarios.

Furthermore multifrequency deep scanning of small sky regions are also useful to understand how the polarized foregrounds contaminate a possible detection of CMBP. BaR-SPOrt is designed to carry on both these tasks.

Both in the Northern and in the Southern hemisphere sky patches of interest are available. The ideal regions, from the CMB point of view, are the ones at high galactic latitude, far from local cirrus, where the foreground should be minimal. In the Southern sky, the patch observed by BOOMERanG [9] ($\alpha = 5^h$, $\delta = -45°$) is the ideal target, being opposite to the Sun during Summertime. Scaling from the Jonas map [23] at 2.3 GHz with a synchrotron spectral index $\gamma = 3$ ($T^{synch} \propto \nu^{-3}$) and with a ratio of $P^{synch}/T^{synch} = 0.1$, we can derive a level of polarization at 32 GHz (one of the channel of BaR-SPOrt) $\Delta P < 1\mu K$. The same thing can be done at 90 GHz starting from the DIRBE 240 μm map [24] and scaling the dust temperature as $T^{dust} \propto \nu^{2.7}/(e^{h\nu/kT_D} - 1)$ [25] with $P^{dust}/T^{dust} = 0.05$. The result is $\Delta P < 0.15\mu K$. For the region centered at ($\alpha = 10^h \delta = +35°$) in the Northern hemisphere, it is possible to derive, from the Reich [26] map at 1.4 GHz for the synchrotron contribution, $\Delta P < 0.3\mu K$ @ 32 GHz and $\Delta P < 0.15\mu K$ @ 90 GHz. In figure 2 the three maps [26, 23, 24] cited above have been scaled to 32 GHz and 90 GHz for the evaluation of synchrotron and dust contribution respectively; the selected patches in the Northern and Southern sky are shown in figure 2.

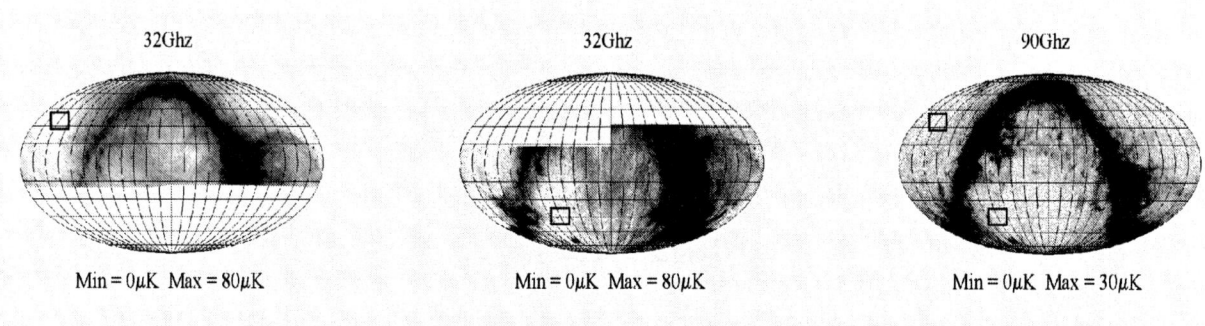

32Ghz   32Ghz   90Ghz

Min = 0μK Max = 80μK   Min = 0μK Max = 80μK   Min = 0μK Max = 30μK

FIGURE 2. Foreground maps computed from the original public data in the radio [26, 23] and far infra-red [24] ranges. The first two have been rescaled to 32 GHz to evaluate the synchrotron component, while the last, related to the dust contribution, has been scaled down to 90 GHz. The two target patches have been superimposed.

From a logistical point of view, both patches are effectively observable. The one in the Southern Sky is accessible with a long duration balloon flight from the Mc Murdo Station in Antarctica, while the one in the Northern hemisphere can be observed during a flight launched from available facilities both in Norway [27] and in Sweden [28]. At the moment no final decision about the scanning strategies has been taken. Simulations are under development to optimize the patch dimension, scanning speed and path for maximizing the final full patch sensitivity. Expected raw sensitivities are reported in table 2.

TABLE 2. BaR-SPOrt expected sensitivities : σ_{1s} is the instantaneous sensitivity, σ_{PX} and σ_{FP} are the final per pixel and the full patch *rms* sensitivities for a flight of two weeks and a patch $20° x 20°$ wide.

Frequencies (GHz)	Bandwidth	Beamsize	$\sigma_{1s}[mKs^{1/2}]$	$\sigma_{PX}[\mu K]$	$\sigma_{FP}[\mu K]$
32	10%	0.5°	0.5	18	0.4
90	10%	0.2°	0.7	64	0.6

THE INSTRUMENT

The BaR-SPOrt payload houses correlation microwave polarimeters (32 & 90 GHz) for the direct measurement of the Q and U Stokes parameters with HPBW=$0°.5$ @32 GHz and $0°.2$ @90 GHz (see figures 3 and 4). In the current baseline the two radiometers will be operated during different campaigns because only one feed at a time can be housed on the optical axis of the telescope (at the moment we are oriented toward a Cassegrain scheme) to meet the very stringent requirements of extremely low spurious polarization (fractions of μK) necessary for such measurements (for a general discussion see [22]). The detection of signals as low as those expected from CMB polarisation ($\leq 1\,\mu$K) implies the use of extremely sensitive and stable radiometers. BaR-SPOrt has been designed to minimize instrumental effects and to reduce $1/f$ noise, thereby increasing the long term stability. Great care has been taken in the realisation of the antenna system to control the spurious polarisation. Correlation techniques are widely adopted in high sensitivity measurements because of their capability to reduce the effects of gain fluctuations. Residual instabilities are recovered using destriping techniques [29, 30, 31, 32, 33], which require the radiometer to be stable only over a single scan period (the scanning time of BaR-SPOrt is between 30 and 60 seconds).

The main instrumental characteristics are:

- direct amplification architecture: no down conversion to avoid possible phase error;
- low cross-polarisation optics providing HPBW of $0°.5$ ($0°.2$) at 32 GHz (90 GHz);
- correlation unit based on a custom design waveguide Hybrid Phase Discriminator (HPD), with unpolarised component rejection ≥ 30 dB [34];
- custom design Orthomode Transducer (OMT) with high isolation between channels (> 60 dB) to limit contaminations from the unpolarised component [32, 34];
- phase modulation (lock-in system) and correlation providing > 70 dB of total rejection to the unpolarised component;
- a cryostat (see figures 5, 6) to cool to $T < (80.0 \pm 0.1)$ K the Low Noise Amplifiers, the circulators, the polariser and the OMT by a closed loop cryocooler. The horn, at present designed to be kept at ≈ 300 K, might be cooled as well. A thermal shield stabilised at temperature $T \cong (300.0 \pm 0.1)$ K, is foreseen to increase the thermal stability;
- custom design internal calibrator to inject reference polarised signals.

The block diagram of the radiometer is reported in figure 3.

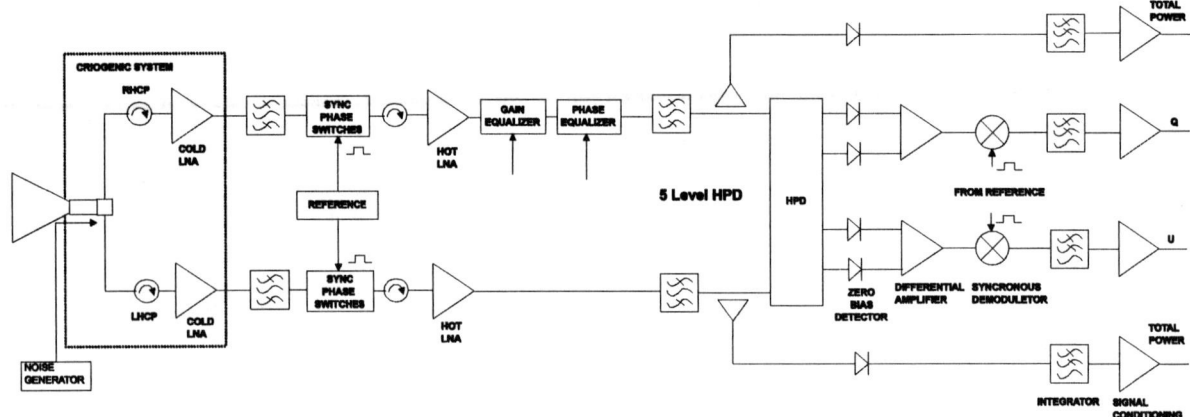

FIGURE 3. Block diagram of the BaR-SPOrt radiometers.

The antenna system collects the incoming radiation and transforms the linearly polarized components (E_x, E_y) of the electric field into the circularly ones (E_R, E_L) which are picked up by the Ortho-Mode Transducer (OMT) [35] (see figure 4).

A two positions ($0 - \pi$) phase shift after the first stage of amplification (LNA) provides phase modulation, followed by synchronous detection, in order to reduce the instability of the components inside the lock-in loop. Just before the Hybrid Phase Discriminator, a fraction of the signal is picked-up and fed into two total power detectors to record the Sky temperature and for monitoring of the System temperature too. The heart of the correlation unit is the HPD that processes the signal in order to have four outputs proportional to:

$$\vec{E}_R - \vec{E}_L \qquad \vec{E}_R + \vec{E}_L \qquad \vec{E}_R + j\vec{E}_L \qquad \vec{E}_R - j\vec{E}_L \tag{1}$$

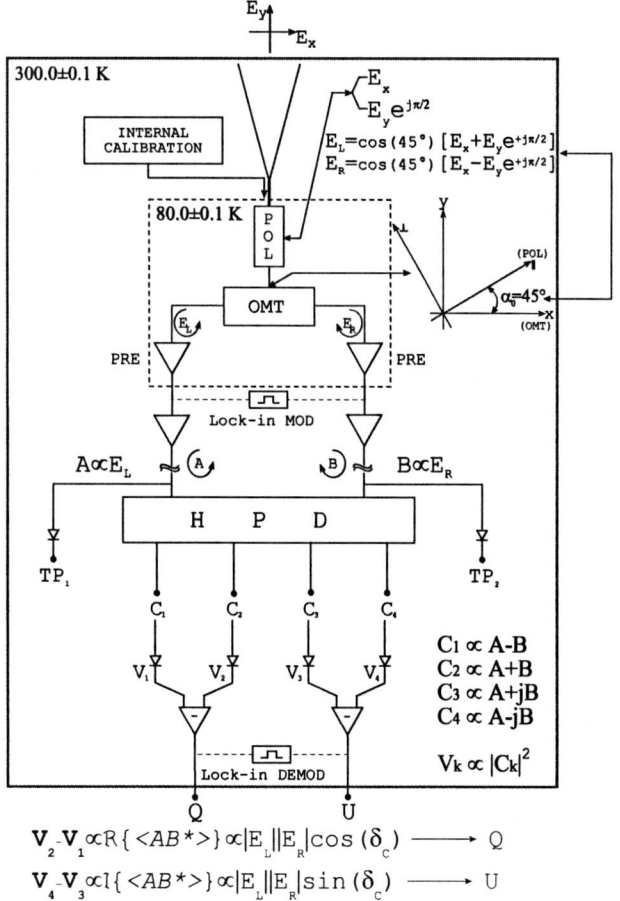

$$E_L = \cos(45°) \, [E_x + E_y e^{+j\pi/2}]$$
$$E_R = \cos(45°) \, [E_x - E_y e^{+j\pi/2}]$$

$C_1 \propto A-B$
$C_2 \propto A+B$
$C_3 \propto A+jB$
$C_4 \propto A-jB$

$V_k \propto |C_k|^2$

$$V_2\text{-}V_1 \propto \Re\{\langle AB*\rangle\} \propto |E_L||E_R|\cos(\delta_c) \longrightarrow Q$$
$$V_4\text{-}V_3 \propto \Im\{\langle AB*\rangle\} \propto |E_L||E_R|\sin(\delta_c) \longrightarrow U$$

FIGURE 4. Scheme of the radiometers with the propagation of the fields collected by the feed horn.

After square law detection the four HPD outputs are:

$$V_1 \propto [|\vec{E}_L|^2 + |\vec{E}_R|^2 - 2\Re(\vec{E}_L \cdot \vec{E}_R^{\,*})] \tag{2}$$
$$V_2 \propto [|\vec{E}_L|^2 + |\vec{E}_R|^2 + 2\Re(\vec{E}_L \cdot \vec{E}_R^{\,*})]$$
$$V_3 \propto [|\vec{E}_L|^2 + |\vec{E}_R|^2 + 2\Im(\vec{E}_L \cdot \vec{E}_R^{\,*})]$$
$$V_4 \propto [|\vec{E}_L|^2 + |\vec{E}_R|^2 - 2\Im(\vec{E}_L \cdot \vec{E}_R^{\,*})]$$

which are properly differentiated to get as final outputs the two quantities:

$$V_2 - V_1 \propto |\vec{E}_L||\vec{E}_R|\cos(\delta_c) \longrightarrow Q \tag{3}$$
$$V_4 - V_3 \propto |\vec{E}_L||\vec{E}_R|\sin(\delta_c) \longrightarrow U$$

where δ_c is the phase delay between the two (L&R) circular components of the electric field. After integration, these provide time averaged values proportional to the Q and U Stokes parameters.

Since the BaR-SPOrt performances strongly depend on the temperature stability, particular care has been put in the thermal design. A mechanical Stirling cryocooler with closed loop control will provide the cooling down to 80 K with stability better than 0.1 K over the period of the flight (for some preliminary results on the performances of the cooler see [36]).

The vacuum needed for thermal isolation implies a large window in front of the feed which is housed inside the cryostat to keep stable its temperature (an active thermal control is also present in warm parts of the radiometer). Concerning the problem of the optical beam propagation through the window, great attention must be paid to the

FIGURE 5. The cryostat housing the 32 GHz radiometer of BaR-Sport with the closed loop cryocooler visible in the foreground.

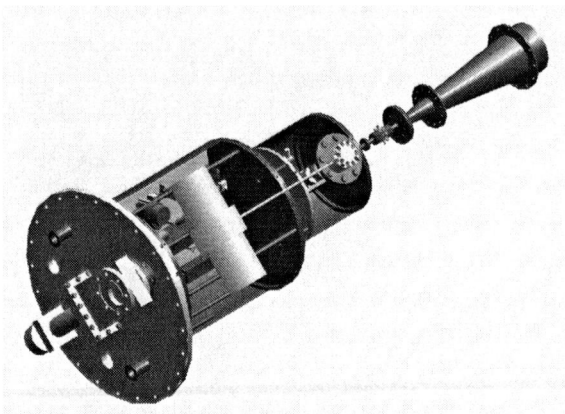

FIGURE 6. The radiometer inside the cryostat.

selection of the material of the vacuum window, even in the case of axial symmetry, which should ensure a null induced polarization. Some polymers have the necessary strength to sustain the pressure gap with very low internal stress [37], but show dichroism and birefringence [38]. This means that they are not suitable for polarization measurements. For many others materials there are no "optical" data in the microwave domain and our experimental team has started a parallel activity [39] to fully characterize a set of materials looking for the smallest spurious polarization.

CONCLUSIONS

The BaR-SPOrt experiment represents a good opportunity for testing in operative conditions state of the art technological solutions to be used in the incoming space mission SPOrt. It is also one of the first instruments with the potentiality to measure CMB polarization in small sky patches on sub-degree angular scale.

ACKNOWLEDGMENTS

Authors wish to thank the organizers of the Workshop for providing a great opportunity to show the BaR-SPOrt experiment. This work is partially supported by ASI.

REFERENCES

1. Jungman, G., Kamionkowski, M., Kosowsky, A., & Spergel, D. N. *PhysRevD*, 54, 1332-1344, 1996.
2. Zaldarriaga, M., Spergel, D. N., & Seljak, U. *ApJ* , 488, 1, 1997.
3. Efstathiou, G. & Bond, J. R. *MNRAS*, 304, 75, 1999.
4. Penzias, A. A. and Wilson R. W. *ApJ*, 142, 419-221, 1965.
5. Smoot, G. F. et al. *ApJL*, 291, L23, 1985.
6. Fixsen, D. J., Cheng, E. S., Gales, J. M., Mather, J. C., Shafer, R. A., & Wright, E. L. *ApJ*, 473, 576, 1996.
7. Zannoni, M. et al. in proceedings of the *International Conference on 3K Cosmology EC-TMR, Roma 1998*, edited by L. Maiani, F. Melchiorri and N. Vittorio, AIP Conference Proceedings 476, 1999, p. 165.
8. Bennett, C. L. et al. *ApJ*, 464, L1, 1996.
9. de Bernardis, P. et al. *Nat*, 404, 955, 2000.
10. Hanany, S. et al. *ApJ*, 545, L5, 2000.
11. Carretti, E. et al., this volume
12. Caderni, N., Fabbri, R., Melchiorri, B., Melchiorri, F., & Natale, V. *PhysRevD*, 17, 1901, 1978.
13. Nanos, G. P. *ApJ*, 232, 341, 1979.
14. Lubin, P. M. & Smoot, G. F. *ApJ*, 245, 1, 1981.
15. Partridge, R. B., Nowakowski, J., & Martin, H. M. *Nat*, 331, 146, 1988.
16. Wollack, E. J., Jarosik, N. C., Netterfield, C. B., Page, L. A., & Wilkinson, D. *ApJ*, 419, L49, 1993.
17. Netterfield, C. B., Jarosik, N., Page, L., Wilkinson, D., & Wollack, E. *ApJ*, 445, L69, 1995.
18. Sironi, G. et al., *NewA*, 3, 1, 1998.
19. Subrahmanyan, R., Kesteven, M. J., Ekers, R. D., Sinclair, M., & Silk, J. *MNRAS*, 315, 808, 2000.
20. Hedman, M. M., Barkats, D., Gundersen, J. O., Staggs, S. T., & Winstein, B. *ApJL*, 548, L111, 2001.
21. Keating, B. G., O'Dell, C. W., de Oliveira-Costa, A., Klawikowski, S., Stebor, N., Piccirillo, L., Tegmark, M., & Timbie, P. T. *ApJL*, 560, L1, 2001.
22. Carretti, E., Tascone, R., Cortiglioni, S., Monari, J., Orsini, M., *NewA* 6, 173–187, 2001.
23. Jonas, J. L., Baart, E. E., Nicolson, G. D. *MNRAS*, 297, 977, 1998.
24. Hauser, M. G., et al, *ApJ*, 508, 25, 1998.
25. Tegmark, M., Eisenstein, D. J., Hu, W., & de Oliveira-Costa, A., *ApJ*, 530, 133, 2000.
26. Reich, P. & Reich, W., *A&AS* 63, 205
27. Boen, K. this volume.
28. Baldemar, P., Widell, O. this volume.
29. Sbarra, C. et al. this volume.
30. Delabrouille, J. *A&AS*, 127, 555, 1998.
31. Wright E. L., *astro-ph*/9612006.
32. Carretti E., et al., 2001 in preparation.
33. Revenu, B., Kim, A., Ansari, R., Couchot, F., Delabrouille, J., & Kaplan, J. *A&AS*, 142, 499, 2000.
34. Tascone R. et al. *AIP 2K1BC Conference Proceeding, Cervinia (Ao), 2001 in press.*
35. Macculi C. et al. *AIP 2K1BC Conference Proceeding, Cervinia (Ao), 2001 in press.*
36. Macculi, C. & Zannoni, M. this volume.
37. J. R. White, Origin and Measurement of Internal Stress in Plastics, Polymer Testing, 4, 165–191, 1984.
38. E. D. Palik, Handbook of optical constants of solid II, Academic Press, 1991.
39. Macculi C., Spurious polarisation from dielectrics, *BaR-SPOrt Int. Tech. memo*, 2000.

B2K: the polarization-sensitive BOOMERanG experiment

S. Masi[1], P.A.R. Ade[2], J.J. Bock[3], A. Boscaleri[4], P. de Bernardis[1], G. De Troia[1], G. Di Stefano[5], V.V. Hristov[6], A. Iacoangeli[1], W.C. Jones[6], T. Kisner[7], A.E. Lange[6], P.D. Mauskopf[2], C. Mac Tavish[8], T. Montroy[7], C.B. Netterfield[8], E. Pascale[4], F. Piacentini[1], F. Pongetti[5], G. Romeo[5], J.E. Ruhl[7], E. Torbet[7], J. Watt[8]

[1] *Dipartimento di Fisica, Universitá La Sapienza, Roma, P.le A. Moro, 2, 00185, Italy.* [2] *Dept. of Physics and Astronomy, Cardiff University, Cardiff CF24 3YB, Wales, UK.* [3] *Jet Propulsion Laboratory, Pasadena, CA, USA.* [4] *IROE-CNR, Firenze, Italy.* [5] *Istituto Nazionale di Geofisica, Roma, Italy.* [6] *California Institute of Technology, Pasadena, CA, USA.* [7] *Dept. of Physics, Univ. of California, Santa Barbara, CA, USA.* [8] *Depts. of Physics and Astronomy, University of Toronto, Canada.*

Abstract. We describe the new BOOMERanG payload, which is being prepared for a new circum-antarctic flight, with the aim to detect the linear polarization of the Cosmic Microwave Background (CMB). In addition to polarization capabilities, obtained by means of special bolometers, the instrument has been improved in the attitude reconstruction system and in the calibration system.

INTRODUCTION

The BOOMERanG payload [1], [2], [3] has been flown in 1997 and 1998/99 and has produced high resolution maps of the microwave sky at 90, 150, 240, 410 GHz [4], [5]. The angular power spectrum of the maps features peaks at multipoles $\ell \sim 210, 540, 840$, [5], [6] as expected in all cosmological models with phase-coherent pressure waves in the photon-baryon fluid at recombination. These are expected in adiabatic models of structure formation, where fluctuations are generated in the inflation process. Independent evidence for these features comes from the DASI [7] and MAXIMA [8] experiments.

One strong prediction of this scenario is that the CMB should have a small level of linear polarization. In fact, in Thomson scattering, any quadrupole anisotropy in the incoming photons results in a degree of linear polarization in the scattered photons [9]. The main terms of the local anisotropy due to density (scalar) fluctuations are monopole and dipole, while the quadrupole term is much smaller: for this reason the expected polarization is quite weak [12], [10]. It is possible to compute the power spectra of the polarization expected for the same cosmological scenario best fitting the current anisotropy data. This scalar component of the CMB polarization is expected to be curl-free, and is called the **EE** component of the polarization. Moreover, in the inflationary scenario, tensor fluctuations (gravitational waves) are generated in the very early Universe, thus producing a different pattern of polarization in the CMB. These fluctuations introduce both curl-free and curl (labeled **BB**) components in the polarization pattern. There are thus several good reasons to search for such small polarization signature in the CMB sky. A cross correlation spectrum between temperature and scalar polarization (labeled **TE**) is expected as well (see e.g. [11]), with amplitude larger than the pure scalar and tensor polarization spectra. This is in the range of sensitivity of the MAP satellite, and will probably not require the development of a dedicated experiment. Typical levels of the rms polarization fluctuations in a 0.5^o beam are $\sim 1\mu K$ for the **EE** component and $\sim 0.2\mu K$ for the **BB** component. Compared to the temperature fluctuations (of the order of $70\ \mu K$ rms in the same beam), these numbers are definitely challenging. Changing the cosmological parameters within the ranges consistent with the BOOMERanG measurements of the anisotropy does not change much the resulting polarization. The polarization signal is very small with respect to the CMB anisotropy itself, and even smaller with respect to many instrumental and astrophysical contaminants. Despite very intense and long lasting efforts [13], [14], [15], [16], [17], [18], [19], [21], [20], CMB polarization has not been detected yet. The best upper limits for the polarization power spectrum are still of the order of 5-10 μK.

CP609, *Astrophysical Polarized Backgrounds,* edited by S. Cecchini et al.

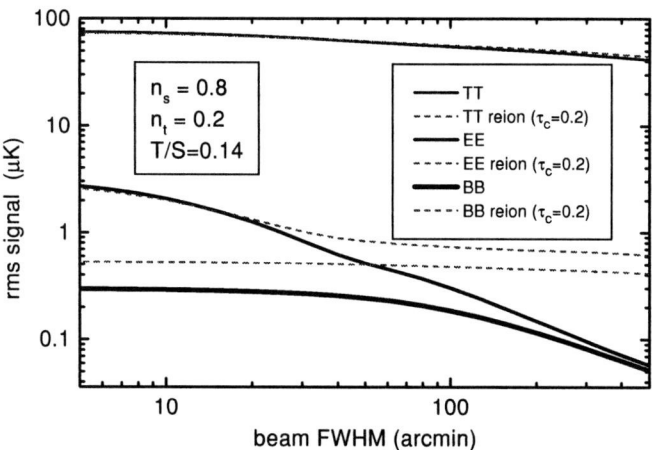

FIGURE 1. rms signal detected by a polarization experiment as a function of the beam FWHM. The temperature anisotropy (TT), gradient polarization (EE) and curl polarization (BB) signals are plotted. A model with a significant tensor component has been selected, for illustration purposes.

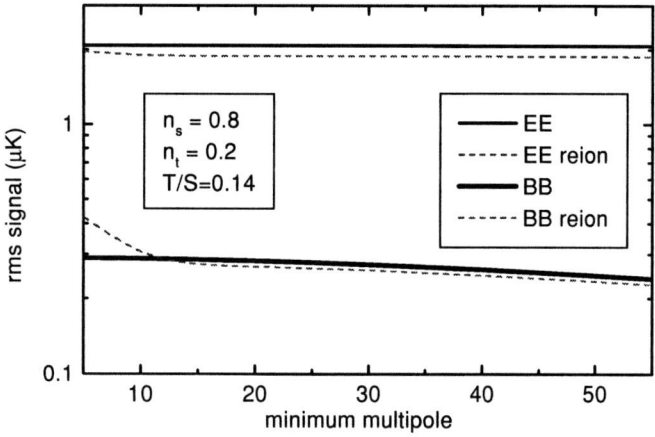

FIGURE 2. rms signal detected by a polarization experiment as a function of the minimum multipole to which the survey is sensitive. This is if the order of π/θ, where θ is the maximum size of the survey.

We have a quite detailed idea of the signal we expect in the standard scenario. The angular power spectrum can be computed using CMBFAST, and the rms signal expected for a polarimeter covering a given multipoles range can be computed by integrating the power spectra. In fig.1 we show how the detected rms signal depends on the angular resolution of the polarimeter, while in fig.2 we show how the rms signal depends on the minimum multipole to which the polarimeter is sensitive. This, in turn, depends on the area of the sky survey (a size of the survey of θ allows detection of multipoles $\ell > \pi/\theta$). It is evident that, for a polarization experiment oriented to a first detection of **EE**, a good angular resolution $\sim 1^o$ to $\sim 10'$ should be preferred, while the size of the survey can be quite limited. The detection of the reionization signal and the measurement of τ_C would require, instead, a very wide survey.

EXPERIMENTAL APPROACH

The experimental approach to the problem is to find an instrument setup able to extract the small CMB polarization signal in the presence of overwhelming CMB anisotropy, astrophysical and instrumental foregrounds.

For this purpose most polarimeters use a modulating analyzer to extract, by means of synchronous demodulation techniques, the small polarized component. Rotating wire grids, half wave plates, K-mirrors, Faraday rotators, Fresnel rombs, have been used or proposed as polarization analyzers (see e.g. [22]). All these techniques suffer for the need of long integration time: one needs to point to the same sky pixel during many cycles of the analyzer. It is thus difficult to produce extended maps of the CMB polarization.

In BOOMERanG we use the sky scan to modulate the signal. We want to map the two orthogonal components of the linear polarization with two separate bolometers, $B1$ and $B2$, each sensitive to one polarization direction, and combine the two signals to retreive the linear polarization Stokes parameter Q. The U parameter can be measured as well by means of an identical device rotated by 45°. (see fig.4). Polarization sensitive bolometers (PSB) of this kind have been prepared for the coming flight of BOOMERanG and for the High Frequency Instrument of the Planck satellite [23].

Very little is known about the polarization of CMB foregrounds. Synchrotron radiation is strongly polarized, and microwave emission from spinning dust grains is also expected to be significantly polarized. For this reason, the best spectral region for this search is in the 150 GHz range, where thermal emission from interstellar dust (ISD) is low enough [24], and the two above mentioned polarized foregrounds are not an issue. If we assume a conservative 5% polarization for the dust (see e.g. [25]), the resulting polarized signal at 150 GHz is lower than the CMB polarization to be detected (see fig.3).

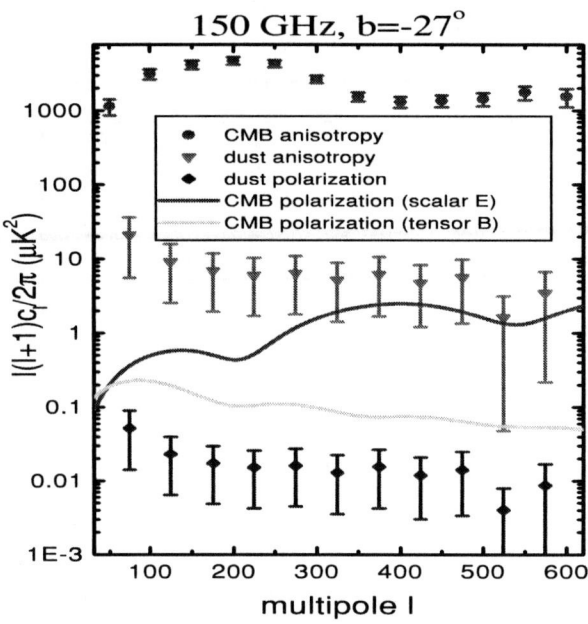

FIGURE 3. Comparison between anisotropy and polarization of the CMB and anisotropy and polarization of interstellar dust in high latitude cirrus clouds. The anisotropy measurements are from the BOOMERanG experiment at 150 GHz. The CMB polarization power spectra are computed assuming the best fit parameters obtained fitting the anisotropy power spectrum; the interstellar dust anisotropy has been extrapolated to 150 GHz using the measurements at 410 GHz, and the polarization of interstellar dust has been estimated simply assuming a conservative 5% of the anisotropy.

B2K

With the BOOMERanG payload, recovered and refurbished after the 1998 flight, signal modulation is obtained by scanning the sky at a constant rate v. Polarization signals are detected at a frequency f far from the 1/f knee of the noise and far from the effects of instrumental drifts. This sky scan approach is similar to the anisotropy measurements performed with BOOMERanG in its long duration flight in 1998. In fact $f = v\ell/\pi$; with $\ell = 300 - 1000$ and $v = 1^o/s$ we get $f = 1.7 - 5.5 Hz$, well inside the useful bandwidth of the 0.3 K bolometers used in the BOOMERanG instrument.

In BOOMERanG an off-axis telescope close to the Dragone configuration [26] has been used. We have analyzed in detail its polarization properties on a wide region of the focal plane. The telescope is an off-axis gregorian, with a 1.3 m diameter aluminum primary enclosed in a low emissivity cavity. Secondary and tertiary are inside the cryostat, and reimage the primary focal plane inside the dewar. The cold tertiary acts as the Lyot stop of the system, and is surrounded by a 2K absorber, thus improving sidelobes rejection and significantly reducing the straylight on the detectors. The spurious polarization induced by the BOOMERanG off-axis telescope at the edge of the used focal plane is $\sim 10^{-4}$, well suited for this purpose.

Combining signals from two different detectors instead of using always the same detector to measure the two polarization components is a significant concern (see e.g. [27]). One has to demonstrate that this procedure does not introduce significant systematics. Spurious polarization induced by the optics or by the bolometers polarizes any unpolarized input signal. Since the spurious polarization is constant during the scans, only sky brightness gradients are a concern, while the constant instrumental and astrophysical backgrounds do not affect the measurements. Cross polarization induced in the optics or in the PSBs can be treated in a similar way. We have carried out detailed simulations to investigate the problem in a quantitative way . We find that, if the dominant unpolarized signal is CMB anisotropy, spurious polarization and unbalanced cross polarization up to a few % are acceptable . A careful pre-flight calibration is required, to measure the cross-polarization of the system to a few % [28].

The B2K1 focal plane design was optimized for sensitivity per pixel and foreground discrimination. It consists of 8 close-packed corrugated feed elements arranged in 2 rows separated by an elevation of 0.5° (see fig.4). Each row contains 4 elements separated in cross-elevation by 0.5°. In the upper row corrugated shaped horns feed 2-color photometers which operate at 245 GHz and 345 GHz. Polarizing grids are placed on the front of photometer feed horns. The photometers use spider web bolometers similar to the ones used in B98. The horns in the lower row feed

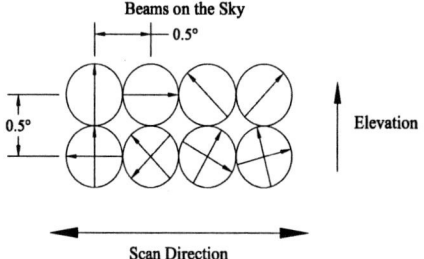

FIGURE 4. Left: Photo of three corrugated feedhorns used in the 150GHz channels of B2K. Right: sketch of the B2K focal plane. The upper row of feeds has grid polarizers at the entrance, and 2-color photometers at 245 and 345 GHz at the output. The lower row of feeds has 150 GHz polarization sensitive bolometers (PSBs) at the output.

pairs of PSB operating at 150 GHz . The PSB's consist of 2 square mesh bolometers placed 65 μm apart in the same groove of a circular corrugated waveguide. Each bolometer is metallized along one direction and each element in the pair is sensitive to a different polarization (see fig.5). The PSBs provide an instantaneous measurement of the Stokes I and Q parameters. The relative rotation of each pixel allows the determination of the U parameter as well as I and Q. Each bolometer has an in-flight NET of $170\mu K_{CMB}\sqrt{s}$. The instrumental sensitivity to temperature anisotropies will

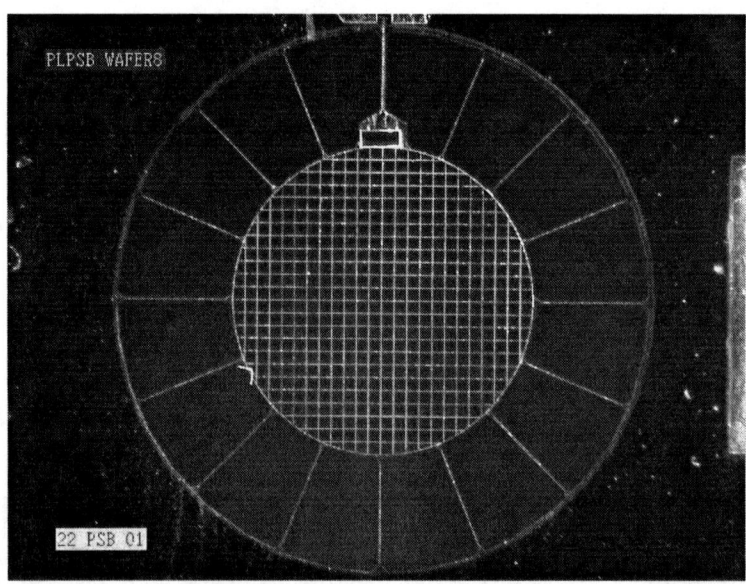

FIGURE 5. Photograph of a PSB used in the B2K focal plane. The two grids of the two separate bolometers are overlap in the picture. The grids have a diameter of 2.6 mm and are made out of 2 μm thick Si_3N_4 wires metallized only in one of the two orthogonal directions.

be of the order of $170/\sqrt{8} = 60 \mu K_{CMB}\sqrt{s}$, with a sensitivity to Q and U of $\sim 170 \times \sqrt{2}/\sqrt{4} = 120 \mu K_{CMB}\sqrt{s}$.

The frequency bands are determined using metal mesh filters for the low pass and waveguide cut-offs for the high pass. The PSB's are expected to have a beam FWHM of 11 arcmin and the photometer channels have a 6 arcmin FWHM. The feed elements are placed as close as mechanically possible to optimize the overlap of maps made by different channels. This is especially critical for the photometers, since Stokes Q and U can be determined only by differencing maps made by different photometers.

POLARIMETER CALIBRATION

We have seen above how important it is to characterize the polarization properties of CMB polarimeters. The pre-flight calibration of a polarimeter is more difficult than the usual photometer calibration. In particular it is very important to study the co-polar and cross-polar response (beam and integral) of the polarimeter. We have developed a polarized, far field, sine-modulated source filling the beam of the instrument. This will allow us to carry out a through polarization characterization of all the detectors in the instrument. A 10' diameter beam is defined by a stop in the focus of a 1.3 m off-axis paraboloid (the spare of the BOOMERanG mirror). Placing a small size source in the center of the focal plane we limit the spurious polarization induced by the off-axis mirror to $< 10^{-4}$. Behind the stop there are two slant wire grid polarizers (P1 and P2), and a 77K black-body source. Rotating P2 at constant speed we modulate the signal, producing a sine wave modulation at twice the rotation frequency. Rotating P1 (in steps) we change the illuminator from co-polar to cross-polar (and all intermediate directions). With this source we plan to measure the co-polar and cross-polar response for each detector in the BOOMERanG focal plane with better than 1% accuracy.

THE NEW ACS SENSORS

In addition to the Attitude Control System (ACS) sensors used in the previous flight (see [30],[2]) we are developing for the new payload two daytime star trackers. This kind of imaging sensors -ccd based in our case- gives full attitude reconstruction, with an accuracy that can be as good as few arc-seconds, if at least two stars are framed at one time. To run the star tracker in daytime, we have to deal with the diffused sky background [31, 32, 33] that, even at balloon altitude, is not negligible. As a general rule, this background is proportional to the atmospheric pressure [34] and, hence, it is few thousandths than at sea level. In order to be able to detect a faint star, we must require that the CCD sensor will not saturate and that it has enough dynamic range to separate the star signals from the bright background. One of our star trackers is based on the Kodak KAF1401 sensor, featuring 70 dB of dynamic range and a well capacity of $45ke^-$. We found that, to be able to have at least one star framed at one time, we need to detect magnitude 4 stars, with a field of view (FOV) of 8 degrees. Since the KAF sensor has a sensitive area of $8.9 \times 7.04 mm^2$ we need to use a $50mm$ lens. A 4th magnitude star focused on 9 pixels at $f = 0.8$ gives a signal that is roughly 30 times fainter than the background. This can be detected by taking advantage of the wide dynamic range of the sensor, that allows us to use a 14 bit ADC (see [34]). This sensor will not be moving with respect to the payload during the scans. The wide FOV assures detection of the same stars during the scans.

FIGURE 6. Left: The integrated B2K payload (telescope earth shield removed). Right: The tracking star camera mounted on the B2K payload.

The other camera (see fig.6) will have a smaller FOV (about 0.5^o) and will track the reference star during the scans. The star position will be recovered combining the position of the star on the CCD array and the position angles of the camera with respect to the gondola, measured by two 16 bit absolute encoders.

The combined use of these two new sensors will allow sub-arcmin attitude reconstruction of the payload.

CONCLUSIONS

The BOOMERanG payload has been refurbished and improved in critical areas like calibration and pointing reconstruction. The addition of PSBs in the focal plane allows the search for the elusive CMB polarization. The system is scheduled to fly from Antarctica at the end of 2002.

ACKNOWLEDGMENTS

This activity is being supported by Universita' di Roma La Sapienza, PNRA and ASI in Italy, NSF and NASA in the USA, PPARC in UK and CIAR and NSERC in Canada.

REFERENCES

1. Masi S., et al., in "3K cosmology", AIP Conf. Proc. 476, 237, (1999); astro-ph/9911520

2. Piacentini F., et al. 2001, astro-ph//0105148 (Ap.J. in press).
3. Crill B.P., et al., in preparation (2002). See also Crill B.P., Ph.D. thesis, Caltech, 2000.
4. de Bernardis, P. et al. 2000, Nature, 404, 955
5. Netterfield C.B., et al. (2001), astro-ph/0104460
6. de Bernardis, P., et al., 2002, Ap.J. in press; astro-ph/0105296
7. Halverson N.W. et al. 2001, astro-ph/0104489
8. Lee A.T. et al. 2001, astro-ph/0104459
9. Rees M., Ap.J., 153, L1, (1968)
10. Hu W., White M., New A. 2, 323, (1997)
11. Kaminokowsky M., Kosowsky A., Ann.Rev.Nucl.Part.Sci. 49, 77-123, (1999)
12. Kaiser N., MNRAS, 202, 1169, (1983)
13. Caderni N., Phys.Rev.D., 17, 1901, (1978); Phys.Rev.D., 17, 1908, (1978)
14. Nanos G.P., Ap.J., 232, 241, (1979)
15. Lubin P. and Smoot G., Ap.J., 245, 1, (1981)
16. Partridge B., et al., Nature, 331, 146, (1988)
17. Wollack et al., Ap.J., 476, 440, (1997)
18. Netterfield B., et al., Ap.J., 474, 47, (1997)
19. Hedman M.M. et al., Ap.J., 548, L111, (2001); astro-ph/0010592
20. Subrahmanyan R., et al., MNRAS, 315, 808 (2001)
21. Keating B., et al., 2001, "A Limit on the Large Angular Scale Polarization of the Cosmic Microwave Background", astro-ph/0107013
22. Pisano G., New Astron. Reviews, 43, 329, (1999)
23. Bock J., et al., in "2K1BC Workshop : Experimental Cosmology at Millimetic Wavelenghts" de Petris and Gervasi editors, AIP.
24. Masi S., et al., Ap.J., 553, L93, (2001A)
25. Tegmark M., et al., Ap.J., 530, 133, (2000)
26. Dragone M., IEEE Trans. AP-30, 331, (1982); IEEE Trans.AP-22, 472, (1974)
27. Carretti E., et al., 2001, "Limits Due to Instrumental Polarisation in CMB Experiments at Microwave Wavelengths", astro-ph/0103318
28. Masi S., et al., 2001, "Scanning polarimeters for measurements of CMB polarization", in "2K1BC Workshop : Experimental Cosmology at Millimetic Wavelenghts" de Petris and Gervasi editors, AIP.
29. Philhour B., et al., 2001, "The Polatron: A Millimeter-Wave Cosmic Microwave Background Polarimeter for the OVRO 5.5 m Telescope", astro-ph/0106543
30. Romeo G., et al., 2001, "Three Sun Sensors for Stratospheric Balloon Payloads", in "2K1BC Workshop : Experimental Cosmology at Millimetic Wavelenghts" de Petris and Gervasi editors, AIP.
31. Yves André "Conception et developpement de senseurs stellaires diurnes embarques a bord de nacelles de ballons stratospheriques" P.h.D thesis
32. E. Rossi et Al., "The in-flight performance of the zebra day-time star sensor" Advance in space research, Vol.13, No.2, 1993, Pag. 159-163
33. E. Pascale, C. MacTavish, "Daytime Sky Brightness at Stratospheric Balloon Altitude", 2002, In preparation.
34. E. Pascale, A. Boscaleri, "A daytime bore-sight star tracker for balloon experiments", 2002, In preparation

ELISA: A small Balloon Experiment for a Large Scale Survey in the Sub-millimeter

J.-Ph. Bernard[1], I. Ristorcelli[2], B. Stepnik[1], A. Abergel[1], F. Boulanger[1], M. Giard[2], G. Lagache[1], J.M. Lamarre[1], C. Meny[2], J.P. Torre[3], M. Armengaud[2], J.P. Crussaire[1], B. Leriche[1], Y. Longval[1]

IAS, UMR-CNRS, Bât.121, Université Paris XI, 91405 Orsay Cedex, France
CESR,FRE-CNRS-UPS, 9 av. du colonel Roche, BP4346, F-31028 Toulouse cedex 04, France
SA, UPR-CNRS, Verrières-le-Buisson, France

Abstract. This paper presents the technical aspects and scientific objectives of the balloon-borne Experiment for Large Infrared Survey Astronomy (ELISA). The emphasis is put upon the synergies existing between the ELISA project and future space missions, both with respect to technical and scientific aspects. ELISA is a small balloon project for an experiment dedicated to measure the Far-Infrared to Sub-millimeter continuum emission of dust over a large fraction of the sky, with unprecedented sensitivity and angular resolution. The primary mirror of the telescope, similar to the one used for the Top-Hat mission, will have a diameter of 1m, ensuring an angular resolution of about 3.5'. PACS-type bolometer arrays will be used in four photometric bands centered at 170, 240, 400, and 650 μm and providing a 22'x45' instantaneous field of view per channel. A liquid He cryostat will host the cold optics, including the secondary mirror of the telescope, as well as the detectors, which will be cooled to 0.3 K using an He3 close-cycle fridge. Mapping of the sky will be accomplished by rotating the gondola over a large azimuth range (up to 60 degree amplitude). The pointing of the experiment will be maintained to a constant elevation during the azimuth scans through a feed back loop using the signal from a large format, fast stellar sensor, operating day and night. The scientific goal of the experiment is to map the diffuse Sub-millimeter emission along a large fraction of the Milky Way. The astronomical data obtained will be used to derive the emission properties of the dust grains in the Interstellar Medium (ISM), such as their temperature and emissivity. It will also allow to systematically measure the polarization of the dust emission. It should also lead to the detection of a few thousand point sources such as newly formed stars and distant galaxies. In addition to these goals, the ELISA project will serve as a test bed for the detector technology that will be used for the HERSCHEL and the PLANCK space missions to be launched in 2007. The ELISA data will also be usable to help calibrate the observations of HERSCHEL and PLANCK and to plan the large-scale surveys to be undertaken with HERSCHEL. Owing to these objectives, 3 flights of the ELISA experiment, including one from Southern hemisphere, are foreseen in the period from 2004 to 2006. The ELISA project is carried out by an international collaboration including France (CESR, IAS, CEA, CNES), Netherlands (SSD/ESTEC), Denmark (DSRI), England (QMW), USA (JPL/Caltech), Italy (ASI).

INTRODUCTION

Interstellar dust plays a key role in the process of star formation and in the energetic equilibrium of the Galaxy. It seems then essential to understand its physical and chemical properties, and their evolution, from the most tenuous to the densest phases of the interstellar medium (ISM). The current dust models consider three main components, differing in composition, size, structure and emission mechanism (e.g. Désert 1990). The smallest particles (PAHs) and the very small grains (sizes < 15 nm) are transiently heated and emit in the infrared range, which has been extensively studied with ISO. The largest grains are in thermal equilibrium, dominating the far-infrared and Sub-millimeter spectrum with a continuum emission following a gray body emission law of the form $I_\nu = \kappa \, (\nu/\nu_0)^\beta \, B_\nu(T)$, where T is the average temperature, β the dust emissivity spectral index, and κ is proportional to the column density. In order to constrain both parameters T and β, Sub-millimeter multi-band observations are needed, including the range 100-300 μm (inaccessible from the ground), since it contains the peak of the emission spectrum for dust at T <

CP609, *Astrophysical Polarized Backgrounds*, edited by S. Cecchini et al.
© 2002 American Institute of Physics 0-7354-0055-5/02/$19.00

30 K. The emission from large grains is so far poorly constrained by observations since the Far-IR and Sub-millimeter domain is still almost quasi unexplored, excepted at the large angular scale of COBE (7°) (e.g. Boulanger 1996, Lagache 1998). The PRONAOS balloon-borne experiment (Lamarre 1994) has mapped the dust emission in four photometric channels from 200 to 650 μm, toward a few selected regions of the ISM. These observations have brought new insights about the nature of large grains in the ISM, and have raised a number of questions about the nature and evolution of dust: PRONAOS has directly revealed the existence of cold condensations (T around 12 K) in different sites of star forming regions (Ristorcelli 1998, Dupac 2001), but also in translucent and optically thin dust clouds at high galactic latitude. This cannot be explained by the standard dust models currently used and can be interpreted as the existence of porous dust aggregates (Bernard 1999, Stepnik 2001). In addition, the observations have also evidenced a significant correlation between the dust equilibrium temperature and the spectral index, which may reflect new quantum processes within the grains, specific to low temperatures. However, these observations are limited to a very small fraction of the sky (a few square degrees), and a full characterization of the dust emission clearly calls for more statistics and a larger survey. Due to a limited spectral range (λ > 350 μm) and the necessary subtraction of atmospheric emission and low spatial frequencies, ground observations cannot constrain both temperature and spectral index of dust, nor measure the low brightness extended emission.

MAIN OBJECTIVES

In that context, the ELISA experiment objective is to survey the Galactic emission arising from different sites of the ISM. For that purpose, two specific observing modes will be used. A large survey will enable to map the galactic plane emission, building a catalog of young stellar objects, and a deep survey at high galactic latitudes will be performed to study the emission from more diffuse clouds. The expected sensitivities for each observing mode are given in Table 1. A large census of the dust emission will be deduced from those surveys. From the temperature and spectral index maps, we will statistically study their correlations. Spectral and spatial variations of the large grain emissivities will be analyzed in relation with the physical conditions (density, turbulence), the other dust components (VSGs and PAHs), and the molecular abundances. A specific study of the Galactic cold component will be performed. The grain properties in cold core are expected to be very different from those in the diffuse medium, and ELISA observations should enable to characterize both spatial and spectral changes of dust properties (due to molecular ice mantles, molecular and/or grain coagulation leading to aggregates...).

TABLE 1. Sensitivity of the ELISA experiment in the two main observing modes: Galactic Survey (700 square degree per hour) and Deep Survey (10 square degree per hour), for both extended emission and point sources. The quoted values are for S/N ration of 3. The Hydrogen column density and visual extinction corresponding to the sensitivity figure is also given

Channel	1	2	3	4
Wavelength (μm)	170	240	400	650
Frequency (GHz)	1760	1250	750	545
Beam Size (arcmin)	3.5	3.5	3.5	3.5
Galactic Survey:				
I (Mjy/sr)	12.6	9.5	2.1	1.7
S (Jy)	10.2	7.8	1.7	1.4
N_H (10^{21} H/cm^2) (Av)	1.5 (0.8)			
Deep Survey:				
I (Mjy/sr)	1.5	1.13	0.25	0.2
S(Jy)	1.23	0.93	0.2	0.16
N_H (10^{21} H/cm^2) (Av)	0.18 (0.1)			

Galactic Plane: Large Scale Survey

The understanding of the very early phases toward star formation requires a better knowledge of the composition and physical and chemical properties of the pre-stellar cold and dense cores. In particular, the opacity, which is dominated by dust grains, must play a predominant role in the phase preceding the collapse, and may influence the fragments sizes. It is therefore expected that global quantities such as the Initial Mass Function and Star Formation Efficiency should be strongly linked to the dust absorption and emission properties. In addition, since dust grains participate directly to most of the physical processes influencing the gas ionization stage, their properties could significantly impact the ambipolar diffusion efficiency, which is thought to be one of the major regulating processes of star formation.

With ELISA, we plan to conduct a large scale survey along the Galactic plane, at |b|<20°, down to 2 MJy/sr (3-σ) at 650 μm. The famous nearby star forming regions (such as ρ-Ophiuchi, Taurus, Orion, Serpens) will be mapped, as well as more distant large molecular complexes (e.g. Cygnus). Such a large scale survey will also cover a large fraction of the Galactic ridge and therefore allow to derive dust properties as a function of the distance to the Galactic center toward the inner regions of the Galaxy, using the Galactic rotation curve and correlation with velocity information in the HI and molecular large scale surveys.

This survey will allow a systematic search and study of very cold condensations. A large number is expected, including at least the thousands of cold cores detected in extinction against the infrared background with the MSX (Egan 1998) and the Isogal (Hennebelle 2000) surveys.

High Latitude Clouds: Deep Survey

The ELISA experiment will aim at limited observations (typically a few ten square degree per cloud, due to the low level emission) of a representative sample of cirrus clouds. This will allow, for the first time, a complete census of dust temperature and emissivity properties of dust toward diffuse regions (typically Av < 0.1) at the angular scale of a few arcminutes. The size of the maps, the number of different cirrus and their careful selection should guarantee minimum observational bias and statistical relevance of the results. For these observations, comparison to HI and eventually molecular line observations measurements will be crucial, in particular to derive the dust emissivity. We plan to include several Intermediate and High Velocity HI Clouds (IVCs and HVCs respectively) in the cirrus sample, in order to search for their dust emission - if any - in the Sub-millimeter. Detection of cold dust and derivation of dust temperature in those clouds may be the only way to derive their distance to the Galactic plane and help elucidate their origin.

The results obtained toward diffuse clouds by ELISA should be very useful to anticipate the methods to be used to subtract Galactic foreground contribution in the PLANCK data. They will be particularly relevant to the problem of the separation between thermal dust emission and grain rotation, which is likely to dominate the spectrum at longer wavelengths.

Polarization Measurements

Our current plan is to equip each photometric channel of ELISA with a set of polarizers, which will allow to measure the degree of polarization of the radiation, and to derive the polarization direction. Up to date, very little measurements of the polarization have been obtained on large scale in the Sub-millimeter at the angular resolution of a few arcminutes. We expect that most of the polarization in the ELISA channels will be coming from polarized dust emission. Although polarization from dust is known to be fairly low (<10 %), the high sensitivity of ELISA should allow detection in clouds with Av ≥ 25 mag and Av ≥ 3 mag in the Galactic and deep survey modes respectively. We expect that these measurements will strongly help constraining the nature of dust and the efficiency of its alignment with respect to the magnetic field. In particular, we expect that fractal grain aggregates should have different polarization and alignment properties than classical dust grains. The correlation between the dust temperature and the observed polarization should therefore shed a new light on the nature of cold dust. In dense star forming regions, such observations should reveal the orientation of the magnetic field, which is of prime importance to understand star forming mechanisms. Finally, the ELISA polarized large scale maps will allow a global study of the polarized galactic foregrounds, which will have to be carefully taken into account when measuring CMB polarization with future space missions such as Planck.

Statistical Study of Young Stellar Objects

The IRAS Point Source Catalog (PSC) remains the reference database for statistical study of star formation in our Galaxy. The IRAS wavelength coverage has allowed to evidence the evolutionary sequence from cold and deeply embedded young stellar objects to progressively hotter and older stars (class I, II, III). However, the coldest and youngest sources (class -I, 0), the protostars in the isothermal contraction phase and in the phase preceding gravitational collapse can only be evidenced in the Sub-millimeter, because their emission peak is expected in the 100-300 μm region. Similarly, the study of the early stages of the formation of massive stars suffers from the lack of good candidates (which are expected to be rare, due to the very short time scales involved) that cannot be readily identified in the IRAS PSC. Sub-millimeter measurements seem to be the best way to distinguish them from the far more common solar mass Young Stellar Objects (YSO). Yet, no extensive point source catalog similar to the IRAS PSC exist in the FIR and Sub-millimeter. Generally speaking, the short characteristic lifetime of YSOs and the necessity of unbiased studies again calls for large surveys. An angular resolution better than that of DIRBE or FIRAS is critical in order to separate efficiently individual sources.

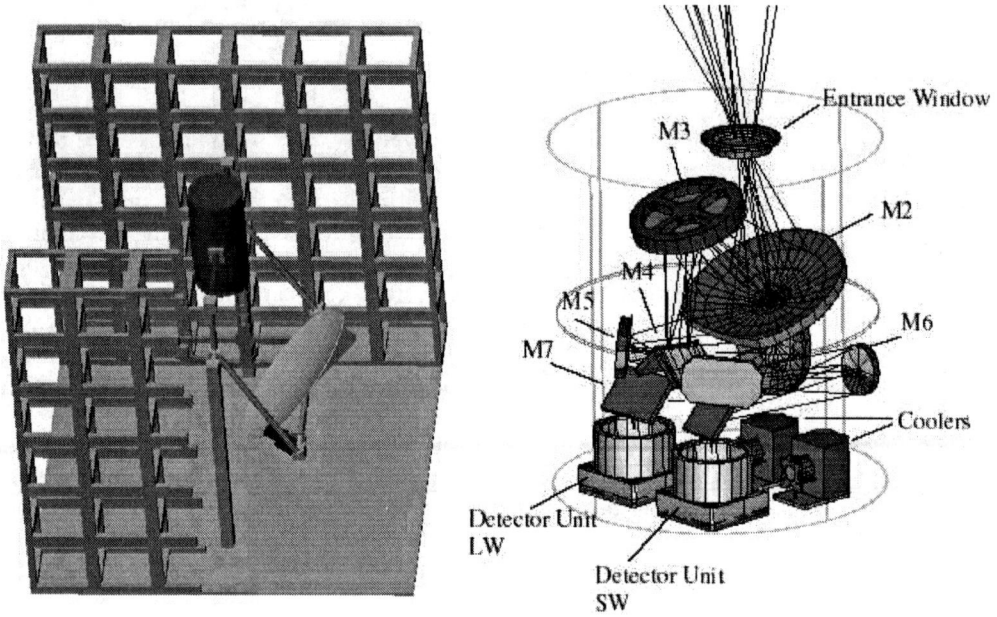

FIGURE 1. Left: Schematic view of the ELISA experiment. The gondola used is a general use gondola provided by CNES. The main dish , stellar sensor and Cryostat can be pointed at various elevations. Baffles, which will surround the pointed subsystems, are not shown here. Right: Implementation of the cold optics in the ELISA cryostat. Orientation is upside-down compared to the left pannel. Each detector unit contains bolometer arrays of 2048 elements.

INSTRUMENT MAIN CHARACTERISTICS

The ELISA experiment is being designed to fly at a ceiling altitude of around 4 mbar (37 km) in the stratosphere, for a large galactic survey to be performed in 3 flights (10 to 20 hours each, including one flight from the Southern hemisphere). An angular resolution of 3.5' has been chosen as a compromise between the need of having a resolution better or similar to IRAS and the aim of surveying a large galactic fraction per flight, with a small project carrying a 1m-diameter telescope and a payload lighter than 500 kg. The telescope is a 1m-diameter off-axis Gregorian, which primary is the carbon-fiber mirror of the TOPHAT experiment, provided by DSRI. The secondary mirror is integrated within the cryostat, and the equivalent focal distance is about 2m (see the preliminary scheme on Figure 1). One of the photometric channels will coincide with the DIRBE band at 240 μm in order to monitor and correct for possible drifts due to remaining atmospheric emission or parasitic signal of instrumental nature. A preliminary study to optimize the best determination of T and β leads to 4 Sub-millimeter large band channels (Δλ/λ about 30 %) centered at 170, 240, 400 and 650 μm.

The institutes taking part to the project main developments are : CESR in Toulouse, IAS in Orsay, CEA in Saclay, CNES in Toulouse, DSRI in Copenhagen, SSD/ESTEC in Noordwijk, QMW in London, LAT in Toulouse, La Sapienza U. in Roma, together with the scientific collaboration of Toronto, Nagoya and Tokyo Universities.

In the nominal configuration, we will use the PACS-type bolometers arrays for the four channels, developed by CEA and LETI. An adequate cold optics scheme will allow to split spatially the beam in onto two sets of 16 * 16 bolometers arrays per channel. Specific care is given to minimizing internal stray-light levels. The large focal plane to be used with the arrays (50 mm for two bands), induces a sophisticated cold optics layout, which is shown in Figure 1. The liquid He cryostat holding the cold optics and the detectors will be cooled down to 0.3 K using an He^3 closed cycle fridge.

Mapping of the sky will be accomplished by rotating the gondola over a large azimuth range ($\pm$ 30°) at constant elevation, in order to reduce the residual atmospheric contribution. A servo-control loop ensures the stabilization in elevation with a 15" accuracy, by means of a reaction wheel, a magnetometer and a fast and large field (15°) stellar sensor, operating day and night. The elevation can range from 15° to 60°. The selected rotation speed of the gondola (1.2°/s) is a compromise between, on the one hand, the need to cover a large amplitude (integrating the sky rotation) and reduce the instrument drifts, and, on the other hand, the need to distinguish point sources detection from parasitic "spikes", and respects both the detectors and stellar sensor response times.

OBSERVING STRATEGIES AND SIMULATIONS

The scientific objectives previously described require two opposite needs in term of observing strategy: the need to reach a sufficient integration time per beam for the (T, β) determination even for low level brightness emission, and the need to survey a large fraction of the galactic plane. Those constrains lead to the trade-off of using two observing modes: a large survey mode corresponding to a sky coverage of ≈ 700 square degree per hour, and a deep survey corresponding to mapping faint regions at a rate of a few 10 square degree per hour. These different coverage rates will be reached by adapting the elevation step between scans and/or repeating the observations in order to increase the signal to noise ratio. A detailed modeling will allow to optimize the observing strategy, and to adapt the parameters such as scan speed, scan length and amplitude of the elevation corrections. A simulation of the ELISA observations in the galactic survey mode, is shown on Figure 2. The Sub-millimeter galactic emission has been extrapolated from IRAS with an average dust spectrum corresponding to T=17.5K, β = 2. A cold condensation with a typical spectrum T=13K, β = 2 has been superposed, scaled to the 100 μm emission. Such cold component, undetected at 100 μm, and too diluted in the DIRBE beam could then be revealed with ELISA during the large galactic survey.

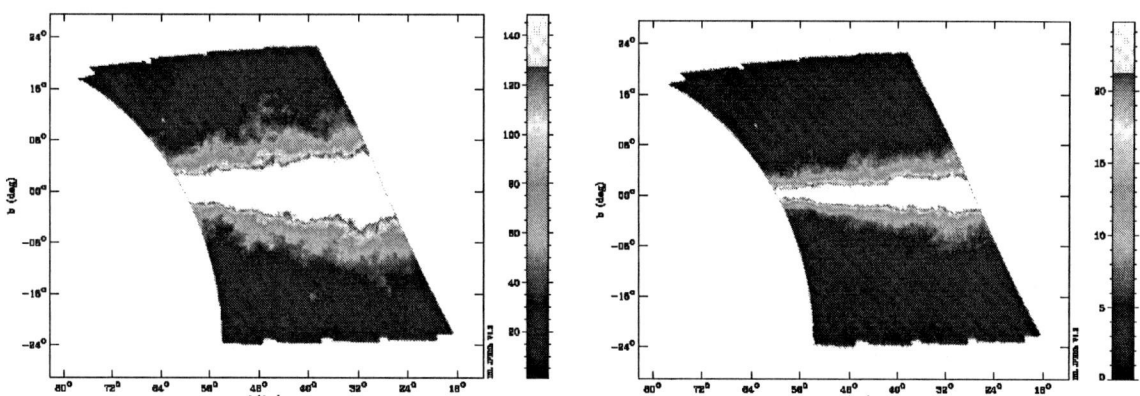

FIGURE 2. ELISA simulated observation at 240 μm (left) and 650 μm (right) corresponding to 2 hours, in the large survey mode (700 square degree per hour). A typical cold condensation has been added at (l,b)=(64°, 11°) to illustrate detection of such structures. The intensity scale is in Mjy/sr.

SYNERGIES WITH PLANCK AND HERSCHEL, SCHEDULE

HERSCHEL instruments will allow to analyze the dust grain emission and the gas-grain interaction with an unprecedented sensitivity, spectral coverage and angular resolution. However, as for ISO, the constraints of an observatory mission will lead to a limited number of observed regions, which will be selected in advance, on the basis of the current data-sets and models. The PRONAOS experiment has confirmed that extrapolation from IRAS emission is very uncertain. Most of the cold condensations discovered with PRONAOS are undetected at 100 µm, and a significant grain emissivity variation has been shown in the Sub-millimeter range, at the angular scale of a few arcminutes. Both are very difficult to study with DIRBE and FIRAS measurements due to an insufficient angular resolution (1° and 7°). The ELISA survey will be very useful to prepare and optimize observations with HERSCHEL, and identify new targets (in particular cold condensations). It will play a role similar to that of the IRAS survey for ISO. The PLANCK "early release" catalog will contain mainly point sources and galaxies clusters. The extended emission maps of PLANCK will probably not be delivered in time for follow-up observations with HERSCHEL. The FIRAS instrument concept has provided the best absolute calibration in the far-infrared and Sub-millimeter range. It will be a reference for the flight calibration of SPIRE, in particular on extended sources. However, the angular resolution is very different between the two instruments (30" and 7°), which makes direct and systematic comparisons very difficult. ELISA will provide data at 3.5' resolution cross calibrated with FIRAS, which it will be possible to use as a secondary calibrator for SPIRE. In addition, the results obtained with ELISA will be very useful to anticipate on the methods to be used to subtract Galactic foreground contribution in the PLANCK data. It will be particularly relevant to the problem of separation between dust emission and grain rotation, which is likely to dominate the spectrum at longer wavelength.

Thanks to the use of the PACS-type bolometer arrays, we can expect that the main objectives of ELISA could be reached with 2 flights (15 to 20 hours each), one in each hemisphere. Due to HERSCHEL calendar constraints, the supplying of these arrays cannot be expected before the early 2005. In that context, it remains realistic and desirable to perform two flights in a nominal configuration with the bolometers arrays, before PLANCK and HERSCHEL observing time starts (June 2007). In order to reduce a risk of additional delay for ELISA scientific return, we propose to prepare and validate the experiment with a first flight in 2004. For this first flight (called 'technological'), we propose a configuration exactly similar to the nominal one (especially the same optics), except for the detectors and its associated electronics. Individual bolometers will be used (3 per band), with a 'replica' of the Planck electronic readout, offering both a cost reduction and a first end-to-end test 'on the sky'. The ELISA project is currently submitted to CNES.

ACKNOWLEDGMENTS

We are deeply indebted to Guy Serra who initiated the concept of the ELISA experiment. Guy died in August 2000. He was scientifically renowned as a pioneer of infrared and Sub-millimeter astronomy. He will remain in our mind both as a great scientist and a generous humanist.

REFERENCES

1. Bernard J.P., Abergel A. et al., A&A 347, 640 (1999).
2. Boulanger F., Abergel A. et al., A&A 312, 256 (1996).
3. Désert F.X., Boulanger F., Puget J.L., A&A 237, 215 (1990).
4. Dupac X., Giard M., Bernard J.P. et al., ApJ, in press (2001).
5. Egan M.P., Shipman R.F. et al., ApJ 494, L199 (1998).
6. Hennebelle P., Teyssier D. et al., ESA-SP 455, p.125 (2000).
7. Lagache G., Abergel A. et al., A&A 333, 709 (1998).
8. Lamarre J.M. et al., IR Phys. Technol., Vol.35, 277-289 (1994).
9. Ristorcelli I., Serra G. et al., ApJ 496, 267 (1998).
10. Stepnik B., Abergel A. et al., A&A, in press (2000).

Measuring CMB polarisation with the Planck HFI

J. Delabrouille, J. Kaplan
on behalf of the Planck HFI consortium

PCC, Collège de France, 11 place Marcelin Berthelot, 75231 Paris cedex 05

Abstract.
The Planck High Frequency Instrument (HFI) is the most sensitive instrument currently being built for the measurement of Cosmic Microwave Background anisotropies. In addition to unprecedented sensitivity to CMB temperature fluctuations, the HFI has polarisation-sensitive detectors in 3 frequency channels (143, 217 and 353 GHz), which will constrain full-sky polarised emission of the CMB and foregrounds at these frequencies. The sensitivity of the instrument will allow a clear detection of CMB polarisation signals and should yield a precise measurement of its power spectrum at all angular scales between $\ell = 50$ and $\ell = 1000$, as well as constraints on the polarised emission at larger scales where a polarised signal from inflationary gravity waves or from reionisation is expected in many cosmological scenarios.

INTRODUCTION

The COBE-DMR [1], Boomerang [2, 3], DASI [4, 5] and Maxima [6, 7, 8] experiments together now have yielded strong constraints on the Cosmic Microwave Background (CMB) anisotropy power spectrum [9], and in particular a convincing detection of the first three acoustic peaks, providing compelling evidence that indeed the primordial fluctuations were produced during an inflationary phase. The next great challenge in the the study of the statistical properties of the CMB is to measure the polarisation signal, both that due to acoustic oscillations in causally connected regions of the Universe before decoupling and, even more challenging, the primordial polarisation spectrum due to gravity waves generated during inflation. Measurement of the correlated spectrum of polarisation and temperature at sub-degree scales will provide yet another test of the basic acoustic oscillation scenario and of the global paradigm, while measurement (or absence of detection) of polarisation signals due to tensor modes could strongly constrain inflationary models, as well as yield direct observational evidence for the existence of primordial gravity waves [10, 11, 12, 13].

This has been widely recognised by the CMB community, and a large number of experiments dedicated to detecting the CMB polarisation are currently in operation or being planned. In this paper, we review the overall design of the Planck High Frequency Instrument (HFI) as a polarisation sensitive instrument, and discuss its capabilities in terms of measuring CMB polarisation.

THE PLANCK HIGH FREQUENCY INSTRUMENT

The Planck mission, to be launched by ESA in spring 2007, is the third generation satellite dedicated to observing CMB anisotropies. The DMR instrument on COBE yielded the first detection of CMB anisotropies on large angular scales. The MAP mission, launched by NASA in June 2001, will provide a measurement of anisotropies at 15 arcminute resolution with a sensitivity of a few tens of μK per resolution element. The Planck mission will measure CMB anisotropies with a sensitivity yet an order of magnitude better, a few μK over 7 arcminute resolution pixels.

The HFI is one of the two instruments on board Planck. It is a 48–detector instrument, using bolometers cooled to 100 mK by an open–cycle spatial dilution fridge. Its 48 detectors are distributed into 6 frequency channels ranging from 100 to 850 GHz. Originally designed as a temperature-sensitive instrument only, it has been modified for polarisation sensitivity in three of its frequency channels. In the current (final) design, half of the detectors of the Planck HFI are polarisation sensitive (see table 1).

The other instrument on the Planck spacecraft, the Low Frequency Instrument (LFI), polarisation-sensitive as well, is described by Villa et al. in these proceedings [14].

TABLE 1. Summary of Planck HFI main characteristics. Sensitivities for intensity and polarisation are given per square pixel with a beam size side, for a 1 year mission.

Central Frequency (GHz)	100	143	217	353	545	857
Beam size (arcmin)	9.2	7.1	5.0	5.0	5.0	5.0
N det. unpolarised	4	4	4	4	4	4
N det. polarised	-	8	8	8	-	-
I sensitivity (μK/K)	2.2	2.4	3.8	15	80	8000
U and Q sensitivity (μK/K)	-	4.8	7.6	30	-	-
Flux sensitivity (mJy)	9.0	12.6	9.4	20	46	52

Polarisation-sensitive bolometers

The Planck HFI Polarisation Sensitive Bolometers (PSB) are built so that two sensors using absorbers made with parallel wires [15], coupled to orthogonal polarisation modes, are located in the same integration cavity. The two sensitive devices coupled in this way share the same optics (horns, filters, telescope), but are different detectors (with each its own coupling efficiency, time constant, sensitivity). Each of them is read out by its own read-out electronics chain, as illustrated in Figure 1.

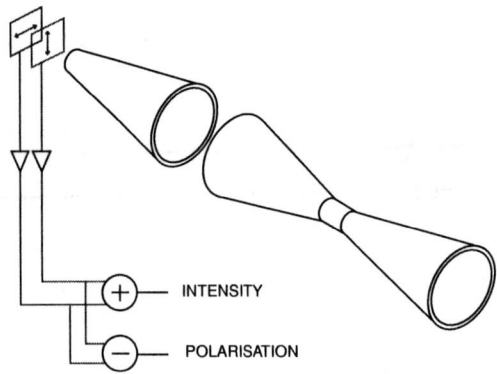

FIGURE 1. Schematic view of the Planck HFI polarisation dispositive using Polarisation Sensitive Bolometers.

With this design, the optical coupling to the sky should be identical for the two elements of the PSB (same beam shape, same response as a function of the wavelength), except for the sensitivity to orthogonal polarisation directions. Numerical simulations of the instrumental beam show that differences in beam shapes between the two polarisation directions of a single PSB are at negligible level (typically one per cent effect or less) [16, 17]. Noise levels, time constants of the detection chain (or more generally its impulse response), however, may differ slightly between the two elements, although at the manufacturing level it is hoped to match as well as possible the time constants and noise levels at the optimum (minimum) value.

The amount of depolarisation achieved with this design has been measured in-lab to be less than a few percent [18], with a target of about 3 %, which should negligibly impact the sensitivity to polarised signals.

Each PSB, by differencing its two data streams, directly provides a measurement of the Q Stokes parameter (*i.e.* a pure polarisation signal) in its own reference system (on an axis system where the x and y coordinates are measured along the two orthogonal polarisation-sensitive directions).

PSB layout

As emphasized in [19], at least three (linear) polarisation sensitive measurements with different orientations are needed to fully measure I, Q, and U in a given direction. For perfectly matched, uncorrelated noise in the data, the minimal error box volume, as well as uncorrelated noise between the measurements of these three Stokes parameters, are realised when the polarimeter orientations are evenly distributed between 0 and 2π. Couchot et al. [19] denote such configurations (using $n \geq 3$ evenly distributed polarimeters) *optimised configurations*.

Then, a single PSB yielding two orthogonal polarisation measurements is not by itself sufficient to measure the three Stokes parameters I, Q, and U. The layout of the detectors for the Planck HFI is such that every PSB has a companion oriented at 45 degrees, with relative locations in the focal plane such that during scanning, it measures polarisation on the same pointing trajectory a fraction of a second before or after the first one (see Figure 2). Such a pair of companion PSBs yields four timelines which correspond effectively, after rephasing the measurements to get co-extensive pointings, to a polarisation measurement with a 4-polarimeter *optimised configuration*.

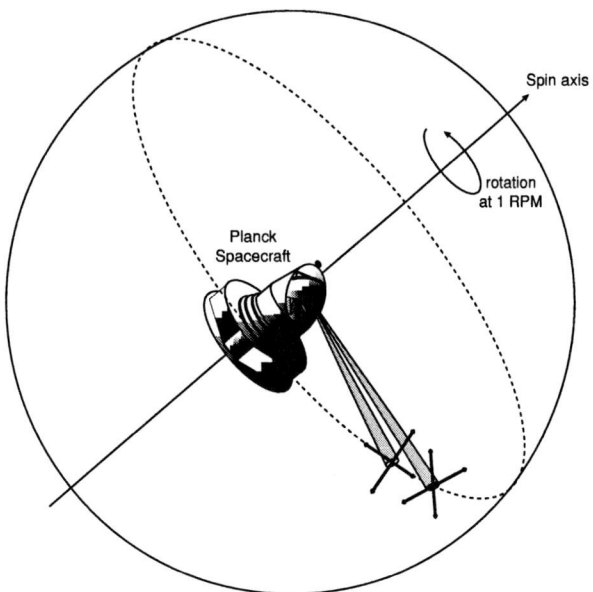

FIGURE 2. Polarisation measurement in the context of the Planck scan-strategy. For each PSB-pair detector set, the 4 measurements correspond to polarimeter orientations at 0, 45, 90 and 135 degrees with respect to the direction defined by the scanning.

Such a configuration has several advantages. The matching of the orientations with the scanning direction ensures that one of the PSBs permits (by simple difference) to measure Q and the second U in the reference system where one axis is parallel to the scanning and the other one perpendicular. The fact that two polarimeters share the same horn permits the best rejection of leakage of intensity I signals into polarisation data, as discussed later. In addition, there is in principle built-in redundancy, as if even one detector fails, the three remaining timelines are sufficient to obtain the three Stokes parameters of interest (even if they do not then constitute an *optimised configuration*). This built-in redundancy permits an internal consistency check of the data (and thus of systematic effects) if all detectors work properly (in terms of sensitivity).

For each polarisation-sensitive frequency channel (at 143, 217 and 353 GHz), there are two such companion PSB pairs, which provides yet another redundancy level, as well as better overall sensitivity.

The HFI detector layout in the focal plane is shown in Figure 3.

POLARISATION DATA PROCESSING

Whereas a lot of experience has been already gathered by the scientific community in the analysis of CMB anisotropy data, no such know-how has yet been acquired for the specific measurement of polarisation. Special effort has been undertaken in the HFI Data Processing Center (DPC) to develop data reduction tools specifically tailored and optimised for Planck HFI polarisation data. In this section, we discuss (non exhaustively) some of the issues adressed by the

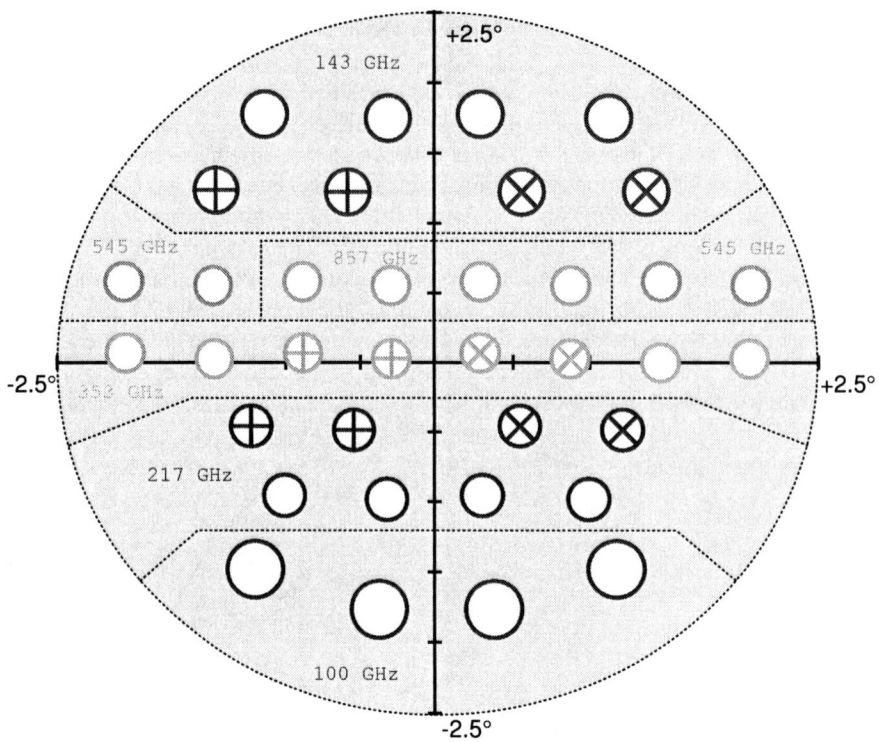

FIGURE 3. Planck HFI detector layout. In this figure, the horizontal and vertical axis are parallel and perpendicular to the line of scanning respectively. Polarisation bolometer fields of view are represented with a cross inside, showing the orientations of polarisation sensitivities.

Planck HFI DPC so far, although part of the discussion applies (and we believe can be useful) to other instruments as well.

Modelling the measurement

There are several particularities to the measurement of polarisation, as compared to temperature, that will require the implementation of specific data processing software for the processing of such data. The specifics of polarisation signals come from two origins:

1. particularities of the signals we want to measure and which impact the data processing: very weak signals, poorly known galactic foregrounds and systematics;
2. specifics of the instrumental set-up for polarisation measurements.

Neglecting for the moment instrumental imperfections, each polarimeter of the Planck HFI, sensitive only to one linear polarisation, measures in each sample a linear combination of I, Q and U (integrated over the detector beam):

$$d(\alpha) = \frac{1}{2}[I + Q\cos 2\alpha + U\sin 2\alpha] \qquad (1)$$

where α is the angle between the polarimeter orientation and the x-axis used for defining the Stokes parameters.

Combining data samples

At least three independent samples, with different angles α but coextensive beam pointings, are required to measure I, Q and U. For an ideal measurement with two PSBs, assuming perfectly coincident beam pointings, the data from

138

the four timelines are given by:

$$
\begin{bmatrix} d_1 \\ d_2 \\ d_3 \\ d_4 \end{bmatrix} = \begin{bmatrix} d(0) \\ d(\pi/2) \\ d(\pi/4) \\ d(3\pi/4) \end{bmatrix} = \frac{1}{2} \begin{bmatrix} 1 & 1 & 0 \\ 1 & -1 & 0 \\ 1 & 0 & 1 \\ 1 & 0 & -1 \end{bmatrix} \begin{bmatrix} I \\ Q \\ U \end{bmatrix} + \begin{bmatrix} n_1 \\ n_2 \\ n_3 \\ n_4 \end{bmatrix} \tag{2}
$$

where $n_1, \ldots n_4$ are noise terms, and I, Q, U denote *the same* beam-integrated Stokes parameters on the sky in the pointing direction. This equation can be recast in the matrix form:

$$
d = AS + n \tag{3}
$$

where d is the data, A a 4×3 matrix, S the three-element Stokes-parameter vector, and n the noise.

With this ideal setup, for well balanced, uncorrelated noise, the best Stokes parameters estimates in the PSB-pair frame are obtained as

$$
\begin{aligned}
I &= \tfrac{1}{4} \times (d_1 + d_2 + d_3 + d_4) \\
Q &= \tfrac{1}{2} \times (d_1 - d_2) \\
U &= \tfrac{1}{2} \times (d_3 - d_4)
\end{aligned} \tag{4}
$$

For unbalanced and/or correlated noise, the optimal least square solution is

$$
\widetilde{S} = [A^t N^{-1} A]^{-1} A^t N^{-1} d \tag{5}
$$

where N is the noise covariance matrix, $N_{ij} = \langle n_i n_j \rangle$.

Monitoring systematic effects

So far, we have not said anything about unavoidable noise correlations along the time streams, we have just considered the measurement at a single given pointing direction. In addition, all four detectors of a PSB pair have been assumed to have coextensive beams, which is actually not the case. The actual data processing for polarised map-making has to take into account these imperfections of the measurement. We concentrate on two major issues, which are low frequency drifts and pointing mismatch, and show how these imperfections impact polarised data processing for the Planck HFI.

Low frequency drifts

Low frequency drifts, due both to $1/f$ noise in the detection chain and to thermal fluctuations of payload elements detected by the sensors, are expected to be present in the detector timelines. This will be the case for PSB timelines as much as for unpolarised bolometers timelines. For each polarimeter i, writing explicitly the summmation over pixels to avoid confusion on other repeated indices, the data stream can be modelled as:

$$
d_{it} = \sum_p M_{itp}[I_p + Q_p \cos 2\alpha_{itp} + U_{itp} \sin 2\alpha_{itp}] + n_{it}. \tag{6}
$$

where d_{it} is the data of polarimeter i at time t, (I_p, Q_p, U_p) are the Stokes parameters in pixel p, M_{itp} is the pointing matrix for polarimeter i, telling how much pixel p contributes to the signal of polarimeter i at time t. α_{itp} is the angle, in pixel p, between the co-polar direction of polarimeter i at time t and the reference direction (e.g. parallel and perpendicular to longitude and latitude lines), and n_{it} is the noise timestream for detector i. Usually for temperature measurements, the pointing matrix is modelled as a sparse matrix containing only one non-vanishing element per line, which just tells which pixel of the sky is pointed at at time t.

Reordering the four timelines of a PSB pair d_{it} into a single data vector d, the four noise streams into one single noise vector n, and the three maps of Stokes parameters I_p, Q_p and U_p into a single vector S, we can recast the four equations in eq. 6 (one for each detector) into one single linear equation:

$$
d = MS + n. \tag{7}
$$

Denoting as N the noise autocorrelation (which now encompasses both correlations *along* timelines and correlations *between* timelines), the best estimates for the Stokes parameters, $\widetilde{S}$, can in principle be optained by a Global Least Square (GLS) inversion of the linear equation, $\widetilde{S} = [M^t N^{-1} M]^{-1} M^t N^{-1} d$. This however is a formidable task for Planck HFI, even for temperature maps, because of the size of the system. It is even more so with polarisation data.

A first-order map-making algorithm which approximates the GLS solution has been developed in [20] to construct polarisation maps from all timelines of an *optimised configuration* in the context of the Planck HFI. The method simplifies the resolution of the linear system under the assumption that noise correlations on timescales less than 60 seconds are negligible. Although a useful step towards a global solution, this simplified solution (which does not take into account imperfections of the model as it assumes a perfect knowledge of the pointing) can not be used for Planck in its present implementation. For polarisation mapping indeed, special care has to be taken in the implementation of the inversion of the linear system, as discussed in the next paragraphs.

Pointing mismatch

In the formal GLS solution, it is implicitly assumed that the matrix M, which encompasses both the pointing of the detectors and the mixing of the Stokes parameters, is known and perfectly describes the actual detector pointing. The fact that this is not exactly true complicates the processing of polarisation data in a somewhat tricky way.

Let us assume that a number m of polarised data samples, all corresponding to a pointing towards the same given pixel p, are to be combined to recover a best estimate of I, Q and U at that point. For a pixel size θ_p, the measurements are actually not pointing all to the exact same place, as illustrated in Figure 4. Any gradient of the temperature through the pixel will generate a difference in the readout of two polarimeters which do not integrate signal from perfectly co-extensive beams. For mismatched pointings, if the two polarimeters do not have the same orientation, a temperature gradient term will leak into the estimated polarisation (obtained through difference terms in the inversion of the linear system). This pointing error effect, even if essentially small or even negligible for temperature measurements, can nonetheless be important for polarisation because of the relative levels of I, Q and U on the sky.

For each pixel, the fake polarisation generated in this way will depend on the exact distribution of all pointings inside the pixel and on the distribution of corresponding polarimeter orientations. Relative pointings, therefore, are to be monitored very carefully for a polarisation-measuring experiment using single polarisers.

The naive model of the very sparse mixing matrix with one single non-vanishing element per line can be refined. The exact pointing, however, has to be perfectly well known, which is not the case in any of the present instruments (and not the case for Planck). The order of magnitude of the pointing reconstruction accuracy required for this effect to be smaller than the expected CMB polarisation for Planck is about 30 arcseconds. For the effect to be completely negligible, a relative pointing reconstruction of typically 5 arcseconds or better is needed. This requirement becomes less stringent in the limit of many measurements per pixel, as the effect of random uncorrelated errors in pointing reconstruction will tend to cancel out.

Planck HFI polarised map-making

The use of Polarisation Sensitive Bolometers (or of any setup in which two orthogonal polarimeters share the same optics) provides an elegant solution to the pointing reconstruction accuracy problem, as beams are in principle perfectly co-extensive. Then, instead of implementing a GLS map-making on single polarimeter timelines, the map-making can be implemented on differences between two orthogonal polarimeter readouts. If the transfer functions of the two polarimeters in each PSB (time constants and readout impulse response) are well matched (which can be done by numerical post-filtering the two timelines if not built in hardware) the temperature gradient leakage problem is solved. This method is currently the baseline polarised map-making solution for nominal HFI performance (co-extensive PSB beams, nominal balanced noise levels, pointing reconstruction accuracies of a few tens of arcseconds or worse).

Additionnal discussion on polarisation systematics can be found in [21].

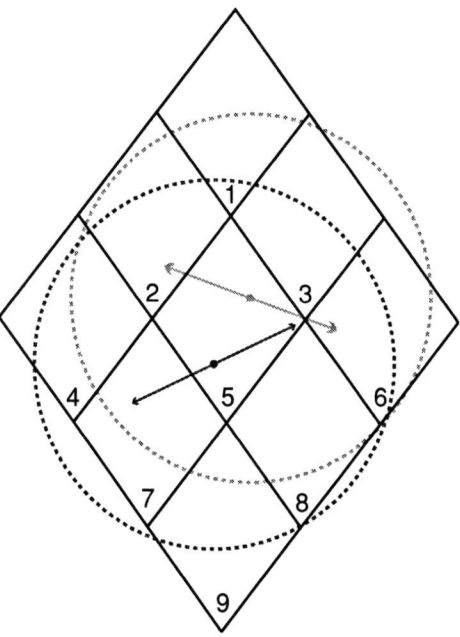

FIGURE 4. Illustration of the effect of single pixel pointing approximation or of pointing mismatch due to improper knowledge of the exact pointing. Here, in pixel 5, two polarisation measurements with different orientations are not pointed at the exact same place. Even if the beams are much bigger than the pixel, as should be for proper sampling, the difference between the two measurements includes a term proportional to the gradient of the beam-smoothed temperature map, which may be larger than the actual average polarisation in that pixel. For a careless implementation of a GLS map-making solution directly from polarimeter data streams, this may become a "killer–effect" for polarisation measurements.

Polarised component separation

Polarisation maps obtained with Planck in general, and the Planck HFI in particular, will contain at each frequency a mixture of the polarised emission of several astrophysical components. Although the component separation can be made based on the same general principles as for temperature maps, the overall performance of the separation is still unclear. First order methods assuming prior knowledge of the emission spectra and the spatial spectra have been developed [24] to generalise the Wiener separation method first implemented by Bouchet & Gispert [22] and Tegmark & Efstathiou [23]. However, as little prior information on the polarised emission of foregrounds is actually available, blind separation methods [25, 26, 27] must be investigated in this particular context.

Note still that for the Planck HFI, contrarily to the temperature case where several astrophysical foregrounds are expected to have temperature contribution larger than, or of the order of, the noise level, foreground polarised emission is expected to be well below both the polarisation sensitivity and the CMB polarisation, making component separation less critical for CMB polarisation measurements than for temperature anisotropies mapping (fig. 5).

THE SENSITIVITY OF THE PLANCK HFI

A detailed modelling of the performance of the HFI bolometers onboard Planck in the background conditions at the L2 Sun-Earth Lagrange point permits to predict the sensitivity of the HFI in each channel. For this estimation, it is assumed that the observation time is evenly shared between all the sky pixels. Corresponding estimated sensitivites on I, Q and U per resolution side square pixel, for all HFI frequency channels, are given in table 1.

The sensitivity of the Planck HFI to intensity and polarisation is shown in Figure 5. Galactic foreground emission level estimates are shown for high galactic latitudes ($b \simeq 70°$). Sensitivity levels per resolution element (1σ) are shown as horizontal lines for six unpolarised frequency bands and three polarised frequency bands. The polarisation sensitivity is not sufficient to detect CMB polarisation at the level of individual pixels, nor to map foreground polarised emission at high galactic latitudes. Still, the power spectrum sensitivity on the 143 GHz channel alone (assuming the other channels are used as foreground and systematics monitors) is good enough to measure the $E - T$ and $E - E$ spectra quite well (Figure 6).

Most interestingly, the sensitivity of the Planck HFI is at a level which should allow to put strong contrains on B modes, which are expected in a large range of cosmological scenarios to be much larger at low ℓ values than the E modes of Figure 6, and thus within the detection reach of Planck (especially after combining the data from all HFI and LFI channels).

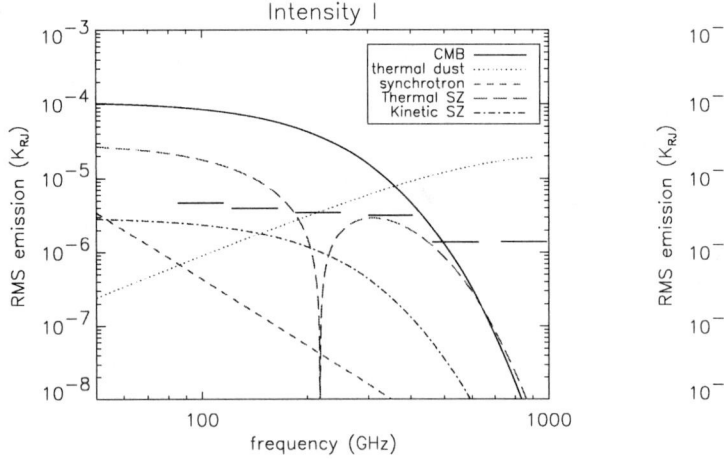 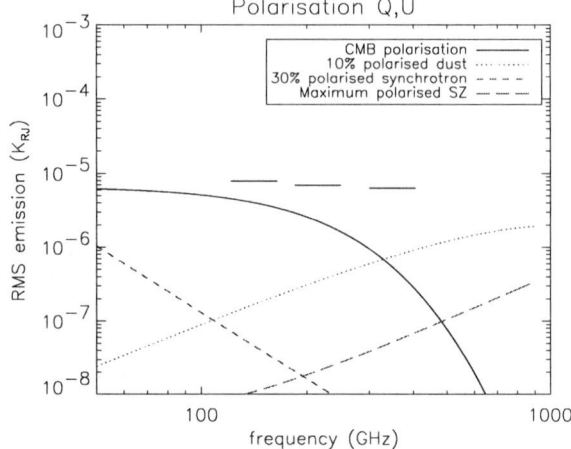

FIGURE 5. RMS intensity (left) and polarised emission (right) of various astrophysical components at high galactic latitude ($b \simeq 70°$). Planck HFI Intensity and Polarisation sensitivities per resolution element in all frequency bands appear as horizontal bars on the plots.

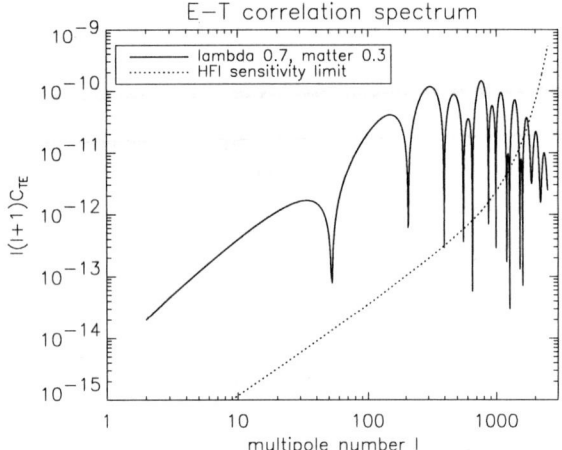 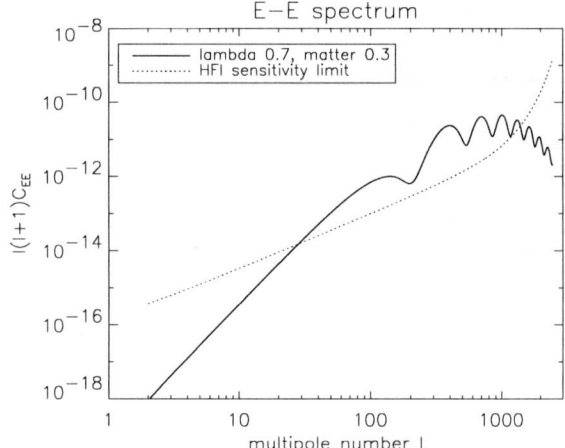

FIGURE 6. HFI sensitivity to $E - T$ cross correlation (left) and to $E - E$ spectrum (right) for the 143 GHz HFI channel only. Power spectra are plotted for scalar modes only in a CDM-like cosmological model with $\Omega_b = 0.045$, $H_0 = 65 \, \text{km/s/Mpc}$, $\Omega_\Lambda = 0.7$ and $\Omega_m = 0.3$, and have been obtained using the CMBFAST software [28].

CONCLUSION

Originally planned and proposed solely as a CMB temperature anisotropy sensitive instrument, the Planck HFI has been revised to become as well a CMB polarisation sensitive instrument. Unprecedented sensitivity at high resolution is achieved through the combination of the use of new polarisation sensitive bolometers, cooled to 100 mK on a spaceborne mission with low background, and of the selection of observing frequency bands where diffraction limits the resolution at the 5 to 7 arcminute level. In addition, the selected frequency range is at the expected minimum of the polarised emission from galactic foregrounds and extragalactic compact sources.

A substantial effort is being made to understand the impact of all possible systematic instrumental effects throughout the detection process and the data reduction pipeline. The minimisation of such systematic errors through a rigorous choice of the instrumental setup, careful on-ground testing, and continuing dedicated effort to the development and optimisation of data reduction methods for polarisation measurements with the Planck HFI, give us confidence that the instrument can meet its ambitious objectives.

ACKNOWLEDGEMENTS

We thank all the people who have contributed in a way or an other to the preparation of the measurement of polarisation with the Planck HFI. Special thanks to Yannick Giraud-Héraud, Jean-Michel Lamarre, Michel Piat, Cécile Renault and Cyrille Rosset for useful discussions and dedicated work. Thanks also to Jim Bartlett for useful suggestions towards improving the manuscript, and to Radek Stompor for interesting discussions about polarised map-making.

REFERENCES

1. Smoot, G. F. et al., 1991, ApJ letter, 371, L1
2. de Bernardis, P. et al., 2000, Nature, 404, 955
3. Lange, A. E. et al., 2001, Physical Review D, vol. 63, Issue 4, id. 042001
4. Halverson, N. W. et al., 2001, submitted to ApJ (astro-ph/0104489)
5. Pryke, C. et al., 2001, submitted to ApJ (astro-ph/0104490)
6. Hanany, S. et al., 2000, ApJ letter, 545, L5
7. Lee, A. et al., 2001, ApJ letter, 561, L1
8. Stompor, R. et al., 2001, ApJ letter, 561, L7
9. Jaffe, A. H. et al., 2001, Physical Review Letters, vol. 86, Issue 16, 3475
10. Zaldarriaga. M. & Seljak, U., 1997, Physical Review D, vol. 55, Issue 4, 1830
11. Kamionkowski, M., Kosowsky, A. & Stebbins, A., 1997, Physical Review Letters, vol. 55, Issue 12, 7368
12. Seljak, U. & Zaldarriaga. M., 1997, Physical Review Letters, vol. 78, Issue 11, 2054
13. Kamionkowski, M., Kosowsky, A. & Stebbins, A., 1997, Physical Review Letters, vol. 78, Issue 11, 2058
14. Villa, F., on behalf of the Planck LFI, 2001, these proceedings.
15. developped at Caltech/JPL by J. Bock and collaborators.
16. Yurchenko, V., 2001, private communication
17. Fosabla, P., 2001, private communication
18. Maffei., 2001, private communication
19. Couchot, F., Delabrouille, J., Kaplan, J., and Revenu, B., 1999, A&A Supplement Series, 135, 579
20. Revenu, B., Kim, A., Ansari, R., Couchot, F., Delabrouille, J., and J., K., 2000, A&A Supplement Series, 142, 499
21. Kaplan, J. & Delabrouille, J., 2001, these proceedings.
22. Bouchet, F.R., Gispert, R., 1999, NewA, 4, 443
23. Tegmark, M. and Estathiou, G., 1996, MNRAS, 281, 1297
24. Bouchet, F.R., Prunet, S. and Sethi, Shiv K., 1999, MNRAS, 302, 663
25. Baccigalupi, C. et al., 2000, MNRAS, 318, 769
26. Maino, D. et al., 2001, submitted to MNRAS (astro-ph/0108362)
27. Snoussi, H. et al., 2001, to appear in MaxEnt 2001 proceedings (astro-ph/0109123)
28. Seljak, U. & Zaldarriaga, M., 1996, ApJ 469, 437

The Low Frequency Instrument of the Planck Mission

F. Villa[1], N. Mandolesi[1], M. Bersanelli[2], R.C. Butler[1], C. Burigana[1], A. Mennella[3], G. Morgante[1], M. Sandri[1], L. Valenziano[1]

On behalf of the LFI Consortium

[1]TESRE-CNR – Bologna, Italy
[2]Università di Milano, Dip. Di Fisica, Milano, Italy
[3]IFC – CNR, Milano, Italy

Abstract. The Low Frequency Instrument (LFI) is one of the two instruments onboard the ESA Planck satellite. LFI will image the Cosmic Microwave Background anisotropies and the polarization status in four different bands, from 30 GHz to 100 GHz, with an array of 54 radiometers. The characteristics of the instrument are reported, underlining the ability of LFI to perform polarization measurements.

INTRODUCTION

The prediction of CMB polarization anisotropy comes from the well known physics of Thomson scattering and the assumption that the temperature anisotropies are the result of fluctuations at last scattering (see, e.g. [1]). The detection of the (partial) polarization of the CMB at different scales would represent a fundamental test of basic cosmological assumptions. Moreover, polarization measurements will provide independent estimates of the cosmological parameters and will help breaking degeneracies in their determination [2].

The ratio between the polarization and temperature anisotropies is small, since only those photons that last scattered in an optically thin region could have possessed a quadrupole anisotropy (multiple scattering causes photon trajectories to mix and hence erases the anisotropies). This fraction depends on the duration of last scattering: for the standard thermal history, it is 5-10% and shows its main peaks on characteristic scales of tens of arcminutes, slightly smaller than the scales where the temperature power spectrum exhibits its main doppler peaks.

The significant experimental challenge lies in the very small level of the cosmological polarization signal, requiring both great sensitivity and deep control of potential systematic effects of instrumental or astrophysical origin. The Low Frequency Instrument (LFI) is one of the two instruments onboard the ESA's Planck satellite, and will map the intensity and the polarization of the sky anisotropy at four frequency channels between 30 and 100 GHz by a set of 54 HEMT–based radiometric detectors intrinsically sensitive to the linear polarization of the received radiation [3]. Because of its extremely advanced technology, LFI has a sensitivity adequate to produce statistical information on the CMB polarization.

We describe in this paper the Planck mission and the Low Frequency Instrument, with particular emphasis to the polarization measurement capabilities.

THE PLANCK MISSION

Planck is the third medium–sized mission of the Horizon 2000 program of the European Space Agency (ESA). It will measure and accurately characterize small fluctuations in the Cosmic Microwave Background (CMB) temperature, which are only about 10^{-5} of the CMB temperature, as well as the polarization properties of the CMB, expected at the 10^{-6} level.

CP609, *Astrophysical Polarized Backgrounds,* edited by S. Cecchini et al.

Planck is a scanning mission that will be launched in 2007 together with the Herschel Space Observatory (HSO) by an Ariane 5 ESV launcher from Kourou (French Guiana, South America). The spacecraft will travel for approximately three months before reaching a small Lissajou orbit around the second Lagrangian point (L2) of the Earth–Sun system, at about 1.5 millions of kilometers from Earth. The line of sight of the Planck Telescope will observe great circles separated by 2.5 arcmin one from each other, performing a full sky coverage in 6 months and thus allowing two full sky surveys during 1 year of nominal mission lifetime.

For several reasons Planck is considered the third generation of space missions dedicated to imaging the sky fluctuations at mm–wavelength after the NASA's COBE and MAP missions: (i) the active cooling of the two instruments which drastically reduces the noise of the detectors; (ii) the unprecedented wide frequency coverage and angular resolution; (iii) the payload and the instruments configurations which are designed to minimize all the potential sources of systematic effects; (iv) the redundancy of the most important channel for Cosmology, the 100 GHz, which is covered both by radiometric and bolometric detectors.

The payload is constituted by an off axis dual reflector telescope [4] and by two instruments which share the focal region of the telescope: the radiometric Low Frequency Instrument (LFI) and the bolometric High Frequency Instrument (HFI) [5].

The telescope is a 1.5 meter optimized aplanatic telescope which permits to accommodate the large Focal Plane Unit box inside which the two instruments are assembled and integrated and will be passively cooled at about 50K by a dedicated set of thermal shields. The HFI will cover the 100 – 857 GHz frequency range in six observing bands and it is based on advanced bolometers cooled to 100 mK. Half of the channels at 143 GHz, 217GHz and 353 GHz will be equipped with Polarization Sensitive Bolometers (PSB) [6] in order to detect the polarization of the incoming radiation. The LFI is working in four bands between 30 GHz and 100 GHz. The HEMT–based radiometric detectors of LFI are intrinsically sensitive to the polarization allowing it to be able to measure the polarization at all frequencies and for all detectors.

THE LOW FREQUENCY INSTRUMENT

Overview and scientific capabilities

The Low Frequency Instrument is a multi-frequency radiometer array, located in the focal surface of the Planck telescope. It is designed to produce images of the sky in the 30 – 100 GHz spectral range, with an unprecedented combination of sky coverage, calibration accuracy, control of the systematic errors, stability and sensitivity, including the polarized component (see Table 1 for an overview of the LFI characteristics and performances).

Its primary scientific goal is to measure the Cosmic Microwave Background (CMB) anisotropies at high angular resolution, with an accuracy set by astrophysical limits. Moreover, it will measure the polarization characteristics of the CMB with an accuracy of a few μK in regions near the ecliptic caps. The LFI will produce full sky maps at 30, 44, 70 and 100 GHz, with angular resolution of 30', 24', 14' and 10', respectively, and with an average sensitivity per resolution element $\Delta T/T \sim (2 \div 7) \, 10^{-6}$. These unprecedented angular resolution and sensitivity will uncover the wealth of cosmological information encoded in the anisotropy pattern at degree and sub-degree angular scales.

The LFI measurements will determine the primary cosmological parameters (Hubble constant, deceleration parameter, curvature of space, baryon density, dark matter densities including neutrinos, amplitude and spectral index of the primordial scalar density perturbations, and the gravity wave content of the Universe) to an accuracy of a few percent. The LFI data can test models for the origin of primordial perturbations, i.e. whether they are due to topological defects or to quantum fluctuations, and to constrain the global properties of the Universe (topology, rotation, shear, etc.). Polarization measurements will independently confirm these findings and help in breaking the degeneracies in the determination of cosmological parameters.

The secondary scientific objective of LFI is to build a map over the whole sky of all major Galactic and extragalactic sources of emission at the four frequency channels, including both extended and compact sources, as well as the SZ effect. LFI is expected to provide rich samples of sources for astrophysical studies. At the same time, the LFI data are also essential to accurately subtract low-frequency foregrounds (such as radio sources, which are the main astrophysical limitation for CMB measurements up to 150-200 GHz).

TABLE 1. LFI main characteristics and goal performances

	30 GHz	44 GHz	70 GHz	100 GHz
Noise per 30' pixel (mK)	6	6	6	6
1–sec sensitivity (mK s $-\frac{1}{2}$)	179	219	310	552
Number of Feeds	2	3	6	16
System Temperature (K)	7.5	12	21.5	39.6
Effective Bandwidth	20%	20%	20%	20%
Angular Resolution	30'	24'	14'	10'
Average $\Delta T/T$ (10-6) per 30' of angular resolution	2.2	2.2	2.3	2.2
Average $\Delta T/T$ (10-6) per $FWHM^2$ resolution element	2.2	2.8	4.9	6.6
ΔT (μK) at the poles per $FWHM^2$ resolution element *	0.41	0.45	0.58	0.68
ΔT (μK) at the poles per $FWHM^2$ resolution element **	1.0	1.1	1.5	1.7

$\Delta T/T$ and ΔT are expressed in thermodynamic temperatures
* Corresponding to the two ~ 4°☐ regions near the ecliptic poles
** Corresponding to the two ~ 30°☐ regions near the ecliptic poles

Instrument description

LFI consists of four main units: the Front End Unit (FEU) reported in figure 1; the Back End Unit (BEU); the Radiometer Electronics Box Assembly (REBA); the Sorption Cooler System (SCS). The FEU is located at the focus of the telescope, as one component of the joint LFI/HFI focal assembly. It is composed by 27 corrugated feed horns, 27 Ortho Mode transducers, 27 Front End Modules (FEM) each containing two receivers. The BEU is mounted on the equipment platform and includes 27 Back End Modules (BEM).

The FEU is cooled to 20 K by the vibration–less hydrogen sorption cooler and is connected to the BEU, which is at 300 K, by 102 waveguides. With this configuration, the power dissipation for the Front End Unit can be kept below 0.6 W, a sufficiently low value to enable active cooling of the HEMTs to 20 K.

LFI takes full advantage of the dramatic progress of transistor amplifier technology achieved over the last decade, particularly for low noise performance and reliability of cryogenically cooled indium phosphide (InP) high-electron-mobility transistors (HEMTs). These devices exhibit the best noise performance ever reached in the LFI frequency range.

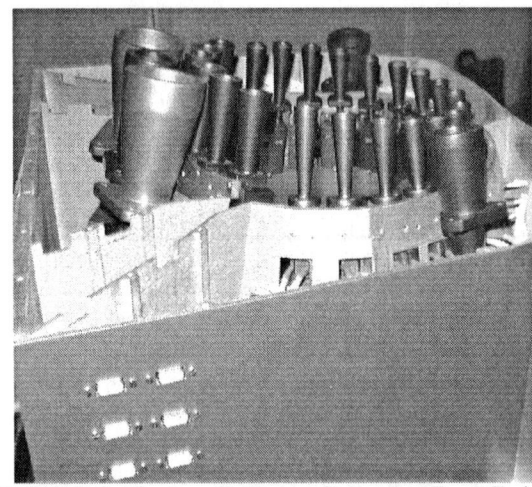 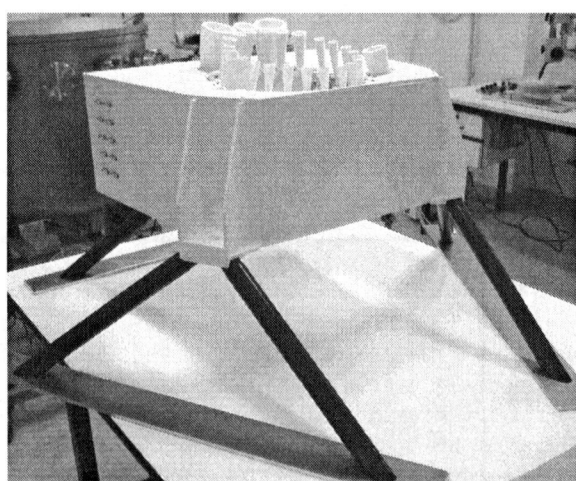

FIGURE 1. The Low Frequency Instrument Front End Unit. The picture shows the mockup which has been built to check the integration and assembling procedures of the LFI (the cover plate of the main frame has been removed on the left picture). The array of 27 feed horns of which the biggest are the two 30 GHz ones is clearly visible. The 100 GHz and the 70 GHz horns are located around the circular hole into which the HFI will be located. The 44 GHz horns are located at approximately 120° one from each other. The front end modules are not visible since fixed into the main frame structure which guarantees the mechanical stiffness and the appropriate thermal characteristics. On the right picture, note the bipods which support the Focal Plane Unit and interface with the telescope structure.

In Figure 2 the radiometer block diagram is reported. The signal entering the corrugated feed horn is divided by the Ortho Mode Transducer into two orthogonal linearly polarized signals that propagate through two independent pseudo-correlation receivers. This solution maximizes the number for channels per available focal plane area, and, at the same time, adds polarization capability to the instrument.

The signal entering one receiver is then mixed, by an hybrid, with the signal coming from the reference load which is attached to the HFI 4K shield [7]. The outputs of the hybrid are amplified by the cryogenically cooled HEMTs low noise amplifiers. The amplified mixed signals are thus shifted in phase appropriately (+/− 180°) and synchronously (~8 KHz) with the data acquisition, decoupled by another hybrid, and sent to the back end section. The signals are further amplified in the back end, then filtered, and finally detected by two diodes. This scheme guarantees a very low 1/f noise both in the front end section and in the back end section.

The connection between each FEM and each BEM is performed by four high performance 1.5 meter long bent twisted composite (copper – stainless steel – gold plated stainless steel) rectangular waveguides (102 waveguides for the whole array). The waveguides are thermally connected to the three V–grooved shields of the payload at 50, 90 and 140 K from the FEU to the BEU respectively.

For all frequencies, about 35 dB of front-end gain is sufficient to guarantee that the overall noise is dominated by the low-noise front-end amplifiers. In this way, active cooling of the LFI front end reduces the noise temperature substantially, roughly a factor of 3 compared to an optimized passive cooling design. The low-noise amplifiers use InP HEMTs in cascaded gain stages. The amplifiers at 30 and 44 GHz will use discrete InP HEMTs incorporated into a microwave integrated circuit (MIC). At these frequencies, cryogenic MIC amplifiers have demonstrated noise temperatures approaching 10K, with 20% bandwidth. At 70 and 100 GHz, MMICs (Monolithic Microwave Integrated Circuits) architectures, which incorporate all circuit elements and the HEMT transistors on a single InP chip, are used. The LFI will fully exploit both MIC and MMIC technologies at their best.

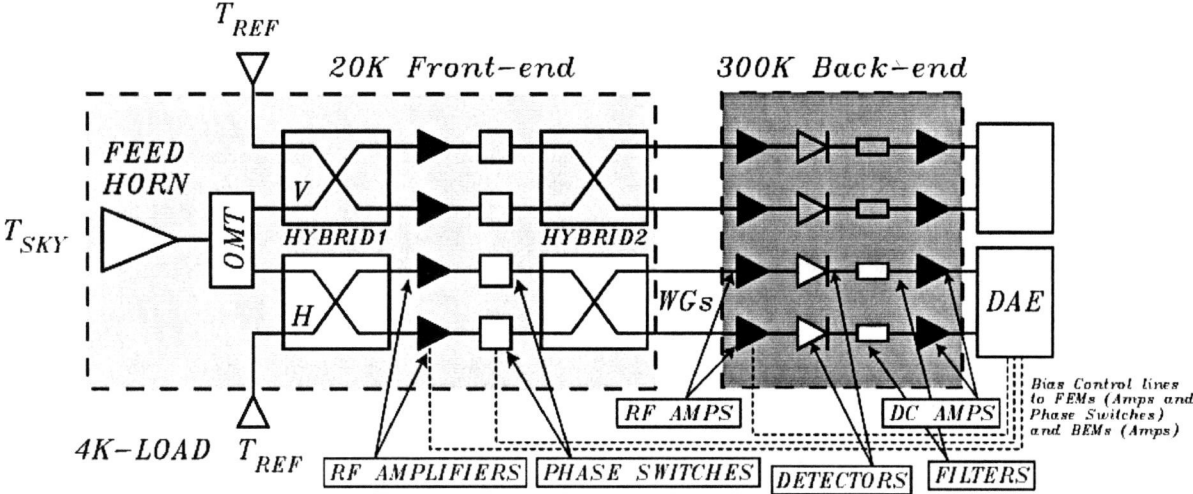

FIGURE 2. LFI radiometer block diagram. In the figure we show the 20 K front end and the 300 K back end of the radiometer chain assembly which are connected by waveguides. The figure also shows the electronic boxes (DAE) and the bias controls of the radiometer active devices.

Optical Interfaces

Although LFI has been designed to measure primarily the anisotropies of the CMB, a strong effort has been made to optimize the focal plane for the sky polarization reconstruction with no impact on the instrument design. The changes have been done at the Front End Unit level by modifying the polarization orientations of the intrinsically polarized receivers.

To recover the linear polarization information of the incoming radiation, the detectors of the LFI have been optimized in orientation according to figure 3 which reports the polarization axes of the detectors in the focal plane and the corresponding beam polarization directions on the sky. Each feed horn is identified by two orthogonal axes, aligned with the electric field coupled to the two arm of each corresponding OMT. The projection of these axes on the sky through the optics defines the two polarization axes of the main beam. Except for the 44 GHz #24, the horns have been paired off along the scan direction, aided by the symmetry of the focal plane. Each scan–aligned pair, except the (8,11) and the (5,14) are oriented at 45° in order to get the linear polarization parameters (however these two are 45° tilted with respect to each other for different scan directions). Each horn pair will observe the same spot on the sky during the same scan along the great circles.

One of the major contribution to the spurious polarization is expected to be produced by the optics, i.e. the cross–polar level of the telescope (for the dual profiled feed horns the cross–polarization is at a level of –40dB). The off–axis location of all the LFI feed horns degrades the polarization purity of the main beams. Nevertheless, our optical simulations show, as reported for example in Figure 4, that the peak of the cross–polarized field is still 25 dB below to the co–polar peak, for all the LFI beams [8]. For a more detailed analysis of the systematic errors in LFI polarization measurements see [9].

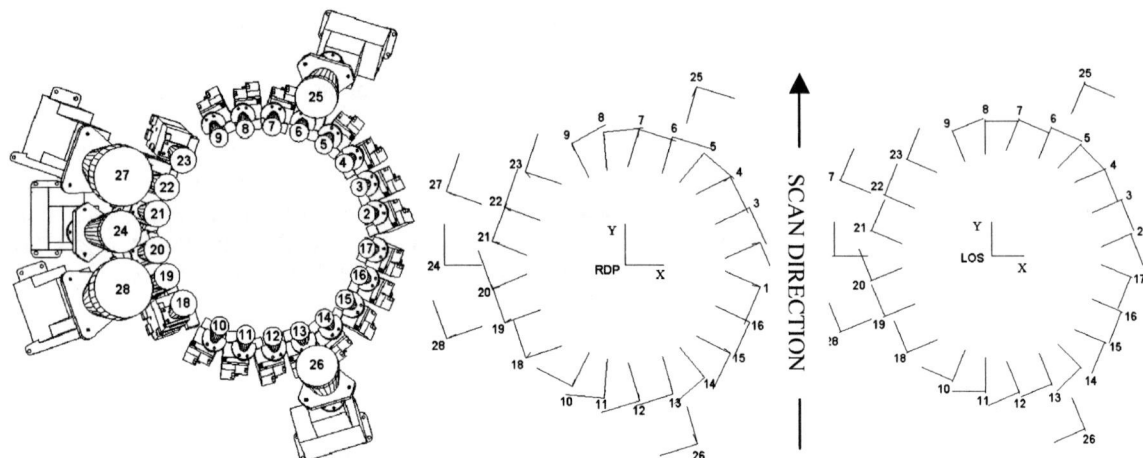

FIGURE 3. Left panel: Planck/LFI Baseline Focal Plane Unit configuration with the adopted numbering. Center panel: Polarization alignment for the LFI detectors on the focal plane. Right panel: Polarization alignment of the corresponding LFI beams. The Main Beam coordinate systems are referred to the Line Of Sight. It should be noted that for convenience the origin of each coordinate system has been taken such that (X,Y) is proportional to the angular separation of each beam pointing direction with respect to the line of sight (LOS). The polarization orientations have been optimized on the sky to recover the stokes parameters relative to the linear polarization. The scan direction is vertical in this figure (aligned with the Y_{RDP} and Y_{LOS} axes).

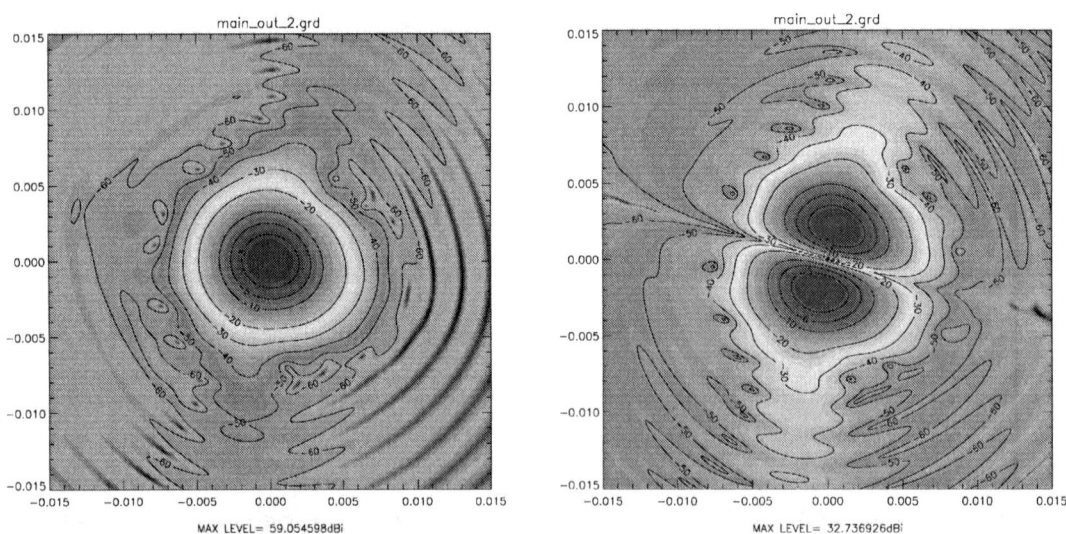

FIGURE 4. Example of main beam simulations for the LFI feed #2. The simulations have been performed with a Physical Optics code both in co– polar (left plot) and in cross– polar (right plot) components according to the Ludwig third definition of polarization [10]. The co–polar component peaks at 59.05 dBi while the cross–polar peaks at 32.74 dBi.

CONCLUSIONS

The Planck Low Frequency Instrument is conceived to obtain high resolution, high sensitivity, full-sky imaging of the Cosmic Microwave Background including polarization. Although not specifically designed for polarimetry, the LFI Focal Plane Unit layout has been optimized also for linear polarization measurements of the radiation coming from the sky, taking the advantage of its intrinsically polarization sensitive radiometers. It is expected that LFI will measure the statistics of the predicted and to be detected polarized E– component of the CMB at sub–degrees angular scales with excellent sensitivity, especially in the regions near the ecliptic caps.

ACKNOWLEDGMENTS

This paper represents the work carried out by a large number of people who are working on the development of the Planck Low Frequency Instrument. We wish to thank the instrument consortia and in particular it is a pleasure to acknowledge Paddy Leahy and Vladimir Yurchenko for the useful discussion about the polarization topic.

REFERENCES

1. Kosowsky, A., *New Astronomy Reviews,* **43**, Issue 2-4, 157-168 (1999).
2. Hu, W., Seljak U., Withe, M., Zaldarriaga, M., *Phys.Rev. D,* **57**, 3290- (2001).
3. Mandolesi, N. et al., "The Low Frequency Instrument, a Proposal Submitted to the ESA in response to the Announcement of Opportunity for the FIRST/Planck mission" (1998)
4. Villa, F., Bersanelli, M., Burigana, C., Butler, R.C., Mandolesi, N., Mennella, A., Morgante, G., Sandri, M., Terenzi, L., and Valenziano, L., "The Planck telescope", in *2K1BC Workshop "Experimental Cosmology @ mm – waves"*, AIP Conference Proceeding, American Institute of Physics, New York, in press
5. Puget, J-L., et al., "The High Frequency Instrument, a Proposal Submitted to the ESA in response to the Announcement of Opportunity for the FIRST/Planck mission" (1998)
6. Delabrouille, J., Kaplan, J., these Conference Proceedings
7. Valenziano, L., Bersanelli, Butler, R.C., Cuttaia, F., Mandolesi, N., Mennella, A., Morigi, G., Morgante, G., Sandri, M., Terenzi, L., and Villa, F., "The 4K reference load for the Planck Low Frequency Instrument", in *2K1BC Workshop "Experimental Cosmology @ mm – waves"*, AIP Conference Proceedings, American Institute of Physics, New York, in press
8. Sandri, M., Villa, F., "Planck/LFI: Main Beam Locations and Polarization Alignment for the LFI Baseline FPU", Planck LFI Technical Note, *PL-LFI-PST-TN-027*, Issue/Rev. 1.0 (2001)
9. Leahy, J.P., Yurchenko, V., and Morag A. Hastie, these Conference Proceedings.
10. Ludwig, A. C., *IEEE Trans. Antennas Propagat.* **21**, 116-119 (1973).

Precision and widefield polarimetry with the Australia Telescope Compact Array

R.J. Sault*, D.P. Rayner* and M.J. Kesteven*

*ATNF, P.O. Box 76, EPPING, NSW, 1710, Australia

Abstract. This paper describes some of the polarimetry done with the Australia Telescope Compact Array (ATCA). Comparatively low and stable instrumental polarization allows high quality polarimetric images to be formed. The instrument has proven particularly successful in studying low level circular polarization. The instrument has also proven itself excellent for widefield imaging of galactic polarization. We consider some of the techniques used in widefield polarimetry of an interferometer.

INTRODUCTION

The Australia Telescope Compact Array (ATCA) is an interferometer array located near the township of Narrabri in eastern Australia. It consists of 6 antennas (5 movable) on an east-west track. It currently observes at 20, 13, 6 and 3 cm wavelengths routinely. Commissioning observations using new 12 and 3.5 mm wavelengths receivers are also now possible. With low noise receivers ($T_{\mathrm{sys}} \approx 40K$ at centimetre wavelengths), 22m diameter dishes and an effective bandwidth of 200 MHz (split into two independently tunable bands), sensitivities of $50 - 100\mu\mathrm{Jy}$ are achieved in a 12 h synthesis. A detailed description of the array has been given in Frater and Brooks [1].

Although it is a general purpose instrument, achieving high precision polarimetry was an important issue in the design of the ATCA. The array simultaneously measures two orthogonal linear polarizations (the X and Y feeds), and forms the four possible polarization cross products between a pair of antennas. With an alt-az mount, the dishes rotate relative to the sky during the course of a long synthesis. The antennas have cassegrain optics, with the feeds being on-axis (the feeds are located on a rotating turret, which is used to change between some observing bands). The feeds are wide bandwidth corrugated horns. The 20 and 13 cm bands share one feed, as do the 6 and 3 cm bands. A switched noise source, injected into each horn, is used to continuously measure receiver gains. As this noise is injected symmetrically into the X and Y feeds, it is also used to track changes in the phase difference between the X and Y signal paths for each antenna (the phase difference typically varies by about a degree during a 12h synthesis).

This paper consists of three somewhat distinct parts. Firstly we consider the calibration issues relevant when observing "small" objects (i.e. when off-axis polarimetric response can be ignored). We demonstrate the successfulness of this, and the stability of the ATCA, with circular polarisation images with dynamic ranges of greater than 100,000. Secondly we consider widefield issues. We consider models of the off-axis response of the ATCA, and discuss techniques to eliminate it. Finally we consider mosaicing and deconvolution of large polarimetric images from the ATCA,

PRECISION POLARIMETRIC CALIBRATION

The basic way to describe the polarimetric response of a phase-coherent linear system is by way of Jones matrices. These are complex-valued 2×2 matrices which give the change of the two polarization states as they pass through a system. An exhaustive description of using Jones matrices in the context of radio interferometry is given by Hamaker et al. [8]. A Jones matrix can be used to describe the response of a single element in the signal path, with the composite Jones matrix of the signal path, **J**, being the cascaded matrix multiplication of the Jones matrices. The two Jones

CP609, *Astrophysical Polarized Backgrounds,* edited by S. Cecchini et al.

matrices typically determined in the calibration process describe receiver gain, **G**, and polarization leakage, **D**:

$$\mathbf{G} = \begin{pmatrix} g_X & 0 \\ 0 & g_Y \end{pmatrix}, \tag{1}$$

$$\mathbf{D} = \begin{pmatrix} 1 & d_X \\ -d_Y & 1 \end{pmatrix}. \tag{2}$$

$$\mathbf{J} = \mathbf{GD} \tag{3}$$

For an interferometer, the Kronecker (or tensor) product of the Jones matrices of the two signal paths results in a response matrix which relates the measured correlations, **V**, with those from an ideal system, $\hat{\mathbf{V}}$.

$$\mathbf{V} = (\mathbf{J}_i \otimes \mathbf{J}_j^*)\hat{\mathbf{V}} \tag{4}$$

To some extent Jones matrices and Kronecker products are a matter of notation. But it is the compactness of the notation, and its reflection of the signal path, which makes them an elegant tool to describe polarimetric systems. Our main interest in them is that they provide a useful abstraction through which we have analysed the issue of calibration of a polarimetric interferometer. In practice a real interferometer has to be calibrated to determine, for example, gains and leakages. In practice we use observations of point sources to determine these calibrations. Jones matrices and Kronecker products provide a formalism through which we can determine the number and types of calibration observations required to sufficiently characterise the Jones matrices of our system. An analysis of this is described in [14]. Note that to polarimetrically calibrate some observations, the Jones matrices of the system do not need to be fully determined (there are some degrees of freedom in the Jones matrices which have no effect on the calibration of some observation types). This is an important point: to get well calibrated data, we do not necessarily need to fully determine the Jones matrices which describe a system.

An important tool in calibrating an antenna with an alt-az mount is that we can use parallactic angle rotation of the sky during a long synthesis to our advantage. Whereas polarization leakages are fixed to the antenna, source polarization rotates with the sky. In this way, it is possible to decouple instrumental and calibrator parameters. This means that we can use a calibrator without prior knowledge of its polarization properties.

Our motivation for studying the finer details of polarimetric calibration issues of an interferometer results from out interest in measuring high precision circular polarization in a study of radio galaxies. Circular polarization is typically very weak (a source which is 0.3% circularly polarized is considered to be strongly circularly polarized). To work at fractional polarizations of tenths or hundredths of a percent, second order effects (e.g. leakage of linear polarization into circular) needs to be accounted for in the calibration. A review of this work is given in [12]. Figure 1 gives two examples. The left panel shows a radio galaxy with a core/lobe structure. The both the core and lobes are linearly polarized, whereas only the core shows circular polarization. As calibration artifacts would be expected to produce spurious circular polarisation emission either coincident with the linear polarisation or total intensity, the lack of circular polarisation in the lobes is fairly conclusive that the circular polarisation that we measure is real. The right panel shows a blank field in circular polarization at the 30μJy level. This field is centred on a 5 Jy source, suggesting that the systematic errors in this image are probably smaller than 1 part in 160,000. An exhaustive evaluation of all the errors considered in forming images such as this is given in [11].

OFF-AXIS POLARIMETRY

When making polarimetric images of large objects, the off-axis response of the antennas must be considered. Again this is most conveniently described by a Jones matrix of the primary beam voltage response (e.g. [13]), **P**, where the matrix coefficients are a function of the position of a source within the primary beam. If the X and Y feeds of an array respond identically (apart from a 90° rotation, then the form of the Jones matrix will be

$$\mathbf{P} = \begin{pmatrix} A(r,\psi) & B(r,\psi) \\ -B(r,\psi+\frac{\pi}{2}) & A(r,\psi+\frac{\pi}{2}) \end{pmatrix}. \tag{5}$$

Here r is the distance from the pointing centre and we define ψ as the angle from the axis of the X probe to a point in the primary beam.

To good approximation, all ATCA antennas show the same off-axis response and the X and Y feeds appear identical. The response, however, is a function of frequency. The ATCA's off-axis polarization is mainly a characteristic of the

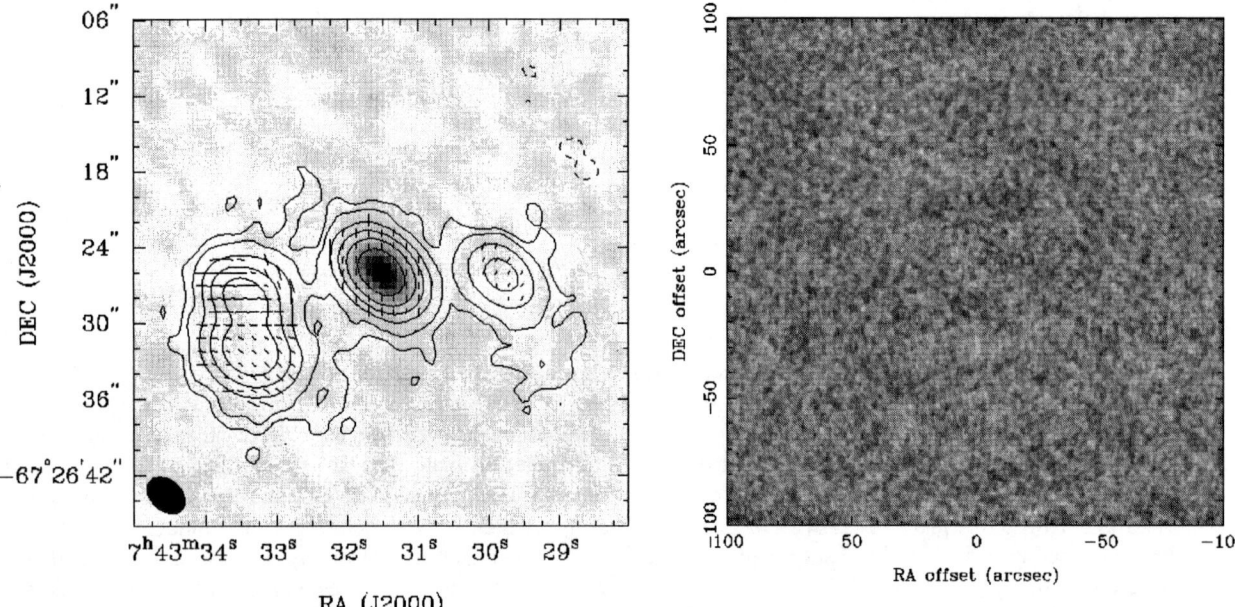

FIGURE 1. Circular polarization in radio galaxies. Left panel: a core/lobe source showing total intensity (contours), linear polarization (vectors) and circular polarization (grey scale). Right panel: an essentially artifact-free circular polarization image despite being centred on a 5 Jy radio source. The rms of the image is 30μJy.

feed assembly. Figure 2 shows the off-axis polarimetric response, in fractional spurious linear polarization, for the ATCA antennas at its four main bands. The grey scale saturates at 7%, and the diameter of the region shown is about twice the FWHM of the primary beam response (the spurious polarization is worst at about the half power points). Whereas the off-axis response at 20 and 6 cm is good (it peaks at about 0.7%), the 3cm response is poorer and the 13cm response is quite poor. The poorer response is a result of two factors. Firstly both 13 and 3 cm are the higher frequency bands in their feed horns, and the feed horns have been optimized for the lower band. Secondly, at 13cm, there was an error in the design of the 20/13cm feed assembly which makes the polarimetric response much narrower band. An engineering correction to the 13 cm system is planned to eliminate the latter problem.

For a snapshot observation, or for an observation with an equatorial telescope (where the dish does not rotate relative to the sky during an observation), this polarimetric beam response can be corrected in the image domain (e.g. [3] and [4]). This is possible as the effect is not time varying in the image (there is no parallactic angle rotation). This approach has been used for both VLA snapshots and long syntheses with the WSRT. However this is not generally possible with a long synthesis for an alt-az mount (or any other mount where the sky and dish rotate relative to one another). In this case an iterative algorithm to form a correct image is needed. This would involve steps of estimating the polarized images, correcting the visibilities for the off-axis errors, and then re-imaging etc. This is a very slow operation when there are many visibilities and pixels with emission – as there generally will be in a widefield image. In practice, very few ATCA observations have attempted to use this approach.

An interesting special case presents itself when the beam matrix produces to the spurious linear and circular polarization showing circular symmetry. This is the case when the beam matrix is of the form

$$\mathbf{P} = \begin{pmatrix} a + b\cos(2\psi) & b\sin(2\psi) + c \\ b\sin(2\psi) - c & a - b\cos(2\psi) \end{pmatrix}. \tag{6}$$

Here a, b and c can be functions of radial distance, but not ψ. With the circular symmetry, parallactic angle rotation is irrelevant, and the effect can, once again, be corrected in the image plane. This case is of interest as it is this form that will result from a perfect (circularly symmetric) dish (e.g. [7]). Although this form poorly approximates the response of the ATCA dishes at 13 and 3cm, it is a tolerable approximation to the 20 and 6 cm response. This can form the basis of much more computationally cheap algorithm.

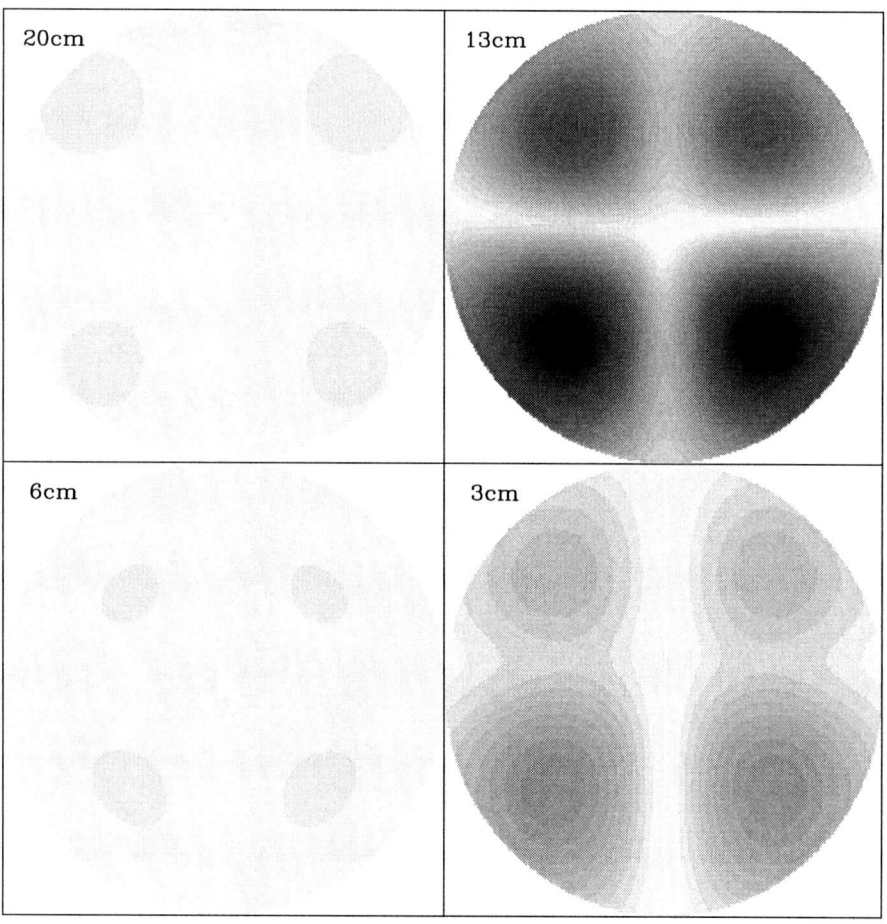

FIGURE 2. The spurious fractional linear polarization caused by off-axis polarimetric response. The area plotted is approximately twice the FWHM of the primary beam at each wavelength band. The grey scale saturates at 7%.

It is worth noting that off-axis polarimetric response can even be an issue for precision measurements of compact objects. For example, for the ATCA at 3cm, we have found that off-axis polarimetric response and small ($\sim 5-10''$) time-varying pointing errors produce a quite detectable signature in precision circular polarization measurements of point sources.

In practice, ATCA observers tend not to correct for off-axis polarimetric response. Properly correcting in the 13 cm band is too time consuming relative to the extra science that is achieved relative to the other bands, whereas the 20 and 6 cm bands are sufficiently good to ignore the imperfections.

POLARIMETRIC MOSAICING

One of the shortcomings of a single-pointing interferometric observation is its inability to measure large-scale structure where this structure corresponds to baselines shorter than the minimum separation of antennas. This problem can be partially solved by the technique of mosaicing (see [5] and [2]). This is the process of using pointings from many pointings to both form an image that is much larger than the antenna primary beam, and to simultaneously recover shorter spacings than are possible with a single pointing. While mosaicing has been applied predominantly only to total intensity, exactly the same principles apply to polarized emission. The mosaicing process usually includes a non-linear deconvolution step, which is needed to extrapolate unsampled spacings in the Fourier plane: maximum entropy deconvolution algorithms are probably the most popular. In general, non-linear algorithms work best when they are presented with all the data simultaneously (i.e. deconvolving mosaic data will be better behaved if all the pointings are

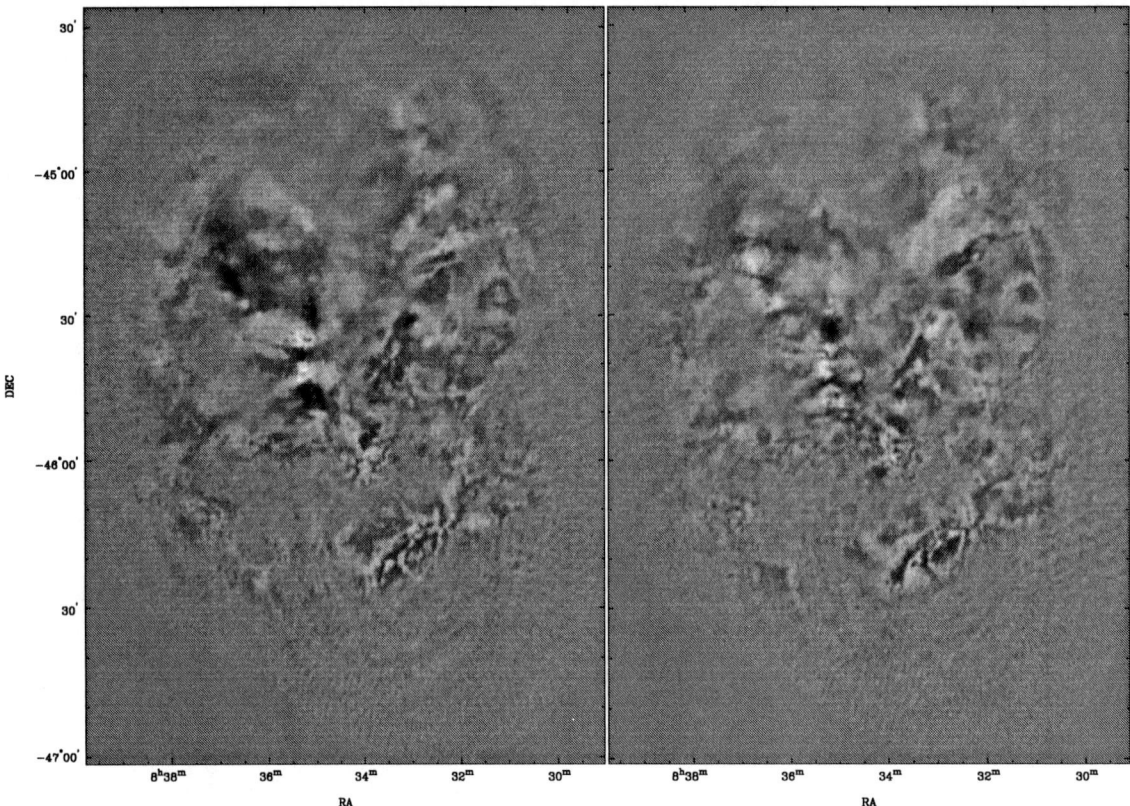

FIGURE 3. Linear polarization emission of the Vela-X region. The left and right panels are Stokes Q and U respectively.

jointly processed), and so it would appear better to deconvolve total intensity and polarized emission images jointly.

The formalism for using maximum entropy algorithms for joint polarization has been known for many years (e.g. [10]). Following the lead of Holdaway et al. [9] in VLBI, we have used the entropy measure

$$H = -\sum_i I_i \left(\log(\frac{2I_i}{M_i e}) \quad + \quad \frac{1+p_i}{2} \log(\frac{1+p_i}{2}) \right.$$
$$\left. + \quad \frac{1-p_i}{2} \log(\frac{1-p_i}{2}) \right). \qquad (7)$$

Here I_i and p_i are the total intensity and fractional polarization of the ith pixel respectively, and M_i is the so-called "default" image. In this entropy measure, the two terms containing the fractional polarization, p, will always contribute to reducing the total entropy, and the minimum reduction is achieved when p is zero. More details about polarimetric mosaicing can be found in [15].

There is an intriguing advantage to joint polarimetric deconvolution in recovering short spacing information. At centimetre wavelengths, polarized emission tends to have smaller scale sizes than the total intensity. Positionally varying Faraday rotation, for example, will impose small scale structure on a uniformly polarized background. When observed with an interferometer, the total intensity can be resolved out even when the polarized emission is not. This leads to sources with apparent polarized emission of greater than 100%. In this case, a joint maximum entropy deconvolution will produce a better (though possibly still quite unreliable) estimate of the total intensity.

Figure 3 shows an example of an ATCA polarimetric mosaic of the Vela-X region observed at 20 cm. The fractional polarization in this region is particularly strong, reaching 60% in some region. This image has not included any compensation for off-axis polarization. The good polarimetric purity of the 20 cm system makes the off-axis response comparatively small to start with. Additionally an analysis shows that the combination of parallactic angle rotation and mosaicing (covering each part of the sky with multiple overlapping primary beams) significantly averages down the impure response. Another excellent example of an ATCA polarimetric mosaic can be found in [6].

CONCLUDING COMMENTS

Its good engineering design, coupled with appropriate functionality in the on-line and off-line systems makes the ATCA an excellent polarimetric interferometer at centimetre wavelengths. Regardless of whether the observer is interested in polarimetry or not, calibrated polarimetric data is a normal outcome of a continuum observation. Apart from "bread and butter" polarization studies, the ATCA is able to exploit niches in high precision circular polarization measurements and in widefield polarimetric imaging.

REFERENCES

1. Frater, R.H., Brooks, J.W., (eds.), J. Electr. Electron. Eng. Aust., Special Issue, 12, no. 2 (1992)
2. Cornwell T.J., A&A, 202, 316 (1988)
3. Cotton, W.D., AIPS Memo No. 86, National Radio Astronomy Observatory (1994)
4. Cotton, W.D., in G.B. Taylor, C.L. Carilli, R.A. Perley (eds.), Synthesis in Radio Astronomy II, ASP Conf. Series 180, 111 (1999)
5. Ekers, R.D., Rots, A.H., in: van Schooneveld C. (ed.), Proc. IAU Col. 49, Image Formation from Coherence Functions in Astronomy. D. Reidel, p. 61 (1979)
6. Gaensler, B.M., Dickey, J.M., McClure-Griffiths, N.M., Green, A.J., Wieringa, M.H., Haynes, R.F., ApJ, 549, 959 (2001)
7. Ghobrial, S.I., IEEE Trans. AP, 24, 418 (1976)
8. Hamaker, J.P, , Bregman, J.D., Sault, R.J., A&AS, 117, 137 (1996)
9. Holdaway, M.A., Wardle, J.F.C., in: Proceedings of the Meeting on digital image synthesis and inverse optics, Proc. SPIE 1351, 714 (1990)
10. Ponsonby, J.E.B., MNRAS, 163, 359 (1973)
11. Rayner, D.P., Ph.D. thesis, University of Tasmania, Hobart, Australia (2000)
12. Rayner, D.P., Norris, R.P., Sault, R.J., MNRAS, 319, 484 (2000)
13. Sault R.J., Ehle M., ATNF Technical Document Series No. 39.3/088, Australia Telescope National Facility (1996)
14. Sault, R.J., Hamaker, J.P, , Bregman, J.D., A&AS, 117, 149 (1996)
15. Sault, R.J., Bock, D.C.-J., Duncan, A.R., A&AS, 139, 387 (1999

AMiBA - array for microwave background anisotropy

M. Kesteven for the AMiBA team

ATNF, CSIRO, Australia

Abstract. AMiBA is a 90 GHz interferometric array of the ASIAA (Academia Sinica, Institue of Astronomy and Astrophysics). It will make a detailed study of the polarisation of the CMB anisotropy; it will also undertake a survey of Sunyaev-Zel'dovich clusters. It is under construction at present, with an expected completion date of late 2003.

INTRODUCTION

As this conference has amply demonstrated, the search for (and the study of) the polarisation in the CMB anisotropy is one of the last of the CMB frontiers. AMiBA is another of the bold experiments hoping to make a contribution to this field. In 2000 the ASIAA and the Department of Physics, NTU (National University of Taiwan) received a grant for the construction of a CMB experiment: AMiBA [1]. Construction has now started on a dual-purpose telescope - a compact interferometric array to study the polarisation of the CMB anisotropy, and a larger array to survey large areas of the sky for SZ clusters.

THE EXPERIMENT GOALS

AMiBA has been designed to address two particular CMB goals:

Polarisation

The aim is to detect the polarisation, and to determine its power spectrum, over the l-space range of approximately 500 to 2000.

The telescope for this mode will be a close-packed, high-brightness-sensitive array, with peak sensitivity around l=1000.

This will be the initial configuration of the array.

Sunyaev-Zel'dovich Effect

The initial task here will be to make a large area survey for SZ clusters.

In this mode the resolution of the telescope will be increased - with larger diameter antennas, more widely spaced.

THE INSTRUMENT SPECIFICATIONS

- 19 antennas (30 cm diameter for polarisation; 150 cm for the SZE).
- F_c = 90 GHz; bandwidth = 20 GHz.
- Cooled Receivers, 100 K Tsys

- Dual Circular Polarisation System
- Four channel (all stokes parameters) 8 spectral channel correlator, covering all 172 baselines.

THE PLATFORM

We require an array mount with three degrees of freedom - that is, azimuth, elevation and rotation about the optical axis (the "z-axis"). A conventional azimuth/elevation mount modified to provide the third axis is clearly a possibility; however, we incline for the moment to a hexapod mount.

This is unconventional, in that such a mount has not yet been used for a radio telescope, although a number of telescopes have used such a system for the subreflector. A detailed hexapod design was prepared for the ALMA prototype [2]. We find that the hexapod meets our pointing/rotation requirements without trouble (although the third axis capability does rely on a three-fold symmetrical antenna configuration to compensate for the limited z-axis rotation). The hexapod is currently more expensive that a similarly sized conventional mount, on the factory floor. However, we find that the hexapod does become competitive when one compares the as-installed costs and benefits :

- Access to the receivers mounted below the array antennas is much better than for a conventional mount - we have clear access to the under-side of the array platform.
- The stowed position is more compact, with all 6 jacks fully compressed. This makes for greater safety, and a simpler weather-proof housing.
- The shipping and assembly costs are reduced, since the mount dismantles to quite modest-sized components.

THE OPTICS

The antennas will be simple cassegrains, with a primary F/D of $\simeq 0.3$. We plan to use cylindrical shrouds for the polarisation experiment in order to reduce the mutual coupling between the close-packed antennas. The secondary focus is close to the antenna vertex, so that the same feedhorn and dewar assembly can be used on both antenna sizes.

THE RECEIVERS

The receivers will be cooled (12 K) InP MMICs. The entire frontend will be cooled - feedhorn, $\lambda/4$ polariser and Orthomode Transducer.

A single down-conversion stage will be provided to translate the 20 GHz bandwidth to baseband. Phase-switching is necessary in order to counter any offsets within the correlator; it will be implemented via the conversion system.

The array will be calibrated with a broad-band noise source, coherent over the entire array, which will be injected into the throat of the feedhorns.

The proposed observing mode (drift scans) means that some form of fringe-tracking is required. The fringe rates will be small, as the array is small - we therefore propose to perform the phase tracking in software. This means that we must extract the visibility data from the correlator at a relatively rapid pace, in order not to blur the fringes.

THE CORRELATOR

The correlator is a complex XF correlator, and is an interesting analog-digital hybrid.

The correlator will produce four products (from which we can derive the four stokes paameters) for every baseline. Each product is the fourier transform of a 17-lag cross-correlation spectrum.

Two 17-tap delay lines are used in each module, with 20 GHz multipliers at each tap. We enter the digital domain after the multipliers.

A similar design has been deployed with success by A. Harris [3]

THE OBSERVING MODE

We plan to use a drift scan technique to combat the systematics - the mutual coupling between antennas and ground pickup in particular.

The basic target field is a small number of beamwidths long (along the HA axis), and one beamwidth wide. The array is positioned ahead of the field, and held stationary while the target field drifts by; the array is then repositioned, and the cycle continues. The essential point is that this scheme establishes a clear signature difference between sky (drifting at the sidereal rate) and the systematics (stationary).

The subsequent processing will key to these signatures to remove the systematics.

SCHEDULE TARGETS

Construction of the prototype has begun; the prototype integration tests will take place during january-june 2002. A trial polarisation experiment will start in late 2002. The full instrument is expected in 2003

REFERENCES

1. Lo,K.Y, Chiueh,T.H, Martin,R.N, Ng Kin-Wang, Liang,H, Pen Ue-Li & Ma Chung-Pei, 2000, IAU Symposium 201 - New Cosmological Data and the Values of the Fundamental Parameters, eds Lazenby,L and Wilkinson,A
2. Kingsley,J.S, Martin,R,N & Gasho,V.L, 1999, ALMA memo # 263
3. Harris,A.I & Zmuidzinas, 2001, J. Rev. Sci. Inst. **74**, 1531

QUEST — A 2.6–m mm–wave telescope for CMB polarization studies

L. Piccirillo*, P. A. R. Ade*, J. J. Bock[§†], M. Bowden*, S. E. Church**, K. Ganga[‡], W. K. Gear*, J. Hinderks**, B. G. Keating[§], A. E. Lange[§], B. Maffei*, O. Mallié*, S. J. Melhuish*, J. A. Murphy[¶], G. Pisano*, B. Rusholme**, A. Taylor[‖] and K. Thompson**

*Department of Physics and Astronomy, 5 The Parade, Cardiff, CF24 3YB, U.K.
[†]Jet Propulsion Laboratory, 4800 Oak Grove Dr., Pasadena, CA 91109
**Department of Physics, Stanford University, Stanford, CA 94305-4060
[‡]Infrared Processing and Analysis Center, California Institute of Technology, Pasadena, CA 91125
[§]Division of Physics, Math, and Astronomy, California Institute of Technology, Pasadena, CA 91125
[¶]Experimental Physics Department, National University of Ireland, Maynooth. Co. Kildare., Ireland
[‖]Institute for Astronomy, University of Edinburgh, Royal Observatory, Blackford Hill, Edinburgh, EH9 3HJ, U.K

Abstract. We describe QUEST (**Q** and **U** **E**xtra–galactic **S**ub–mm **T**elescope), a CMB polarimeter, operating at millimetre wavelengths. Interesting features of its design are outlined.

INTRODUCTION

Studies of the Cosmic Microwave Background (CMB) have demonstrated their worth in improving our understanding of the early Universe. Measurements of its black–body spectrum and spatial variation of its temperature have helped to constrain cosmologies to an ever–tightening parameter space.

Since the COBE DMR detection of CMB anisotropy [7] and the direct observation of individual structures (for example Hancock et al. [3]), sensitivities have improved to the extent that the first peak of the CMB spatial frequency (C_ℓ) spectrum may be identified with confidence, and there is increasingly strong evidence for second and third peaks (see for example results from DASI [2], MAXIMA–1 [4] and BOOMERANG [5]). The spectrum will be even better determined by forthcoming measurements from new balloon experiments and satellite missions such as *MAP* [1] and *Planck* [8].

With the anticipated improvement of data quality from new and revised ground–based, balloon and satellite experiments we expect better to determine values such as the total energy density and curvature of the Universe. However, despite improved statistics, degeneracy in parameter determination based on temperature anisotropy data will continue to be a problem. But by measuring the third characteristic of the CMB radiation field, its polarization, this problem may be addressed.

THE INSTRUMENT

The following paragraphs describe QUEST as it is currently being designed and built at institutions in Cardiff and California.

CP609, *Astrophysical Polarized Backgrounds*, edited by S. Cecchini et al.
© 2002 American Institute of Physics 0-7354-0055-5/02/$19.00

FIGURE 1. QUEST Mechanical Layout. A cryostat holding re–imaging optics, the wave plate and the focal plane assembly is mounted behind the primary mirror. This assembly may rotate about the 'Z' axis. This in turn is mounted on an az. / el. platform.

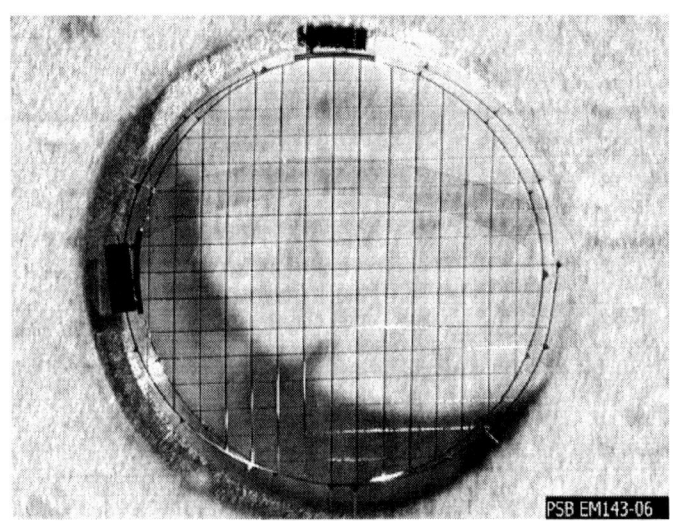

FIGURE 2. Polarization–Sensitive Bolometer

Optical Layout

An entirely on–axis optical design has been chosen. Millimetric radiation falling on the primary mirror is reflected onto a secondary mirror and thence to a Cassegrain focus. Lens elements are used to form a 1°5 image at the focal plane, with a resolution of 5′.

A disadvantage of the classical Cassegrainian layout is the blockage caused at the centre of the primary by the secondary mirror. This 'hole' in the field distribution will result in an increased side lobe level aside the primary beam. However, removing this blockage by adopting an off–axis layout has its own problems; such a system has an in–built

160

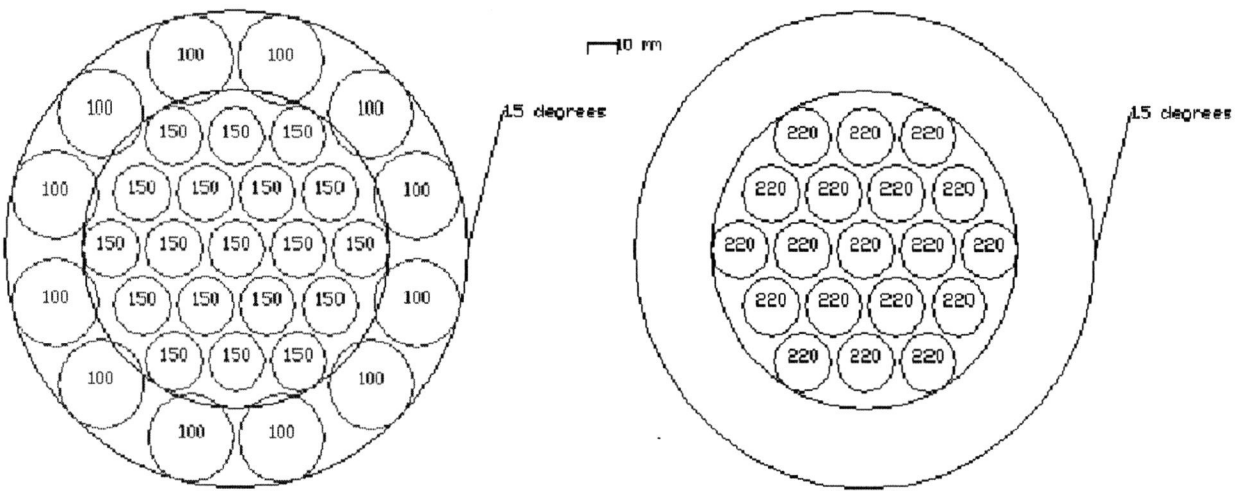

FIGURE 3. QUEST Focal Plane

standing offset in polarization due to the loss of circular symmetry in the geometry of the reflecting surfaces.

We avoid problems of blockage and reflection from mirror supports by suspending the secondary above the primary on a cone formed of expanded polystyrene foam. This absorbs very little mm–wave radiation (attenuation less than 1 per cent), so the degradation of system noise temperature is not significant. A similar arrangement is in use on COMPASS [6] at Ka band, for which the contribution of a few Kelvin has been measured.

Wave Plate

Immediately ahead of the focal plane is a half wave plate. This rotates incoming linear–polarized radiation. For each rotation of the wave plate the plane of polarization is rotated twice; so the signals at the polarization–sensitive detector outputs are modulated at four times the rotation rate. This modulation will be used to permit instrumental offsets to be distinguished from sky signals.

Currently we are studying the feasibility of building a cooled wave plate mechanism. In case the difficulties inherent in such a device prove insurmountable, we retain the option with our cryostat design to have the wave plate at ambient temperature. Also the re–imaging lenses may or may not be cooled.

Mechanical Structures

The mirror structure is mounted on a typical elevation above azimuth mount; but in addition, we add a third rotation about the 'Z' or 'field' axis of the antenna — see Figure 1. This will introduce another means of modulating the polarization of signals falling on the detectors, to better distinguish instrumental from sky signals. Also, it allows greater flexibility in our scanning strategy.

Wavelengths

QUEST will be a ground–based telescope. Therefore the wavelengths used for observation must be chosen to fall within the windows of low atmospheric absorption. These are at 100 GHz and 150 GHz, with an option for future upgrade to add detectors for the 220–GHz window. Band–pass filters will define a fractional band–width of approximately 20 per cent.

Detectors

Rather than using polarizers and separate detectors for each polarization, QUEST will employ *polarization sensitive bolometers* (PSBs) — see Figure 2. These comprise two thermometers in a single module, each with its own absorbing mesh, perpendicular to the other. The two outputs correspond to orthogonal linear polarizations. Since both detectors share a single module, coupled to the telescope optics through the same feed horn, they look at the same sky pixel through a common column of atmosphere. Therefore after differencing the two outputs to measure the level of polarization, there is less sensitivity to sky conditions than with a beam–switching temperature anisotropy experiment.

The arrangement of detectors in the focal plane is shown in Figure 3. Our primary science band is 150 GHz, so the nineteen detectors for this frequency are grouped in the central region where the spurious instrumental polarization signals are lowest. The twelve 100–GHz detectors are placed in a ring around the outside edge of the focal plane. The field of view is 1°.5. A second focal plane consisting of a further nineteen PSBs at 220 GHz may be added as a future upgrade.

SURVEY DESIGN AND EXPECTED SCIENCE RESULTS

We have optimised the design of the QUEST survey to (i) detect EE and ET modes, and (ii) measure in detail the polarization power spectra C_ℓ^{EE} and C_ℓ^{ET}. The first of these aims, the detection of polarization, suggests the strategy of integrating over a small patch of sky, a few degrees, until polarization is detected. This will be easiest in the ET cross-polarization, combining the QUEST data with the MAP temperature field, but a similar strategy will provide a first early detection of EE modes. To satisfy the second requirement, the survey area is increased to a few hundred square degrees. Figures 4 and 5 show the expected signal and noise properties of the EE, ET and BB spectra. The B-modes are not expected to be detected from this type of survey, although interesting limits may be set. In addition to a detection and measurement of the E-mode polarization power spectra, the signal/noise ratio per pixel will allow us to make maps of the polarized sky, which will help point source extraction.

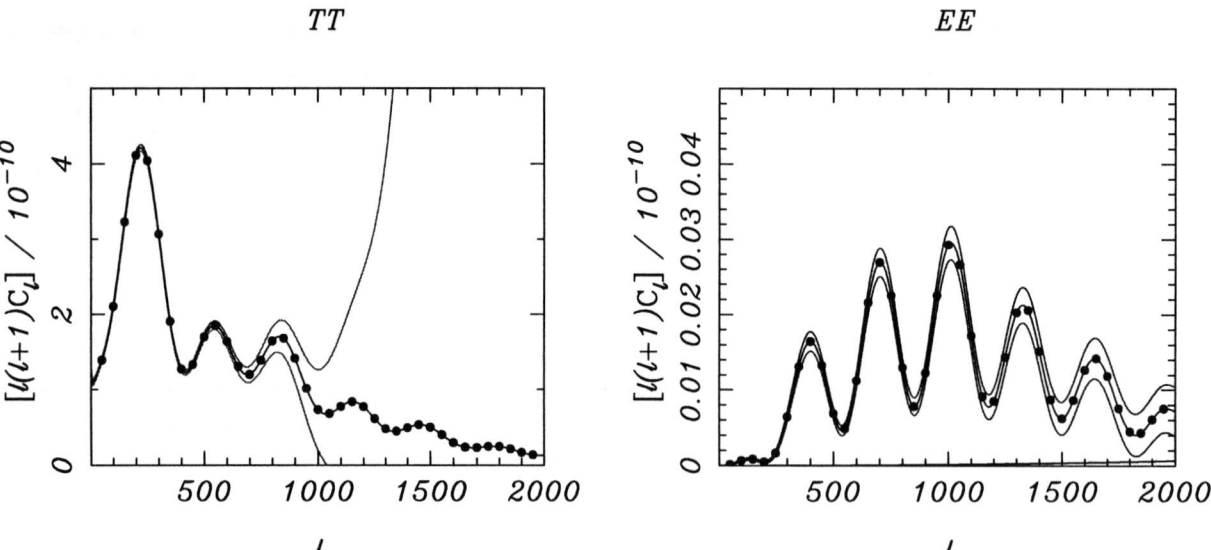

FIGURE 4. *LHS: TT power spectrum expected from MAP with expected measurement errors. RHS: Expected EE polarization power spectra and uncertainties from QUEST, for a survey with 5' pixels, covering an area of 300 sq. deg. and sensitivity of* $\sigma = 3.2\mu K$ *per pixel.*

Measurement of the EE polarization power provides new information on the epoch of reionization, and helps break the degeneracy between the magnitude of the mass power spectrum and the optical depth to ionization. In addition the peak oscillations of the polarization spectra provide additional information on the baryon abundance and Hubble parameter. While these are the major improvements on measurements of the temperature field, they help break a

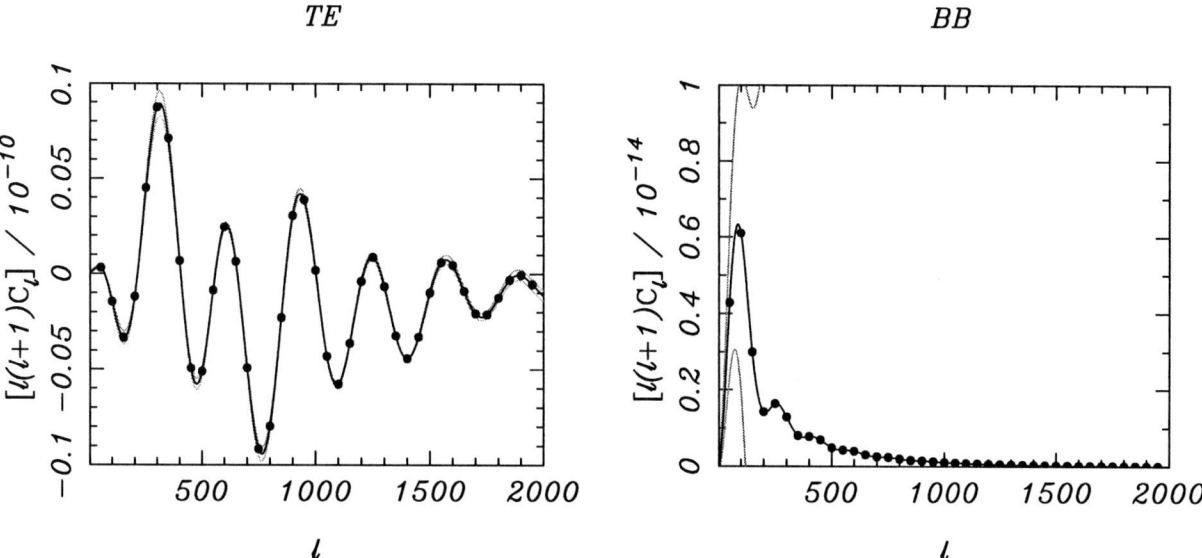

FIGURE 5. *LHS: ET power spectrum expected from cross correlating with MAP. RHS: BB power spectrum, for a survey with 1' pixels, covering an area of 300 sq. deg. This would suggest that B-type polarization would be only marginally detectable, but interesting limits may be set.*

number of major degeneracies in the parameter estimation, substantially reducing the uncertainties on parameter estimation as a whole. For example we expect a factor 5 improvement in the estimation of the combination $\Omega_B h^2$.

As well as parameter estimation, measurement of the polarization signature in the CMB will also allow us to test the paradigm of adiabatic perturbations, and probe the epoch of reionization in detail. The latter should allow us to investigate such questions as when reionization occurred and whether the reionization was homogeneous or inhomogeneous.

REFERENCES

1. Bennett, C. L., et al., The Microwave Anisotropy Probe (MAP) Mission, in American Astronomical Society Meeting, **191**, 8701 (1997)
2. Halverson, N. W., et al., Submitted to Ap. J., `astro-ph/0104489` (2001).
3. Hancock, S., Davies, R. D., Lasenby, A. N., Gutiérrez, C. M., Watson, R. A., Rebolo, R., Beckman, J. E. , Nature, **367**, 333 (1994).
4. Lee, A. T., et al., Ap. J., **561**, L1 (2001).
5. Netterfield, C. B., et al., Submitted to Ap. J., `astro-ph/0104460` (2001).
6. Piccirillo, L., et al., in Proceedings of 2K1BC Workshop "Experimental Cosmology @ mm-waves" Breuil–Cervinia, in press (2001).
7. Smoot, G. F., et al., Ap. J., **396**, L1 (1992).
8. Tauber, J.A., The Planck Mission, in IAU Symposium, **204**, Martin Harwit and Michael G. Hauser, eds. (2000).

The Milano Polarization Experiment Devoted to the Study of the Cosmic Microwave Background

M. Gervasi*, E.S. Battistelli*†, G. Boella*, F. Cavaliere**, A. Passerini*, G. Sironi*
and M. Zannoni*

*Dipartimento di Fisica - Università di Milano Bicocca - Piazza della Scienza 3, 20126 Milano, Italy
†Dipartimento di Fisica - Università "La Sapienza" - Piazzale A. Moro 2, 00185 Roma, Italy
**Dipartimento di Fisica - Università di Milano - Via Celoria 16, 20133 Milano, Italy

Abstract. In the last years several experiments have been devoted to the search for CMB polarization, but no positive detection has been obtained so far. In fact the expected degree of polarization is very low and its detection is extremely difficult. Very special design and system stability are necessary to reach sensitivities of one part in 10^6 or greater. Here we present the Milano polarimeter, a 33 GHz correlation system which can be used for detecting both linear and circular polarization. It has been designed to operate at large angular scale ($7° - 14°$) from ground. In this contribution we describe the experience acquired so far during observations and the solutions adopted to improve the sensitivity.

INTRODUCTION

The study of the polarization properties of the Cosmic Microwave Background (CMB) is very important in order to have a complete picture of the Universe. The positive detection of the temperature fluctuations in the last years [1, 2, 3] is giving us a large amount of information on many parameters governing the Universe. Nevertheless, several cosmological parameters can not be unambiguously determined by using only anisotropy measurements. In fact there is a degeneracy which can be removed only through the measurement of the CMB polarization power spectrum.

The polarization of the CMB has been originated by the anisotropic scattering of photons with matter during the recombination era [4]. Therefore a linear polarization (ΔT_p) correlated with the spatial fluctuations of the CMB temperature (ΔT_a) is expected. In particular the maximum correlated signal should be found to be $\alpha = \Delta T_p / \Delta T_a \leq 0.1$ at the horizon scale at recombination ($\Delta \theta \sim 1°$). α should rapidly decrease at larger angular scale [5] if no reionization occurred in the primordial Universe.

Among the information we can get from CMB polarization measurements there is the reionization optical depth, and therefore the reionization epoch and the involved fraction of matter. This information is carried by large angular scales ($\theta > 1°$) and can tell us the history of the Universe in the post-recombination era. A measurement of the polarization power spectrum, removing some degeneracies, will help to distinguish among scalar, vector (vortices) and tensor modes of matter density perturbations. Finally, deviations from a uniform and isotropic expansion of the Universe as well as the existence of rotational modes associated to primordial magnetic fields have CMB polarization as a signature. In this case also circular polarization can be detected [6].

No experimental evidence of CMB polarization has been obtained so far. In fact, only upper limits have been set, mainly for linear polarization. The most recent measurements of CMB linear polarization are shown in Table 1. The best upper limits (few parts in 10^6) are still higher than the expected polarization level, but the sensitivity of ongoing experiments is now approaching that required for a detection.

CP609, *Astrophysical Polarized Backgrounds*, edited by S. Cecchini et al.
© 2002 American Institute of Physics 0-7354-0055-5/02/$19.00

TABLE 1. Recent CMB linear polarization upper limits.

Resolution (deg)	Wavelength (mm)	Sky Coverage	Upper Limit (μK)	Reference
14	9.1	*SCP*	226	[7]
7	9.1	*SCP*	267	[7]
7	8 − 12	$\delta = +43$	10	[8]
2.9	10 − 7.5	$\delta = +63$	37	[9]
2.1	10 − 7.5	$\delta = +63$	54	[9]
1.6	10 − 7.5	$\delta = +63$	28	[9]
1.2	10 − 7.5	$\delta = +63$	41	[9]
0.9	10 − 7.5	$\delta = +63$	79	[9]
1.2	8 − 12	*NCP*	30	[10]
0.24	3.3	$\delta = +89$	13	[11]
0.038	340	$\delta \simeq -50$	16	[12]

THE MILANO POLARIMETER

The instrument design

We have developed an instrument able to measure both the linear and the circular polarization components of the incoming radiation. The polarimeter is a correlation receiver designed for ground based observations. The radiometer is sensitive to large angular scales, therefore it is able to investigate a possible reionization occurred in the post-recombination universe [13]. The main characteristics of the polarimeter are summarized in Table 2. For a complete description of the instrument see [14].

TABLE 2. Main characteristics of the Milano polarimeter.

Detection technique	Etherodyne Correlation receiver
Observing frequency	$\nu_{rf} = 33$ GHz
Intermediate frequency	$\nu_{if} = 3$ GHz
Frequency band	$\Delta\nu = 1.5$ GHz
Beamwidth FWHM	$\Theta_{ant} = 7°\,\&\,14°$
Polarization mode	Linear & Circular
Outputs	U, Q, V Stokes parameters

- The antenna beam pattern is defined by a corrugated conical horn aimed directly at the sky. The angular resolution can be switched from 14° to 7° by adding an extension to the horn.

- The polarization mode can be switched from circular to linear by adding an iris polarizer (Pol) before the separation of the two components by means of an orthomode transducer (OMT).

- The two components are therefore processed in two separated amplification chains. All the front-end components are cooled down to ~ 20 K by a mechanical cryogenerator. The front-end amplifiers are low noise cryogenic HEMTs.

- The signals are then phase-modulated, before being downconverted from ν_{rf} to ν_{if}, using the same low phase noise oscillator (LO) for the two channels.

- The two signals pass through a correlation unit made by a phase discriminator (PhD), and are finally demodulated and integrated by a Phase Sensitive Detector (PSD).

The outputs of the system give directly the Stokes parameters. Using the iris polarizer we get the parameters U and Q describing the linear polarization:

$$O_1 = K_1 < E_A E_B > \cos(\delta + \phi) = K_1' \left(Q\cos(\phi) - U\sin(\phi) \right) \tag{1}$$

$$O_2 = K_2 < E_A E_B > \sin(\delta + \phi) = K_2' \left(Q\sin(\phi) + U\cos(\phi) \right) \tag{2}$$

While in the alternative set-up, i.e. removing the iris polarizer, we get the parameters U and V describing the circular polarization:

$$O'_1 = K_1 < E'_A E'_B > \cos(\delta + \phi) = K'_1 \left(U \cos(\phi) - V \sin(\phi) \right) \qquad (3)$$

$$O'_2 = K_2 < E'_A E'_B > \sin(\delta + \phi) = K'_2 \left(U \sin(\phi) + V \cos(\phi) \right) \qquad (4)$$

Here δ is the polarization angle, while ϕ is the phase difference between the two channels introduced by the instrument. Adjusting the instrumental phase by a phase shifter (PhS) we can separate Q and U (or U and V) in the two outputs.

The Milano polarimeter has been used for sky observations in Antarctica. In 1994-95 we used the prototype (**Mk1**) from Terra Nova Bay, and got the upper limits shown in Table 1 [7]. In 1998-99 we repeated the observations from Dome Concordia, placed at 3233 m a.s.l., using an updated version of the instrument (**Mk2**), which had the front-end cooled down to ~ 20 K. The instrument has now been improved (**Mk3**) and is ready for a new observing campaign [15].

MK3 model performances

- We perform a square wave phase modulation $[0, \pi]$ with a synchronous detection (PSD) of the observed signal, at a frequency of $\nu_m = 781$ Hz. This technique helps in minimizing both the $1/f$ noise and the instrumental off-set. The modulation frequency ν_m is chosen far enough from both the low frequency part of the spectrum, dominated by the $1/f$ noise, and the network frequencies ($\nu_{net} = 50$ Hz and its harmonics). A measurement of the power spectrum of the output signal is shown in Figure 1.

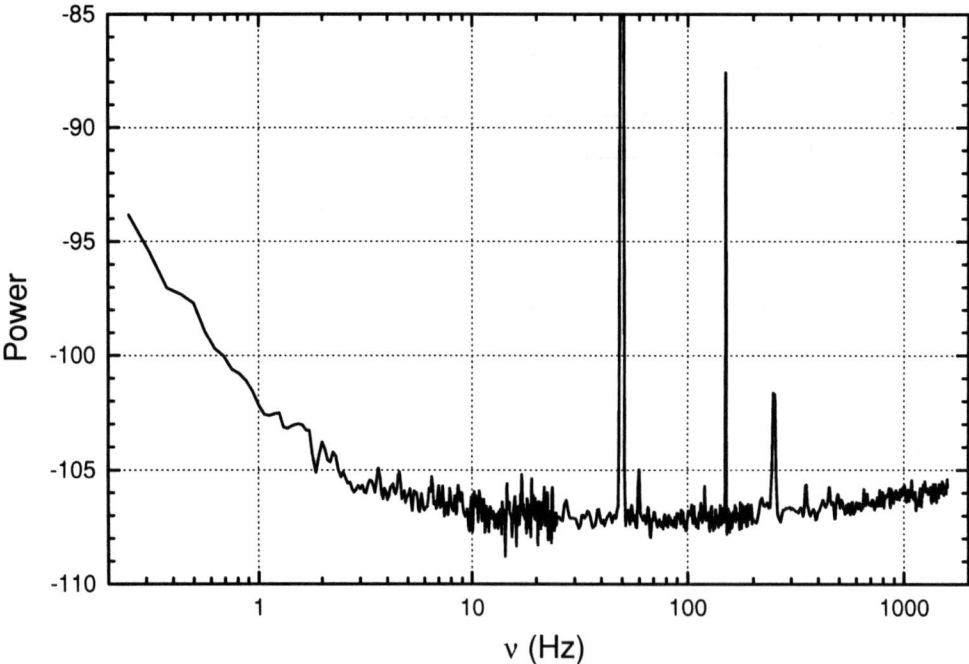

FIGURE 1. Measured power spectrum of the output signal. The units of the power are dB. The $1/f$ noise is effective up to few Hz.

- We improved the long term stability of the instrument by taking care of the thermal control of both the sensitive components of the instrument and the environment where the polarimeter is operated. In fact we use an insulated tent surrounding the instrument. An active loop of temperature control completes the system. This system is also effective for protecting the external electronics and the mechanic cryocooler in case of large external temperature variations: During preliminary tests we had temperature fluctuations of the HEMT amplifiers $\Delta T_{HEMT} \leq 0.5$ K over several days of operation.
- New low noise InP preamplifiers have been implemented.

The current version of the polarimeter is more stable and effective in cancelling out a large part of the instrumental off-set. It also allows us to obtain a gain large enough to approach the needed sensitivity. With the new preamplifiers the system temperature is $T_{sys} \sim 40 - 50$ K and the noise equivalent temperature $\Delta T_{min} \sim 1 \text{mK}/\sqrt{\text{Hz}}$.

During the 1998-99 Antarctic observations we tested the calibration technique. We used a stable noise generator as internal reference mark in order to measure and monitor the system gain stability. An external grid calibrator gave us information on the response of the system to an external polarized signal. A total power monitor helped the complete characterization of the signals. Combining these information we got a complete calibration of the system [16, 17].

OBSERVATION PROGRAMS

The instrument has been designed as a complete observatory that can be moved and operated in sites with different characteristics. The polarimeter is operated as a semi-automatic machine, i.e. drift observations are completely automatic, while manual intervention is needed for calibration with external sources and for maintenance. These features have been selected in order to have an experiment qualified to be operated in Antarctica during the winter season, and able to be easily installed in different environmental situations. In fact the observation during the Antarctic night is only the final destination of the experiment, after a series of deep tests of its sensitivity and, possibly, after independent observations of the sky polarization from different sites.

- The polarimeter will start observations from the Testa Grigia site [18] on the Italian Alps, at 3480 m a.s.l.. We will address our investigation at large angular scales ($7° - 14°$) first. Observations will be carried out in transit mode. Then we will place the receiver at the focal plane of the MITO telescope in order to observe with an angular resolution of 10-20 arcmin ([18], [19]).
- Later on the polarimeter will be operated from the Antarctic Plateau in order to take advantage of the long Antarctic night. In fact, long duration observations in quasi static weather conditions are possible, avoiding solar radiation disturbances. Natural targets are the South Celestial Pole (SCP) regions $7°$ and $14°$ wide.

Dome Concordia will become a permanent station in the next years and could be an ideal place to operate the polarimeter. It is placed at an altitude of 3233 m a.s.l. and latitude $75°$ South. Very stable weather conditions are common in winter, so that an observing efficiency approaching 100% can be obtained. When observing at the zenith, a circle around the pole with radius of $15°$ is described. Besides, when observing the SCP, in case of linear polarization a typical signature is expected:

$$O_1 \propto \sin(4\pi\frac{t}{T} + \phi_\circ) \tag{5}$$

$$O_2 \propto \cos(4\pi\frac{t}{T} + \phi_\circ) \tag{6}$$

Here t is the sidereal time, and T a sidereal day, while $\phi_\circ$ is an arbitrary phase angle.

ACKNOWLEDGMENTS

This research was carried out within the Concordia Project (supported by IFRTP and PNRA), and has been supported by PNRA and CSNA, by CNR, by MURST and Universities of Milano and Milano-Bicocca.

REFERENCES

1. de Bernardis, P., et al., *Nature*, 404, 955, 2000.
2. de Bernardis, P., et al., *this volume*, 2001.
3. Hanany, S., et al., *ApJ*, 545, L5, 2000.
4. Rees, M.J., *ApJ*, 153, L1, 1968.
5. Hu, W., and White, M., *New Astronomy*, 2, 323, 1997.
6. Kosowsky, A., *New Astronomy Rev.*, 43, 157, 1999.
7. Sironi, G., et al., *ASP Conf. Ser.*, 141, 116, 1998.
8. Keating, B.G., et al., submitted to *ApJL*, 2001 (*astro-ph/0107013*).
9. Torbet, E., et al., *ApJL*, 521, L79, 1999.
10. Wollak, E.J., et al., *ApJL*, 419, L49, 1993.
11. Hedman, M.M., et al., *ApJL*, 548, L111, 2001.
12. Subrahmanian, R., et al., *MNRAS*, 315, 808, 2000.
13. Gervasi, M., et al., *AIP Conf. Proc.*, 476, 154, 1999.
14. Sironi, G., et al., *New Astronomy*, 3, 1, 1998.
15. Zannoni, M., et al., proceedings of the *2K1BC Workshop:Experimental Cosmology at millimetre wavelengths*, Breuil-Cervinia, July 9-13, 2001, *AIP Conf. Proc.*, in press, 2002.
16. Zannoni, M., *PhD Thesys*, University of Milano, 1999.
17. Gervasi, M., et al., *SIF Conf. Proc.*, 68, 165, 2000.
18. De Petris, M., et al., *New Astronomy*, 1, 121, 1996.
19. De Petris, M., et al., *Applied Optics*, 28, 1785, 1989.

The Polarimetric Observation Facility at the Medicina 32 m Parabolic Antenna

Alessandro Orfei, Mauro Roma

Istituto di Radioastronomia - CNR, Via Gobetti 101 40129 Bologna (Italy)

Abstract. We describe the wideband analog polarimeter made at the Medicina radio observatory usable with the receivers of our 32m parabolic antenna. The principle under which the electronics was designed as well as the circuit realization will be showed, togheter with the description of the software made to manage the observations and the off-line processing of the data acquired.

INTRODUCTION

The design and realization of a polarimetric facility for the Medicina radiotelescope was proposed years ago in collaboration with the Istituto TeSRE-CNR in order to make available a capability to measure, at as many frequencies as possible, the linear polarization of the galactic background in terms of Stokes parameters. The facility was inteded to also provide the amount of instrumental polarisation due to receiving system.

At the start of the project two possibilities were considered. One was to use commercially available products that process the signal at RF level and make available a correlation of the two channels coming from our receivers. This would need one device per receiver and, above all, the rearrangement of each receiver.

The other possibility was to build an IF polarimeter, i.e. a device that works in the down converted band 100-500MHz, so that it can process the signals coming from all our receivers.

The latter was chosen and in the last few years we have been working out the analog electronics that processes the left and right polarisation coming from the receivers placed on the parabolic antenna and gives four outputs, the LHCP (Left Hand Circular Polarisation) and RHCP (Right Hand Circular Polarisation) total powers (I Stokes parameter) and the Q,U Stokes parameters. The V parameter can also be derived by off line data processing .

THE HARDWARE OF THE POLARIMETER

The following relationship among Stokes parameters and electric fields of the RHCP and LHCP receiver outputs hold

$$I = <E^2_r(t)> + <E^2_l(t)> = I_p + I_n \tag{1}$$

$$Q = <E_r(t)*E_l(t)> = I_l\cos2a \tag{2}$$

$$U = <E_r(t)*E'_l(t)> = I_l\text{sen}2a \tag{3}$$

$$V = <E^2_r(t)> - <E^2_l(t)> = I_p\text{sen}2\beta \tag{4}$$

The symbol $<>$ indicates a time integration and:

- E_r, E_l are the electric fields of RHCP and LHCP, E'_l is the electric field of the LHCP 90° out of phase.
- I_p, I_n are the polarized and non polarized power.
- I_l is the linearly polarized power and Q,U the relative Stokes parameters, a is the polarisation angle.
- $I_p\text{sen}2\beta$ (or V) is the circularly polarized power.

CP609, *Astrophysical Polarized Backgrounds,* edited by S. Cecchini et al.

The relationships (1) to (4) say the electronics tasks are:
a) realizing two total powers, one for the LHCP input the other one for the RHCP
b) realizing two multipliers, one having RHCP and LHCP inputs, the other one having RHCP and LHCP 90° out of phase. After each multiplier a low pass filter is needed to realize the operation of symbol < >. In fig. 1 the block diagram of the polarimeter is showed. For more details see [1].

All cables have the same length to minimize the degradation of the polarimeter axial ratio. For the same reason a second Hybrid 90 has been inserted, in principle not strictly necessary. A multiplexer is necessary because the A/D converter board have one input only. So doing the four outputs are not contemporary acquired: this is done at time spacing of the order of a second. A modem for RS232 link is necessary because of the long distance between the polarimeter (located in the vertex room of the antenna) and the computer placed in the control room: locating the polarimeter as near the receiver as possible avoids detection of phase fluctuation due to long coaxial cables.

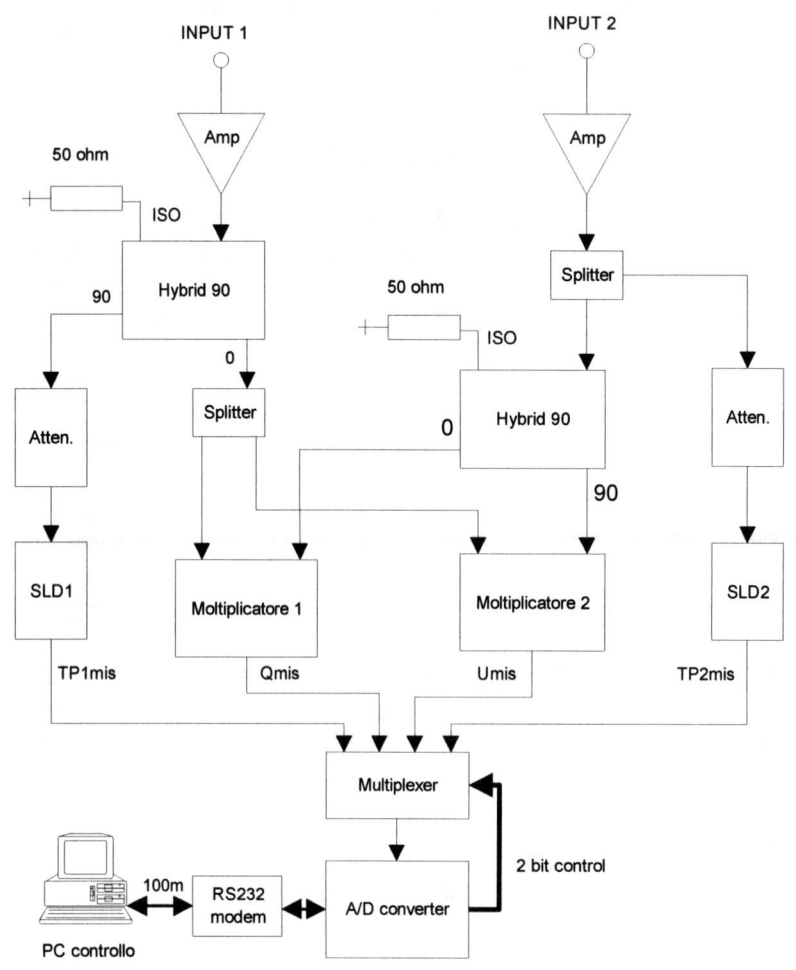

FIGURE 1. Block diagram of the polarimeter

FUNDAMENTALS FOR THE POLARIMETRIC OBSERVATION

The outputs of the hardware are not directly the Stokes parameters, because the data are not calibrated and because of the instrumental polarisation due to the whole receiving system. The effects are that the polarized part of the incident wave is partially depolarized and, inversely, the unpolarized part is partially polarized: the observation strategy, together with the mathematical model, gives the possibility to calibrate I,Q,U,V and to cancel out the contamination.

There are various references to polarimetric systems, for example [2,3,4] using interferometers, or [5] for single dish measurements but using two horns technique.

The mathematical model of our system is treated in [1]. The raw data acquired by the polarimeter, TP1mis, TP2mis, Qmis, Umis (fig. 1), are time labelled and acquired in an observation sequence that provides onsource data, offsource data, and data after injecting a noise calibration signal. The off line software makes true the model showed in equations (5) to (8)

$$I' = I + P\Sigma\cos(2\chi + 2a + \sigma) \tag{5}$$

$$V' = V - P\Delta\cos(2\chi + 2a + \delta) \tag{6}$$

$$Q' = P\cos(2\chi + 2a + \gamma_q + \psi_R - \psi_L) + I\Sigma\cos(\sigma + \gamma_q + \psi_R - \psi_L) + V\Delta\cos(\delta + \gamma_q + \psi_R - \psi_L) \tag{7}$$

$$U' = P\sin(2\chi + 2a + \gamma_u - \psi_R + \psi_L) - I\Sigma\sin(\sigma - \gamma_u + \psi_R - \psi_L) - V\Delta\sin(\delta - \gamma_u + \psi_R - \psi_L) \tag{8}$$

$\mathbf{P} \equiv Q + jU \equiv Pe^{j\theta}$ is the source polarisation vector (P magnitude, θ polarisation angle). The Medicina parabolic antenna is an alt-az mount antenna so when the source is tracked the receiving system sees it at variable angles, thus θ is a sum of a constant term (twice the source polarisation angle) and twice the parallactic angle χ. $\psi_R - \psi_L$ is the phase difference between left and right channels in the receiver, γ_q and γ_u are the phases due to the electric path inside the polarimeter, the Σ,Δ vectors are a combination of the previous quantities, σ and δ their angles, and they represent the contamination. Some considerations on equations (5) to (8):

- The ideal case should be that the polarimeter gives us $Q'=P\cos(2\chi + 2a)$, $U'=P\sin(2\chi + 2a)$, $I'=I$, $V'=V$
- Receiving a polarized signal gives I',V' different from the true I,V Stokes parameters due to instrumental polarisation ($\Sigma,\Delta \neq 0$), depolarising part of the incoming power. Further, I,V pollutes also the true Q,U parameters giving Q',U'; again this happens for the presence of the instrumental polarisation.
- Observing non polarized signal should give true I,V and polluted Q,U.

The observing polarimeter software [6,7] consists of two parts. First of all the user is enabled to setup the observation by time scheduling the source(s) and interfacing with the control system of the antenna. Once the observation is started the second part of the software monitors the acquisition. The user interfaces are showed in fig. 2 and 3

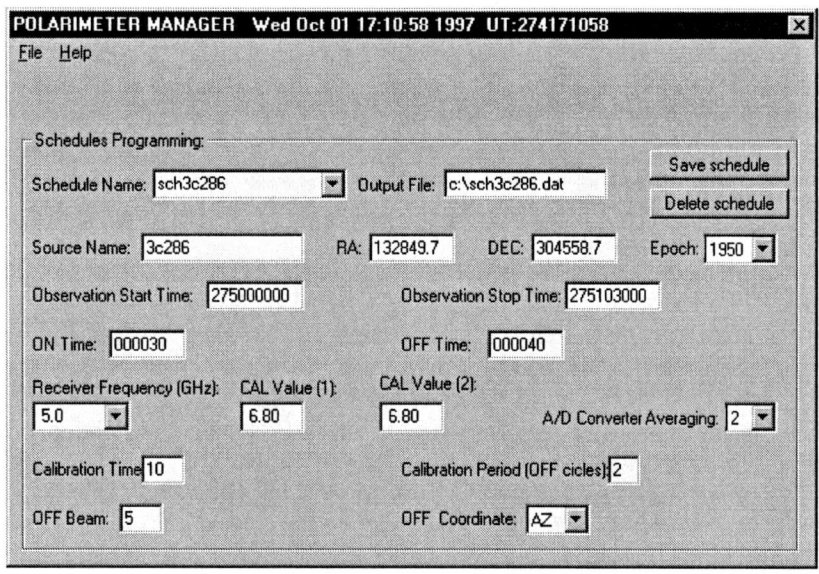

FIGURE 2. User interface to setup the observation

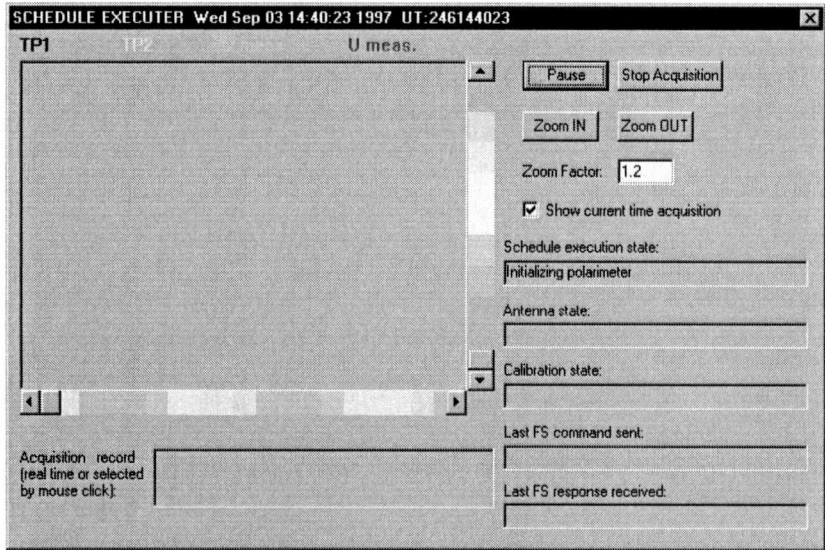

FIGURE 3. User interface during observation

In the setup menu, receiving frequency, sources, duration of on-off and calibration cycles, number of beams to go off together with which coordinate, are selectable. In the observing menu, flowing of data acquired, both numerically and graphically, as well as the antenna status are shown, together with the possibility to stop or to pause the acquisition .

Raw data Tp1mis, Tp2mis, Qmis, Umis are transformed in I', Q', U', V' by the post processing software and then a sinusoidal fitting versus parallactic angle is performed. Results of this fitting are the offsets, the amplitude and the phase of the sinusoids. Moreover, the software can show the polarization circle showing Q' versus U'. In any case Q, U, P, a, I, V and instrumental quantities can be derived.

TEST MEASUREMENTS WITH THE MEDICINA POLARIMETRIC FACILITY

First tests were performed at 5GHz on known polarized (3C286, 3C138) and non polarized sources (3C84). Complete results are in [1] and [6] and we could say here that the measurements fit the model. In the following figures 4 to 6 reduced data for 3C286 and 3C84 are reported. As foreseen, the polarized source observation gives a sinusoidal trend and a polarisation circle with a center offset (effect of the instrumental polarisation). The non polarized source observation, instead, gives only a cloud of points around a centre different from (0,0). Recall that the amount of offset is dependent on the total power of the observed source. You can disregard the outlier points present in each of the four picture of fig. 4.

Fig. 5 also allows to calibrate unknown sources because at 5GHz 3C286 has a flux of 7.48Jy and a polarisation percentage of 11.5%. This means that 1100 counts (the radius of the circle) correspond to 0.823 Jy.

During these years, tests have been repeated many times on different atmospheric conditions showing, as predictable, a very low dependance of the polarised outputs, on the weather conditions, compared to the total power outputs. At 8 GHz a factor 16 in the rms noise of the total power outputs was detected when passing from a rainy day to a clear sky day, while only a factor of 3 was observed in the Q, U outputs. The polarimetric facility was used for source as well as planet observation. A development of the software is in progress to make available sky measurement facilities by making a raster scan of the antenna. A rough estimation of the polarized output sensitivity at 8 GHz is $10mK/s^{1/2}$ by using a bandwidth of 80 MHz. This narrower band was used to avoid a known interference presents in the lower part of the IF band.

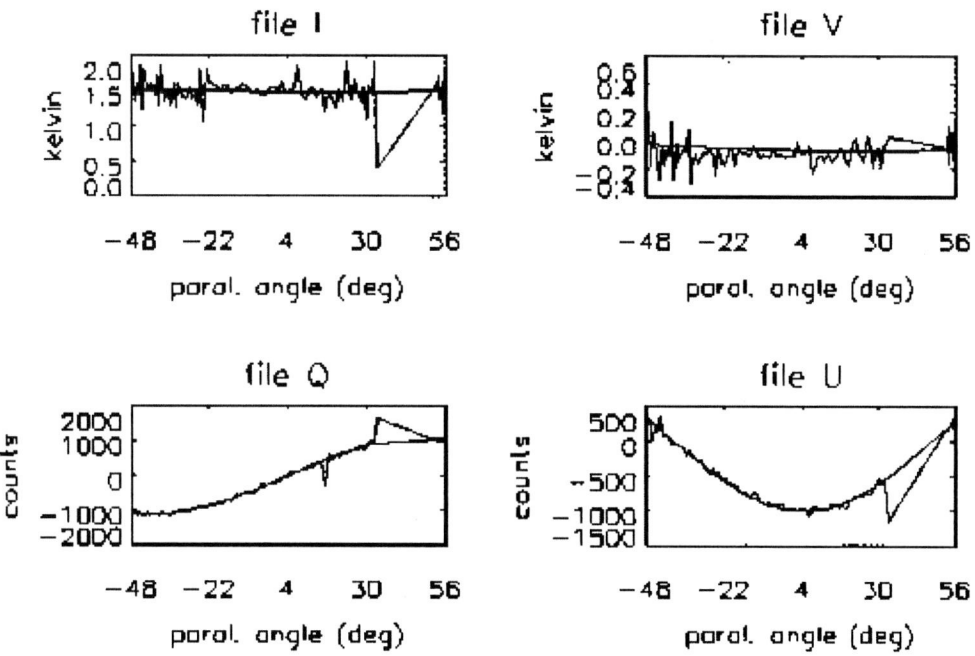

FIGURE 4. Stokes parameters for 3C286

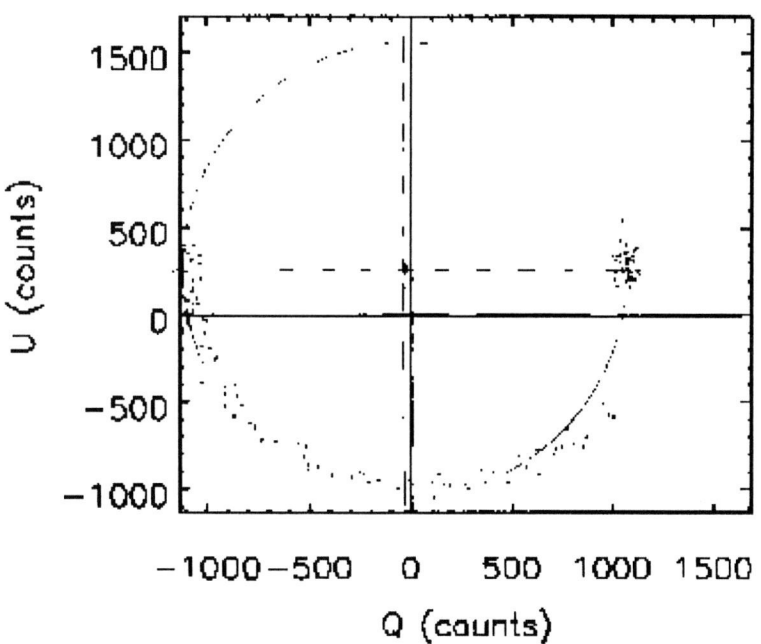

FIGURE 5. Polarisation circle for 3C286

173

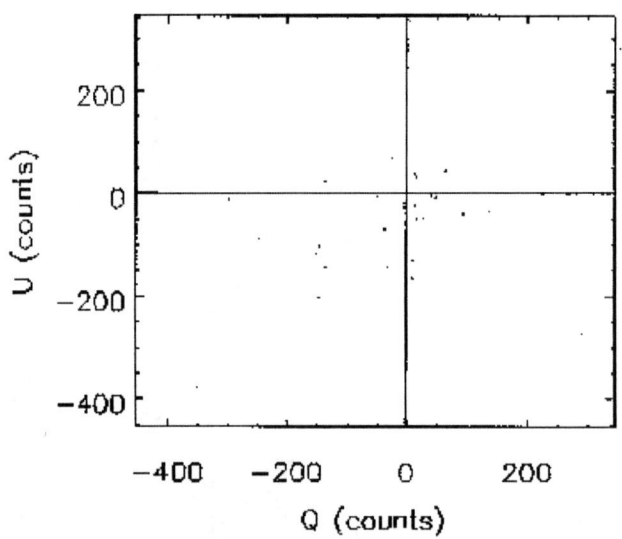

FIGURE 6. Polarisation circle for 3c84

REFERENCES

1. A. Orfei, M. Roma Rapporto Interno IRA 258/98
2. D.H Roberts, J.F.C. Wardle, L.F. Brown *Astr. Journal* **427**, 718-744, (1994)
3. R.G. Konway, P.P. Kronberg, *Mon. Not. R. Astr. Soc.* **142**, 11-32 (1969)
4. M.M. McKinnon, *Astr. & Astroph.*, **260**, 533-542, (1992)
5. Z. Turlo, T. Forkert, W Sieber, W. Wilson, *Astr. & Astroph.* , **142**, 181-188, (1985)
6. T. Dal Pozzo, Tesi di laurea, (1997)
7. A. Orfei, T. Dal Pozzo, Rapporto Interno IRA 238/97

DEVICES, ANALYSIS TOOLS AND SYSTEMATICS

Millimeter Wave Passive Components for Polarization Measurements

O. A. Peverini*, M. Baralis*, R. Tascone*, D. Trinchero*, A. Olivieri*, E. Carretti†
and S. Cortiglioni†

*IRITI-CNR, Politecnico di Torino, Torino , Italy
†ITESRE-CNR,Bologna, Italy

Abstract. The Stokes parameters of the polarized sky emission are detected by a correlation unit called Hybrid Phase Discriminator (HPD), which uses signals obtained by an Ortho-mode Transducer (OMT). In the millimeter wave range and for rather large bandwidths, heterodyne receivers are not applicable, and the correlation units have to work at the frequency of the radiometer. This contribution deals with a Ka-band prototype of a new configuration of waveguide HPD, which presents a high degree of sensitivity for the detection of linearly polarized radiation.

INTRODUCTION

The construction of multifrequency polarization maps of the galactic emission and the Cosmic Microwave Background investigation require the availability of radiometers with a high level of rejection for the non polarized components [1]. In the Astrophysical experiment SPOrt (Sky Polarization Observatory) onboard the International Space Station (ISS) [2], the instrumentation will consist of four correlation radiometers. In the past, the correlation process was performed after a down conversion by using a local oscillator [3]. Apart from stability problems, this is possible only for narrow band investigations. In the SPOrt project the bandwidths are of 10% centered at 22, 32, 60 and 90 GHz. Obviously, the absolute bandwidths are so large that it is not possible to process the signals at lower frequencies, hence the correlation must be performed at the antenna frequency. The polarized emission will be detected by simultaneously measuring the Q and U Stokes parameters, corresponding to:

$$Q = |E_x|^2 - |E_y|^2 \quad \text{and} \quad U = 2\Re\{E_x E_y^*\} \tag{1}$$

These parameters will be detected by means of correlation units (HPD) whose input signals come from an antenna with double circular polarization [2]. The antenna system consists of a corrugated horn, a polarizer which converts the circular polarizations into the linear ones and an ortho-mode transducer (OMT), which separates the two polarizations. Hence the two outputs of the OMT, under ideal conditions, are:

$$A = \frac{1}{\sqrt{2}}(E_x + jE_y) \quad \text{and} \quad B = \frac{1}{\sqrt{2}}(E_x - jE_y) \tag{2}$$

The correlation product between the two signals A and B yields the Q and U Stokes parameters of the polarized emission:

$$AB^\star = \frac{1}{2}(|E_x|^2 - |E_y|^2 + 2j\Re\{E_x E_y^*\}) = \frac{1}{2}(Q + jU) \tag{3}$$

This paper deals with a new configuration of waveguide Hybrid Phase Discriminator which yields at its four output ports the sum and the difference in phase and quadrature of the two input signals. The analog operations are performed at the antenna frequency by means of 3 dB directional couplers and phase shifters. Moreover, two input waveguide filters for the definition of the operative band, and two additional directional couplers for monitoring the total power of the emission are integrated in the HPD. The configuration of this device extends over five levels using E-plane and H-plane discontinuities to form the various components. In this way, a very compact configuration has been obtained which was called FL - HPD (Five Level Hybrid Phase Discriminators, patent pending)

CP609, *Astrophysical Polarized Backgrounds*, edited by S. Cecchini et al.
© 2002 American Institute of Physics 0-7354-0055-5/02/$19.00

HYBRID PHASE DISCRIMINATORS

The scheme of the correlation units (HPD) used in the radiometers is shown in Fig. 1. The two input signals A and B, attributed to the double circular polarization, are filtered by the two 13 cavity waveguide filters (Fa and Fb), which present a high rejection in the stop band with the purpose of defining the measuring band. Subsequently a fraction of the two signals is drawn off by two 3 dB directional couplers (Hya and Hyb) and is detected by two diodes with quadratic characteristic, to monitor the signal level in the two channels. The remaining fractions (A_0 and B_0) are divided into identical parts along two branches. For this operation, two 3 dB directional couplers ($Hy1$ and $Hy2$) are used instead of two power splitters because a high level of decoupling between the two branches is required. These are the first components of the correlation block. In the upper branch the signals A_1 and B_1 are combined by the 3 dB coupler $Hy3$ to produce the output signals C_3 and C_4, proportional to $A + jB$ and $A - jB$, respectively. In the lower branch, before the combination performed by the 3 dB coupler $Hy4$, the signal B_2 undergoes a 90°phase shift with respect to the signal $A2$. In this way, the output signals C_1 and C_2 are proportional to $A - B$ and $A + B$, respectively. The detection of the signals C_k, with $k = 1, \cdots, 4$, by quadratic characteristic diodes and the subsequent difference by means of two differential amplifiers yields the real and imaginary parts of the average of the correlation product $< AB^\star >$. The real and imaginary parts of $< AB^\star >$ correspond to the Q and U Stokes parameters of a linearly polarized radiation, respectively.

$$Q = \tfrac{1}{2} < |A + B|^2 - |A - B|^2 >$$

$$U = \tfrac{1}{2} < |A + jB|^2 - |A - jB|^2 >$$

(4)

In (4) the quantities $|A|^2$ and $|B|^2$ are eliminated by cancellation. Moreover, the level of these two quantities is practically defined by the non polarized component, which can be 40-60 dB higher than the polarized one. Hence, the device must present a very high rejection for the auto-correlation terms. This can be obtained by imposing very severe specification to the various components, even if the admitted error for the detection of the Stokes parameters can be within 10%. The problem was attacked from two sides: first, the development of specialized synthesis and analysis techniques able to guarantee a high level of accuracy in such a way that experimental tuning is not necessary. Second, the study of a compact configuration particularly robust with respect to the mechanical uncertainties.

CONFIGURATION OF THE FL-HPD

The device was designed in rectangular waveguide. The dimensions of the internal waveguides are chosen in order to minimize the dispersion effects of the directional couplers within the corresponding operative bands. By means of appropriate waveguide transitions, integrated in the device, the external connections use standard rectangular

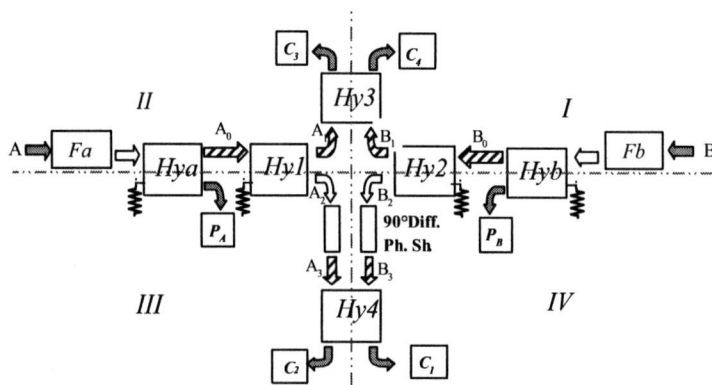

FIGURE 1. Scheme of the HPD where the two input filters and the two directional couplers for monitoring the total power are integrated.

waveguides: WR42 for K-band (22GHz), WR28 for Ka-band (32 GHz), WR15 for Q-band (60 GHz) and WR10 for W-band (90 GHz).

The direct implementation of the circuit scheme of Fig. 1 involves a cross-like geometrical configuration, which requires volume for the K-band realization, and with accuracy problems especially for the higher bands. It is important to define a geometrical configuration whose manufacturing process can be done in a unique phase, by avoiding movements of the sample during the manufacturing. In this way, positioning errors are eliminated. Moreover, within the limits of the mechanical uncertainties, it is important to guarantee a high level of symmetry in order to obtain a high rejection for the non polarized radiation. The selected configuration, called FL-HPD (Five Level Hybrid Phase Discriminator) is a *sandwich* structure, closed by two covers where the input and output standard flanges are placed [6]. The waveguides are obtained within two plates of constant thickness separated by a 0.1 mm thick layer placed in the middle of the device (the thickness of this layer is the same for all the bands). The directional couplers are obtained by coupling two parallel waveguide lengths by means of H-plane rectangular apertures realized on the central layer. The phase shifter consists of a cascade of H-plane stubs obtained by rectangular holes on the adjacent waveguide plate. In order to maintain a high level of integration, the two 13 cavity waveguide filters are obtained by a cascade of E-plane discontinuities. All components form a snake-like geometry which expands over five levels in the E-plane. All the mechanical parts present only through holes, hence they can be manufactured by wire electric discharge machines. Moreover, apart from a 180° rotation, the two waveguide plates are identical and can be manufactured at the same time by overlapping the two pieces. The matched loads of the directional couplers are inside the FL-HPD and are made of ECOSORB MF-190 material. Two kinds of matched transitions are present in the device: L-shaped junctions, to obtain the eight lateral input/output ports in standard waveguide; C-shaped junctions to connect different levels in the waveguide plates. With reference to Fig. 1, the shaded arrows correspond to the L-shaped junctions, whereas the dashed arrows are related to the C-shaped junctions. The white ones represent direct connections. All the devices were designed and analyzed with the methods used in [6].

Q-U COMPENSATION TECHNIQUE AND RESULTS

Based on the previously described architecture, a prototype of the complete HPD in the Ka-band was designed and manufactured. The 8 port-device was characterized by measuring its scattering parameters. In Fig. 2 the reflection and transmission scattering coefficients are reported. In order to evaluate the performance of the FL-HPD the measurements of the scattering parameters were elaborated to obtain the transfer functions which yield the Stokes parameters. If an ideal behavior of the diodes and of the differential amplifiers is assumed, it is possible to define a spectral distribution of the Stokes parameters whose integration yields the relevant data. To this end, one can consider the quantities:

$$C_k = |S_{ka}A + S_{kb}B|^2 \text{ with } k = 1, 2, 3, 4 \tag{5}$$

where, with reference to Fig. 1, S_{ka} and S_{kb} are the scattering parameters of the FL-HPD. By subtracting C_1 from C_2 and C_4 from C_3, one obtains:

$$\begin{bmatrix} Q_m \\ U_m \end{bmatrix} = \begin{bmatrix} |C_2|^2 - |C_1|^2 \\ |C_3|^2 - |C_4|^2 \end{bmatrix} = \underline{\underline{H}} \begin{bmatrix} \Re\{AB^\star\} \\ \Im\{AB^\star\} \end{bmatrix} + \underline{\underline{T}} \begin{bmatrix} |A|^2 \\ |B|^2 \end{bmatrix} \tag{6}$$

where

$$\underline{\underline{H}} = \begin{bmatrix} H_{qq} & H_{qu} \\ H_{uq} & H_{uu} \end{bmatrix} \quad \text{and} \quad \underline{\underline{T}} = \begin{bmatrix} T_{qa} & T_{qb} \\ T_{ua} & T_{ub} \end{bmatrix} \tag{7}$$

Moreover, the real elements of matrices $\underline{\underline{H}}$ and $\underline{\underline{T}}$ are defined in terms of the scattering parameters of the device:

$$H_{qq} = 2\Re\{S_{2a}S_{2b}^\star - S_{1a}S_{1b}^\star\} \tag{8}$$

$$H_{qu} = -2\Im\{S_{2a}S_{2b}^\star - S_{1a}S_{1b}^\star\} \tag{9}$$

$$H_{uq} = 2\Re\{S_{3a}S_{3b}^\star - S_{4a}S_{4b}^\star\} \tag{10}$$

$$H_{uu} = -2\Im\{S_{3a}S_{3b}^\star - S_{4a}S_{4b}^\star\} \tag{11}$$

$$T_{qa} = |S_{2a}|^2 - |S_{1a}|^2 \tag{12}$$

$$T_{qb} = |S_{2b}|^2 - |S_{1b}|^2 \tag{13}$$

$$T_{ua} = |S_{3a}|^2 - |S_{4a}|^2 \tag{14}$$

$$T_{ub} = |S_{3b}|^2 - |S_{4b}|^2 \tag{15}$$

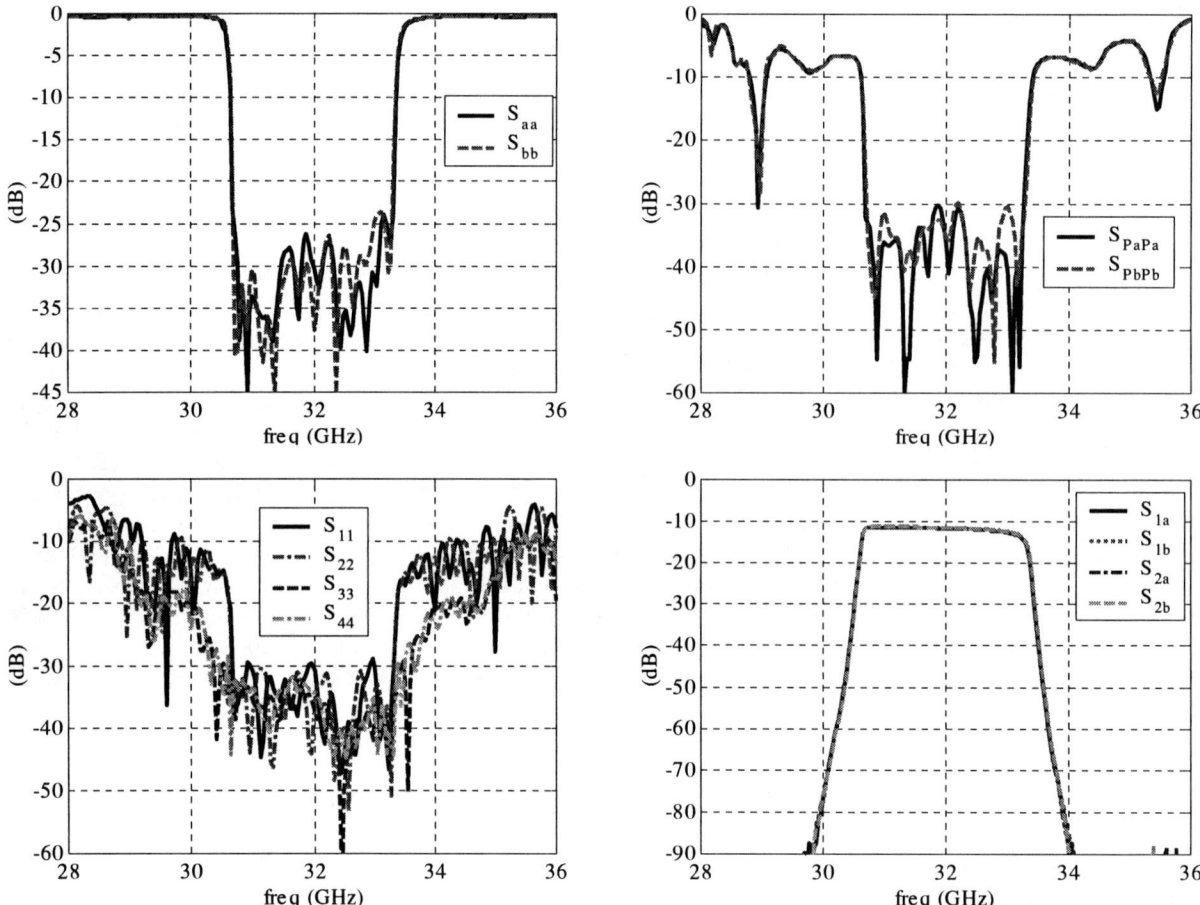

FIGURE 2. Measured reflection and transmission scattering coefficients of the Ka-band FL-HPD.

The eight transfer functions defined by the previous equations are shown in Fig. 3. In the case of a linearly polarized emission, the cross product $AB^\star$ can be written as:

$$AB^\star = \frac{1}{2}\, |\underline{E}|^2 \exp(2\,j\theta) = \frac{1}{2}(Q + jU) \tag{16}$$

where $\underline{E}$ is the spectral distribution of the electric field of the emission and θ is the angle with respect to a principal direction of the polarizer. In the Q-U real plane this polarization is described by the vector:

$$\underline{P} = \frac{1}{2}|E|^2[\cos(2\theta)\hat{q} + \sin(2\theta)\hat{u}] \tag{17}$$

where $\hat{q}$ and $\hat{u}$ are the unit vectors of the Cartesian reference system in the Q-U plane. It describes a circle, when the polarization angle θ ranges from 0 to π. This circle is transformed into a rotated ellipse by the real matrix $\underline{H}$. As far as the auto correlation terms $|A|^2$ and $|B|^2$ are concerned, they produce a shift of the ellipse in the measured Q-U plane. However, the HPD presents a rejection of the auto correlation terms of about 30 dB, which can be further improved by adopting lock-in techniques. Once the measured Stokes parameters are obtained, equation (6) can be inverted in order to obtain the Stokes parameters of the sky polarized components. To this end, the evaluation of the matrices $\underline{H}$ and $\underline{T}$ and of the auto correlation terms is necessary. This can be accomplished by adopting the online calibration procedure described in [5]. Moreover, one can introduce a ratio between the two conversion factors Q_m/Q and U_m/U, which accounts for the different level of losses between the two channels Q and U. Using a fasorial representation with respect to the variable θ, the vector $\underline{P}$ can be expressed as:

$$\underline{P} = \Re\{\underline{P}e^{j2\theta}\} \tag{18}$$

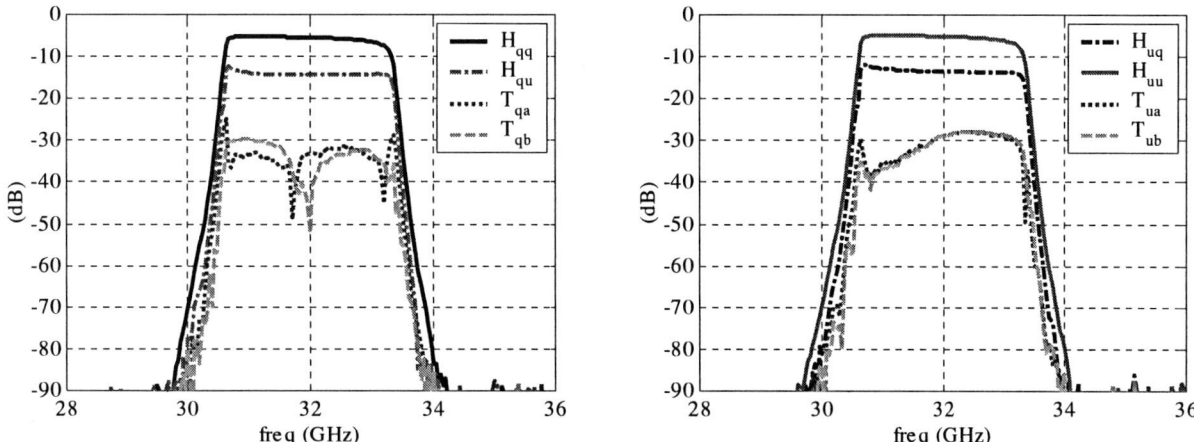

FIGURE 3. Transfer functions corresponding to the evaluation of the Q and U Stokes parameters obtained from the measurements of the Ka-band FL-HPD prototype.

where $\underline{P}$ is a complex vector

$$\underline{P} = \frac{1}{2}|E|^2(\hat{q} - j\hat{u}) \tag{19}$$

According to this notation, the measured Stokes parameters are described by the vector $\underline{P_m}$:

$$\underline{P_m} = \frac{1}{2}|E|^2[(H_{qq} - jH_{qu})\hat{q} + (H_{uq} - jH_{uu})\hat{u}] \tag{20}$$

which can be written as

$$\underline{P_m} = \frac{1}{2}|E|^2[\mathcal{H}_q\hat{q} + \mathcal{H}_u\hat{u}] \tag{21}$$

where

$$\mathcal{H}_q = H_{qq} - jH_{qu} \qquad \text{and} \qquad \mathcal{H}_u = H_{uq} - jH_{uu} \tag{22}$$

are complex quantities. In order to equalize the two channels Q_m and U_m, it is convenient to introduce the two amplification factors D_q and D_u, so that eq. (21) is transformed into:

$$\underline{\tilde{P}_m} = \frac{1}{2}|E|^2[\mathcal{H}_q D_q\hat{q} + \mathcal{H}_u D_u\hat{u}] \tag{23}$$

Since the amplification factors are applied to the detected D.C. signals, the functions $\mathcal{H}_q$ and $\mathcal{H}_u$ have to be thought as their averages. Projecting $\underline{\tilde{P}_m}$ onto the orthogonal basis $\underline{P}$ and $\underline{P}_\perp = \hat{v} \times \underline{P}$ ($\hat{v}$ is the unit vector perpendicular to the Q-U plane):

$$p_\parallel = \underline{\tilde{P}_m} \cdot \underline{P}^* = \frac{1}{4}|E|^4[D_q\mathcal{H}_q + jD_u\mathcal{H}_u] \quad \text{and} \quad p_\perp = \underline{\tilde{P}_m} \cdot \underline{P}_\perp^* = \frac{1}{4}|E|^4[D_q\mathcal{H}_q - jD_u\mathcal{H}_u] \tag{24}$$

and enforcing $p_\parallel = \frac{1}{2}|E|^4$ and $p_\perp = 0$ one obtains:

$$D_q = \frac{1}{\mathcal{H}_q} = \frac{1}{|\mathcal{H}_q|e^{j\varphi_q}} \qquad \text{and} \qquad D_u = \frac{1}{j\mathcal{H}_u} = \frac{1}{j|\mathcal{H}_u|e^{j\varphi_u}} \tag{25}$$

With these two complex amplification factors, the elliptical deformation would be completely recovered. However D_q and D_u must be real, a part from a common phase $\varphi = (\varphi_q + \varphi_u)/2$, which represents the offset of the detection. Hence, enforcing these conditions:

$$D_q = |\frac{1}{\mathcal{H}_q}|e^{-j\varphi} \qquad \text{and} \qquad D_u = |\frac{1}{\mathcal{H}_u}|e^{-j\varphi} \tag{26}$$

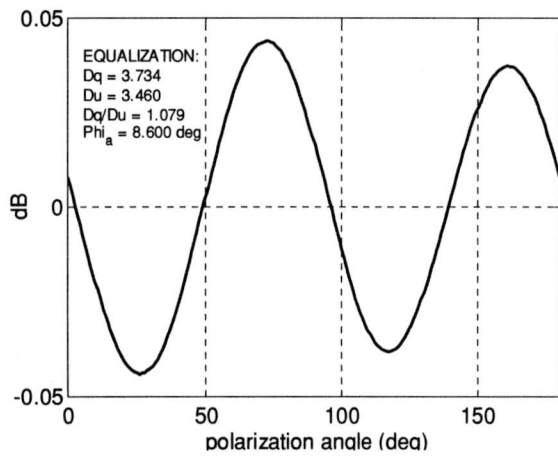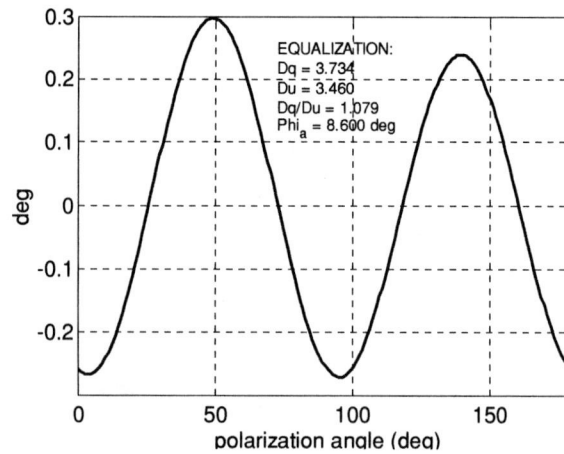

FIGURE 4. Detection error as a function of the polarization angle for a linearly polarized radiation. The angle is detected with an error less the 0.28 degrees. As for the amplitude, the maximum error is 0.04 dB.

the projections (24) become:

$$p_{\parallel} = \frac{1}{4}|E|^4(e^{j\Delta} + e^{-j\Delta}) = \frac{1}{2}|E|^4\cos(\Delta) \quad \text{and} \quad p_{\perp} = \frac{1}{4}|E|^4(e^{j\Delta} - e^{-j\Delta}) = \frac{j}{2}|E|^4\sin(\Delta) \quad (27)$$

where $\Delta = (\varphi_q - \varphi_u)/2$. By these projections it is possible to evaluate the axial ratio of the ellipse, which defines the maximum error $(1 + \tan\Delta)/(1 - \tan\Delta)$. Finally, in order to recover the right level of the Q and U Stokes parameters, the factors D_q and D_u are divided by $\cos(\Delta)$. By applying these amplification factors to the measured data of Fig. 3, the detection error of the linearly polarized radiation is reduced from 0.17 dB to a value better than 0.04 dB for the amplitude, whereas for the direction the error of 1.14 deg decreases to 0.28 deg. Fig. 4 shows the amplitude and phase error as a function of the polarization angle when the compensation is applied.

ACKNOWLEDGMENTS

This work has been financially supported by the Agenzia Spaziale Italiana. The authors would like to thank M. Franciotti for the prototype manufacturing.

REFERENCES

1. E.Carretti, R. Tascone, S. Cortiglioni, J. Monari, M.Orsini, *New Astronomy*, Vol. **6/3**, pp. 173-188 (2001).
2. S. Cortiglioni, S. Cecchini, M. Orsini, G. Boella, M. Gervasi, G. Sironi, R. Fabbri, J. Monari, A. Orfei, Ng. Kin-Wang, L. Nicastro, U. Pisani, R. Tascone, L. Popa, I. A. Strukov, " Sky Polarization Observatory (SPOrt): a Project for the International Space Station," *Proc. of ESA Workshop on Space Exploration and Resources Exploitation* , Cagliari, 1998, edited by Mauro Novara-ESTEC.
3. G. Sironi, G. Boella, G. Monelli, L. Brunetti, F. Cavaliere, M. Gervasi, G. Giardino, A. Passerini, *New Astronomy*, Vol. **3**, pp.1-13 (1998).
4. R.Tascone, P.Savi, D. Trinchero, R. Orta, *IEEE Trans. Microwave Theory Tech.*, Vol. **48**, No. 3,pp 423-430 (2000).
5. M. Baralis, O. A. Peverini, R. Tascone, D. Trinchero, V. Niculae, A. Olivieri, E. Carretti, S. Cortiglioni, C. Macculi, C. Sbarra, J. Monari, A. Orfei, G. Sironi, M. Zannoni, "Calibration techniques and devices for correlation radiometers used in polarization measurements," these proceedings.
6. R. Tascone, D. Trinchero, M. Baralis, O. A. Peverini, A. Olivieri, E. Carretti, S. Cortiglioni, "Millimeter wave passive devices for measurements of the polarized sky emission," in the proceedings of *2klBC Workshop Breuil-Cervinia*, Italy, 2001, edited by M. De Petris and M. Gervasi.

Calibrating CMB Polarization Telescopes

S. T. Staggs[*], D. Barkats[*], J. O. Gundersen[†], M. M. Hedman[*], C. P. Herzog[*], J. J. McMahon[*] and B. Winstein[**]

[*]*Physics Department, Princeton University, Princeton, NJ, 08544*
[†]*Department of Physics, University of Miami, Coral Gables, FL, 33146*
[**]*Department of Physics and the Enrico Fermi Institute, University of Chicago, Chicago, IL, 60637*

Abstract. Instruments for measuring the polarization of the cosmic microwave background (CMB) must be designed for accuracy as well as precision. The requirement for precision translates into a need for detectors with unprecedented sensitivity. Accuracy requires good methods for calibrating the response of the instrument to small polarized signals superimposed on large unpolarized signals. Since well-characterized polarized astrophysical millimeter sources are in short supply, we present an alternative method here. A flat metal plate is mounted in front of the telescope and nutated about a vertical axis, providing a varying polarized signal of amplitude near 30 mK.

INTRODUCTION

The polarization of the cosmic microwave background promises to provide new and interesting information about the structure and dynamics of the early universe. There has been a recent surge of interest in measuring this elusive polarized signal, and a suite of highly sensitive polarimeters are now being used to search for the cosmological polarization. Calibrating these polarimeters is a significant challenge. In the absence of bright, well-characterized astrophysical point sources, specialized techniques are needed to generate polarized signals. This paper describes the use of a nutating metal plate to calibrate the Princeton IQU Experiment (PIQUE). Related techniques using dielectric sheets and wire grids are described in Keating et al., 2001 [5].

STOKES PARAMETERS

In order to calibrate a polarimeter, the polarization state of the incident electromagnetic radiation must be consistently defined using the Stokes parameters I, Q and U (a fourth parameter V measures the circularly polarized component of the radiation). These parameters are defined in terms of two orthogonal components of the electric field E_x and E_y:

$$I = <E_x^2> + <E_y^2>, \tag{1}$$

$$Q = <E_x^2> - <E_y^2>, \tag{2}$$

$$U = <2E_x E_y>. \tag{3}$$

These parameters are the basis of all recent treatments (e.g. [1],[2],[4]) of the theoretical predictions of the CMB. As defined above, the Stokes parameters have units of intensity/c, where c is the speed of light. Because the CMB is observed to be an excellent blackbody, its intensity is uniquely determined by its temperature. Therefore, the convention in CMB experiments is to express measurements in units of temperature. The Stokes parameters with units of temperature are denoted here T, Q_T and U_T. The parameter T is related to I by a units-conversion factor: $T = \hat{k}I$. To maintain consistent definitions of the Stokes parameters, the same factor $\hat{k}$ must be used to define Q_T and U_T: $Q_T = \hat{k}Q$ and $U_T = \hat{k}U$.

We note in passing one instance where care needs to be taken in keeping the Stokes parameters consistent. Many CMB receivers operate with scalar feed horns coupled to waveguide and are thus sensitive to only one polarized

CP609, *Astrophysical Polarized Backgrounds*, edited by S. Cecchini et al.
© 2002 American Institute of Physics 0-7354-0055-5/02/$19.00

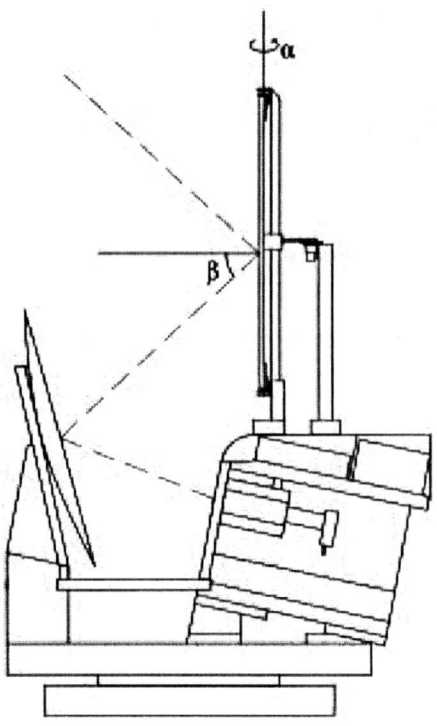

FIGURE 1. A figure showing the geometry of the calibrator plate used for PIQUE. The PIQUE optics comprise a 1.2 m off-axis parabola coupling to a scalar feed horn. The experiment is designed to operate at fixed elevation to observe near the North Celestial Pole (NCP). For calibration, the instrument is rotated by 180° in azimuth; then the additional reflection off the chopping plate results in observations once again near the NCP.

component of the radiation field at a time. It is common to calibrate each single-polarization receiver by viewing an unpolarized source of known temperature T_{cal} and equating the output of the receiver (which is, e.g., $\propto |E_x|^2$) with $T_{cal} + T_{rec}$, where T_{rec} is the (known) receiver temperature. Then a single receiver can be used to estimate I or T in CMB experiments, since the CMB is largely unpolarized. Frequently, an experiment comprises pairs of receivers, so that two receivers measure two orthogonal components of the electric field (e.g., E_x and E_y) in a given pixel, yielding two different estimates of the effective temperature, T_x and T_y. One then estimates the temperature T as $T = (T_x + T_y)/2$, and the correct estimate for Q is $Q_T = (T_x - T_y)/2$, not merely the difference between T_x and T_y.

CALIBRATION WITH A NUTATING METAL PLATE

PIQUE is a correlation polarimeter that observed the sky around 90 GHz with a quarter-degree beam [6]. In its second year, PIQUE made observations with two receivers, one at 90 GHz and the other at 40 GHz. The polarimeter is calibrated using a nutating aluminum plate to provide a controlled polarized signal for the radiometer. The thermal emission from this plate and the radiation from the sky reflected off this plate are weakly polarized because of the finite conductivity of aluminum. This effect has been described in several places; a recent treatment may be found in Cortiglioni, 1994 [3].

From classical electrodynamics, it can be shown that the reflection coefficients of radiation polarized parallel ($R_{//}$) or normal ($R_\perp$) to the plane of incidence is given by (assuming the conductivity is high):

$$R_{//} = 1 - 2\delta/\lambda \sec\theta, \qquad (4)$$
$$R_\perp = 1 - 2\delta/\lambda \cos\theta, \qquad (5)$$

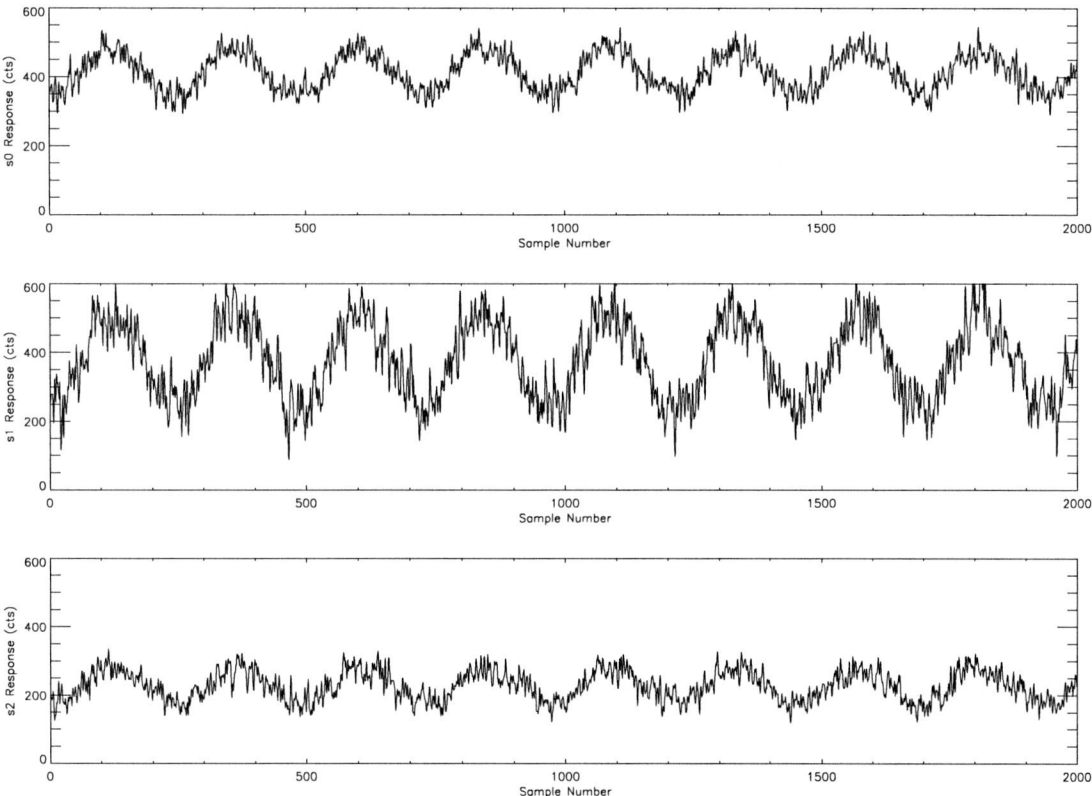

FIGURE 2. Response of PIQUE polarimetry channels to the nutating plate. The output of the three polarimetry channels are displayed as a function of time – data are sampled at 62.5 Hz. The oscillations have a period on the order of seconds.

where θ is the angle of incidence, δ is the skin depth, λ is the wavelength of the radiation, all in MKS units. The fact that these two reflection coefficients have different dependencies on the incident angle means that initially unpolarized radiation reflected from the plate will acquire a polarized component dependent on the orientation of the plate. It follows from Kirchoff's law that the thermal emission of the plate will also be polarized, and that the polarized component will have the opposite sign (i.e. if the reflected radiation has a positive Q, the emitted radiation will have a negative Q).

The size of the polarized signal from the plate depends on the orientation of the plate with respect to the detector. The geometry of the PIQUE calibrator is shown in Figure 1. The ray from the feed horn at the focus of the parabola arrives at the plate at an elevation angle $\beta \simeq 41°$ (after reflecting once off the parabola). The vector normal to the plate makes an angle α with the vertical plane bisecting the telescope. The plate nutates about a vertical axis, so α oscillates sinusoidally in time as $\alpha = \alpha_o \cos(\omega_o t + \phi)$, where typical values for the amplitude and the frequency are $\alpha_o \simeq 5°$ and $\omega_o \sim 1$ Hz.

Given this geometry, one can calculate how the polarized signal Q_T introduced into the receiver will vary as a function of α (assuming α small):

$$Q_T(\alpha) = \kappa \frac{\alpha}{\alpha_o}(T_{plate} - T_{sky}), \tag{6}$$

where T_{plate} and T_{sky} are the physical temperature of the plate and the effective noise temperature of the sky (including the atmosphere, the CMB, and any Galactic signals). The dependencies on the composition and the geometry of the plate are included in the dimensionless parameter κ:

$$\kappa = 2\frac{\delta}{\lambda}\gamma\frac{\sec\beta - \cos\beta}{\sin\beta}\alpha_o. \tag{7}$$

185

The factor γ in this expression is a factor that accounts for how much of the telescope beam intersects the plate. If the plate is small and uniformly illuminated $\gamma = \sec \beta$, but if the plate is infinitely large $\gamma = 1$. For PIQUE at 90 GHz, only 3% of the beam spills past the chopping plate, so γ can be reasonably approximated as unity.

Inserting the correct expressions for the skin depth, wavelength and relevant angles, one finds that $\kappa \simeq 10^{-4}$. Therefore, if $T_{plate} \simeq 300K$ and $T_{sky} \simeq 50K$, the polarized signal injected into the telescope varies with an amplitude of about 30 mK.

When the plate is installed and begins nutating, the outputs from the polarimetry channels show clear sinusoidal oscillations, as shown in Figure 2. By comparing the amplitudes of these oscillations with the calculations for the polarized signals introduced by the nutating plate, the responsivity of the telescope can be determined. This requires estimates of T_{plate} and T_{sky}. The plate temperature is measured using a thermometer attached to the plate itself, while the sky temperature must be derived from the output of the total power channels in PIQUE. The accuracy of the nutating plate method has been confirmed by independent tests using calibrated cryogenic loads. These tests are tedious and require partial disassembly of the instrument. The leading uncertainty in the nutating plate calibration is the DC conductivity σ of the aluminum; a conservative estimate of 20% uncertainty on σ gives rise to an overall calibration uncertainty of 10%. We plan to measure the conductivity in separate tests and to replace the aluminum plate with a stainless steel plate as final checks. In summary, the nutating plate provides a dependable means of calibrating PIQUE at regular intervals in the field.

REFERENCES

1. Zaldarriaga, M., and Seljak, U., *Phys. Rev. D*, **55** (1997).
2. Kamionkowski, M., Kosowsky, A., and Stebbins, A., *Phys. Rev. D*, **55** (1997).
3. Cortiglioni, S., *Rev. Sci. Instrum.*, **65** (1994).
4. Zaldarriaga, M., *ApJ*, **503**, 1 (1998).
5. Keating, B. G., O'Dell, C. W., Gundersen, J. O., Piccirillo, L., Stebor, N. C., and Timbie, P., An instrument for investigating the large angular scale polarization of the cosmic microwave background (2001), `astro-ph/0111276`.
6. Hedman, M. M., Barkats, D., Gundersen, J. O., Staggs, S. T., and Winstein, B., *ApJL*, **548**, L111–L114 (2001).

The Moon as a Calibrator of Linearly Polarized Radio Emission for the SPOrt Project.

S.Poppi*, E.Carretti†, S.Cortiglioni†, V.D.Krotikov** and E.N.Vinyajkin**

*IRA/CNR, via Gobetti 101 Bologna (ITALY)
†TeSRE/CNR, via Gobetti 101 Bologna (ITALY)
**NIRFI Institute 25 B.Pecherskaya st.Nizhnij Novgorod (RUSSIA)

Abstract. The Moon could be the best external calibrator for the Sky Polarization Observatory (SPOrt) experiment, providing the highest polarized signal at large angular scales ($\geq 7°$) in the 22-90 GHz range. Maps of linearly polarized lunar radio emission have been realized at 8.3 GHz with the 32-m radiotelescope of IRA-CNR (Medicina-Italy) at full Moon, new Moon, first and last quarter. We derived estimates of spectral and time properties of both the intensity and the linear polarization of the Moon radio emission, taking into account the radiative transfer of heat in lunar soil and the surface roughness. A comparison between predictions of the theory and observations is presented.

INTRODUCTION

Nowadays, several experiments have been planned to measure Galactic and cosmological linearly polarized emission at microwave frequencies, starting in the very close future. Among these, SPOrt-ISS will observe at 22, 32, 60 and 90 GHz with HPBW=7° [1]. At these frequencies and angular scales, few sources are strong and extended enough to be used as calibrators. The Moon, which is one of the most powerful sources at microwave wavelengths, could satisfy the appropriate requirements. The polarization characteristics of the radio emission of the Moon are determined by the value of the dielectric constant of its soil and by the difference of Fresnel coefficients for linear polarization in the plane of incidence of the radio wave and in the orthogonal plane [2]. Roughness of the lunar surface and an averaging effect produced by the antenna diagram lead to a decrease of the value of the linearly polarized signal. If the HPBW is much smaller than the angular size of the Moon ($2R_M$), the observed values of the Stokes parameters are expected to be close to the local true values. Otherwise, (HPBW$\gg 2 R_M$) the integrated values of the Stokes parameters are measured. The reasons for a nonzero integrated value of the linearly polarized component of the Moon radio emission and for its phase dependence are the spatial and time dependence of the thermal regime of the lunar surface due to the spatial and time variations of solar illumination [2]. The only map present in the literature has been made at 1.4 GHz [3]. Though the brightness temperature phase variation is near 1% at 21 cm, it was ignored.

The Moon is a discrete source for the SPOrt antennae, so it is necessary to know integrated values of the Stokes parameters (I, Q and U) of the Moon radio emission at the SPOrt frequencies for any lunar phase. There are no sufficient and reliable data in the literature on these values.We were thus forced to develop a theory for the linear polarization of the lunar radio emission by taking into account new data on the lunar soil physical properties and to carry out new polarimetric measurements at 8.3 GHz.

OBSERVATIONS

Observations were carried out at 8.3 GHz (FWHM=4.8$'$) with 80 MHz bandwidth, using the polarimetric facility of the Medicina 32m Radiotelescope [4].The calibration of the linearly polarized flux has been performed observing the source 3C286 (5.0 Jy, 11.8% polarized, PA=34° [5]). The Moon has been observed performing right ascension scans across its disc, with a drift speed of 15$''$/s, and repeating the scans for several declinations at equal intervals

CP609, *Astrophysical Polarized Backgrounds*, edited by S. Cecchini et al.
© 2002 American Institute of Physics 0-7354-0055-5/02/$19.00

of 1.6$'$. The Stokes parameters I, Q, U have been obtained for each scan (Fig. 1), and corrected for the parallactic angle variation during the observations. From these, maps of the linearly polarized radiation brightness temperature ($T_b^p = \sqrt{Q^2 + U^2}$) have been reconstructed as well as the distribution of the polarization vectors over the Moon disc (Fig. 2). The integrated values of the Q and U parameters have been calculated and collected in table 1.

TABLE 1. Summary of the observations: columns represent the date of the observation (day of the year 2001), the lunar phase ($\phi = 0$ is Full Moon), the angular diameter of the Moon and the integrated values of the Q and U parameters respectively.

Day	Phase	Diameter	Q^I (K)	U^I (K)
98	10.8 $^\circ$	32.0$'$	- 0.43 (0.02)	0.47 (0.02)
113	177.3°	30.8$'$	-0.38 (0.02)	-0.14 (0.01)
135	87.8 $^\circ$	29.6$'$	- 0.51 (0.03)	0.41(0.02)
242	327.0 $^\circ$	29.4$'$	-0.23 (0.01)	0.42 (0.02)
244	349.4 $^\circ$	29.4$'$	-0.54 (0.03)	0.37(0.02)
245	0.8 $^\circ$	29.4$'$	-0.49(0.02)	0.52(0.03)

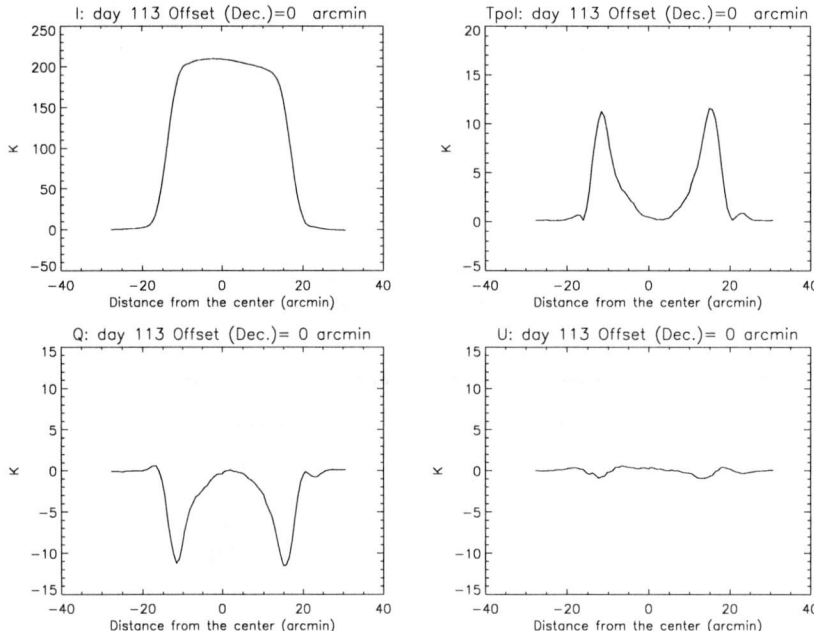

FIGURE 1. A right ascension scan across the center of the Moon, at $\phi = 177.3°$: parameters I (top left),Q (bottom left), U, (bottom right) and polarized brightness temperature ($T_b^p = \sqrt{Q^2 + U^2}$)

THE MODEL

We calculated both the local Stokes parameters, using a model resolution of $0.1'$, and their integrated values over the lunar disc (Q^I, U^I), according to [2, 6] but by also taking into account the dependence of the thermal conductivity of the lunar soil on the temperature [7, 8].

The brightness temperature of a point over the Moon surface having selenographic longitude φ and latitude ψ for linear polarization in the plane of incidence of the radio wave (v) or in the orthogonal plane (h) has been represented as a Fourier series [9]:

$$T_{bv,h} = [1 - R_{v,h}(\varphi,\psi)] \left[T_0(\psi) + \sum_{n=1}^{\infty} (-1)^{\alpha_n} \frac{T_n(\psi)\cos(n2\pi t/t_0 - \varphi_n - n\varphi\xi_n(\varphi,\psi))}{\sqrt{1 + 2\delta_n\cos(\rho') + 2\delta_n^2\cos^2(\rho')}} \right] \tag{1}$$

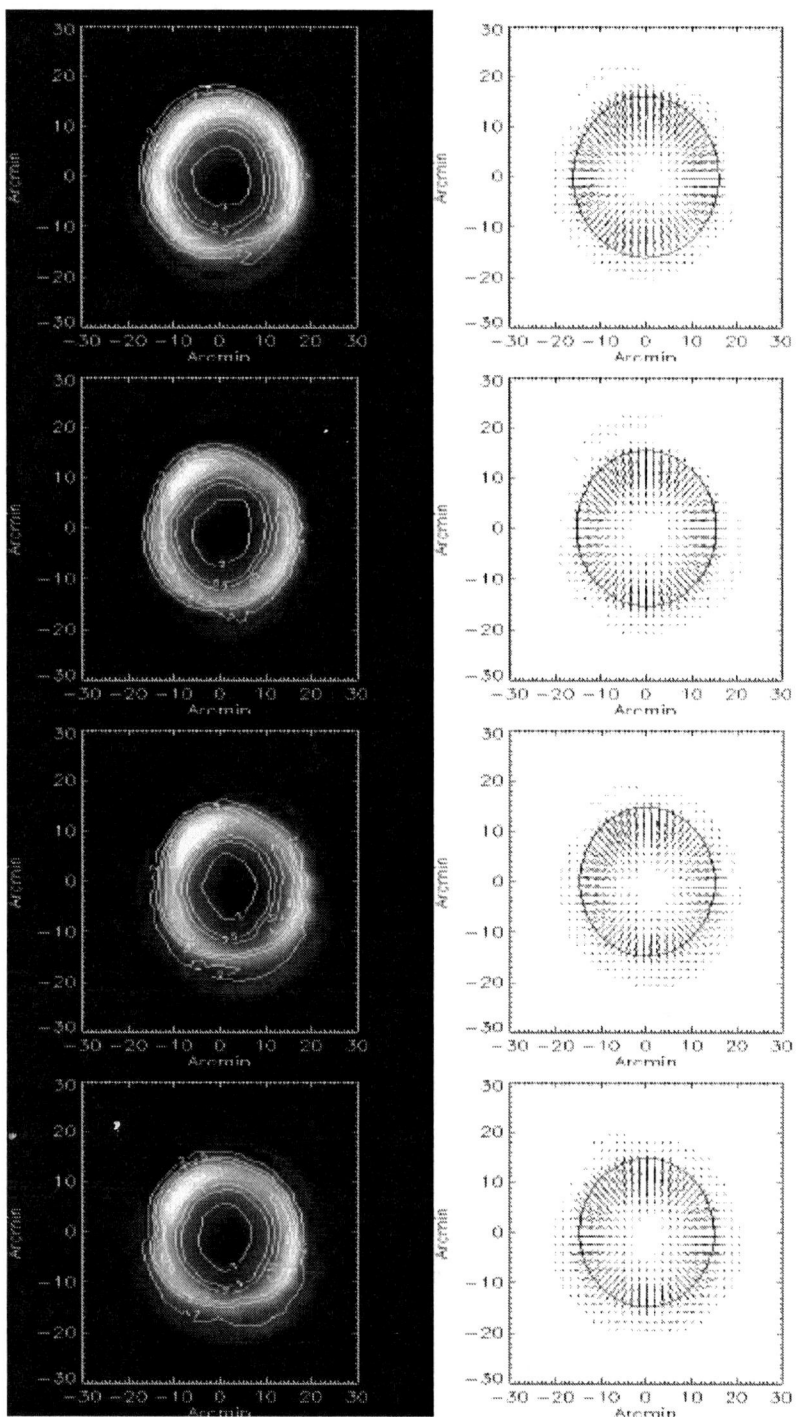

FIGURE 2. T_B^p distribution maps (left) and polarization vectors (right) for different lunar phases: last quarter, full Moon, new Moon, between new moon and first quarter, respectively from top to bottom.

where $R_{v,h}(\varphi,\psi)$ is the Fresnel coefficient for v and h linear polarization respectively, $T_0(\psi), T_n(\psi)$ and φ_n are the constant, the amplitude and the phase of the daily variation of the surface temperature respectively, t_0 is the synodic month, $\xi_n(\varphi,\psi)$ is the phase shift of the $n-th$ brightness temperature harmonic with respect to the surface temperature harmonic, δ_n is the ratio of penetration of the electrical to and thermal wave into the lunar soil for the $n-th$ harmonic, ρ' is the refraction angle, $\alpha_n = (n-1)(n-2)/2$. It can be demonstrated that it is enough to take into account only five harmonics [9].

The values of the physical parameters of lunar soil were chosen from the results of numerous radio astronomical, radar and direct studies [9, 10, 11, 12]. We adopted dielectric constant $\varepsilon=2$, standard deviation of surface slopes $\sigma=10°$ and parameter $\delta_n = 2\lambda n^{1/2}$, where λ is the observing wavelength and n is the number of harmonics of brightness temperature phase dependence. Amplitudes and phases of harmonics of surface temperature correspond to a thermal inertia parameter $I_s=0.001$ for night surface temperature and to a ratio of radiative to contact parts of thermal conductivity $\alpha=1$ at 350 K. A full description of all the parameters we used can be found in [9]. From this model we calculated integrated values of the Stokes parameters I, Q and the degree of linear polarization P=Q/I as a function of the lunar phase ϕ (according to [6] we set $\phi=0$ for the full Moon) at the frequencies 22, 32, 60, 90 and 8.3 GHz (integrated value of U are zero for this homogeneous model). Figure 3 contains the phase dependence of Q (in K) and P (in percent) at five frequencies. It is seen from fig. 3 that the phase variations decrease with decreasing frequency. Maxima and minima of these curves correspond to those phases for which the brightness temperature distribution over lunar disc has the largest East-West asymmetry. With further decrease of frequency limit values of P and Q are reached, whose value are mainly determined by the dependence of the constant component of the surface temperature on the latitude. Analysis with $\varepsilon=2$ has revealed that the influence of roughness with $\sigma=10°$ is insignificant and becomes appreciable when $\sigma=20°$.

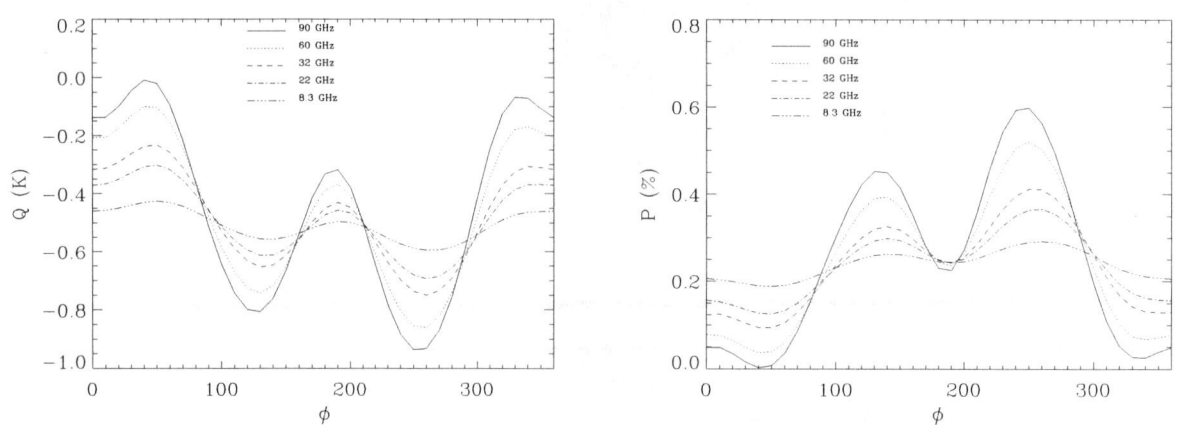

FIGURE 3. Q parameter and percentage of polarization calculated at $\nu =90, 60, 32, 22$ and 8.3 GHz

RESULTS

As a first result, it can be seen that the polarized emission is not homogeneous. In fact the measured maps show an emission maximum in the north-east border which is present for all the lunar phases: this could be due to an heterogeneous distribution of the dielectic constant of the regolith over the surface. In order to perform a comparison with the model predictions at 8.3 GHz, we calculated the distributions of brightness temperature of total and linearly polarized T_b^p radio emission across the disc of the Moon for some fixed selenographic latitudes. Figure 4 (left plot) contains the values of T_b^p calculated from the model and smoothed by 4.8' beam (dotted line) as well as the observed values (dashed line) across the equator of the Moon for $\phi= 177.3°$. The solid line represents the model calculations for an infinite resolution and shows that the antenna beam is smoothing the Stokes parameters distribution only for distances larger than 0.6 R_M.

For integrated polarization the comparison between the observations and the model shows partial agreement too. In fact, from fig. 4 (right plot) it can be seen that the observations follow the model forecast for the Stokes parameter Q at 8.3 GHz. Non zero values of U prove that there are spatial large scale inhomogeneities of lunar regolith physical properties.

190

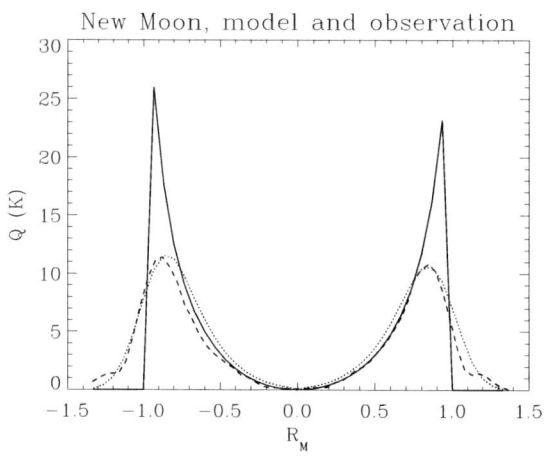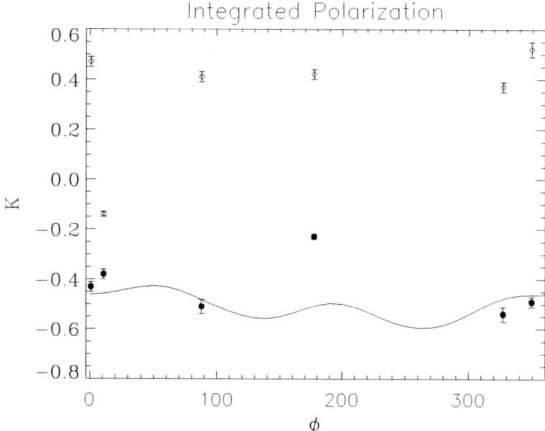

FIGURE 4. Left: expected local Q parameter at full angular resolution (solid) as well as after including the beam effect (dotted) compared to the measured values (dashed). Right: measured integrated Q^I (filled circles) and U^I (empty squares) values compared to the model predictions for Q^I

This allow us to make a preliminary evaluation of the lunar antenna temperature for the SPOrt beam. As it can be seen from Fig.3 the absolute value of Q has a maximum at $\phi = 250°$. Assuming an average moon radius $R_M = 15'$ it is possible to evaluate the antenna temperature from:

$$T_a^Q = T_b^Q \left(\frac{2R_M}{HPMW} \right)^2 \qquad (2)$$

where T_b^Q is the brightness temperature of the integrated Q parameter calculated at $\phi = 250°$.

These results, showed in table 2, allow us to use the Moon as a calibrator for all SPOrt-ISS channels which have instantaneous sensitivity $\sigma_{IS} = 1\ mK$. Moreover, the signal is more powerful at high frequencies, where the lack of strongly polarized sources is a major problem.

TABLE 2. Evaluation of integrated brightness and antenna temperatures of the Q parameter for SPOrt-ISS.

$\nu(GHz)$	$T_b^Q(K)$	$T_a^Q(K)$
22	0.69	$3.5 \cdot 10^{-3}$
32	0.75	$3.8 \cdot 10^{-3}$
60	0.86	$4.4 \cdot 10^{-3}$
90	0.94	$4.8 \cdot 10^{-3}$

CONCLUSIONS

Preliminary analysis has revealed that the Moon might be the best external calibrator at centimetric and millimetric wavelenghts, providing the most powerful linearly polarized signal, for large angular scale experiments measuring polarization of CMB and Galactic background radio emission.

Satisfactory agreement between calculated and observed polarized signals as well as brightness temperatures allows us to suggest the Moon as a linearly polarized calibration source for observations by antennae with $HPBW \gg 2R_M$. Future development of this work will involve new observation at 8.3 GHz in order to collect data at as many lunar phases as possible. Moreover, the model will be improved by taking into account a non homogeneous dielectric constant distribution. Local and integrated calculations of the U parameter will be performed.

Additional polarized observations at higher frequencies for various lunar phases are necessary for further improvement of the model.

ACKNOWLEDGMENTS

This work was made in the framework of the SPOrt Project, an ASI funded program. Partial support comes from CNR-RAS agreement.

REFERENCES

1. Carretti, E., "The SPOrt project", *this volume*
2. Alekseev V.A., Krotikov V.D.,*Radiophysics and quantum electronics (Izvestiya vuz. Radiofizika)* 1968, August, p.647. (Russian original 1968, V.11, N8, P.1133).
3. Moffat,P.H, *MNRAS*,1972,**160**,139.
4. Orfei, A., "The Polarimetric observation facility at the Medicina 32-m Parabolic Antenna" , *this volume*.
5. Tabara H.,Inoue M., *A&AS*,**39**,379-393.
6. Troitskij,V.S., *Astronomicheskij zhurnal* (in Russian), 1954, **V.31**, N6, P. 511.
7. Troitsky V.S. , Burov A.B. , Alyoshina T.N.,*Icarus*, 1968,**8**, N7 ,p. 423-433.
8. Troitsky V.S. , *Philosophical Transactions of the Royal Society of London*, A, 1969, **264**, pp. 145-149.
9. Troitskii V.S., Tikhonova T.V., *Radiophysics and quantum electronics (Izvestiya vuz. Radiofizika)*. 1973, May 15, p.981 (Russian original 1970, v.13, p.1273).
10. Muhleman, D.O. "Microwave emission from the Moon", in *Thermal characteristics of the Moon, Progress in Astronautics and Aeronautics*, MIT Press, Cambridge, Massachusets, 1972, v.28, pp51-81.
11. Troitsky V.S., Krotikov V.D. Astronomicheskij vestnik (in Russian). 1989,**23**, N1. P.78.
12. Krotikov V.D., Pelyushenko S.A. Astronomicheskij zhurnal (in Russian). 1987,**64**, P.417.

A Destriping Technique for SPOrt Polarization Data

C. Sbarra*, E. Carretti*, S. Cortiglioni*, M. Zannoni†, R. Fabbri**, C. Macculi* and M. Tucci†

*I.Te.S.R.E./CNR, Via P. Gobetti 101, I-40129 Bologna, Italy
†Dipartimento di Fisica, Universitá di Milano-Bicocca, Piazza della scienza 3, I-20126 Milano, Italy
**Dipartimento di Fisica, Universitá di Firenze, via Sansone 1, I-50019 Sesto Fiorentino, Italy

Abstract. A new destriping technique suitable for polarization data is described. The performances of the algorithm are studied in the frame of the SPOrt Experiment onboard the International Space Station.

INTRODUCTION

Low frequency noise affects most radiometers inducing correlations between successive samples of the measured signal. In a spin-scanned system it can lead to striping in the projected maps, compromising the result of the experiment. However, it has been shown [1] that this noise is only important if the scanning frequency f_s is smaller than the so-called knee frequency f_k appearing in the noise power spectrum equation:

$$F(f) = \sigma^2 \left[1 + \left(\frac{f_k}{f} \right)^\alpha \right] \qquad (1)$$

where σ^2 is the variance of the white noise component and α depends on the radiometer features.

The SPOrt experiment [2] onboard the International Space Station (ISS) will be measuring the Q and U Stokes parameters for 18 months with an angular resolution of 7°. The accessible fraction of sky, about 80%, will be covered every 70 days and will be viewed under precessing circles (scans) with a scanning frequency $f_s = 1.8 \times 10^{-4}$ Hz. For the correlation polarimeters of the SPOrt-ISS experiment the dominant contribution to the low frequency noise comes from amplifier gain fluctuations, for which $\alpha \simeq 1$ [3]. The goal knee frequency of the SPOrt polarimeters is $f_k = 1.8 \times 10^{-5}$ Hz, largely sufficient to ensure stability over the time needed for a scan. The istantaneous sensitivity is expected to be 1 mKs$^{1/2}$.

THE ALGORITHM

Our proposed destriping algorithm assumes the only effect of the low frequency noise is to add a different constant offset to each scan. A simple iterative procedure can then be applied to remove the offsets from the Time Ordered Data (TOD) before map-making: we start from a null map as guessed sky emission, use it to get a zero order estimate of the offsets, subtract them from the TOD, project the result on a map, corresponding to our first order sky emission estimate, and iterate the cycle.

Total Intensity Case

We detail the destriping algorithm for the simpler total intensity case, and extend the computation to polarization data in the next section.

Let's define the TOD vector **Y** as

$$\mathbf{Y}^M = \mathbf{A}^{M \times N} \cdot \mathbf{X}^N + \mathbf{B}^{M \times R} \cdot \mathbf{OFF}^R + \mathbf{N}^M \qquad (2)$$

where M is the number of observations, N is the number of pixels in the projected map, R is the number of scans, $\mathbf{X}$ represents the true sky emission map, $\mathbf{A}$ is a pointing matrix relating observations to pixels [4, 5], $\mathbf{B}$ is another pointing matrix relating observations to scans, $\mathbf{N}$ is a vector containing only white noise and $\mathbf{OFF}$ is a vector containing the scan offset. The average value of the white noise vector being zero, the offset vector estimate of order zero can be evaluated with the following steps:

$$
\begin{aligned}
\mathbf{B} \cdot \mathbf{OFF}^{(0)} &= \mathbf{Y} - \mathbf{A} \cdot \mathbf{X}^{(0)} \\
\mathbf{OFF}^{(0)} &= \mathbf{N_{obs}}^{-1} \cdot \mathbf{B}^{\mathrm{T}}(\mathbf{Y} - \mathbf{A} \cdot \mathbf{X}^{(0)})
\end{aligned}
\tag{3}
$$

where $\mathbf{N_{obs}} = \mathbf{B}^{\mathrm{T}}\mathbf{B}$ is diagonal and contains the number of observations in each scan. In fact, in the first cycle when we set $\mathbf{X}^{(0)} = 0$, this simply corresponds to taking the average of the measurements in each scan. We can now subtract the estimated offsets from the TOD:

$$
\begin{aligned}
\mathbf{Y}^{(0)} &= \mathbf{Y} - \mathbf{B} \cdot \mathbf{OFF}^{(0)} \\
&= \mathbf{A} \cdot \mathbf{X} + \mathbf{B} \cdot (\mathbf{OFF} - \mathbf{OFF}^{(0)}) + \mathbf{N}
\end{aligned}
\tag{4}
$$

and project the cleaned data on a map by averaging the measurements corresponding to the same pixel, obtaining our first estimate of the true sky emission map (we can in fact write $\mathbf{Y}^{(0)} = \mathbf{A} \cdot \mathbf{X}^{(1)} + \mathbf{N}$):

$$
\mathbf{X}^{(1)} = \mathbf{N_{pix}}^{-1} \cdot \mathbf{A}^{\mathrm{T}} \cdot \mathbf{Y}^{(0)}
\tag{5}
$$

where $\mathbf{N_{pix}} = \mathbf{A}^{\mathrm{T}}\mathbf{A}$ is diagonal and contains the number of observations in each pixel. $\mathbf{X}^{(1)}$ can in turn be used as a guess map for another iteration and the full cycle can be repeated until the error introduced by interrupting the iteration is negligible when compared to the pixel variance. A rule of thumb may be:

$$
\sqrt{\left\langle (\mathbf{X}^{(i+1)} - \mathbf{X}^{(i)})^2 \right\rangle} < \frac{1}{3}\sigma_{pix}
\tag{6}
$$

For SPOrt-ISS, 10 iterations have proven to be largely sufficient.

Polarization Case

In order to extend the previous algorithm to the polarization case, we replace the scalar elements of the previously defined vectors with (Q,U) pairs referred to a fixed reference frame. The scalar elements (0 or 1) of the pointing matrix $\mathbf{A}$ of the total intensity case are multiplyed by rotation matrices:

$$
\mathcal{R}(\alpha) = \begin{pmatrix} \cos 2\alpha & \sin 2\alpha \\ -\sin 2\alpha & \cos 2\alpha \end{pmatrix}
\tag{7}
$$

where α represents the angle between the absolute reference frame and the instrument reference frame at the time of each measurement.

For both polarization and total intensity data the average signal over the covered sky is lost in the destriped map.

SIMULATIONS AND TESTS

The performances of the destriping algorithm have been investigated using numerical simulations of the SPOrt-ISS mission and producing several maps with HEALPix [1]. We start by simulating noise maps, containing either white or white plus 1/f noise, and compare the power spectra measured from the corresponding maps. The average result is shown in Fig. 1 for two different values of the knee frequency: even when f_k is one order of magnitude larger than

---

[1] http://www.eso.org./science/healpix

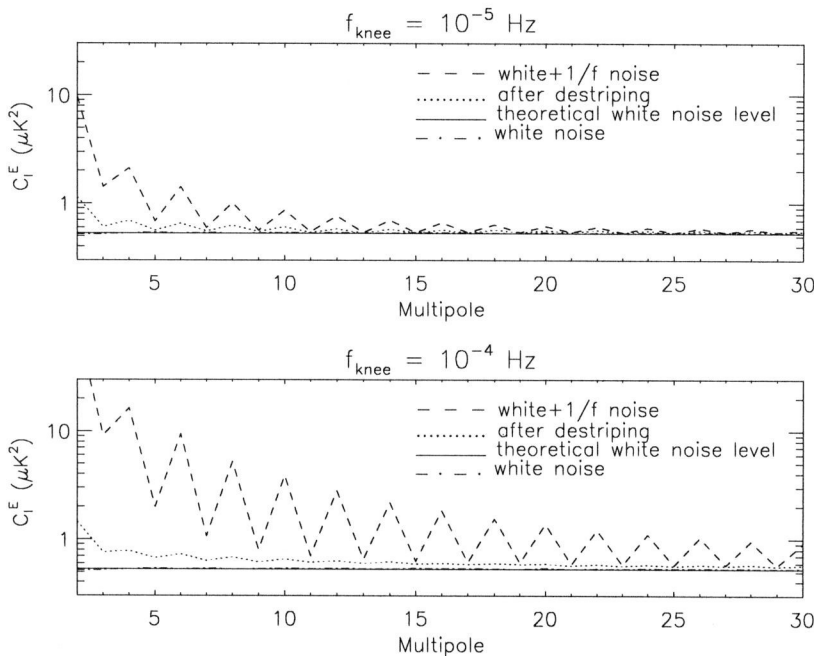

FIGURE 1. Average of measured noise power spectra, before and after destriping, for two values of the knee frequency f_k. The measured power spectrum of purely white noise is included for comparison and cannot be distinguished from its theoretical level.

TABLE 1. Excess pixel noise due to the presence of low frequecy contributions, with respect to the ideal white noise case

f_k (Hz)	Before Destriping	After Destriping
10^{-4}	32%	3%
10^{-5}	3%	$\leq 1\%$

expected, the noise power spectrum after destriping is close to that of purely white noise. A small residual is however present at low values of ℓ. To asses quantitatively the efficiency of the destriping algorithm we measure the excess pixel noise induced by the low frequency component with respect to the case of purely white noise. The result is shown in table 1.

To check for residual noise correlations after the destriping algorithm has been applied we also measure the two-point correlation functions $C^{Q,U}(\theta)$ from the simulated Q and U noise maps. In the computation, the Stokes parameters are properly rotated so that Q is defined with reference to the great circle connecting the points in each pair. The averages of the measured correlation functions $C^Q(\theta)$, before and after destriping, are shown in Fig. 2 together with their one sigma bands, for the pessimistic case $f_k = 10^{-4}$ Hz. Even in this case residual correlations are small. The spread of the measured correlation functions after destriping is increased by roughly 15% with respect to the case of purely white noise. This figure can be taken as a conservative upper limit on the possible SPOrt-ISS sensitivity worsening due to the presence of instrumental low frequency noise. In fact, for Gaussian fluctuations, the correlation functions carry the same information as the power spectra.

A complete analysis is finally performed on simulated maps including both a CMB polarization signal (ΛCDM, $\tau = 0.2$, $\Omega_m = 0.3$, $H_0 = 65$, B-mode=0), generated with CMBFAST [2], and white plus 1/f noise ($f_k = 10^{-4}$ Hz). The

---

[2] http://www.sns.ias.edu/ matiasz/CMBFAST/cmbfast.html

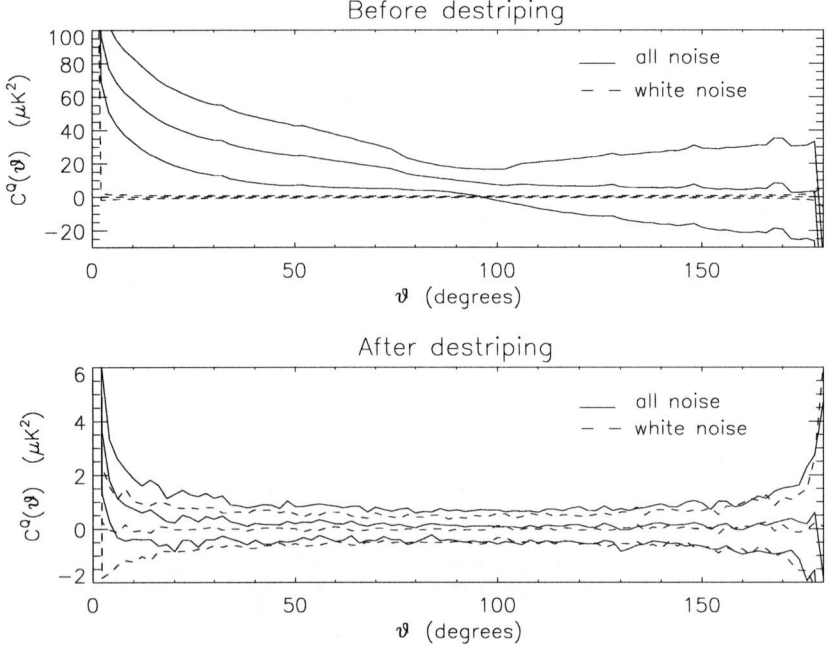

FIGURE 2. Average of measured two-point correlation functions for $\simeq 100$ simulated Q maps including only instrumental noise: white (dashed) or white plus 1/f (solid), before and after destriping, for $f_k = 10^{-4}$ Hz. Measured one σ bands are shown as well. The noise correlation functions for U are similar and are not shown.

$C_\ell^{\mathrm{E,B}}$ power spectra are obtained by integration from the measured correlation functuions $C^{\mathrm{Q,U}}(\theta)$:

$$C_\ell^{\mathrm{E}} = \int_0^\pi [C^{\mathrm{Q}}(\theta)F_{\ell,2}^1(\theta) + C^{\mathrm{U}}(\theta)F_{\ell,2}^2(\theta)]\sin\theta d\theta \tag{8}$$

$$C_\ell^{\mathrm{B}} = \int_0^\pi [C^{\mathrm{U}}(\theta)F_{\ell,2}^1(\theta) + C^{\mathrm{Q}}(\theta)F_{\ell,2}^2(\theta)]\sin\theta d\theta \tag{9}$$

where the functions $F_{\ell,2}^i(\theta)$ can be found in [6]. The method is similar to that used by Szapudi et al. [7], though it has been extended to polarization data. It has the advantage of being simple and rather fast ($O(\mathrm{N}_{\mathrm{pix}}^2)$), and takes authomatically into account any edge effects. In presence of noise we use as estimator for the correlation function of the signal the expression:

$$\tilde{C}(\theta) = \sum_{ij} w_{ij}(\Delta_i\Delta_j - \mathrm{N}_{ij}) \tag{10}$$

where Δ_i is the content of pixel i, N_{ij} is the noise correlation between pixels i and j, and w_{ij} is the weight of the pair, which we relate to the sensitivity of both pixels. We check that in case of purely white noise, when N_{ij} is diagonal, the measured power spectra as well as the error on the C_ℓ are as expected, as shown in Fig. 3: the measurement is unbiased and the standard deviation of the distribution of power spectra measured from different maps is in good agreement with the expected value:

$$\Delta C_\ell^{\mathrm{E,B}} = \sqrt{\frac{2}{(2\ell+1)f_{sky}}}(C_\ell^{\mathrm{E,B}} + W_p^{-1} \cdot B_\ell^{-2}) \tag{11}$$

where f_{sky} is the fraction of sky observed, $W_p^{-1} = 4\pi\sigma^2/\mathrm{N}_{\mathrm{pix}}$ and $B_\ell^2 = e^{-\ell^2\sigma_b^2}$ is the beam smearing.

After including the 1/f component of the noise and destriping, if we use $f_k = 10^{-4}$ Hz we find that in order to recover the original CMBP power spectra we cannot simply assume the noise pixel-to-pixel correlation matrix to be diagonal (white noise). Instead, the noise correlation function must be evaluated, by averaging many Monte Carlo simulations, and subtracted from the measured correlation functions before integrating. The average of 60 power spectra measured in this way is shown in Fig. 4 together with the measured $\Delta C_\ell^{\mathrm{E,B}}$.

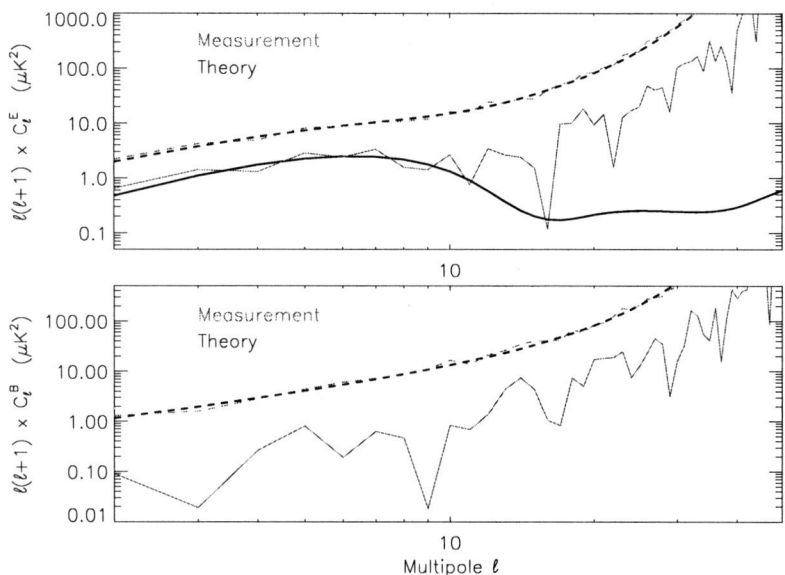

FIGURE 3. E and B-mode power spectra (solid), together with their one sigma errors including Cosmic Variance and white noise (dashed), as predicted by the theory (smooth thick) and as measured from 60 simulated maps including signal and white noise (jagged). The B-mode power spectrum is zero in the theoretical model we used.

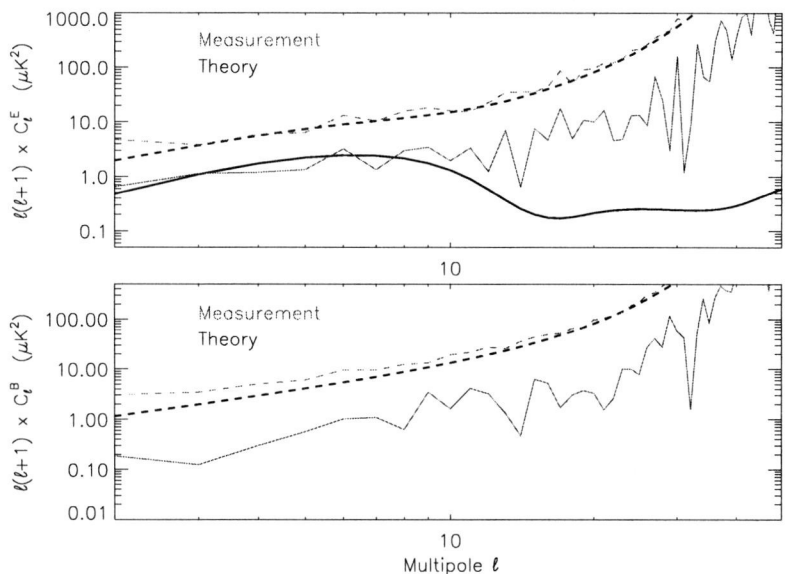

FIGURE 4. Same theoretical power spectra and expected errors as in Fig. 3 (smooth thick), compared to the average and spread of power spectra measured on 60 simulated maps including signal and white plus 1/f noise ($f_k = 10^{-4}$), after destriping (jagged).

The effect of residual pixel noise correlations for the more realistic value $f_k = 1.8 \times 10^{-5}$ Hz is being evaluated but it is not available yet. The analysis performed up to now is meant to provide only a conservative upper limit on the possible effects of low frequency noise on the SPOrt-ISS real performances.

With real SPOrt-ISS data the noise contribution to the measured Q and U maps can be estimated by subtracting two half mission maps, and then checked for compatibility with the expected distribution.

CONCLUSIONS

We have implemented a simple destriping algorithm based on an iterative procedure that is suitable for SPOrt-ISS polarization data. Improvements of the algorithm are still possible and are currently under study. After performing a complete analysis on simulated maps containing both a CMB polarization signal and full instrumental noise, we are able to set a conservative upper limit of $\simeq 15\%$ on the possible sensitivity worsening in SPOrt-ISS measured CMBP power spectra due to the presence of low frequency noise, in the unlikely case of a knee frequency about one order of magnitude larger than expected.

ACKNOWLEDGMENTS

This work has been performed within the SPOrt Collaboration and has been financially supported by the Agenzia Spaziale Italiana (ASI). Authors wish to thank B. Audone and F. Amisano for fruitful discussion and acknowledge use of the CMBFAST and HEALPix packages.

REFERENCES

1. Janssen M.A. et al., Internal Report PSI-96-01, 1996, astro-ph/9602009.
2. Carretti E. et al, "The SPOrt Project", these Proceedings.
3. Carretti E., Tascone R., Cortiglioni S., Monari J., and Orsini, M., *NewA* **6** 173-187 (2001).
4. Wright E. L., "Scanning and Mapping Strategies for CMB Experiments", IAS CMB Data Analysis Workshop, 1996, astro-ph/9612006.
5. Tegmark M., *ApJ* **480**, L87-L90 (1997).
6. Zaldarriaga M., 1998, Ph.D. Thesys, M.I.T., astro-ph/9806122.
7. Szapudi I., Prunet S., Pogosyan D., Szalay A.S., and Bond J.R., *ApJ* **548**, L115-L118 (2001).

Statistical Evaluation and Processing of Uncorrelated and Correlated Outputs of the SPOrt Radiometer

Primo Attinà, Bruno Audone, Franco Amisano

Alenia Spazio S.p.A.,
Strada Antica di Collegno 253, 10146 Turin, Italy

Abstract. The outputs of SPOrt radiometer present different statistical properties. The determination of the statistics of SPOrt radiometer outputs allows the determination of the maximum likelihood estimator for the best possible estimation of the signals. In the data processing 1/f instrumental noise can be eliminated through the technique of circulant matrices.

INTRODUCTION

The SPOrt Experiment has been conceived to obtain a measurement of the polarized component of the sky emissions at microwaves. The measurement of this component has a great importance for scientists and a successful experiment would be an important achievement for radio-astronomical research.

The SPOrt radiometer will provide two different outputs: the total power measurements and the Q and U Stokes parameter measurements. Observations shall be performed at 22, 32, 60 and 90GHz.

In our work the input signal, corresponding to the Cosmic Background Radiation (CBR), is represented as white noise. It means that this kind of signal can be characterized through a normal distribution.

At the end of the Ortho Mode Transducer (OMT), on each channel of the instrument there will be a signal with these characteristics. However, it must be remarked that the signals on the two radiometer channels are not statistically independent because of the presence of polarized CBR. The determination of output signal statistics has been a significant achievement of the SPOrt radiometer analysis.

DETERMINATION OF INSTRUMENT SENSITIVITY

By appropriately defining the functional blocks of the instrument, both for the total power outputs and the Q and U Stokes parameters' outputs, it is possible to define the sensitivity of the instrument for the two different outputs.

The sensitivity is defined by determining the minimum signal power that can be detected by the instrument (*the system noise output power is equal to the output power due to the sky signals*).

By expressing it in terms of equivalent temperature, the sensitivity will be:

Total power $$\Delta T = T_{sys} \cdot \sqrt{\frac{1}{B \cdot \tau}} \qquad (1)$$

CP609, *Astrophysical Polarized Backgrounds*, edited by S. Cecchini et al.

Q and U Stokes parameters
$$\Delta T = T_{sys} \cdot \sqrt{\frac{1}{2 \cdot B \cdot \tau}} \qquad (2)$$

T_{sys} is the system noise temperature, B is the pre-detection bandwidth and τ is the time constant of post-detection low-pass filter. In the equation it has been assumed that no channel gain instability occurs.

STATISTICS OF RADIOMETER OUTPUTS

The outputs of SPOrt radiometer present different statistical properties. The determination of the statistics of SPOrt radiometer outputs is necessary because it allows the determination of the maximum likelihood estimator that will allow the best possible estimation of the signals.

After the band-pass filtering each signal can be represented as a complex narrow-band Gaussian process with zero mean value. The generic signal of this kind will be indicated with x(t). It will be:

$$x(t) = u(t) + j \cdot v(t) \qquad (3)$$

where the signal components u(t) and v(t) are real Gaussian processes. The probability density function (PDF) of x(t) is a complex Gaussian PDF. It can be obtained by assuming that u(t) and v(t) are statistically independent. It is supposed that they are real Gaussian processes with null mean and equal variance (corresponding to the half part of the x(t) variance σ_x^2).

For the total power measurement square law devices are employed for signal detection. The detector output can be defined as follows:

$$y(t) = a \cdot |x(t)|^2 = a \cdot \left[u^2(t) + v^2(t) \right] \qquad (4)$$

In the expression, a is the constant parameter characteristic of the square law device.

With the appropriate mathematical steps the PDF of y has been evaluated. The signal can be expressed as the sum of two terms, corresponding to the square of u and v. It will be:

$$y = w + z \qquad (5)$$

With the appropriate mathematical steps, the PDF of y is obtained.

$$f_y(y) = \int_{-\infty}^{+\infty} f_w(y-z) \cdot f_z(z) \cdot dz = \frac{1}{a \cdot \sigma_x^2} \exp\left(-\frac{y}{a \cdot \sigma_x^2} \right) \qquad (6)$$

In the total power case the density function is exponential. Through its knowledge it is possible to evaluate the mean value of y, representing the total power of the input signal entering a channel. However, the total power measurement does not allow the discrimination of the polarized CBR power from the unpolarised term contribution.

This result can be achieved by evaluating the Q and U Stokes parameters. The corresponding output of the radiometer can be expressed in the following form:

$$y = x_A \cdot x_B^* \qquad (7)$$

The PDF of the random process defined by previous equation can be evaluated by expressing the signals in terms of their real and imaginary parts and by introducing the multivariate complex Gaussian PDF.

$$\mathbf{x} = \begin{pmatrix} \mathbf{x}_A \\ \mathbf{x}_B \end{pmatrix} \tag{8}$$

The expression of the multivariate complex Gaussian PDF is as follows:

$$f_{\mathbf{x}}(\mathbf{x}) = \frac{1}{\pi^2 \det(\mathbf{C}_{\mathbf{x}})} \exp\!\left[-(\mathbf{x}-\boldsymbol{\mu})^{\mathrm{H}} \cdot \mathbf{C}_{\mathbf{x}}^{-1} \cdot (\mathbf{x}-\boldsymbol{\mu})\right] \tag{9}$$

The vector $\boldsymbol{\mu}$ consists of the mean values of the complex processes that compose $\mathbf{x}$. According to the previous assumptions, the two mean values will be null in the present case.

The matrix $\mathbf{C}_{\mathbf{x}}$ is the covariance matrix of $\mathbf{x}$.

$$f_{x_A, x_B}(x_A, x_B) = \frac{1}{\pi^2 \sigma_x^4 \cdot (1 - |r|^2)} \exp\!\left[-\frac{|x_A|^2 + |x_B|^2 - 2\Re\{r^* x_A x_B^*\}}{\sigma_x^2 \cdot (1 - |r|^2)}\right] \tag{10}$$

The term r is the cross-correlation coefficient between the signals on the two channels. It is complex and has a module lower than 1. With the appropriate mathematical steps, the final expression of the PDF is obtained. It is not an exponential PDF as in the total power output case and has a more complex expression.

$$f_y(y) = \frac{2}{\pi \sigma_x^4 \cdot (1 - |r|^2)} \cdot \exp\!\left[\frac{2 \cdot \Re\{r^* \cdot y\}}{\sigma_x^2 \cdot (1 - |r|^2)}\right] \cdot K_0\!\left[\frac{2|y|}{\sigma_x^2 \cdot (1 - |r|^2)}\right] \tag{11}$$

The function K_0 is the modified Bessel function of the second type, of 0 order.

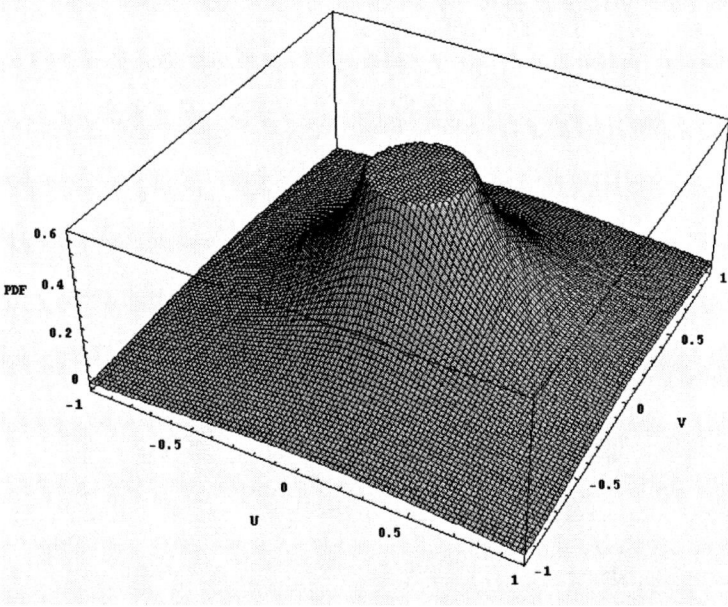

FIGURE 1. PDF of y as a function of u and v, with r amplitude equal to 0.1 and phase equal to 30°

Applying the necessary analytical elaboration, the resulting expression of mean value of y is the following:

$$E\{y\} = r \cdot \sigma_x^2 \tag{12}$$

In Figure 1 the PDF of y for r with amplitude 0.1 and phase 30° has been represented. Variance of x has been assumed equal to 1.

ESTIMATORS

Through the previously determined density functions, it is possible to verify if arithmetical the average of the measurements represents the best possible estimator.

For the measurement of the total power output, the analysis has demonstrated that the arithmetical average is the Minimum Variance Unbiased Estimator (MVUB) for this kind of data. This conclusion depends on the PDF of the radiometer output corresponding to total power measurements, determined as an exponential density function.

However, due to the complexity of PDF it is quite probable that this estimator will not allow the maximum likelihood estimate of Stokes parameters.

Following these considerations, it can be concluded that the arithmetical average is not the MVUB estimator for Q and U Stokes parameters. Assuming that the main goal of the SPOrt experiment is the determination of Q and U, a different estimator should be chosen.

POST-PROCESSING OF MEASURED DATA

Another important topic in SPOrt Experiment is the post-processing of CBR data.

Sky mapping is affected by the long-term effects of 1/f instrumental noise and by the spurious polarisation effects due to the real characteristics of the radiometer.

The removal of the effects of spurious polarisation is based on the data properties depending on orbit characteristics. For the elimination of instrument noise it is possible to develop appropriate signal processing algorithms that allow the determination of the cosmic radiofrequency signal.

The results of the measurements can be represented as M signal levels grouped in a column vector y, that is the TOD (Time Ordered Data) vector:

$$\mathbf{y} = \begin{bmatrix} y_1 \\ y_2 \\ .. \\ y_M \end{bmatrix} \tag{13}$$

It can be expressed as follows:

$$\mathbf{y}(t) = \mathbf{x}(t) + \mathbf{n}(t) = \mathbf{x}(t) + \mathbf{n}'(t) + \mathbf{n}''(t) \tag{14}$$

where $\mathbf{n}'(t)$ is the white noise contribution and $\mathbf{n}''(t)$ is the 1/f noise contribution. The output signal vector $\mathbf{x}(t)$ is related to the vector $\boldsymbol{\theta}$ of the sky pixels CBR contributions through the known matrix $\mathbf{H}$ that depends on the antenna pattern:

$$\mathbf{x} = \mathbf{H} \cdot \boldsymbol{\theta} \tag{15}$$

The aim of signal processing technique employment is the estimation of θ. The measured data are to be processed with Data Adaptive Rank-Shaping Methods, so that it is possible to evaluate the signal contributions.

Assuming the employment of linear methods, the estimator vector may be expressed as:

$$\widetilde{\theta} = \mathbf{W} \cdot \mathbf{y} = \left[\mathbf{H}^{\mathrm{T}} \cdot \mathbf{M} \cdot \mathbf{H} \right]^{-1} \mathbf{H}^{\mathrm{T}} \cdot \mathbf{M} \tag{16}$$

The matrix $\mathbf{M}$ is the inverse of the covariance matrix of the noise vector $\mathbf{n}$, indicated as $\mathbf{N}$. The removal of 1/f noise can be performed assuming that $\mathbf{N}$ is circulant and symmetric. A circulant and symmetric matrix has the following generic form:

$$\mathbf{N} = \begin{pmatrix} c_0 & c_1 & c_2 & .. & c_2 & c_1 \\ c_1 & c_0 & c_1 & .. & 0 & c_2 \\ c_2 & c_1 & c_0 & .. & 0 & 0 \\ .. & & & & & .. \\ c_2 & 0 & 0 & & c_0 & c_1 \\ c_1 & c_2 & 0 & .. & c_1 & c_0 \end{pmatrix} \tag{17}$$

The filtering of 1/f noise is performed by a matrix based on the results of the decomposition of $\mathbf{N}$ (Fourier matrix and eigenvalues matrix).

The estimation of θ has been performed with reduced rank. Different estimators have been considered, depending on the weight coefficients (Abrupt, Unbiased, Maximum Likelihood, Conditional Mean and Wiener estimation).

Software tools in MATLAB[1] have been employed for simulating the previously mentioned steps. The Mean Square Error (MSE) has been evaluated for each kind of estimator depending on the Signal to Noise Ratio (SNR) and assuming different values for both the signal mean value and its standard deviation. It is representative of the estimator effectiveness in achieving the knowledge of signals to be determined.

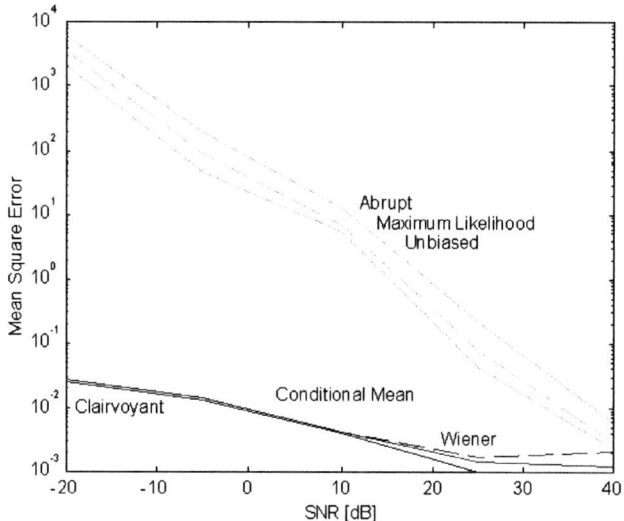

FIGURE 2. MSE for different estimators (Mean Value=0, Standard Deviation=0.1).

---

[1] MATLAB is the software package for mathematical and technical computing by TheMathWorks, Inc.

In the following diagrams, several cases have been reported. Wiener estimator can be employed for null mean value signals only.

The simulation results show that the Mean Square Error is minimum for a very low Signal to Noise Ratio with the Conditional Mean estimator, while the performance of the other estimators (Abrupt, Maximum Likelihood, Unbiased) significantly improves for high SNR levels. However, in SPOrt Experiment very low SNR levels are to be expected, so the Conditional Mean estimation is the best for the filtering of output measurements.

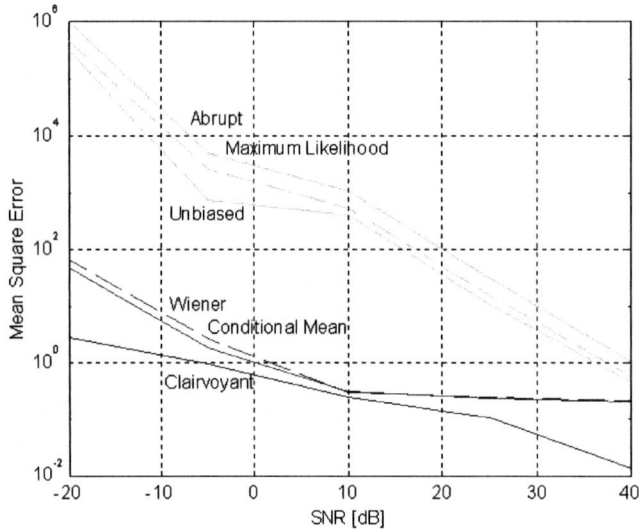

FIGURE 3. MSE for different estimators (Mean Value=0, Standard Deviation=1)

CONCLUSION

The previously reported analysis has shown that the typical considerations about output characteristics of total power radiometers can be applied to SPOrt Experiment too. However, the outputs of Q and U Stokes parameters' measurements present different statistical properties. The involvement of appropriate signal processing techniques may be an useful support for the removal of noise contributions and the cleaning of data.

A Code for Dust Polarized Emission Simulation

N. Ponthieu

ISN, 53 av. des Martyrs, 38026 Grenoble, France, ponthieu@isn.in2p3.fr

Abstract. Visible and infrared radiation polarization by dust grains absorption is a widely debated subject, with theoretical predictions, measurements, and numerical simulation tools. On the other hand, polarization by thermal emission is still waiting for CMB designed experiments for instance, to get its first precise and reliable data. PLEEZ is a numerical simulation code that computes the integration along the line of sight of such a radiation, once the physical parameters of the cloud that yield to polarized emission are provided, such as grain emission properties and alignment characteristics. The first step in PLEEZ implementation is presented here.

PLEEZ MOTIVATIONS

Several authors have proved the asymmetry of galactic dust grains [1], and estimated its influence on polarization by absorption [2, 3]. Concerning polarization by emission, it is clear that the same asymmetry in the grain shape is responsible for asymmetrical radiation, which should lead to polarized radiation. However, the characteristics of emission lobes are not known precisely, and forthcoming measurements with CMB designed experiments like ARCHEOPS, MAP and PLANCK are expected to provide the first precise data in the millimeter range. In the mean time, no simulation code is publicly available to estimate this contribution to infrared polarized backgrounds. PLEEZ aims to provide the line of sight integration of a cloud's radiation, which is one of the steps required for such a simulation.

PHYSICAL PARAMETERS

As PLEEZ is designed to be used once a theory on dust has provided the physical characteristics of the dust cloud, it takes them as input. It will therefore deal with :

1. The local dust density in the cloud.
2. The emission lobe, that is to say the spatial distribution of the intrinsic emission of the grain.
3. The polarization degree of the emission.
4. The grain alignement, that is to say the spatial and dynamical description of the electromagnetic coupling of the grains to the local magnetic field : average direction, dispersion, precession.

By "emission lobe" we usually mean the energy radiated by the system in a given direction of observation. This direction will be denoted by $\vec{e_3}$ in the following. In the case of polarization, we will focus on the electric field of the radiation. To characterize its orientation (polarization), we will distinguish between the direction of the main symmetry axis of the grain projected on the plane of sky (denoted $\vec{e_1}$ hereafter) and a second direction $\vec{e_2}$, orthogonal to $\vec{e_1}$. $(\vec{e_1}, \vec{e_2}, \vec{e_3})$ form an orthonormal basis (cf. fig. 1).

PLEEZ ALGORITHM

To deal with the previous parameters, PLEEZ is implemented in the following way:

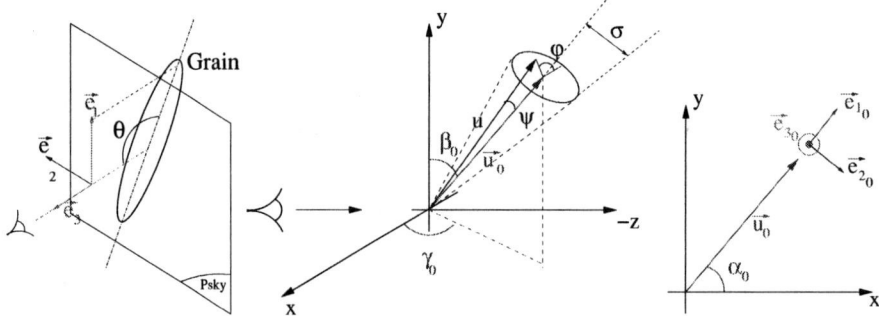

FIGURE 1. 1st *fig.*: Natural frame of a grain. The plane of sky (P_sky) is the plane orthogonal to the line of sight $\vec{e}_3$. The main symmetry axis of the grain is projected on P_sky into $\vec{e}_1$. A third axis is chosen in P_sky to form an orthonormal basis : $(\vec{e}_1, \vec{e}_2, \vec{e}_3)$. 2nd *fig.*: Alignment parameters. $\vec{u}_0$ defines the mean direction with which the grains are aligned in a gaussian cone of variance σ^2. 3rd *fig.*: α_0 is the angle between $\vec{u}_0$ and $\vec{x}$. $\vec{e}_{1_0,2_0,3_0}$ are the directions defined on the first figure and are relative to a grain perfectly aligned with $\vec{u}_0$.

1. **Dust density**. PLEEZ takes this information among the requested parameters that describe the cloud.

2. **Emission lobe.** In order to accomodate any shape of the lobe, and therefore be adaptable to all theoretical predictions on the grain physical properties, PLEEZ takes as input the a_{lm} coefficients of the expansion:

$$E(\theta, \phi) = \sum_{l,m} a_{lm} Y_m^l(\theta, \phi) \tag{1}$$

where E is the electric field radiated by the grain, θ is the angle of sight defined on Fig. 1 and ϕ is the azimuth angle, w.r.t. the grain axis. The user then provides a shape for the two directions $\vec{e}_1$, $\vec{e}_2$, and a phase term.

3. **Polarization degree.** To determine the polarization degree, the relative intensities I_1 and I_2 of the two projections of the electric field on $\vec{e}_1$ and $\vec{e}_2$ must be fixed. In the case of a thermal radiation modelling, such as a grey body, the total emitted intensity I refers to a 4π solid angle. To fix I_1 and I_2 we require : $I = I_1 + I_2$. It also offers the possibility to separate the influence of the shape of the lobes and that of the intensity on polarization.

4. **Alignment and Line of sight integration.** All angular parameters are defined on Fig. 1. The average direction $\vec{u}_0$ is one of the inputs. The grains are then assumed to be aligned with $\vec{u}_0$ with a gaussian dispersion σ (cf. Fig. 1). By this, we mean that ϕ is a uniform random variable that is in $[0, 2\pi]$ and that ψ is gaussian, centered, of variance σ^2.

In order to perform the integration analytically and to give a first order approximation of the expected radiation, it is assumed that σ is small. We thus get, after integration of the statistical distribution of the alignement, e.g. the mean projection on $\vec{x}$ in the observer's frame of the electric field $\vec{E}$:

$$\overline{E_x} = E_x(\vec{u}_0) \times e^{-\sigma^2/2}(1 + \frac{\Lambda}{2} e^{-2\sigma^2}) \tag{2}$$

where $E_x(\vec{u}_0)$ stands for the x projection of the emission lobe if the grain is perfectly aligned with $\vec{u}_0$, and Λ is a constant depending on $\vec{u}_0$ as well. It is now crucial to determine how "small" σ must for the previous approximation to be justified.

To test this, we performed Monte–Carlo simulations. We generated various grain alignments with various emission lobes and compared the numerical integration to the analytical approximation. The results are represented on Fig. 2. This plot shows the comparison for a $\sin^3 \theta$ lobe and for an orientation of $(\beta_0, \gamma_0) = (40, 30)$ (see Fig. 1 for β_0, γ_0 definition). It shows that an expansion until $Y_{3,m}$ can be safely described by our first order approximation, and this, for any σ.

Two other factors must be considered : the Rayleigh factor $3/2(<\cos^2 \psi> -1/3)$ that accounts for depolarization due to grain precession around $\vec{u}_0$, and the factor $F = 3/2(<\cos^2 \alpha_0> -1/3)$ that stands for the depolarization due to $\vec{u}_0$ rotation along the line of sight. Both these factors are directly calculated by PLEEZ, accounting for the alignment input parameters and the cloud's description.

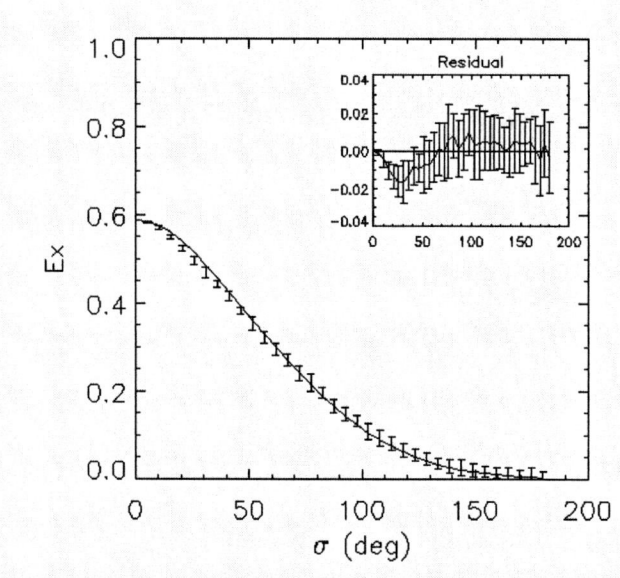

FIGURE 2. Monte–Carlo simulation of the projection on the x axis of the electric field for a $\sin^3\theta$ lobe, $(\beta_0,\gamma_0) = (40,30)$ (see Fig. 1 for β_0,γ_0 definition), as a function of the alignment cone standard deviation σ.

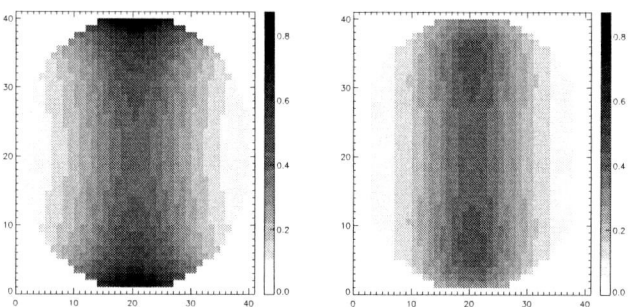

FIGURE 3. Cloud toy model analysis. Diagrams show the polarization degree for an inclination $<\theta>$ of 0 and 30° of the mean alignment direction w.r.t. the line of sight (see Fig. 1 for θ definition). The average degree of polarization are 0.32 and 0.23.

TOY MODEL

In order to show an application of PLEEZ, a toy model is defined as following:

- The cloud is assumed to be spherical, with homogeneous density.
- The alignment is along $\vec{u}_0$, assumed to be "vertical", and σ decreases linearly from 0 to 90° with the distance to the main axis. R is often used to characterize the alignment coherence, and varies here on its whole range, from 1 (perfect alignment) to 1/3 (random alignment).
- The grains are assumed to radiate only along their main axis direction ($\vec{e}_1$), with a dipole lobe. This choice is made to enhance the orientation effect on polarization.

The results are represented on Fig. 3. The cylindrical symmetry along the "vertical" axis is obviously due to the dispersion model adopted. The reader can see that when the line of sight is orthogonal to the mean direction of alignment (left figure), the most polarized pixels are those on the poles. It is because these pixels are on the main axis and therefore have a high level of alignment coherence. The depolarization from the poles to the center of the cloud is due to the damping that results from the integration of edge pixels, far from the central axis, with incoherent alignment. When the alignment direction is tilted, then every line of sight intercepts edge pixels, and the pole–center gradient on polarization maps is damped. It should also be pointed out that the average level of polarization of the cloud is consistent with theoretical predictions for such a perfectly polarized grain emission.

CONCLUSION

PLEEZ has been designed to enable the user to integrate the polarized emission of a dust cloud. It especially offers the possibility of varying the emission lobe properties. Tested on a toy model, it gives consistent results with theoretical predictions. Further implementation should offer the possibility to investigate more sophisticated models of grains distribution and alignment.

REFERENCES

1. R. H. Hildebrand & M. Dragovan, ApJ 450, 663–666, (1995)
2. B. T. Draine, ApJ Suppl. Series 57, 587–594, (1985)
3. B. T. Draine & P. J. Flatau, http://xxx.lanl.gov/abs/astro–ph/0008151v3

Some sources of systematic errors on CMB polarized measurements with bolometers

Jean Kaplan, Jacques Delabrouille

PCC, Collège de France, Paris

Abstract. Some sources of systematic errors, specific to polarized CMB measurements using bolometers, are examined. Although the evaluations we show have been made in the context of the Planck mission (and more specifically the Planck HFI), many of our conclusions are valid for other experiments as well.

THE SPECIFICS OF CMB POLARIZATION SIGNALS

CMB fluctuations are difficult to measure because of their extreme weakness. Systematics must therefore be very well controlled. This is even more true for polarization anisotropies, which are expected to be less than 10% of the temperature fluctuations.

Low frequency noise is a source of troubles for polarization as well as for temperature measurements. However, as the polarized signal depends on the direction of the detector projected in the sky, its suppression, or "destriping" requires a specific treatment. The way in which polarized destriping can be implemented for Planck is outlined in the next section.

A few definitions useful to the following discussion are given in the third section.

Polarizers are never perfect: the unwanted polarization is never totally suppressed and the polarizer direction is never perfectly known. The impact of these uncertainties is evaluated in section four.

In the fifth section, we discuss the crucial difficulties linked to signal differences: calibration, pointing and beam mismatches between different detectors.

We then consider the question of avoiding elliptical error boxes on the Stokes parameters, which might be confused with a polarization signal when the signal to noise ratio is small.

A few concluding remarks are given in the last section.

POLARIZED DESTRIPING

This section is devoted to the elimination of low frequency noises in the framework of Planck. It relies on the Planck scanning strategy, which goes as follows: The telescope beam rotates 60 times around a fixed axis with an opening angle around 85°. Then the axis is shifted by a few arc-minutes and the beam is again rotated 60 times etc... Averaging

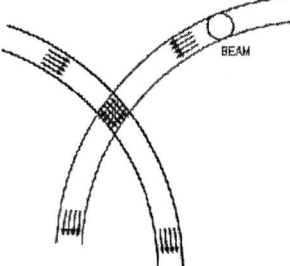

FIGURE 1. Circle crossing

CP609, *Astrophysical Polarized Backgrounds*, edited by S. Cecchini et al.

over the 60 scans of the same circle suppresses most noise with frequencies smaller than the spinning frequency. The remaining noise can be described by 1 offset per ring for each of the 3 Stokes parameters, irrespective of the number of polarized detectors. The circles have many intersections with each other in the sky. At these circle crossings (see figure 1), one can use the redundancy by asking that the sky Stokes parameters be the same along both circles. This results in solving a linear system with the Stokes parameter offsets as variable. The quality of the "destriping" can be

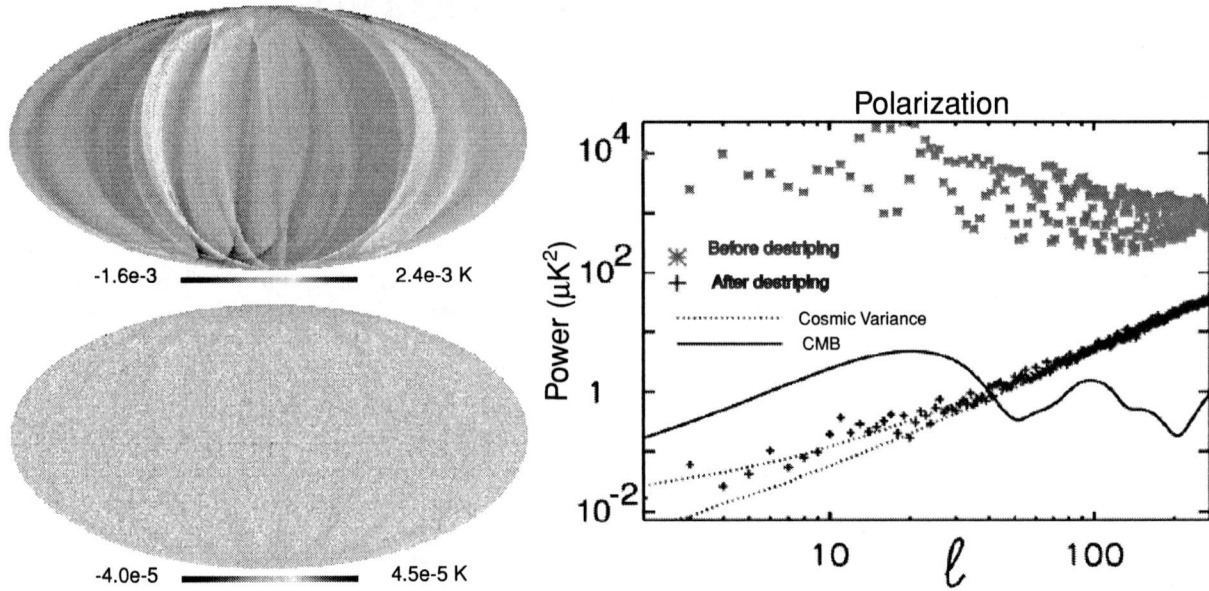

FIGURE 2. Q residual maps before (upper left panel) and after (lower left panel) destriping. The E_l coefficients of the residual maps before and after destriping (right panel)

seen on figure 2. For further details, see [1].

SOME USEFUL DEFINITIONS

Let us consider a "Polarization Sensitive Bolometer (PSB)". The path followed by the incoming radiation is illustrated in figure 3. The far field Stokes parameters $(I, Q, U)_{\text{Far}}(\vec{n})$ are integrated by going through the telescope and the horn

FIGURE 3. The path of a radiation entering a PSB

to give the Stokes parameters on the PSB, $(I, Q, U)_{\text{PSB}}$:

$$
\begin{aligned}
I_{\text{PSB}} &= \int \left(a_I^I(\vec{n}) \, I_{\text{Far}} + a_I^Q(\vec{n}) \, Q_{\text{Far}} + a_I^U(\vec{n}) \, U_{\text{Far}} \right) d\vec{n} \\
Q_{\text{PSB}} &= \int \left(a_Q^I(\vec{n}) \, I_{\text{Far}} + a_Q^Q(\vec{n}) \, Q_{\text{Far}} + a_Q^U(\vec{n}) \, U_{\text{Far}} \right) d\vec{n} \\
U_{\text{PSB}} &= \int \left(a_U^I(\vec{n}) \, I_{\text{Far}} + a_U^Q(\vec{n}) \, Q_{\text{Far}} + a_U^U(\vec{n}) \, U_{\text{Far}} \right) d\vec{n}
\end{aligned}
\tag{1}
$$

In principle the 9 real functions $a_I^I \ldots$ are needed to characterize the beam completely. For a PSB at the center of a perfect instrument, only a_I^I, a_Q^Q and a_U^U are present. The signals measured by the two detectors of the PSB are given by:

$$m_1 = \frac{g_1}{2}\left((1+\varepsilon)I_{\text{PSB}} + (1-\varepsilon)(Q_{\text{PSB}}\cos 2\alpha_1 + U_{\text{PSB}}\sin 2\alpha_1)\right)$$
$$m_2 = \frac{g_2}{2}\left((1+\varepsilon)I_{\text{PSB}} - (1-\varepsilon)(Q_{\text{PSB}}\cos 2\alpha_2 + U_{\text{PSB}}\sin 2\alpha_2)\right) \qquad (2)$$

where ε is the rate of cross polarization leakage: if the incoming radiation is not polarized, ε is the ratio of the transmitted intensity polarized in the wrong direction to that polarized in the right direction. The angles α_1 (α_2) are the angle between the polarised sensitive direction 1 (2) and the x (y) axis of the local reference system. Ideally these two directions are exactly orthogonal and one can choose the local reference frame so that $\alpha_1 = \alpha_2 = 0$ to remove the contribution of U_{PSB}, and make the three coefficients a_U^I, a_U^Q and a_U^U irrelevant.

The gain factors g_1 and g_2 are in general different.

UNCERTAINTIES ON POLARIZATION LEAKAGE AND POLARIZERS ORIENTATIONS

Uncertainties on the characteristics of the polarimeters will generate systematic errors. We focus here on two specific examples:

1. **The cross-polarization leakage** ε in equation (2) is only known up to some uncertainty.

2. **The polarimeter orientation in the sky** The angles α_1 and α_2 in equation (2) as well as the relative orientation of the two PSB's necessary to measure the 3 Stokes parameters are not exactly known either.

We have evaluated these effects as follows

i) Assume a "theoretical setup" of polarimeters with given rates of cross-polarization leakage and given orientations in the sky.

ii) Assume that, due to imperfections in building the instrument, the actual set up is different. An "actual setup" is built by adding random errors to the cross-polarization leakage and to the polarimeter orientations.

iii) Observe a set of Stokes parameters I, Q, U with the "actual setup"

iv) Reconstruct the Stokes parameters using the "theoretical setup"

v) Compare the original and reconstructed Stokes parameters for a random sequence of "actual setups".

The results are displayed in tables 1 and 2 , obtained with 4 polarimeters, $T = 2.73$ K, $Q = U = 1\mu$K. Note that a known rate ε of cross-polarization does not contribute to the systematic error but increases the statistical uncertainty by a factor $1/\sqrt{1-\varepsilon}$

TABLE 1. Errors due to uncertainties on the rates of cross-polarization leakage

RMS error on leakage rates	Relative RMS error on reconstructed polarization	average error on reconstructed polarization direction
0.01	1.4%	0.2°
0.05	7%	2°
0.1	15%	4.5°
0.2	35%	8°

TABLE 2. Errors due the uncertainties on polarizer orientations

RMS error on polarimeters orientations	Relative RMS error on reconstructed polarization	average error on reconstructed polarization direction
0.1°	0.2%	0.1°
0.5°	1%	0.5°
1°	2%	0.9°
2°	5%	2°
5°	12%	5°
10°	24%	10°

In order to check that the above uncertainties do not impact on our ability to measure polarized power spectra, we can test the effect of an imperfect knowledge of polarimetric calibration parameters in yet another way: a sky map simulated from a set of C_l, E_l and B_l spectra is observed with the "actual setup". The map is then destriped as described above and reconstructed using the "theoretical setup". Finally the C_l, E_l and B_l spectra of the reconstructed

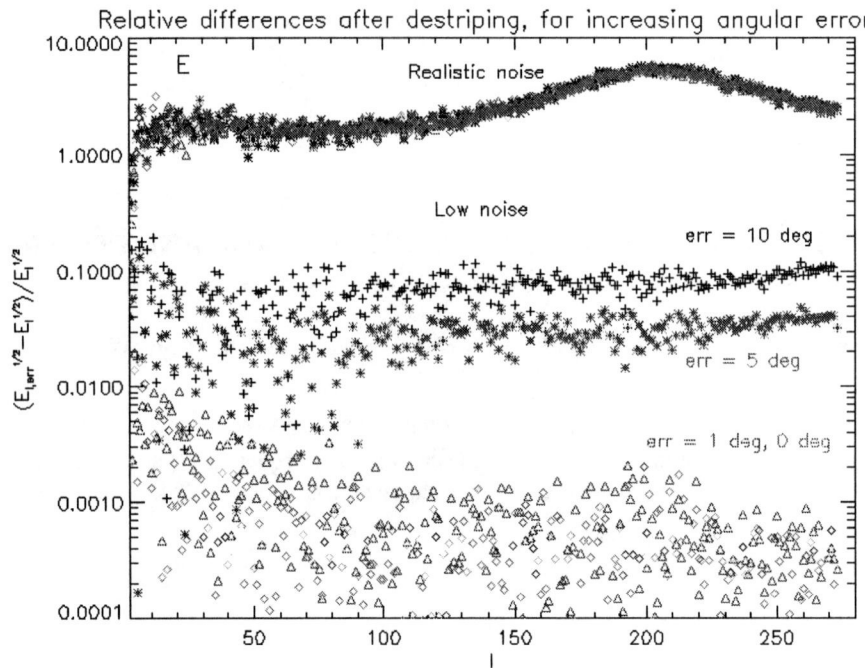

FIGURE 4. The spectra of relative differences between input and output E_l spectra

maps are computed and compared with the inputs. Figure 4 shows the result of this comparison on the E_l spectrum for random orientation errors of order 0 to 10 degrees. With a realistic noise the effects cannot be seen, therefore, the points labelled "low noise" have been evaluated with a noise divided by 10^4. With a $1°$ error, the relative systematic errors on the E_l spectrum remains below 1% (actually closer to 0.1%). The same results apply to the B_l spectrum.

THE CRUCIAL PROBLEM: SIGNAL DIFFERENCES

As the Q and U parameters are computed by differences between detector outputs, any mismatch between the characteristics of the detectors (gain, beam, pointing ...) induces a fake polarization signal.

Relative photometric calibration between polarimeters (cross-calibration)

The gains g_1 and g_2 in equation (2) are in general different. This mismatch should be evaluated and corrected for by cross-calibration. A residual cross-calibration error Δg will result in a spurious polarization signal $\sqrt{Q^2 + U^2} \sim \Delta g \times I$. The Q and U fluctuations induced in this way are strongly correlated to the temperature fluctuations. For the CMB, a 1% calibration error typically induces a 10%-30% systematic error on the polarization fluctuation. A constant calibration mismatch is easily detected and corrected for. The trouble comes from gain variations with time. The time scale for cross calibration have to be very carefully chosen. This question is currently under investigation on the polarized data from the Archeops flight in January 2000. Archeops is a balloon experiment to observe CMB fluctuations with an angular resolution around 10'. It involves 21 bolometers at 143, 217, 353, and 545 GHz. The 6 bolometers at 353 GHz are polarized and arranged in 3 Ortho Mode Transducers. After a technical flight in 1999, the first scientific flight occurred in January 2000 from the ESRANGE base at Kiruna in Sweden, and the data are currently being analyzed. Two more flights are planed in December 2001 and January 2002. (see Ref. [2] for more details).

Pointing and/or beam shape mismatch

The two orthogonal detectors necessary to obtain the Q (or U) parameter from the difference of their outputs should look at the same area of the sky. However, this will in general not be true. Ortho Mode Transducers (OMT) or Polarized Sensitive Bolometers (PSB) (see the contributions of the SPORT and BOOMERANG teams to this workshop) are a nearly perfect answer to this problem as the polarization signal is the difference between the outputs of two detectors sitting behind the same feed. However, even in this case a pointing mismatch can occur if the time constants of the 2 detectors are not the same. This is illustrated in figure 5: in the Planck mission, a 1 ms difference in the time constants

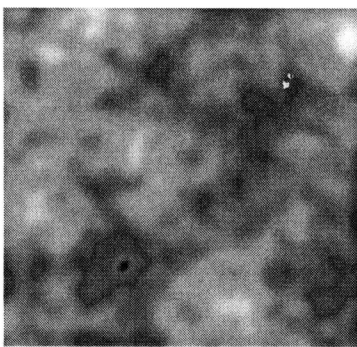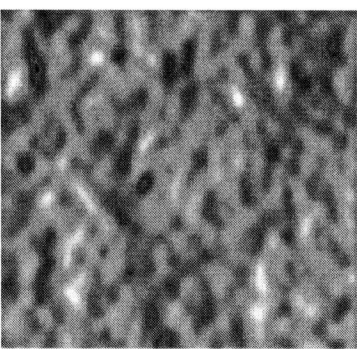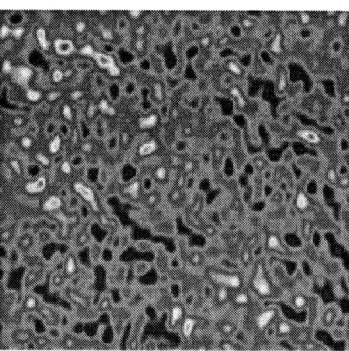

FIGURE 5. The input temperature map (left) and the output Q map with a pointing mismatch of 0.5' (center) and with the two mismatched beams shown in figure 6 (right)

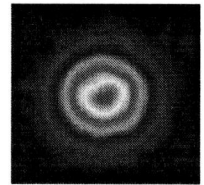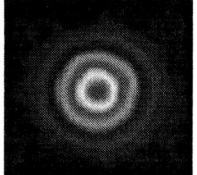

BEAM 1   **BEAM 2**

FIGURE 6. The two mismatched beams leading to the Q map of the right panel of figure 5

of the two orthogonal detectors induces a 0.5' pointing mismatch in the scanning direction. A beam shape mismatch can arise for the same reason and also because the two polarimeters are not oriented in the same way with respect to the horn and the telescope.

In figure 5, a $4.1° \times 4.1°$ temperature map with a dispersion $T_{RMS} = 3.4\,10^{-5}$ and zero polarization (left panel), is observed with two orthogonal polarized detectors. The center panel shows the Q map induced by a pointing mismatch of 0.5' between the two beams, otherwise identical (Gaussian and 7.5' wide). The output map develops Q fluctuations with $Q_{RMS} = 1.2\,10^{-6}$, correlated to the temperature signal. This level is large and only slightly smaller than the expected polarization level of CMB fluctuations. Note that for beams and beam mismatches small compared to the typical CMB structures, the effect grows linearly with the distance between the two beams.

The right panel shows the Q map generated by observing the same input temperature map with the two mismatched beams shown in figure 6. The two beams differ by 2.5% on 1/3 beam size scales. In this case the Q fluctuation has $Q_{RMS} = 3\,10^{-7}$.

OPTIMIZED POLARIMETER CONFIGURATIONS

Because of the low signal to noise ratio, an elliptic error box in the Q,U plane can induce a bias on the polarimeter direction. An elliptic error box means unequal and/or correlated errors on Q and U.

It can be shown [3] that a circular error box is obtained if
i) the polarizer orientations are evenly distributed over 180°, as in figure 7 for 4 polarimeters

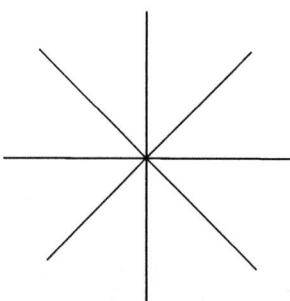

FIGURE 7. An optimized configuration of 4 polarized detectors involving two OMT's or PSB's at 45° from each other. In reality, the two pairs of polarimeters point in the same sky direction at different times.

ii) the noise level should be as homogeneous as possible and uncorrelated among the polarized detectors.

If these conditions are realized, one get as a bonus that the volume of the error box in the I, Q, U Stokes parameter is minimal. Of course, this second condition will be the most difficult to realize in practice.

CONCLUSIONS

In an experiment such as Planck HFI, where polarization measurements are made with detectors sensitive to the total polarized intensity in one direction, the main source of systematic error in polarization measurements is the fact that Q and U Stokes parameters are obtained from signal differences. This can be overcome and even turned into an advantage if the systematics are common to both detectors with the same size and therefore disappear in the difference.

OMT's and PSB's are a partial answer to this requirement as the two detectors have nearly the same lobes and pointings. However the electronic chains (and part of the optics for OMT's) are different. Moreover one still has to combine the signals of two different feeds to get the full polarized information. This latter combination is less dangerous however, as it does not involve intensity differences, and therefore will not generate polarisation where there is none.

A rotating polarizing device in front of *one* bolometer in *one* feed and read by *one* electronic chain may provide a solution to these difficulties, but one has to check that it does not bring new systematics and it may be difficult to implement on satellite or balloon borne experiments.

REFERENCES

1. Revenu, B., Kim, A., Ansari, R., Couchot, F., Delabrouille, J., and J., K., *A&ASS*, **142** (2000), astro-ph/9905163.
2. Benoît, A., et al., *To appear in Astroparticle Physics* (2001).
3. Couchot, F., Delabrouille, J., Kaplan, J., and Revenu, B., *A&ASS*, **135**, 579 (1999), astro-ph/9807080.

Systematics of Microwave Polarimetry with the *Planck* LFI

J. P. Leahy*, V. Yurchenko[†], Morag A. Hastie*, M. Bersanelli** and N. Mandolesi[‡]

*University of Manchester, Jodrell Bank Observatory, Macclesfield, Cheshire, SK11 9DL, England
[†]Dept. of Experimental Physics, National University of Ireland, Maynooth, Co. Kildare, Ireland
**Universita di Milano, Dipartimento di Fisica, via Celoria 16, Milano, I-20133, Italy
[‡]TESRE-CNR, via P. Gobetti 101, Bologna, I-40129, Italy

Abstract. The *Planck* Low Frequency Instrument will recover polarization by differencing the outputs from radiometers sensitive to orthogonal polarizations. We contrast the systematic errors that afflict such a system with those that affect correlation polarimeters; the *Planck* design has some important advantages when measuring very weak signals such as the CMB. We also review systematic effects arising from the choice of scan strategy for all-sky mapping missions like *Planck*. [For the Planck-LFI consortium].

DIFFERENCING VS. CORRELATION POLARIMETERS

Radiometers can be used to measure polarization in two fundamentally different ways.

Differencing polarimeters work in much the same way as polarimeters used in other wavebands; that is, the radiation is filtered to isolate components in several different pure polarization states and these are combined to derive the required Stokes parameters. The *Planck* Low Frequency Instrument (LFI), described in these proceedings by Villa, is of this type. Each feed horn couples the incoming radiation to a waveguide, essentially preserving the polarization state (deviations from this approximation will be discussed later). An OMT separates the radiation into two orthogonal linear components ('X' and 'Y') which are separately amplified and square-law detected. Ideally, the sum of these signals is proportional to Stokes I, while the difference measures one component of the (Q, U) vector. This is formally equivalent to a Wollaston prism system in optical polarimetry. To obtain the full linear polarization vector, more information is needed, and usually a second feed oriented at 45° to the first is used, which also provides a redundant measurement of I. This arrangement leaves V undetermined, but, as we believe there is no significant background of circular polarization, this is not a drawback for *Planck*.

Square-law detection systems have offsets due to receiver noise, noise from lossy components in the signal path, and of course from their sensitivity to the dominant unpolarized emission. The offsets are notoriously variable due to physical temperature changes and also to amplifier gain changes, and they will directly affect the measured Q, U. For this reason most modern polarimeters are based on correlation rather than differencing. Correlation polarimeters make use of phase coherence to derive the full polarization state of the incoming radiation from a single horn. Again an OMT is used to separate two orthogonal polarizations, and each channel is square-law detected. But in addition, the signals from the two channels are correlated, i.e. multiplied together, with and without a 90° phase shift. The two correlation products give the remaining two Stokes parameters. Because all parameters are measured with the same feed, this gives an improvement of $\sqrt{2}$ in sensitivity to linear polarization over a simple differencing system with the same number of feeds; and compared to a more practical null-balanced system (as in the LFI) the correlation receiver has a full factor of 2 advantage. When the main interest is in linear polarization, a polarizer (quarter wave plate equivalent) is usually inserted ahead of the OMT to convert circular to linear polarization, so that the OMT separates polarizations corresponding to right- and left-hand circular on the sky. The correlation products are then the two components of the (Q, U) vector.

Ideally, there would be no offsets in the correlation products, because zero polarization implies zero mean correlation between RHC and LHC. In reality cross-polarization (failure of the OMT to completely separate the RHC and LHC signals) and cross-talk (leakage of noise generated in each amplifier chain into the other) mean that the two channels do have a correlated component, but in conventional systems this polarization offset is much smaller than that on the

CP609, *Astrophysical Polarized Backgrounds*, edited by S. Cecchini et al.

outputs of the square-law detectors. A numerical example is instructive. On a general-purpose telescope, the OMT is required to cover a bandwidth of up to 20 or 30%, which limits the polarization purity to typically 25–35 dB; i.e. the cross-polar voltage is reduced by a factor of ~ 30. As the corresponding voltage in the other channel is unattenuated by definition, the contribution to the correlator offset is $\sim T_{\text{Antenna}}/30$. In total-power systems the offset is dominated by T_{receiver} rather than T_{Antenna}, but in a correlation system we can reduce the cross-talk by more than the cross-polarization, by using circulators or equivalent to provide an additional ~ 20 dB isolation between the channels. Thus we expect an offset of order $T_{\text{Antenna}}/30 + T_{\text{receiver}}/300$. Compared to a simple total power system, with an offset of $T_{\text{Antenna}} + T_{\text{receiver}}$, this is a very big improvement. But consider the goal of measuring μK polarizations in the presence of typically 20 K of receiver noise, and hence a polarization offset of ~ 100 mK. Fluctuations of one part in 10^4 would swamp the wanted signal.

To summarise, in a correlation system, both cross polarization and cross-talk give a false positive signal. Dedicated instruments must therefore be designed to reduce and stabilise these effects. The former is difficult because the critical factor is the voltage, rather than the power, of the cross term. For instance SPOrt, despite an OMT with 60 dB isolation (Peverini et al., these proceedings), still requires the stability offered by a space mission.

In contrast, we will see that cross-polarization and cross-talk are rather unimportant in differencing systems. False positives are instead generated by the total power offsets. But these offsets are also, of course, a serious problem for total power measurements, and practical systems are designed to minimise their effects. When extreme care is taken to do this, as in the *Planck* LFI, the dominant polarization systematic is eliminated at the same time. One is then in a position to make accurate measurements of polarization even when the telescope suffers from cross terms that would be difficult to cope with if it was operated as a correlation polarimeter.

THE LFI RADIOMETERS

As in any space experiment, the LFI is designed to minimise mass, power consumption and complexity consistent with its primary science goal, which is to measure temperature fluctuations in the CMB. For this reason there is no correlator, and the principle reason for measuring both polarizations in each horn is to gain a factor of $\sqrt{2}$ in sensitivity to I. Nevertheless, the potential to measure polarization has always been recognised, and the design has been optimised for this where there is no conflict with the primary goals.

Each feed horn in the LFI feeds two radiometers, one for each polarization. Each radiometer continuously compares the sky brightness with that of an internal 4 K reference load. As described by Villa (these proceedings) the design ensures that the gain (including $1/f$ fluctuations) is identical for the sky and load signals up until detection. After detection, the load gain is adjusted by $\sim$10% to account for the known difference between the load and mean sky system temperatures, and the two signals are differenced to yield a (nominally) zero mean output, whose amplitude is generally dominated by the CMB dipole and therefore is typically $I - T_0 \sim 2$ mK. Despite all these precautions the output is subject to $1/f$ noise mainly due to variations in the amplifier noise temperature, but also due to fluctuations in the reference load temperature, etc. However, the amplitude of of the $1/f$ noise is designed to be below that of the thermal noise for frequencies greater than the 1 rpm spin period of *Planck*. Each feed scans the same circle on the sky, 60 times over between hourly repointings, and stacking the data effectively filters out noise except at frequencies very close to harmonics of the spin frequency. Thus the low frequency $1/f$ noise contributes only to the zero harmonic, a single undetermined offset on each scan circle. Both offset and the average gain can be calibrated by matching to an assumed CMB dipole, after excluding (or modelling) Galactic plane emission [1]. Note that errors in the assumed dipole will cause identical gain errors in the two polarization channels, which will not affect the polarization signal to first order. In practice, residual offsets and gain errors are expected, due to errors in the assumed dipole direction, imperfect Galactic masking, and residual $1/f$ noise at low order harmonics, but the multiply-redundant coverage of the sky means that these can be removed by de-striping techniques [2], which can even be applied directly to the polarization signal [3].

The advantages of the LFI design for polarization measurements could be considerably enhanced if we jettisoned the aim of measuring the total intensity. In this case we could compare the X and Y polarizations with each other directly instead of with the 4 K loads, so that the output of the radiometer was a direct measure of a linear polarization component. This has several advantages. Most importantly the system temperatures of the two channels would be matched at a level of ~ 0.1 K, instead of ~ 1 K, giving better nulling of the residual $1/f$ noise. In addition, the thermal noise is reduced by $\sqrt{2}$ because we have eliminated one level of differencing. Finally, the complexity of the instrument is sharply reduced by eliminating the reference loads, half of the amplifier chains, and half the detectors. Such a design may be appropriate for the post-*Planck* era, when the focus on CMB research will be on polarization.

Polarization response of the LFI radiometers

From the above discussion, we can write the power detected by a given radiometer, say '*P*', as

$$s_P = g_P[T_P - T_0 + R_P + N_P] \tag{1}$$

where g_P is the gain, T_P is the antenna temperature [4] in the polarization matched to this radiometer, T_0 is the temperature assumed for balancing, N_P is white noise and R_P is red (or $1/f$) noise. After averaging the data into rings, N_P is independent at each pixel but R_P is highly correlated.

The polarized brightness can be written in several ways. One of the most useful is in terms of sensitivities to the four Stokes parameters:

$$T_P = \frac{1}{4\pi} \int (I_P I + Q_P Q + U_P U + V_P V) \, d\Omega \tag{2}$$

where S_P is the beam of radiometer P in Stokes parameter S (measured in terms of brightness temperature). Taking $V = 0$, we can re-write this as

$$T_P = \frac{1}{4\pi} \int I_P \{I + \varepsilon_P (Q \cos 2\phi_P + U \sin 2\phi_P)\} \, d\Omega \tag{3}$$

where ε_P is the efficiency of linear polarization response and ϕ_P is the orientation of peak response. Inserting Eq. 3 into Eq. 1, we note the presence of a large total intensity term $I - T_0$, of order 10^3 times the polarized term, as previously discussed.

The two radiometers in each feed are nearly orthogonal so we write $\phi_Y \approx \phi_X + \pi/2$. Thus if we difference the calibrated outputs we get a polarization signal

$$P = \frac{1}{2} \left(\frac{s_X}{g'_X} - \frac{s_Y}{g'_Y} \right) \approx \varepsilon(Q \cos 2\phi + U \sin 2\phi) \tag{4}$$

where g' is our estimate of the true gain g (for simplicity we idealise the beam as a point measurement for now).

More precisely, let $g/g' = 1 + \delta g$, let there be an error $\delta\phi$ in the orthogonality of the two radiometers, and let $\varepsilon_X - \varepsilon_Y = \delta\varepsilon$. Then to second order in the error terms,

$$
\begin{aligned}
P = \ & \langle\varepsilon\rangle (Q \cos 2\phi + U \sin 2\phi) \left[1 + \langle\delta g\rangle - \frac{\delta\phi^2}{2} + \frac{(\delta g_X - \delta g_Y)\delta\varepsilon}{4\langle\varepsilon\rangle} \right] + \frac{\delta g_X - \delta g_Y}{2}(I - T_0) \\
& + \langle\varepsilon\rangle (Q \sin 2\phi - U \cos 2\phi) \delta\phi \left[\frac{\delta g_X - \delta g_Y}{2} + \frac{\delta\varepsilon}{2\langle\varepsilon\rangle} \right] + \frac{R_X - R_Y + N_X - N_Y}{2}
\end{aligned} \tag{5}
$$

Unlike the situation for a correlation polarimeter, cross-polarization (represented here by $\delta\phi$ and $1 - \varepsilon$) only appears in the second order, (except for the overall scaling by $\langle\varepsilon\rangle$), and even then, only affect terms proportional to Q and U, not I. Thus, even if these error terms were completely unknown, we could still detect polarization. In reality we expect them to be very small and calibratable, allowing accurate measurement of the polarized background.

LFI POLARIZATION SYSTEMATICS

Overview

The *Planck* consortium has chosen to analyse systematics in terms of a number of sources, all of which impact on both total intensity and polarization measurements. Many are common in origin and/or effect between HFI and LFI and are being studied jointly. These headings and their principle polarization impacts on the LFI are as follows:

Far Sidelobes: Contamination of the polarized signal by total intensity straylight from the Sun and Galaxy (because the I_P far sidelobes are significantly different for the X and Y polarizations, so fail to cancel); and also by Galactic polarized straylight.

Main Beam: Cross-polarization is discussed in detail in the next section.

Pointing: Alignment errors between different horns corrupt the reconstruction of the (Q, U) vector at each pixel.

Instrument Specific: The principle effect is $1/f$ noise; noise mismatch between X and Y radiometers may also cause minor degradation in sensitivity. Some of these effects are discussed by Kaplan (these proceedings).

Thermal, including internal straylight: The responses of the instrument to many thermal effects are being modelled [5]. Internal straylight affects the polarization signal as for Galactic straylight, except that only fluctuations are important.

Calibration: Kaplan (these proceedings) discusses the effect of inaccurate polarization calibration parameters.

In addition the scan strategy has unique effects on polarization, discussed below.

Detailed studies of these effects are being made as part of the final design optimisation, to ensure that all contributions remain within the top-level error budget. However it is already clear that many will not have a detectable impact on polarization.

As shown in detail in the following section, cross-polarization effects from the main beam appear to be largely negligible, as are the different responses to Stokes I of the X and Y polarizations. Much larger differences are expected in the far sidelobes, but there the largest peaks turn out to be spillover past the subreflector and the main mirror, and initial analysis suggests that the I_P response in these peaks differs by only about 10%; this contributes to $\delta g_X - \delta g_Y$ in terms of the analysis of the previous section. An analysis of total-intensity Galactic straylight [6] showed an expected peak contamination of $\sim 4\,\mu$K at high Galactic latitudes in the worst-case 30 GHz band, corresponding to $\delta T < 1\,\mu$K in the C_ℓ spectrum. Therefore the polarized signal should be less than $\sim 0.1\,\mu$K in the power spectrum, which is just below the low-ℓ white noise at this frequency. At higher LFI frequencies, Galactic straylight will be even lower, because of the falling spectrum of Galactic emission, and also the increased forward gain.

Pointing errors produce error terms in the images proportional to the local gradient. Typical gradients are of order the amplitude of structure on the scale of the beam, divided by the beamwidth. The pointing requirements for *Planck* are set by HFI observations in total intensity, for which the ratio of amplitude/beamwidth will be > 20 times that for LFI observations of Q or U. so for our purposes the pointing is expected to be essentially perfect.

Instrument-specific and thermal effects mainly cause artefacts in the polarization signal through the same mechanisms that affect the temperature signal. These are among the hardest to deal with, and drive many aspects of the LFI and *Planck* mission design. As noted earlier, suppression of these effects in the temperature signal will automatically suppress them in the polarization signal as well, but it is worth noting that often (especially for thermal effects), they produce strongly correlated artefacts in the two channels of each feed and so tend to cancel in the polarization signal.

Cross Polarization

A radiometer operating at a single frequency receiving from a single direction (i.e. a point source) must couple perfectly to some purely polarized signal, that is, in terms of our Stokes response functions, $I_P^2 = Q_P^2 + U_P^2 + V_P^2$. In this sense the concept of 'cross-polarization', which implies that a system always has some sensitivity to the 'wrong' polarization, is rather misleading. In practice what is meant is that we'd like to build a system sensitive only to Q (say), but we find that U_P and V_P are non-zero. Unwanted sensitivity to V is straightforward: it reduces the linear polarization efficiency ε_P but generally creates no false positives, as $V \ll Q, U$. One can also lose efficiency if ϕ varies with frequency or across the beam. However, non-zero response to U at the beam centre (or averaged over the beam) is best described as an error in ϕ rather than as cross polarization in the sense usually intended.

We have been assessing the impact of these effects on the LFI, using physical optics software written by V. Yurchenko [7] and comparing these to simulations with the commercial GRASP8 package described by Villa in his presentation. The two sets of results are in excellent agreement, differing mainly through small differences in the assumed illumination patterns of the feeds.

A priori we do not expect the LFI beams to be very well behaved in polarization as the feeds are located $3°$ to $4°$ away from the centre of the field of view, and the offset pseudo-Gregorian design of the *Planck* telescope breaks circular symmetry. However, the design maximises the effective field of view, while the feeds illuminate the telescope with a strong edge taper, making near-in sidelobes extremely low by the standards of normal radio telescopes; they are at $\lesssim -30$ dB below the peak. Given the absence of sharp spikes in the CMB fluctuations, the total intensity beams are effectively Gaussian.

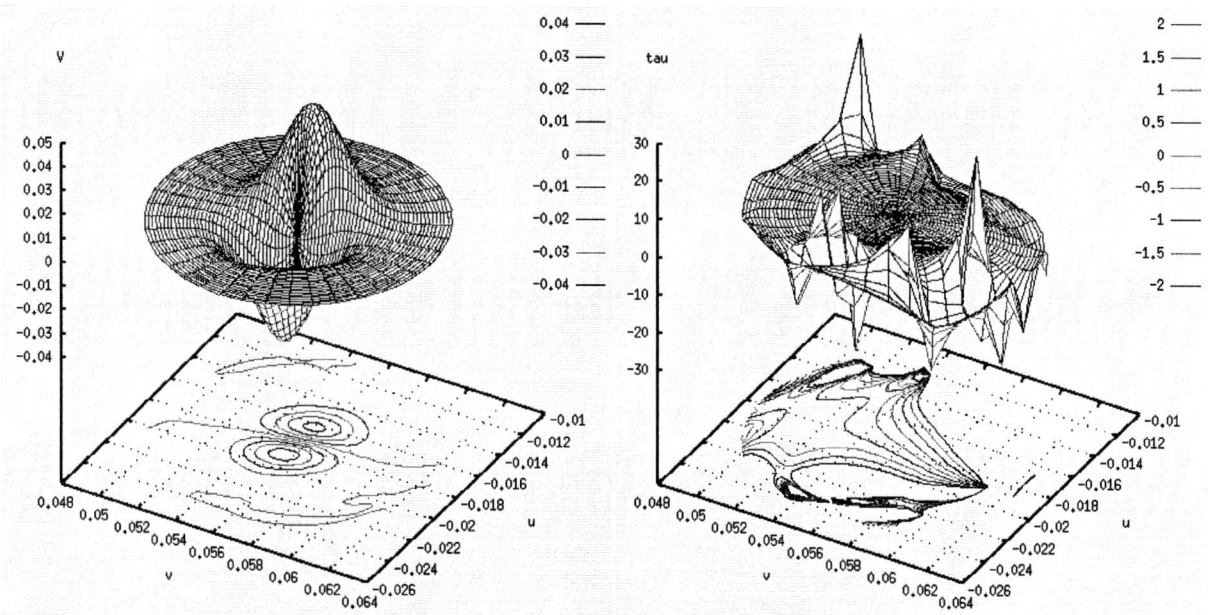

FIGURE 1. Simulations of the polarization response of a 100 GHz feed horn (LFI-9). The coordinates (u, v) are direction cosines with respect to the boresight direction. For comparison, the total intensity beam is centred at $(-0.018, 0.056)$ and the -10 dB contour has a diameter of about 0.005 in (u, v). Left (a): Response to Stokes V, i.e. V_P. Right (b): Variation of orientation angle ϕ_P, in degrees.

The most troublesome property of the beams is that they are elliptical (axial ratio ranging from 1.14 to 1.39), with the horns in each 'Q, U' matched pair differing in the orientation of the ellipse by large angles. Thus deconvolution will be needed to recover the (Q, U) vector at each full-resolution sky pixel; but for background measurements some degradation in resolution is necessary to achieve adequate signal-to-noise in polarization, allowing resolution to be matched through linear techniques.

The total intensity (I_P) beams for the X and Y channels have effective areas identical to better than a few parts in 10^5; mismatches contribute to δg in the analysis of the previous section, but acting only on spatial frequencies near the beamwidth. On this scale amplitudes are $\sim 50\ \mu$K, giving negligible artefacts in the polarization signal.

The major 'unwanted' term turns out to be sensitivity to V, which peaks at several percent of the main beam peak (Fig. 1a) The V_P beams always show a rather symmetric positive-negative structure with the zero line passing almost through the peak of the I_P beam. Integrated over the beam, $|V_P|/I_P \lesssim 0.2\%$ in all cases. Averaged over the beam, the polarization efficiency ranges between 99.1% and 99.7%. If we choose coordinates so the wanted polarization is Q, the U_P beam shows a similar structure to V_P, but with an even lower amplitude. In this case the zero line goes through the beam centre by definition, and the positive-negative structure is inevitable as there is a smooth gradient of ϕ across the centre of the beam. Within the -10 dB contour of the I_P beam, ϕ varies by up to about 2°. Large deviations occur only near nulls in the sidelobe pattern (Fig. 1b). The effective ϕ evaluated from Q_P and U_P integrated over the main beam differs from the value at the peak of I_P by $< 0.15°$.

We also find that the position angle in the far field rotates almost perfectly with the E-field in the focal plane. As a result, if the X and Y responses are orthogonal in the feed, the peak and effective values of ϕ_X and ϕ_Y remain orthogonal in the far field to within $\delta\phi \leq 0.2°$.

These results justify our claim that cross-polarization effects are negligible for the LFI, at least at the pixel level. Kaplan (these proceedings) has followed the effect of miscalibration of these parameters through to the C_ℓ spectra. Here averaging down the thermal noise makes residual polarization systematics more important, but comparison of our results with his findings suggest that even for C_ℓ, the errors induced by the *Planck* telescope optics for the LFI horns will be almost undetectable; presumably even more so for the more favourably placed HFI horns.

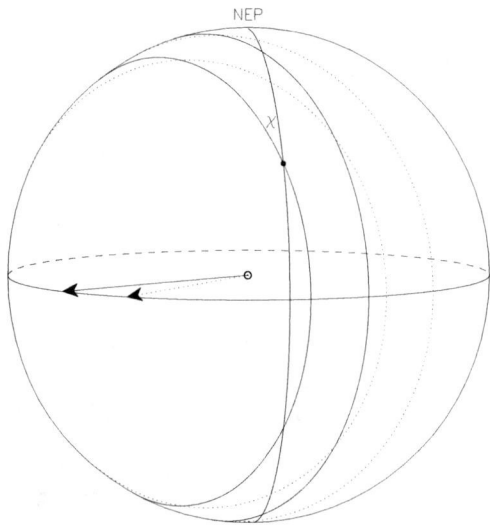

FIGURE 2. Geometry of scan paths. The arrows mark the direction of the spacecraft spin axis at two different times. Scan circles from two horns are shown (the difference in scan radius is exaggerated). NEP is the North Ecliptic Pole. The angle χ is shown for one scan circle at the marked point. Note that at high latitudes scan circles through a given point intersect at significant angles $\Delta\chi$.

FEED ORIENTATION AND SCANNING STRATEGY FOR THE LFI

If several measurements of components (not necessarily orthogonal) of the (Q,U) vector are made, each with a Gaussian error distribution, the 2D error distribution will in general be an elliptical Gaussian [8]. Ideally we would like this distribution to be circular as this minimises the area [9]; furthermore asymmetric errors will certainly complicate the derivation of polarization C_ℓ spectra, and may lead to subtle biases, although no detailed assessment has been made.

It is easy to see that if measurements are made with equally sensitive detectors, with orientations ϕ such that $\{2\phi\}$ are evenly distributed around a circle, then the error distribution will be circular. This is satisfied by the normal set-up of two horns each with orthogonal feeds, oriented at $45°$ to each other; and also by the conventional arrangement in optical polarimetry of $\phi = -60°, 0°, 60°$. Obviously, combining several such sets with arbitrary offsets in ϕ will also yield circular error distributions.

The actual orientations measured on the sky is $\phi_{\mathrm{sky}} = \phi_S + \chi$, where ϕ_S is the orientation of the radiometer polarization relative to the scan direction, and χ is the orientation of the scan circle relative to the sky grid chosen to determine the zero of ϕ. The default strategy for *Planck* is for the spin axis to point in the anti-Sun direction towards the Ecliptic, which makes ecliptic meridians a convenient reference direction (Fig. 2). The figure shows that in this case χ is a function of ecliptic latitude, ranging from $0°$ on the Ecliptic to $90°$ at the maximum accessible latitude.

The radius of the LFI scan circles varies from $81°$ to $89°$. Consequently the values of χ of different detectors measuring the same sky pixel varies, especially near the ecliptic poles. The LFI focal plane has been designed so that most matched pairs of feeds are symmetrically placed on each side of the focal plane parallel to the scan direction (barring small deviations of the spin axis from nominal), so that they share the same scan circle and hence the same relative orientation for all pixels. Unfortunately mechanical constraints prevent this arrangement for four of the sixteen 100 GHz feeds. These are arranged instead as a pair with $\phi_S = 90°$ on one scan circle, and a pair with $\phi_S = 45°$ on another. At 44 GHz there are three feeds, two sharing a common scan circle and the third on one several degrees larger. In these cases some ellipticity in the (Q,U) error distribution is inevitable. Fig. 3 shows the effect on the error ellipse at high latitudes; it is particularly large at 44 GHz because of the large difference in radius between the two scan circles. A second coverage improves the situation slightly since on the second pass pixels are scanned by the other side of the scan circle, hence at angle $-\chi$. Even so, we are left with highly elliptical error distributions near the ecliptic poles, precisely the places where integration times are longest and we have the best chance of detecting polarization at high resolution.

To avoid this problem, and also holes in the coverage at the poles, the spin axis must be moved off the Ecliptic. Several scan strategies are under consideration, including a cycloidal path which maintains a constant angle between the Sun and the spin axis, and a simple sinusoidal oscillation in latitude. In these schemes the polarization error ellipse

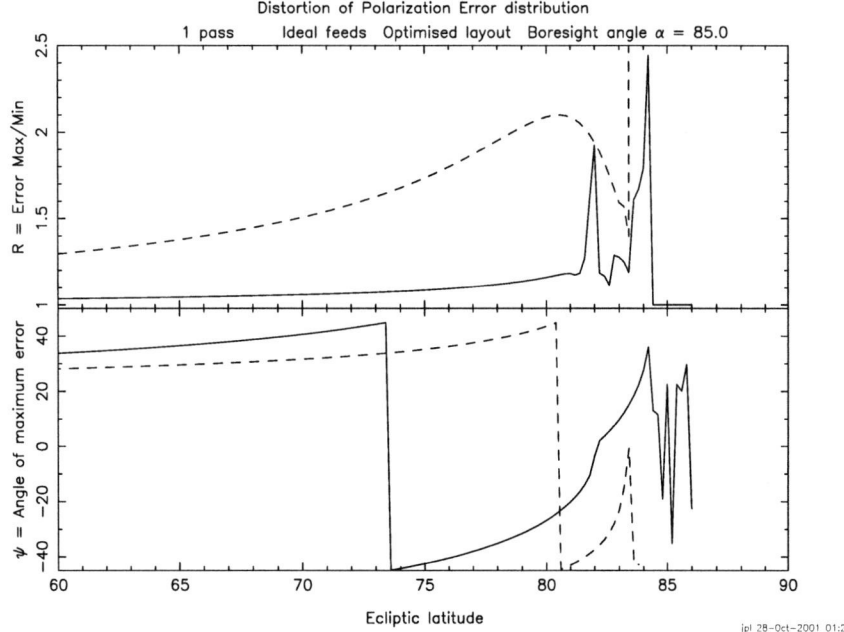

Distortion of Polarization Error distribution

1 pass   Ideal feeds   Optimised layout   Boresight angle α = 85.0

jpl 28-Oct-2001 01:29

FIGURE 3. Axial ratio (top) and orientation (bottom) of the error ellipse as a function of ecliptic latitude, after one sky coverage, for the current LFI focal plane. Solid: 100 GHz: Dashed: 44 GHz, assuming $\phi_S = 0°, 60°, -60°$ for the three horns.

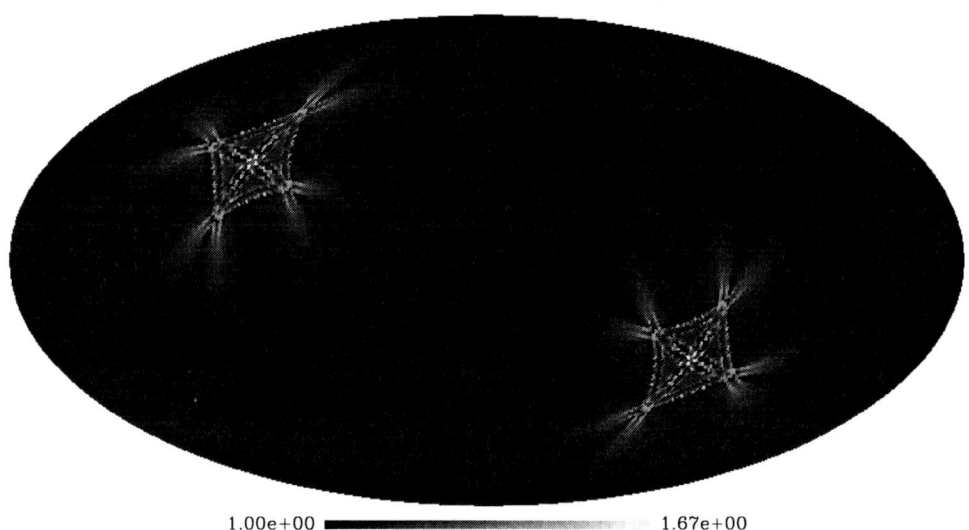

1.00e+00 ▬▬▬▬▬▬▬▬▬▬▬▬ 1.67e+00

FIGURE 4. Sky plot of error ellipticity at 100 GHz assuming a 5° amplitude sinusoidal oscillation in latitude with a wavelength of 90° in longitude. The coordinate system is Galactic, so that the structure around the ecliptic poles can clearly be seen.

depends on both latitude and longitude. Fig. 4 shows an example at 100 GHz. As far as polarization response goes this option is a significant advance over the default strategy, with rather circular error distributions up to the pole except at isolated position on caustics, and even here the axial ratio reaches only 1.7. Similar improvements are seen at 44 GHz.

CONCLUSIONS

We have shown that in a differencing polarimeter the dominant systematics are common to total intensity and polarization signals, except that some cancellation can be expected in polarization. As all LFI horns provide a measurement of I but each only provides one component of the (Q, U) vector, the thermal noise in Q and U will be $\sqrt{2}$ larger than in I. Thus we believe that if we can reach the goal of noise limited sky maps in I, we should be able to do the same for Q and U. Of course, Q and U will not be detected at full resolution in most individual pixels, so the true test of our ability to do useful polarization science is to follow the systematics through to the C_ℓ spectra, and much work is needed to complete this task. But the low level of polarization-specific systematics suggests that here again the hard systematics to beat will be the same ones that affect total intensity, and we know that these can be eliminated with high precision.

Finally we have shown that the polarization response, especially in the most deeply surveyed regions near the ecliptic poles, can be substantially improved by moving the spin axis away from the Ecliptic plane. Further studies of alternative scan strategies will be made to help the project decide on the best option.

ACKNOWLEDGMENTS

We thank our colleagues in the *Planck* Systematics Effects Work Group on polarization for useful discussions, especially Althea Wilkinson, Fabrizio Villa, Jacques Delabrouille and Jean Kaplan. We acknowledge the use of the HEALPix package (www.eso.org/science/healpix/).

REFERENCES

1. Bersanelli, M., Muciaccia, P. F., Natoli, P., Vittorio, N., and Mandolesi, N., *Astron. Astrophys. Suppl. Ser.*, **121**, 393–404 (1997).
2. Maino, D., Burigana, C., Maltoni, M., Wandelt, B. D., Górski, K. M., Malaspina, M., Bersanelli, M., Mandolesi, N., Banday, A. J., and Hivon, E., *Astron. Astrophys. Suppl. Ser.*, **140**, 383–391 (1999).
3. Revenu, B., Kim, A., Ansari, R., Couchot, F., Delabrouille, J., and Kaplan, J., *Astron. Astrophys. Suppl. Ser.*, **142**, 499–509 (2000).
4. Berkhuijsen, E. M., *Astron. Astrophys.*, **40**, 311–316 (1975).
5. Mennella, A., Bersanelli, M., Burigana, C., Maino, D., Ferretti, R., Morgante, G., Prina, M., Mandolesi, N., Butler, C., Valenziano, L., and Villa, F., in *Experimental Cosmology at millimeter wavelengths*, edited by M. D. Petris and M. Gervasi, American Institute of Physics, 2001, in press.
6. Burigana, C., Maino, D., Górski, K. M., Mandolesi, N., Bersanelli, M., Villa, F., Valenziano, L., Wandelt, B. D., Maltoni, M., and Hivon, E., *Astron. Astrophys.*, **373**, 345–358 (2001).
7. Yurchenko, V., Murphy, J. A., and Lamarre, J. M., *Int. J. Infrared and Millimeter Waves*, **22**, 173–184 (2001).
8. Sparks, W. B., and Axon, D. J., *Publ. astron. Soc. Pacific*, **111**, 1298–1315 (1999).
9. Couchot, F., Delabrouille, J., Kaplan, J., and Revenu, B., *Astron. Astrophys. Suppl. Ser.*, **135**, 579–584 (1999).

SPACE AND BALLOON FACILITIES AND AGENCIES

The ASI Science Program

Carlo Musso

Agenzia Spaziale Italiana - Viale di Villa Grazioli, 23 - 00198 Roma, Italy

Abstract. Italy came in the space business in 1963, being the third nation in the world, after the Soviet Union and the United States, to put an artificial satellite into orbit. In 1988 the Italian Space Agency (ASI) was constituted, with the mandate of planning, coordinating and executing civil space activities in Italy.

The core of national space activities is science, for which Italy spends about 25% of the ASI budget, both in national and international programs. The community served by the scientific directorate of ASI is a very wide one, ranging from the science of the Universe and the exploration of the Solar System to life sciences, from Earth observation to the development of new technologies.

The success of Italian space research appears under many different points of view. The national satellite BeppoSAX, named after Giuseppe *Beppo* Occhialini, widely contributed to solve the γ-ray burst puzzle, obtaining the relevant acknowledgment of the 'Bruno Rossi Prize'. Italian researchers kept the PI-ship of various payloads on board ESA missions, such as Epic for XMM-Newton, Ibis for Integral, Virtis and Giada for Rosetta, PFS and Marsis for Mars Express.

Also in the field of the cosmic microwave background (CMB) two important experiments are foreseen in the next future, with Italian PIs: SPOrt on board the International Space Station, dedicated to the polarization of CMB, and LFI (Low Frequency Instrument) on board the ESA Planck satellite, to study CMB anisotropy. Meanwhile, a great success has been obtained with the balloon experiment Boomerang.

Moreover, ASI started a national scientific and technological small mission program. The first three missions are on their way: Agile (a γ-ray observatory), David (an experiment to test very high frequency data transmission), and a third one, devoted to Earth science.

ITALY AND SPACE

In 1962 it was announced that Italy would launch its first satellite within a year. This promise was maintained and so Italy became the third nation, after the Soviet Union and the United States, to put an artificial satellite in orbit.

This was the beginning of a long and very exciting adventure, which in the last years, has brought Italy in an outstanding role in the world panorama of science from space.

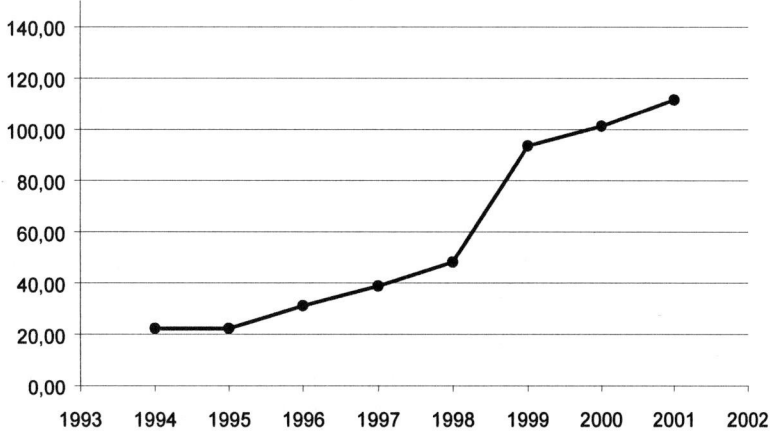

FIGURE 1. ASI budget for science (in million of euros)

The Italian Space Agency, founded in 1988 to promote, fund and manage space activities in Italy, defines through its Science Department the scientific strategies for space research. In the last five years, a serious work has allowed to get an increase of the budget for science of about 50% (see Fig. 1).

This rising of funds has induced a comparable increase of the Italian participation in international scientific programs, not only with ESA, but also with NASA and others agencies and institutions.

SPACE AND SCIENCE

The community served by the scientific directorate of ASI is a very wide one, from the science of the Universe and the exploration of the Solar System to life sciences, from Earth observation to the development of new technologies. This requires the capability of ASI to harmonize the various needs of different communities, with a wide range vision, defining strategies and taking advantages from the opportunities presented by the international scenario. The success of ASI planning is made evident by several outstanding results obtained by Italian researchers.

The national satellite BeppoSAX, named after Giuseppe *Beppo* Occhialini, one of the Italian fathers of high energy astrophysics, deeply contributed to solve the γ-ray burst puzzle. The combined use of its wide-angle and narrow-field cameras allowed the researchers to identify and observe in the X-ray and optical domain the counterparts of these mysterious phenomena (see Fig. 2). The complete solution of the enigma has not been found yet, but great advances have been made thank to BeppoSAX observations. In acknowledgment of this result, the 'Bruno Rossi Prize' was awarded to the BeppoSAX team in 1998.

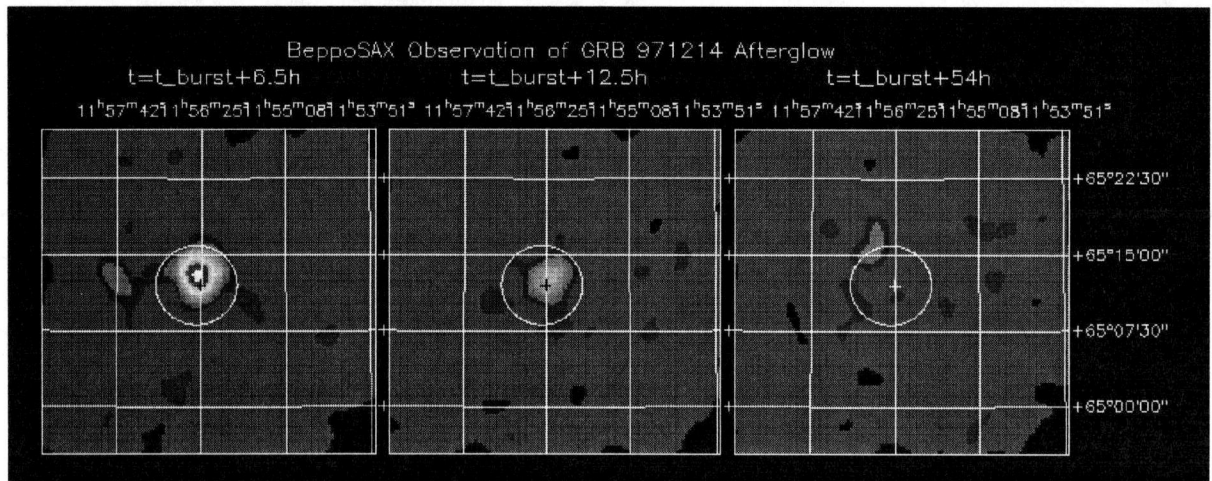

FIGURE 2. Typical BeppoSAX observation of a γ-ray afterglow.

Following a long tradition, starting with COS-B, Italian researchers are very well represented in ESA programs. In particular, we have the PI-ship on Epic for XMM-Newton [1] (X-ray astronomy, launched in 1999), Ibis for Integral (γ-ray astronomy, 2002), Virtis and Giada for Rosetta (to the Wirtanen comet, 2003), PFS and Marsis for Mars Express (mission to Mars, 2003).

Going to the field of the cosmic microwave background (CMB) two important experiments are foreseen in the next future. SPOrt [2], on board the International Space Station, is devoted to study the polarization of the CMB, while Planck is an ESA satellite conceived to solve the problem of CMB anisotropy. One of the two instruments on board Planck, the low frequency instrument (LFI), as well as SPOrt, have Italian PI's, both form the TeSRE Institute of the CNR.

If we add to these future and complementary experiments the great success obtained by Italian scientists with the balloon experiment Boomerang [3], it is clear that a leadership in this sector could be achieved by Italy in the next few years.

In 1997, ASI kicked-off a truly national program for scientific and technological small missions [4]. By the end of 1998, the first two missions of the program were selected: the γ-ray observatory Agile - **A**strorivelatore **G**amma a

Immagini **L**eggero or Extremely Light Imager for **G**amma Astronomy – [5], which has now reached its realization phase for a launch within 2003, and the David satellite - **Da**ta and **Vi**deo **D**istribution -, devoted to the study of data transmission at very-high frequencies, i.e. 90 GHz. The third mission, dedicated to Earth science, is going to be selected by the end of the year 2001[6]. An announcement of opportunity for the fourth mission is expected for the beginning of 2002.

Finally, ASI has been leading a working group to study the feasibility of a nuclear engine of new conception, proposed to the Agency by Carlo Rubbia, since 1998. Such an engine would be one of the crucial issues for allowing the human exploration of the Solar System [7].

REFERENCES

1. Musso C., *L'Astronomia*, **204**, 34-43 (1999).
2. Cortiglioni S. *et al.*, "The SPOrt Program", in *AMiBA 2001: High-z Clusters, Missing Baryons and CMB Polarization*, edited by L-W Chen, C-P Ma, K-W Ng and U-L Pen, ASP Conference Series 999, 2002.
3. De Bernardis P. *et al.*, *Nature*, **404**, 955 (2000).
4. Stalio R. *et al.*, "Technological spin-offs of the PMST program of the Italian Space Agency", in *IEEE Aerospace Conference*, in preparation.
5. Musso C. *et al.*, "The AGILE Mission", in *5th International Symposium on Small Satellites Systems and Services*, edited by CNES, La Baule, 2000.
6. Zoffoli S., Crisconio M., Musso C. & Bignami G.F., "A small glance to Earth from space", in *3rd IAA Symposium on Small Satellites for Earth Observation*, Berlin, 2001.
7. ASI Working Group, "Project 242 - Assessment Report", edited by ASI, 1999.

Luigi Broglio Base Facilities and Activities: the ASI Balloon Program

O. Cosentino and R. Ibba

Base di Lancio Palloni Staratosferici - ASI - S.S. 113 nr 174 Contrada Milo - 91100 Trapani - Italy

Abstract. We briefly summarize the characteristics and capabilities of the "Luigi Broglio" Base for stratospheric balloon launches. Located near Trapani, the geographic position of the base is ideal for stratospheric balloon flights, including those ranging as far as the western coast of Spain. On site services include meteorological readings, remote control of the balloons and real time acquisition of data produced by scientific missions.

GENERAL DESCRIPTION

The "Luigi Broglio" Base was established in 1975, with one Trans-Atlantic successful flight. The payload was recovered in Texas after 5 days of flight. The summer stratospheric circulation pattern from East to West allows payload recovery in Spanish territory after about 20 hours of flight. Weather conditions in the summer, during Azores anticyclone standing on the Mediterranean sea, allow launches early in the morning and at sunset in combination with sea breeze alternation.

The geographical coordinates of the launch site, 38°01′N, 12°35′E, and the predicted trajectory along the 38° northern parallel ensure good conditions for astrophysical research (e.g. very low X-ray background).

The "Luigi Broglio" Base is specialized in launching gross weight payloads heavier than 2000 kg. The average altitude is 40 km and flight duration longer than 20 hours is reached for Trans-Mediterranean flights.

78 experiments have been launched since 1975, with 66 successes and 12 failures. The maximum weight we launched was 3.5 tons, carried to an altitude of 30 km during the flight of a Capsule for an Atmospheric Reentry Demonstrator.

The ASI launching facility offers further interesting opportunities:

- proximity to the sea
- close proximity to a town
- good hotel accommodations
- local industrial support

FACILITIES

The organization and equipment available in the L. Broglio base can fulfil the needs of several experiments. The most important facilities are:

- Hangar for integration of up to 6 scientific payloads
- 500 mt. diameter launching area
- Launching vehicle for gondola weights up to 2500 kg.
- Workshop and electronic laboratory
- Balloon release compact vehicle
- Experimenter's room with telemetry & data acquisition room links
- Gondola suspending amagnetic system

CP609, *Astrophysical Polarized Backgrounds,* edited by S. Cecchini et al.

- Several high precision pointing systems for gondolas (Pivot)
- Unbreakable power system
- UHF telemetry and telecommand ground station
- UHF telemetry and telecommand on board equipment
- S Band telemetry and telecommand ground station
- S Band telemetry and telecommand on board equipment
- Flight control and balloon tracking system
- Real time flight data analog recording (Racal storehouse)
- High precision and stability time reference equipment
- High bit rate real time data acquisition system
- Real time data acquisition, computing and displaying system
- Data elaboration center connected to the network
- Thermo-vacuum test equipment
- Controlled environment for biological research
- Meteorological and sounding station
- Fax, radio facilities, internet
- Balloon management and altitude control

Controlled environment for biological research

Four different incubators were realized in order to satisfy the different requirements of researchers. Electronic control was implemented as to maintain the atmospheric pressure and temperature required by each experiment.

FIGURE 1. The basic incubator

The basic incubator (Fig. 1) is a steel box containing six small racks to allocate the flasks housing the experiments. Two other boxes are equipped with a quartz window to expose the specimen to the solar UV radiations. One of these windows is also enabled various exposure time of the eight specimens contained in the box. The control of the exposure can be remotely operated from ground or pre-programmed.
The fourth box was developed to accommodate a special mannequin designed to simulate the human body. Different dosimeters were placed at the level of the human organs to evaluate the absorbed radiation (Fig. 2).

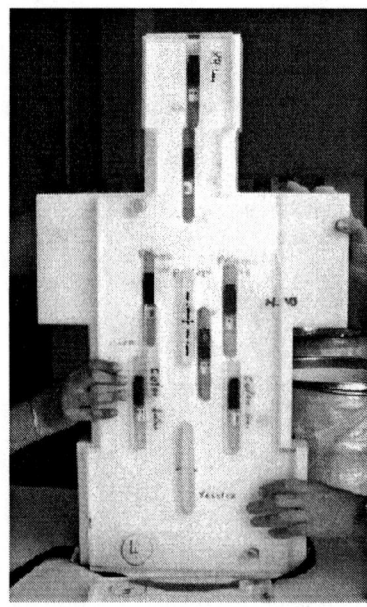

FIGURE 2. Mannequin simulating the human body placed inside the fourth incubator

PAST ACTIVITIES

The Italian Scientific Community has a very strong tradition and several groups are engaged in research through stratospheric balloons, mainly in the astrophysics domain. Important payloads were developed and flown from:

- L. Broglio Base;
- NASA facilities (in U.S. and in Australia)
- ASI/CNES collaboration (Australia)

The most important researches concerns astrophysics subjects, such as γ rays, X rays, IR radiation, Strange Quark matter, Cosmic Ray effect on biological structures etc. The most important experiments launched from the L. Broglio base are listed in table 1.

Technological programs have been carried out as well, such as:

- Preliminary tests of satellite equipment and sub-systems
- A.R.D. (Atmospheric Reentry Demonstrator - capsule rescue qualification)
- HASI (Huygens Atmospheric Structure Instrument)
- Boomerang - Balloon management for pre-determined recovery places

FUTURE ACTIVITIES

Taking into account the most important scientific and technological requirements of nowadays, balloon activity will be developed in the following sectors:

- Long duration flights
- Local technological test
- Technological flights to improve guidance and recovery sub-systems

TABLE 1. Most important experiments launched from the L. Broglio Base

Experiemnt	domain	partners
Milo1	Cosmic particles	Bristol University
Milo2	X Sources and C.B.R.	CNR and Washington University
Milo3	γ Astronomy	IFCTR (CNR) and Max Plank Inst.
Odissea1	Cosmic particles	CONIE SPAIN
Odissea2	γ Astronomy	IFCTR - GIFCO - ITESRE (CNR)
PAF	Radio Astronomy	IROE - GIFCO (CNR)
CAESAR	γ Astronomy	CERS/CEN CNES
CELIMENE	X Astronomy	IAS (CNR)
AGLE	IR Astronomy	CERS - CEN (CNES)
CIRCE	X Astronomy	IAS (CNR)
ULISSE	IR Astronomy	IROE (CNR)
ENEA	X Astronomy	IAS (CNR)
POKER	X Astronomy	IAS (CNR)
TELEMACO	IR Astronomy	IROE (CNR)
ELENA	γ Astronomy	ITESRE (CNR)
FIGARO	γ Astronomy	CERS (CNES) IFCAI/IAS (CNR)
PALLAS	X Astronomy	IAF/IAS (CNR) -Southampton Univ.
ARGO	IR Astronomy	IROE/IFA (CNR)
MINITIR	IR Astronomy	ROME UNIV. IRE/CAISMI/IAS (CNR)
MINIZEBRA	X Astronomy	CNR -Southampton Univ.
PHOSWICH	X Astronomy	ITESRE(CNR) - Univ.
AROME	IR Astronomy	CNES
LAPEX	X Astronomy	ITESRE/IAS (CNR) CERS (FR)
ARD	Technologic	ESA
S.Q.M.	Nuclear research	CNR(TO) TOKIO Univer

Long Duration Flights

Waiting for reliable superpressure balloon technology, the only opportunity for Long Duration Flights is at present North and South Poles circumnavigation during western or eastern stratospheric planetary circulation.

The Antarctica Campaign for 2003 is in progress and ASI is looking forward to performing its launch campaign for heavy payloads in the Arctic Circle.

Antarctica Program

ASI is investigating the possibility to use the Italian Baia Terra Nova base in Antarctica for long duration flights, beginning with weather analysis. The scheduled activities are:

- December, 15, 2001 - January, 15, 2002: Launch Campaign in BTN Base
- Tropospheric and Stratospheric pattern analysis, using zero-pressure small balloons (2000 m^3)
- December 2002: HASI (Huygens Atmospheric Structure Instrument) local flight descent mock-up test
- January 2003: long duration precursor flight and/or Scientific Payload

Local Technological Tests

Picogravity project by stratospheric balloons

Our aim is to arrange a picogravity platform to test high sensitivity scientific instruments in environments free from gravity and dynamic perturbations. A feasibility study for a picogravity facility called "GIZERO" has been completed.

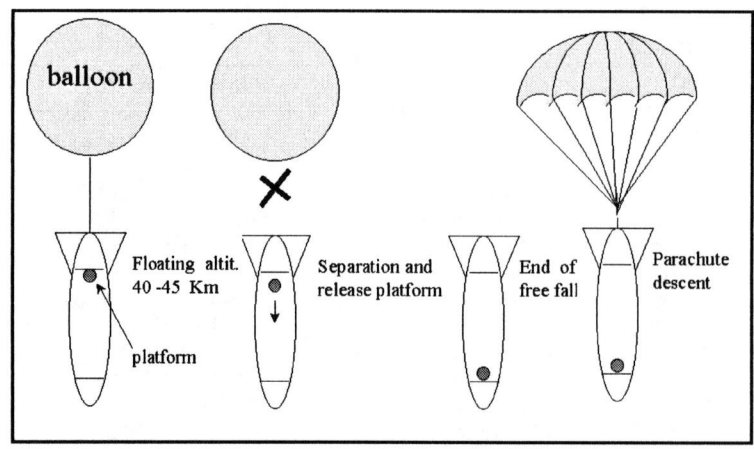

FIGURE 3. Flight steps of GIZERO

This facility consists of a platform in free fall, released from a stratospheric balloon at 42 Km, containing a vacuum capsule (see fig. 3). The characteristics of GIZERO are:

- Vacuum: 10^{-5} Hpa
- Residual G level: 10^{-11} g
- Duration: 25 sec.
- Max P/L weight: 50 Kg
- Angular precision: $3 \cdot 10^{-5}$ r/s

Guidance and recovery sub-systems

Driver Parachute

The goal is to test a particular parachute configuration that would allow us to drive the payload on a predefined target via software and/or TC. The scheduled activities include:

Feasibility project	November 2002
Software definition for management	March 2003
Achievement prototype	June 2004
Guide and recovery strategy definition	September 2004
Technological precursor flight	October 2004

A scheme of the project is shown in Fig.4

Boomerang

Taking into account the fluctuations of the wind, mostly stable during summer time at latitudes around 38 °N, with direction E-W in the upper stratosphere ($28 \div 40$ Km) and W-E direction in the low stratosphere ($16 \div 21$ Km), a driving management system has been proposed that would allow us to launch a balloon and to retrieve it around the same site after a stratospheric flight. The goal is to increase the floating time of the transmediterranean flights by about 8 hours, by carrying balloons over the Atlantic Ocean and by retrieving them in low altitude wind for final recovery in Morocco, Portugal or Spain.

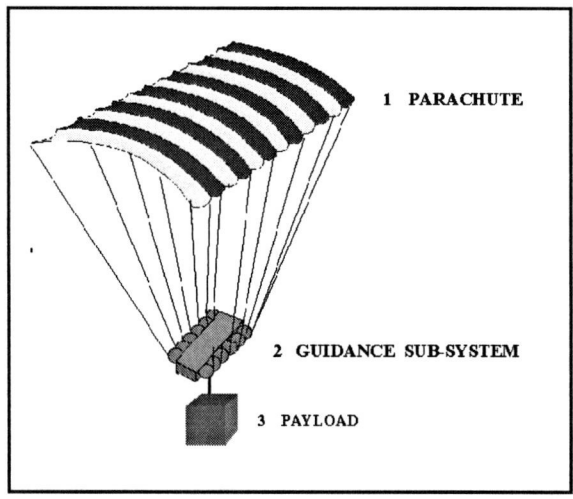

FIGURE 4. Scheme of the Driver Parachute project

High precision pointing systems

A new sub-system based on GPS data acquisition will be ready to be tested during the transmediterranean ASI-INTA Campaign in 2002.

Unmanned Space Vehicles balloon test (CIRA Project)

CIRA (Centro Italiano di Ricerche Aerospaziali) suggested a project to build a space vehicle for the International Space Station (ISS). The program foresees a sequence of big balloon ($\simeq$1200 m^3) flights to lift a space vehicle up to 40 km. After separation from the balloon, the vehicle, equipped with a guidance sub-system, will descend into the atmosphere and will be recovered in the sea. The concept is shown in fig. 5.

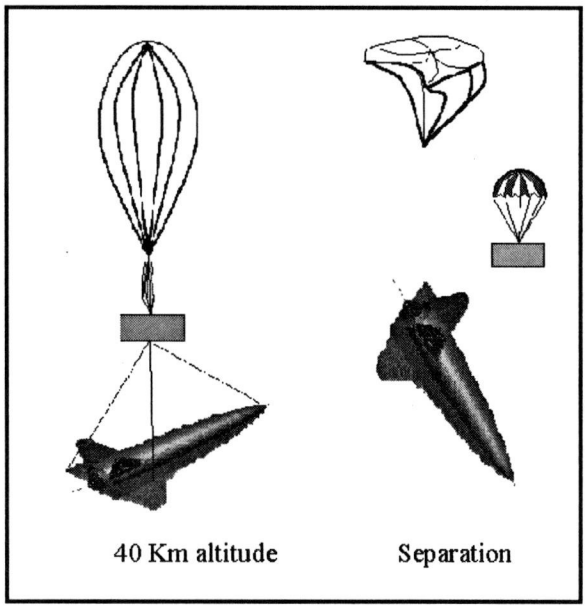

FIGURE 5. Logical scheeme of the CIRA Project

The following Balloon Test Missions have been defined:

1. Dropped Transonic Flight Test (DTFT)
2. Sub-orbital Reentry Test (SRT)
3. Hypersonic Flight Test (HFT)
4. Hypersonic Flight Test (HFT-LP)

Four Test Missions have been scheduled to be executed by three balloon test flights:

Dropped Transonic Flight Test (DTFT)	2003
Sub-orbital Reentry Test (SRT)	2004
Hypersonic Flight Test (HFT)	2005
Hypersonic Flight Test (HFT-LP)	T.B.D

2002 Transmediterranean Campaign

Five launches are expected:

1. Up-grade HASI test
2. Biological payload
3. Technological flight
4. Gamma astronomy
5. Radiometers for Sky polarization

The NASA Balloon Program: An Overview

G. Dwayne Orr[*], Danny RJ Ball[*], I. Steve Smith Jr.[†]

[*]NSBF, P.O. Box 319, Palestine, TX 75802 USA
[†]NASA/WFF, Code 820, Building E107, NASA WFF, Wallops Island, VA 23337

Abstract. The U.S. National Aeronautics and Space Administration (NASA) Balloon Program continues to support the scientific community providing enhanced capabilities across a spectrum of balloon related disciplines. NASA maintains a conventional ballooning capability, which supports three to five campaigns per year. Long duration ballooning (LDB) continues to be a prominent element of the program with a mission model of one campaign per year in both the Northern and Southern Hemispheres. Both polar and mid-latitude LDB capabilities are on-going operational elements of the flight program. The new ultra-long duration balloon (ULDB) project has been progressing with flight tests having been conducted. An overview of the various aspects of the NASA Balloon Program will be presented.

INTRODUCTION

NASA's Balloon Program primarily provides support to the Office of Space Science and the Office of Mission to Planet Earth although it can and does provide support to other NASA elements on a case by case basis. Scientific disciplines supported include cosmic and heliospheric physics, plasma physics, solar physics, high energy astrophysics, infrared astronomy, and upper atmospheric research. Goddard Space Flight Center's Wallops Flight Facility (GSFC/WFF) located at Wallops Island, Virginia provides the program management function.

Flight operations are conducted through an extensive mission support program element that provides the preparation, launch, tracking and recovery of every flight. The Physical Science Laboratory of New Mexico State University has the contract for supporting the NASA Balloon Program including the operation and maintenance of the National Scientific Balloon Facility (NSBF) at Palestine, Texas. The program mission model of 26 flights per year is conducted from various U.S. and foreign sites throughout the world. In addition to operational implementation of the flight program, the mission support activity also provides for the development and implementation of new and enhanced facilities and systems to improve ballooning support. Over the past several years, the NASA Balloon Program has continued to enjoy successful performance across the spectrum of balloon sizes, an accomplishment that is unprecedented in ballooning considering the large, heavy balloons that mostly comprise the program today.

Another important element of the NASA Balloon Program is the on-going balloon research and development (R&D) effort. Since its beginning in 1987, the R&D effort has provided insight and understanding into the performance, limitations, and failure mechanisms of the balloon vehicle and its related technical disciplines, and has acted as the springboard for the advancements made in balloon technology. The R&D effort is directed and conducted from GSFC/WFF, with support for the effort being provided by external organizations possessing the necessary expertise including government, academic, and commercial entities.

FLIGHT PROGRAM

Background

The NSBF at Palestine, Texas serves as the primary facility for the NASA Balloon Program. However, in the recent past its service as a launch site has been limited to meeting only summertime, westerly flight requirements. Today there are an increasing number of flights being conducted from foreign sites and remote U.S.A. locations including Fort Sumner, New Mexico, which has become the U.S. domestic "turnaround" and easterly requirement launch site. In addition, the Program now conducts fewer total flights but of greater scientific and operational complexity. Long Duration Ballooning (LDB) has become an operational capability of the Balloon Program with two two-flight campaigns being planned annually, one each in the Northern and Southern Hemispheres in their respective summer months. The NASA Program has also continued to conduct flight testing and operational qualification of new balloon systems to enhance its capability in support of the scientific community. The requirements of LDB, coupled with the continuing requirements for conventional balloon flights create an on-going challenge to the Balloon Program and the NSBF organization, its people, equipment, and facilities.

Capability

The program presently utilizes five standard balloon designs ranging in volume from 0.33 million cubic meters (Mcm) to 1.12 Mcm. The largest of these will lift more than 3,600 kgs to an altitude of better than 37 kms (See Table 1). The flight program supports both conventional and long duration flights. Typical flight duration for conventional balloons are from a few hours to a couple of days during the "turnaround periods" which occur in the spring and fall. For long duration balloons, flights last for as much as three weeks and are generally flown during the "local summer" easterly wind period.

TABLE 1. NASA Standard Design Balloons.

BALLOON SIZE/# CAPS	MAXIMUM SUSPENDED WEIGHT (KGS)*	FLOAT ALTITUDE (KMS)
1.12 Mcm / 3 cap	3628	36.8
1.12 Mcm / 2 cap	2721	38.7
0.83 Mcm / 2 cap	2948	36.3
0.33 Mcm / 3 cap	3379	29.9
0.33 Mcm / 1 cap	1304	35.5

* Includes Science payload, NASA flight systems, and ballast; some remote launch sites currently have additional launch payload weight limitations.

Recent Campaigns

Launch support requirements continue to dictate a high number of remote campaigns. In the past five years, NSBF has supported campaigns in ten separate locations including: Alice Springs, Australia; McMurdo Station, Antarctica; Juazeiro Do Norte, Brazil; Lynn Lake, Canada; Fairbanks, AK; Wallops Island, VA; Fort Sumner, NM; Kiruna, Sweden; Ottumwa, IA; and Palestine, TX.

In 2000, ballooning activity in support of the Upper Atmosphere Research Program (UARP) consisted of launching the OMS (Observations from the Middle Stratosphere) in-situ gondola from ESRANGE at Kiruna, Sweden. This was the eighth and ninth launch of this gondola in less than four years. Previous launches of this gondola had been conducted from Fort Sumner, NM, Juazeiro Do Norte, Brazil (where it was successfully launched and recovered twice within two weeks), and Fairbanks, AK.

Campaigns continue to be conducted from Lynn Lake, Canada in support of the Cosmic and Heliospheric Physics discipline. This high northern latitude launch site provides the necessary geomagnetic energy cutoffs for conducting the science experiments designed to measure cosmic ray particles entering the Earth's atmosphere. Due to the large detection areas generally required for these experiments, some of the heaviest payloads and largest balloons are flown during these campaigns.

As a demonstration of NASA's operational LDB capability, flights were conducted from McMurrdo Station, Antarctica (See Figure 1) and Fairbanks, Alaska (See Figure 2) during FY 1998. This represented the first time successful long duration balloon launches were conducted from both hemispheres in a given year in support of science. Issues with over-flight of Russian airspace has prevented efforts in conducting additional LDB campaigns from Fairbanks, Alaska. However, recent developments indicate that most of these issues can be resolved and preliminary plans are being made to conduct a 2002 Fairbanks Campaign. Another highlight of the LDB Antarctica Campaigns has been the resounding success of the BOOMERANG experiment launched in December 1998. Data obtained from this Cosmic Microwave Background instrument supports other recent science observations in indicating that the universe is flat.

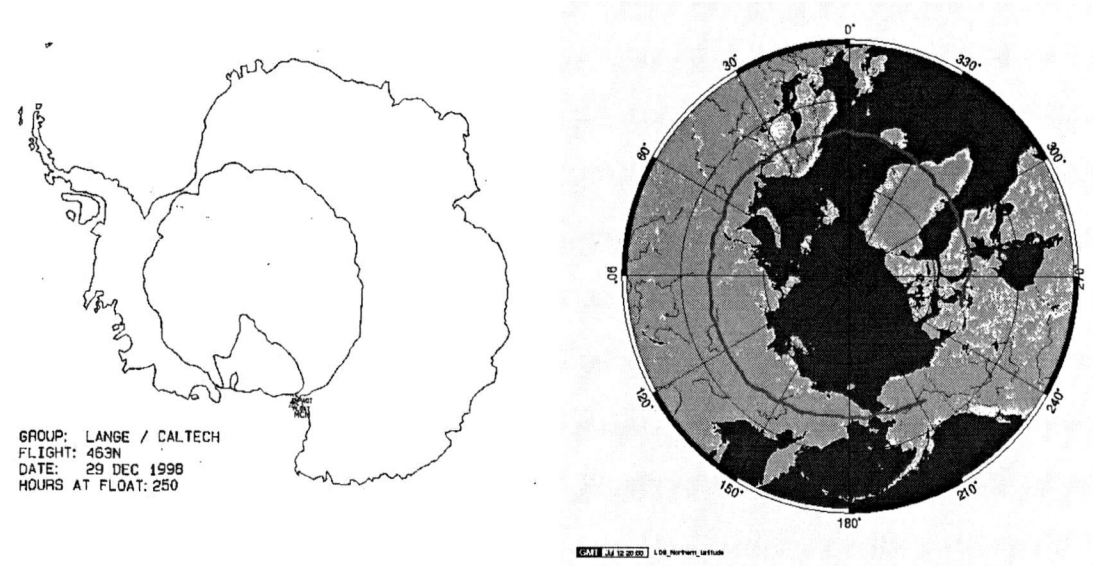

FIGURE 1. '98 Antarctica Campaign – BOOMERANG   **FIGURE 2.** '98 Fairbanks Campaign - Berkeley

ULTRA LONG DURATION BALLOONING

The biggest news in the Balloon Program comes as an output of the Balloon R&D effort. This effort over the past several years has focused on the investigation of an advanced long duration balloon vehicle to provide enhancement of NASA's balloon support capability. In 1997, in response to the advancements made in the R&D activities, a project was approved to develop and demonstrate a new ballooning capability; mission duration lasting up to 100 days above 33.5 km altitude. The objective of the development is to provide a mission capability for supporting approximately 1000 kgs of scientific payload on a 0.6 Mcm super pressure balloon for up to 100 days (~ 5 circumnavigations of the globe). Successful small scale test flights of the new composite material (polyester and polyethylene) super pressure balloon design have been conducted. Two full scale test flights were conducted this past year in Alice Springs, Australia with mixed success. Plans are to continue the development project with a full scale test flight this coming spring in Fort Sumner, NM.

FUTURE

The NASA Balloon Program remains dedicated to excellence in scientific ballooning through the enhancement of existing capabilities and development of new capabilities. Using new designs and materials, we are hoping to establish a new mark in balloon capability. One that will establish the value and contribution that balloons can provide as mission carriers. Already, balloon long duration platforms have been included in NASA's Announcement of Opportunities for UNEX (University Explorers) and SMEX (Small Explorers), along with consideration for inclusion in the next MIDEX (Medium Explorers) Announcement of Opportunities.

REFERENCES

1. Flowers, B.J., and Needleman, H.C. , "An Overview of the NASA Sounding Rockets and Balloon Programs", Proceedings of 14th ESA Symposium on European Rocket and Balloon Programs and Related Research, Potsdam, Germany, ESA SP-437, September 1999.
2. Smith, I.S., Jr., "Overview of the Ultra Long Duration Balloon Project", COSPAR, Warsaw, Poland, 2000.
3. Baldemar, P., Ball, D. RJ, "Medium Duration Heavy Load Balloon Flights from Sweden to Canada – An SSC/Esrange and NASA Joint Effort", Proceedings of 15th Symposium on European Rocket and Balloon Programs and Related Research, Biarritz, France, 2001.

The Esrange Facility in Northern Sweden - your Partner for Successful Aerospace Operations

P. Baldemar, O. Widell

SSC, Esrange, P.O. Box 802, SE-98128 Kiruna Sweden

Abstract. The Swedish Space Corporation, SSC, is internationally recognised as a flexible and successful partner in space operations. The space operations are performed at Esrange in northern Sweden at 68°N and 21°E; an ideal site for launch of sounding rockets, stratospheric balloons and for satellite operations. Balloon flights and aerospace testflights can also be performed from other dedicated ranges in Sweden. Scientific ground-based instruments at Esrange may also be used for coordinated measurements during sounding rocket and balloon flights as weel as data from satellite operated at Esrange. The infrastructure and the skilled personel at Esrange are a guarantee for a successful aerospace mission. SSC has established co-operation with both European and North American partners to accomplish large balloon campaigns with flights as far as into Russia and Canada. The benefit of this co-operation is to ensure handling of very large payloads, reducing the launch readiness time and to be ready for launch during the best scientific conditions.

INTRODUCTION

Esrange located at 67°53′N and 21°04′E is the space operation centre of the Swedish Space Corporation (SSC), a state-owned limited liability company under the Ministry of Industry. The high latitude, the favourable climate and the relative proximity to every modern convenience makes this site unique in the world and an attractive base for international space related projects and activities. The Esrange facility includes a complete satellite station for TT&C (Telemetry, Tracking and Command), reception of data from various satellites, control station for telecommunication satellites, and facilities for launching sounding rockets and stratospheric balloons. Lounching of rockets and balloons is carried out under the ESAP agreement, which is a special project within the European Space Agency (ESA).

Esrange is a well-equipped launching facility and is able to deal with most types of rockets, including large rockets such as Skylark, Terrier Black Brant and Castor 4B. More than 500 rockets have been launched from Esrange since its inauguration in 1966. The impact area of 120 x 75 km on land offers safe landing and recovery. Payloads are speedely recovered by helicopter, normally within one hour after launch. The payload equipment is often re-used on subsequent missions. Most experiments carried out are within the field of microgravity, upper atmosphere chemistry and space physics.

Esrange can launch balloons with a volume of up to 2 million m^3, allowing a payload of 2 metric tonnes to be carried to an altitude of 45 km. The balloons are launched using either auxiliary balloons or a dynamic release technique. A vast area, comprising the whole of northern Scandinavia and European Russia, is available for impact. Payload recovery is a standard procedure carried out by helicopter. Balloons at Esrange are mainly used for atmospheric research, such as studies of the depletion of the ozone layer. Balloons are also used for validation of satellite payloads, astronomical experiments and testing of technical systems to be onboard space probes for planetary missions.

There has always been a need of scientific ground based instruments. The strong points of the ground based remote sensing are the capability for continuous measurements. The ground based instruments at Esrange are operated to support the scientific need during sounding rocket and high altitude balloon flights. Co-ordinated measurements by use of ground based instruments and instruments onboard the flying payloads are normally performed. The Earth's magnetic activity and the aurora activity are more or less continuously monitored. The Esrange MST-radar,

ESRAD, has been in near continuous operation since July 1996, providing information about winds, waves and other phenomena in the lower and middle atmosphere.

A Lidar instrument and metero radar at Esrange, owned by University of Bonn, represents a vital complement to the ground based instrumentation. Authorised users via Internet can easily access data.

For further information please contact: http://www.ssc.se/esrange

SCIENTIFIC INSTRUMENTATION

Developments and improvements

Better performance for ESRAD

The Esrange MST-radar, ESRAD, at 52 MHz (VHF), has been in near continuous operation since July 1996. The possibility to detect very weak echoes from particles and aerosols at altitudes up to 110 km, required an antenna array of 140 Yagi antennas. Different techiques like pulse coding and beam swinging was also implemented to achieve best results.
After 5 years of operation we decided to increase the power and sensitivity of ESRAD.
A large number of additional antennas and new low attenuation feeding cables were installed. The antenna array now comprises 284 pieces of 5-elements Yagi antennas. A significant improvement has been achieved regarding higher radiated power and increased receiving sensitivity. There is still need of an upgraded software package for antenna patching and steering and also a replacement of receivers.

A new optical site

KEOPS is a new concept for multi-disciplinary high-latitude science based on optical measurements and distributed networking. The KEOPS optical facility is located on Pahtavaara mountain at 530 m altitude, 1.5 km west of Esrange at 67°52′N and 21°02′E.
KEOPS mesurements are intended to be unmanned and remote controlled via Internet optimised for telescience applications. The hilltop comprises a 1000 m^2 open area for installation of optical instruments. The field of view of KEOPS is down to 0° for the whole horizon. Current installations contain a set of Fabry-Perot Interferometers providing images of the neutral winds in the thermosphere. A four-channel photometer and a set of all-sky cameras will be installed at KEOPS this year.
Easy access by road, electrical power and the Internet connection offers unique optical non-disturbed site for night-sky observations.

Ground based instruments in operation

Esrange offers a set of basic ground based scientific instruments:

- The magnetometer and the riometers are continuously operated and the photometer is measuring during the period September to April. All measured data are archived at the EGiS system locally at Esrange. Authorised users via Internet can easily access the data.

- A 3-axis flux gate magnetometer measuring variations of the Earth's magnetic field up to ±8000 nT with an accuracy of ±1 nT.

- Two riometers at 27.6 and 35 MHz are equipped with wide- and narrow beam antennas for studies of the Ionosphere

- A four-channel photometer equipped with filters for 427.8 nm, 486.1 nm, 557.7 nm and 630.0 nm.

- A set of night sky cameras are also available for studies of the Aurora Borealis or other light scattered phenomena in the atmosphere.

- Four new Faraday-transmitters for studies of the electron density and radio waves polarisation phenomena in the Ionosphere are available. The radiated power is received by specially designed rocket borne receivers. The frequencies are 1300 kHz, 2200 kHz, 3883 kHz and 7800 kHz with tunable CW output power up to 800 Watts.

User owned groud based instrumentation

RMR Lidar system

A rayleigh-Mie-Raman Lidar owned and operated by Dr. Fricke at the University of Bonn is located at the radar hill at Esrange. The Lidar is normally operated during co-ordinated campaigns involving sounding rockets or stratospheric balloons investigating phenomena in the lower and middle atmosphere. This Lidar is capable of operating in three-colour mode covering 355 nm, 532 nm and 1064 nm in night conditions and near daylight conditions. Different atmospheric phenomena, like PSC and NLC in the range from ground up to 100 km are monitored. Scientists at the Institute of Space Physics (IRF) in Kiruna are trained and capable of operating the Lidar at short notice.

Meteor radar

An All-sky Interferometric Meteor Radar (SKiYMET) owned and operated by Dr. Nick Mitchell, at the University of Wales in Aberystwyth, is located at Esrange. Echoes are received from the Mesopause region by the ionised trail gases when meteors enter at altitudes between 70-100 km. A 6 kilowatt pulsed transmitter is connected to a crossed 4 element Yagi antenna operating at 32.5 MHz. Five receivers are each connected to an individual 2 element crossed Yagi antenna tuned for optimum receiving sensitivity.

Fabry-Perot Interferometers at KEOPS

A set of Fabry-Perot interferometers owned by UCL, UK, and one owned by Dr. David Rees, UK, are located at KEOPS. The FPIs are providing images of neutral wind flows in the thermosphere and nighttime images of the aurora oval. The measurements are unmanned and remotely controlled via Internet. The FPIs are tuned for the red and green lines at 630.0 nm and 557.7 nm.

Meteorological measurements

A great number of balloon borne radio soundings and ozone soundings are carried out every year from Esrange, in order to get information of the current situation in the atmosphere. The measurements are used for scientific purpose or to get necessary wind information for the launch of sounding rockets or stratospheric balloons. The ozone vertical profile is measured by specially designed balloon borne ozonosondes. Standard radiosondes are used for information of wind, position and PTU-data.

Ground winds are measured at several levels up to 100 meters. Troposphere and stratosphere winds are measured with radar or radiosondes. Additional wind information is also measured with ESRAD contributing to the UK meteorological offices operational wind profiler network.

Ground level ozone concentration is measured at Esrange continuously by the Swedish Environmental Research Institute Ltd. (IVL) as part of the national monitoring netwoork. An extensive archive comprising many years of collected PTU, ozone and wind data is available at Esrange.

Onsala Space Observatory and Chalmers University of Technology, Gothenburg, Sweden, are performing seasonal measurements of the meteorological troposphere conditions above Esrange using ground based water vapour microwave radiometer, a celiometer, an IR instrument and a rain radar. Water vapour causes an additional delay (wet delay), which is equivalent to an excess path of radio waves propagating through the atmosphere. The wet delay is a major error contributor in space geodetic techniques such as Very Long Baseline Interferometry (VLBI) and Global Positioning System (GPS).

SPACE OBSERVATORY IN THE KIRUNA AREA

The Swedish Institute of Space Physics (IRF), a space research institute located in Kiruna, was established 1957 and is combining long term observatory activities with advanced space research projects. The IRF of today runs an extensive research programme, carried out by ground based as well as space based means. IRF participates in several international projects and the first Kiruna-designed satellite experiment was launched in 1968. The Auroral Large Imaging System (ALIS), a system for studying the three-dimensional structure of the Northern Lights is developed at IRF. The ALIS system is a multi-station imaging system, which uses sophisticated tomographic reconstruction techniques, artificial intelligence and advanced IT. The system comprises today a network of 6 stations with advanced CCD cameras and a control centre.

During the last ten years IRF has steadily increased its participation in university education in space physics and space technology.

For further information please contact : http://www.irf.se.

SUPPORT

Esrange is a well-equipped range, and with the high-qualified staff we are ready to help the user to perform successful science. The level of support is always adapted to the user's need and confirmed by mutual agreements. Unmanned measurements are normally remotely controlled by Internet.
Requested data like ozone, PTU, telemetry or flight trajectory data will normally be delivered on a CD-ROM media or by FTP, with the help of the Esrange Geophysical information Service (EGiS).

For further information please contact:

    www: http://egis.esrange.ssc.se
    e-mail: egis@esrange.ssc.se

Scientific Balloons From Svalbard

Kjell Bøen

Andøya Rocket Range, P.O. Box 54, N-8483 Andenes, Norway

Abstract. It seems to be a growing demand for launching of scientific balloons from arctic latitudes. Andøya Rocket Range is responsible for scientific-related balloon and sounding rocket operations in Norwegian territory including Svalbard. This paper presents facilities, infrastructure, telemetry and technical possibilities for launch of scientific balloons from Longyearbyen, Svalbard, located at 78 degrees north latitude. Scientific opportunities will also be addressed.

INTRODUCTION

Andøya Rocket Range (ARR) is responsible for scientific related balloon and sounding rocket operations in Norwegian territory, including Svalbard. From its location far north of the Arctic Circle, ARR provides complete services for launching, data acquisition, recovery and ground instrumentation support. The launching services are offered from sites on the mainland of Norway and from Svalbard.

The Andøya launch site has been continuously operational since 1962. More than 760 rockets and 500 balloons have been launched since then. In 1997 ARR established a launch site for sounding rockets in Ny-Ålesund at Svalbard. Since then, rockets campaigns have been performed at Svalbard for institutions from USA, Japan, Germany and Norway.

ARR is planning a new balloon launch site at polar latitude for both short and long duration flights. The launch site is located close to Longyearbyen at Svalbard. The location at 78 degrees north latitude makes the site favorable for circular polar flights driven by the systematic circulation of winds around the North Pole, inside the vortex. Agreements with the Norwegian authorities make ARR able to offer this service.

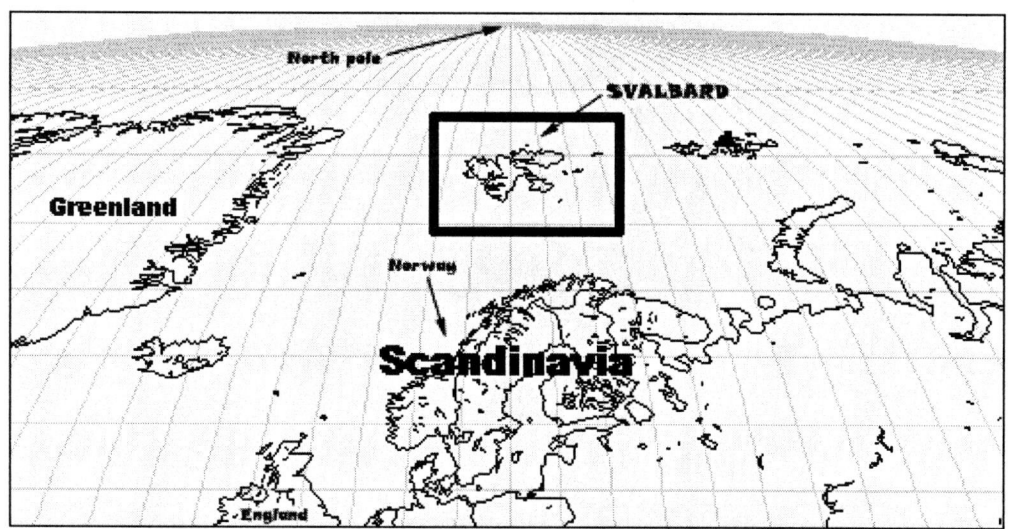

FIGURE 1. Map showing the location of the Svalbard archipelago

SVALBARD - GENERAL INFORMATION

Under the magnificent polar sky, halfway between the North Pole and mainland Norway, lies the island kingdom of Svalbard. The 63 000 square kilometers of Arctic expanse, fjords and glaciers is an exotic, untamed and magnificent destination and there are a lot of tourists visiting the islands during the summer season. The Svalbard archipelago consists of four main islands and about 150 lesser ones. The group of islands cover an area between 71° - 81° N and 10 ° - 35 ° E, see Figure 1. Glaciers cover 60 % of the islands.

Svalbard has settlements in Longyearbyen (Norwegian administrative centre) with approximately 1 400 inhabitants, Barentsburg (Russian mining community) with approx. 850 inhabitants, Ny-Ålesund (Norwegian international research centre) with approximately 40-100 inhabitants, Sveagruva (Norwegian mining community) approximately 90 commuters and Hornsund (Polish research station) approximately 8 inhabitants

Despite Svalbard being so close to the North Pole, the archipelago has a relatively mild climate compared to areas at the same latitude. In Longyearbyen, the average temperature ranges from -14° C during the winter to +6° C during the summer. The very lowest temperature was measured in March 1986 at –46.3° C, while the very highest temperature was measured in July 1979 at +21.3° C. In terms of precipitation, however, Svalbard may be described as an "Arctic desert" with annual rainfall at a mere 200-300 mm. The periods of polar night and midnight sun vary depending on latitude. Longyearbyen enjoys the midnight sun from 19 April to 23 August, while in the period 28 October to 14 February, the sun never peaks over the horizon.

Ny-Ålesund (78° N), a former mining town, is today an international base for different types of arctic research and several countries have scientific and technical personnel permanently staying at Ny-Ålesund. ARR established in 1997 the SvalRak Sounding Rocket Launch Facility in Ny-Ålesund. Since then, several sounding rockets have been launched from this facility; the last one was a large sounding rocket for ISAS, Japan reaching a peak altitude of more than 1100 km. The geographical location, infrastructure and the network of ground-based instruments make the SvalRak launch site ideal for studies of the dayside aurora, magnetospheres boundary layer processes and the magnetic cleft, cap and cusp.

Longyearbyen (see map in Figure2), the largest settlement at Svalbard, is a well-developed modern town, which facilitates all common services and infrastructure. Airport, hotels, guesthouses, flats and rental cars are available, see sketch of Longyearbyen in Figure 3.

FIGURE 2. Map showing the settlements at Svalbard

FIGURE 3. Sketch of Longyearbyen

BALLOON LAUNCH SITES

Longyearbyen airport is located approximately 3 km west of the town Longyearbyen. Balloon launches can take place from the large platforms at the airport. The pre-dominant ground winds (easterly) are favorable for balloon launches bringing the balloons out over the Ice Fjord.

The air traffic control tower is open 24 hours a day, and the air traffic meteorological service is available during the daytime. The air traffic is limited. The airport has two regular airliner flights to the main land a day in addition to a few local flights. Thus, there should be no major local air traffic restrictions to the balloon launches from the airport.

An old unused airfield, located a few km east of the Longyearbyen, may also be used to launch easy-handled small balloons (5-10.000m^3).

PAYLOAD PREPARATION AND OFFICE FACILITIES

A suitable heated storehouse divided into two separate halls can be rented at the airport. The storehouse is suitable for payload preparation, scientific on-line data receiving and the balloon operation and flight control, and it has direct access to the airport apron where the launches can take place.

A number of small offices can be made available in the first floor of the storehouse and there are telephone and ISDN opportunities.

EQUIPMENT AND SUPPORT

A close co-operation with the Norwegian Meteorological Institute ensures that ARR is on-line to the institute database from where digital weather forecasts, analysis and maps are available. Also analyses and prognoses of high altitude conditions are available. The meteorological institute has their own meteorological office with meteorologist located at Longyearbyen airport, primarily to support the aircraft operations at Svalbard.

Digital satellite images, showing cloud cover and also ice-edges can be obtained from Tromsø Satellite Station within 60 minutes after each satellite pass. By combining all the digital meteorological information and by using advanced computer systems, ARR is able to give predictions of the flight path prior to launch. Computer programs are used also for gas filling calculations.

For positioning of the balloon during flight GPS is used. A computer system, also comprising a digital map database, is used for real-time plots of the flight path.

The SvalSat satellite station is located on the 550 meter high Plateau Mountain approximately 10 minutes drive from Longyearbyen airport. The satellite station site has an almost 360 degrees free horizon above 2.5 degrees elevation, except for a small shadow sector in southerly direction, see Figure 4. The SvalSat site is suited for installation of the ARR Mobile Telemetry Station and the Telecommand System. Due to no "line of sight" between the telemetry station and the launch site, a telemetry relay station has to relay the telemetry data during ground tests and the first 500 meters of the flight. This is a technique ARR also have used at other locations in Norway.

Ground support systems necessary for balloon operation summarized:
- On-line meteorological forecasts and prognoses
- Wind profile prognoses
- Satellite images
- Vaisala DigiCora System
- Flight simulation computer programs
- Real-time trajectory information
- Systems for positioning of the payload at ground after impact

ARR have the following flight train modules/systems available for short duration flights:
- House Keeping Module with Telemetry, GPS system, battery packages etc.
- Radio sonde system
- Radar transponder
- Parachute
- Tele Command System
- Pyrotechnics

ARR is currently developing a new Telemetry Storage and Transmission system for balloon payloads. The system will consist of a data storage system, processor and data modem for connection via Iridium Satellite Network to the ARR ground station. The onboard system will store all raw data from the instruments, process and compress data for downloading to the ground station via the Iridium Network. Operation, re-programming and adjustments of the instruments are possible as well.

RECOVERY

Recovery is available on request. The local rescue helicopter (Super Puma) is available for recovery operations. The helicopter company is responsible for the Search and Rescue operations at the islands. Svalbard has polar night 4 months of the year (sun more than 6 degrees below the horizon day and night). Therefore, the helicopters are equipped with advanced navigation systems, communication systems and the crew are equipped with night goggles and trained for operation in dark conditions. It is possible to carry out recovery operations on ice or land even in darkness for payloads equipped with VHF beacon/strobe light and known impact positions from GPS or ARGOS.

The helicopter company has previously successfully recovered payloads up total mass 600 kg. If there are several launches the helicopter can pick up several small payloads in one trip.

TRANSPORT

Air cargo transportation to and from Svalbard is available throughout the year by the daily airliner flight. Typical aircraft size is Boeing 737. Heavy and dangerous equipment should however be shipped by ship. Sea transportation from the Norwegian mainland is possible every 3 weeks from approximately March to December while there is no ice in the fjord outside Longyearbyen.

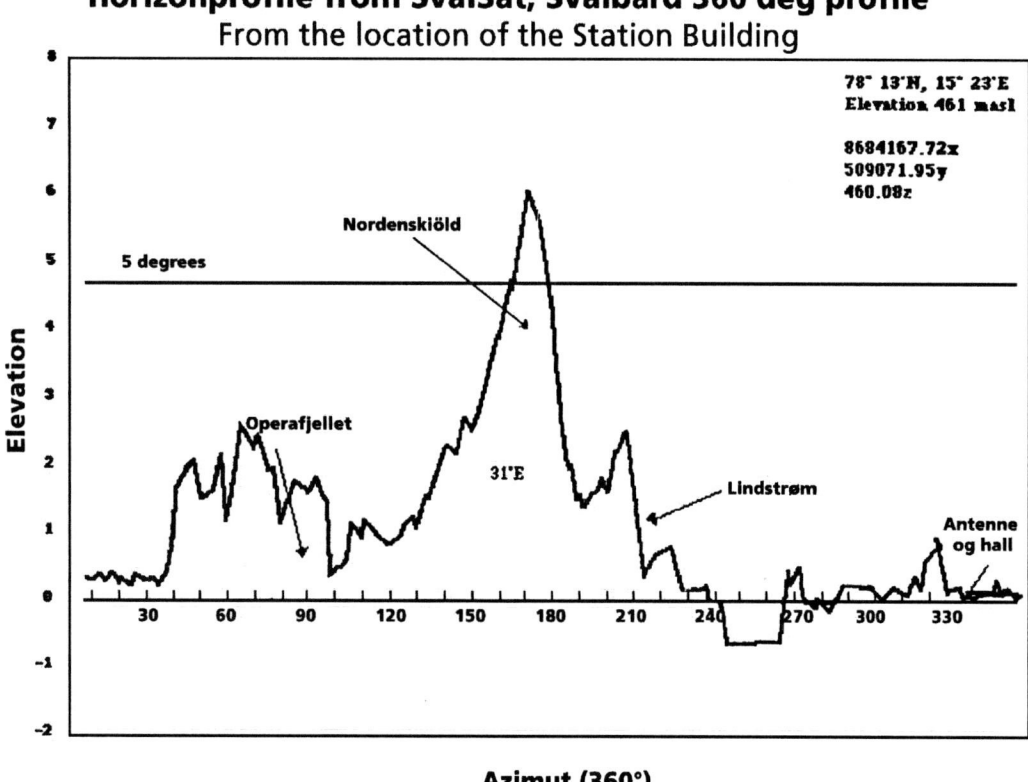

FIGURE 4. Telemetry horizon profile for SvalSat.

CONCLUSION

Longyearbyen would indeed be suitable for launching of stratospheric balloons. Besides the favorable location with respect to the vortex, Longyearbyen offers good facilities, logistics and infrastructure considering its remote location.

Stratospheric balloon operations can be obtained from Longyearbyen by using ARR. The range offers a ready-made flight train concept for the scientists, only the payload itself has to be shipped to the range. ARR staff and ARR contractors perform launch operation, tracking- and data reception, telecommand and recovery. By connecting the telemetry receiving systems to the computer network, the scientists can monitor the payload data during flight from their home office in near real-time.

Balloon operations from Longyearbyen, Svalbard, conducted by ARR can be summarized as:
- Two possible launch sites
- Suitable facilities for preparations and operations
- Long- and short duration flights
- Summer and winter campaigns can be conducted
- Meteorological support available
- Recovery operations are available, also in dark conditions.

ARR is together with CNES planning a demonstration flight from Longyearbyen during the winter 2002/2003. The first scientific campaign is planned for January 2004 (provided that the project will be funded). A total of 15 balloons of different sizes will be launched and with both short and long duration flight time as well. Investigators from France, UK, Germany, Denmark, Italy and USA will participate.

POSTER PRESENTATIONS

The Rapid Prototyping Approach in the Design and Testing of Radiometric Systems

Primo Attinà, Bruno Audone, Demis Boschetti

Alenia Spazio S.p.A.,
Strada Antica di Collegno 253, 10146 Turin, Italy

Abstract. System design and testing activities tend to a higher integration degree. Rapid Prototyping technique allows the automatic design and testing of the instrument, through the employment of appropriate software tools for the verification and refinement of technical specifications before they are frozen in the effective hardware configuration.

INTRODUCTION

System design and testing activities have evolved since the last 30 years toward a higher integration degree suitable to match the needs of cost and time reduction in conjunction with improved quality and performances.

In the early '60s the system design was carried out maintaining a well-defined separation between the system specification definition phase and the system design one. Once the hardware blocks were designed they were manually tested by injecting the signals defined in the applicable specifications at the inputs and measuring the signals at the outputs (Fig.1).

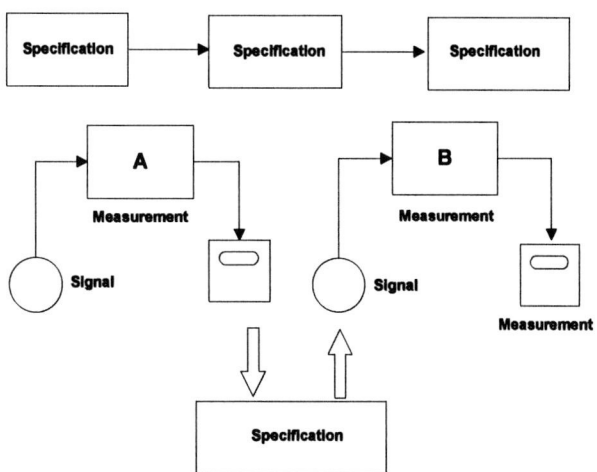

FIGURE 1. Manual Design and Testing

In the '70s the use of stimulation and recording devices driven by computers established an actual improvement with respect to manual techniques (Fig.2).

The actual breakthrough came out when the simulation concept was introduced. The simulation model differs from the mathematical model because it tightly matches the actual hardware input and output parameters. In this manner it is possible to establish an interactive feedback between the hardware models and the software ones (A and B block in Fig.3).

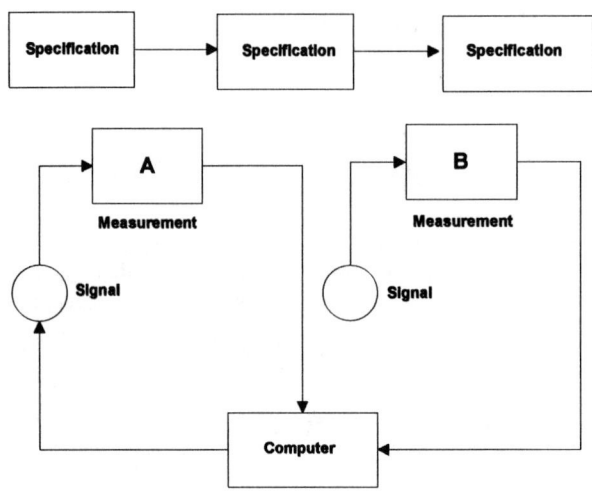

FIGURE 2. Manual Design and Automatic Testing

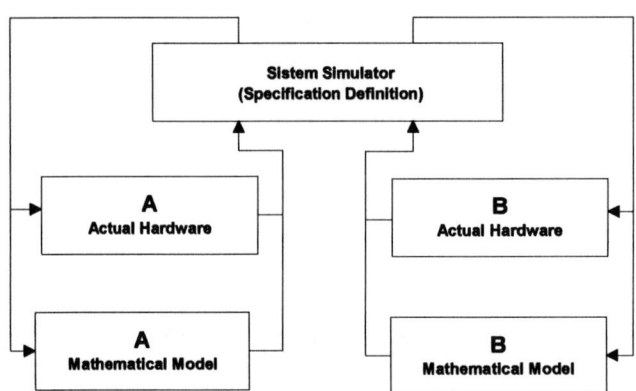

FIGURE 3. Automatic Design and Testing

RAPID PROTOTYPING APPROACH TO SPOrt PROGRAM

In Fig.4 the system definition is shown: it is required to design and test the radiometer installed on the Space Station operating in its intended environment.

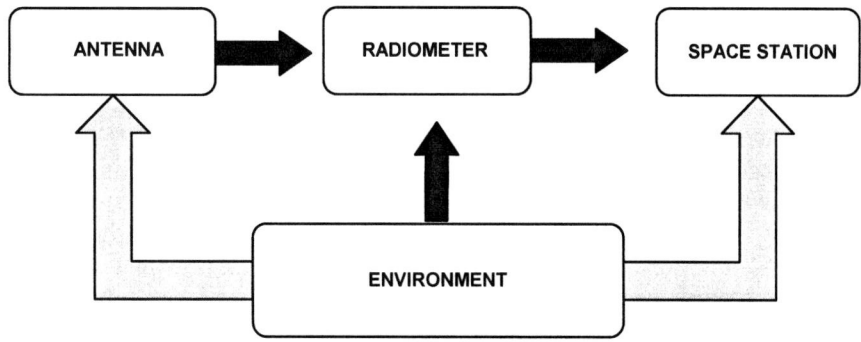

FIGURE 4. SPOrt Experiment

The Radiometric module is simulated by means of ADS (Advanced Design Simulator), a simulation tool developed by Agilent, which has the capability of generating a mathematical model of the radiometer. This model starts from a preliminary definition of the system under consideration. Then gradually its implementation is improved as soon as more information become available from the component/subsystem manufacturer: for example the characteristics of the amplifiers, the S parameter description of the hybrid phase discriminator (HPD) and Ortho Mode Transducer (OMT) and so on. The simulation model plays the role of the breadboard with the additional advantage that at any time it is possible to compare the data measured on the subassemblies of the subsystem with the actual final performances of the system. In this manner it becomes possible to evaluate and grant concessions in case that a specific subsystem component cannot achieve its design aims but the verification at system level shows that concession does not impair the final system performance.

The Rapid Prototyping Approach is also useful in system testing because during the development of the system we have built the model of the subassemblies; therefore it becomes possible to substitute the software modules with the actual components and carry out a step by step integration procedure of the system.

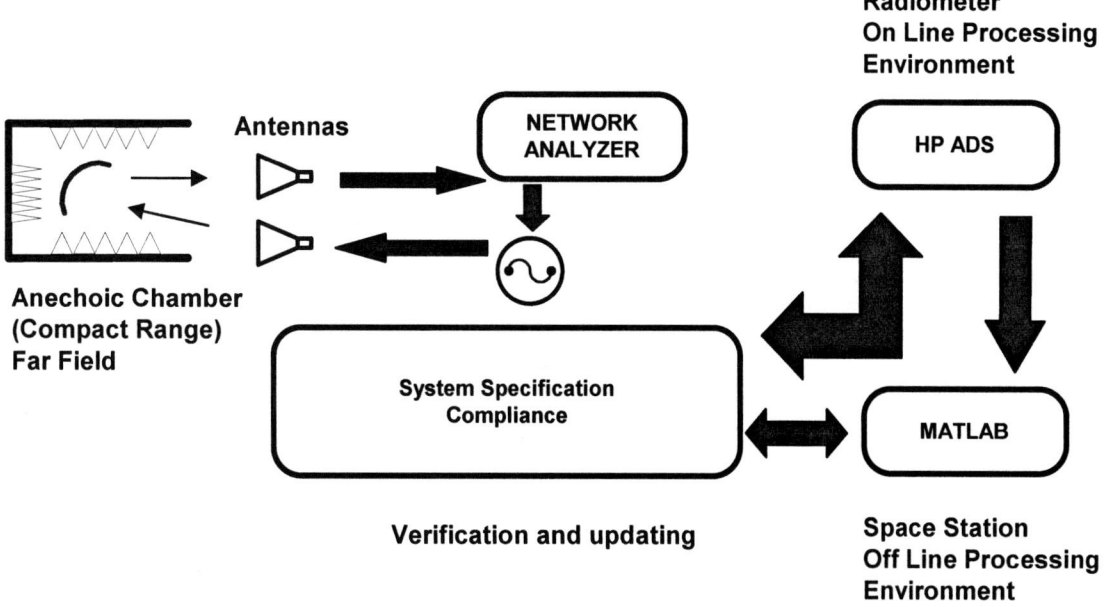

FIGURE 5. Software / Hardware Simulation (First Step)

At the first step the antenna subsystem model is substituted by the actual antenna (Fig.5), which is installed in the suitable test range where the far field test conditions are created. The proposed test range is the Compact Range of Alenia where the antenna radiation tests can be performed in the required frequency range from 1 to 100 GHz. The output signals measured with a network analyzer are sent as inputs to the ADS where they are processed as actual input signals. In this manner it is possible to verify the matching conditions between the radiometer model and the actual antenna hardware. The output data are measured and sent to MATLAB[1], which simulates the actual operation.

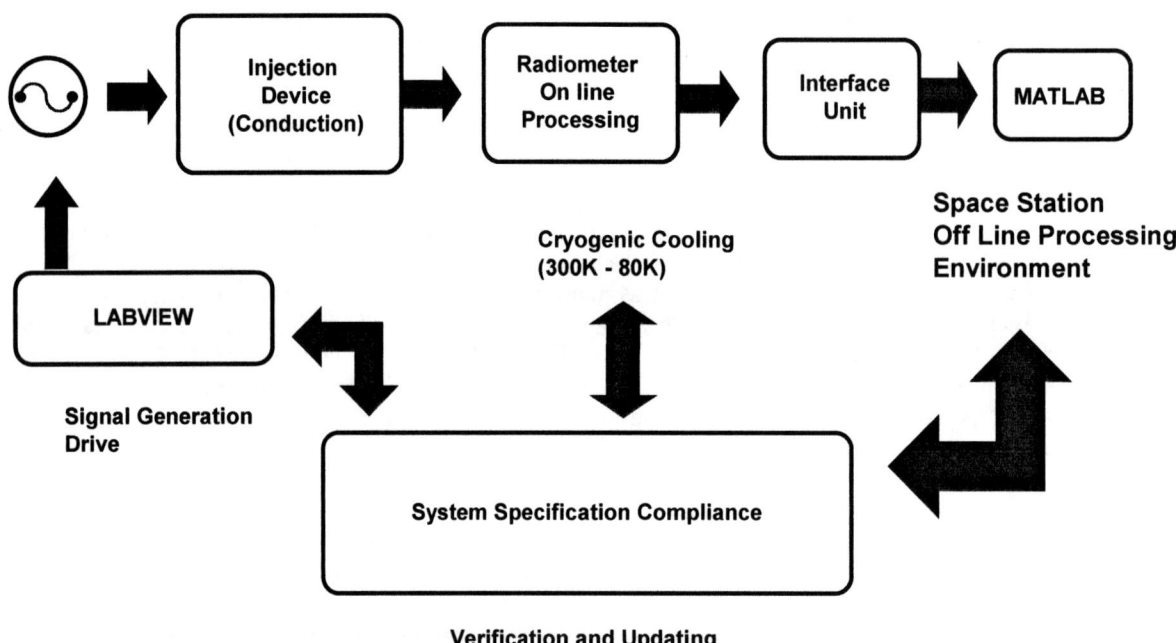

FIGURE 6. Software / Hardware Simulation (Second Step)

At the second step (Fig.6) the antenna is simulated by a suitable injection device compliant with the antenna subsystem mathematical model while the radiometer is substituted by the actual hardware unit installed in a suitable thermo-vacuum chamber, where the actual cryogenic cooling conditions are realized (300K and 80K). The input signals to the radiometer come from suitable generators driven by specific software routines developed in LABVIEW[2] environment. The output data are measured through a suitable interface and sent to MATLAB, which simulates the actual operation environment and the off line processing algorithms.

The third step of the Rapid Prototyping procedure (Fig.7) requires the integration of the hardware antenna and hardware radiometer in conjunction with the simulated environmental conditions and off line processing; this phase may be difficult to be realized because it requires the development of an antenna to antenna coupling device. In the previous phase the signals were injected by conduction; in this phase in order to take into account the hardware antenna it becomes necessary to find a suitable way of injecting signals by radiation. This is only possible by violating the far field conditions. In this final step no specification modification is foreseen because the design was frozen in the previous step; therefore only the verification of the system specification compliance will be carried out. The radiometer output data are sent to MATLAB simulation environment through a suitable interface unit

---

[1] MATLAB is the software package for mathematical and technical computing by TheMathWorks, Inc.
[2] LABVIEW is the software tool for graphical design and engineering developed by National Instrument Corp.

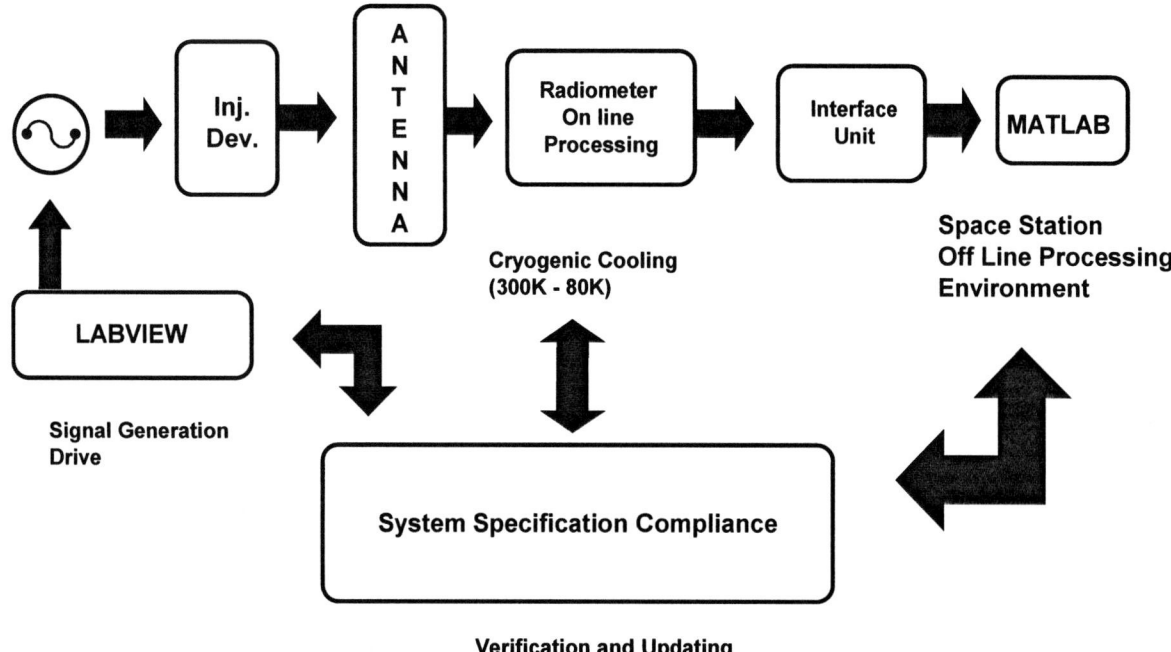

FIGURE 7. Software / Hardware Simulation (Third Step)

The final step is obviously the complete real environment using the actual hardware devices; this is only possible during the operational use of the payload installed on the International Space Station.

SPOrt RADIOMETER SIMULATION WITH HP-ADS

In the schematic (Fig. 8) the radiometer channels are shown from the feed horn input to the HPD outputs (total power and Stokes parameters). The HPD may be in the ideal configuration or correspond to the actual HPD whose parameters were measured. In this manner it is possible to compare the results of the ideal situation with the actual one. This procedure can be used for all components of the radiometer. As soon as the measured parameters become available they are substituted in the simulation model, allowing the evaluation of the impact of the variations induced by actual hardware.

CONCLUSION

The Rapid Prototyping approach represents a fundamental conceptual framework for the modern design and testing activities. The employment of this approach in the SPOrt Radiometer design appears to be a good contribution for the successful performance of the experiment.

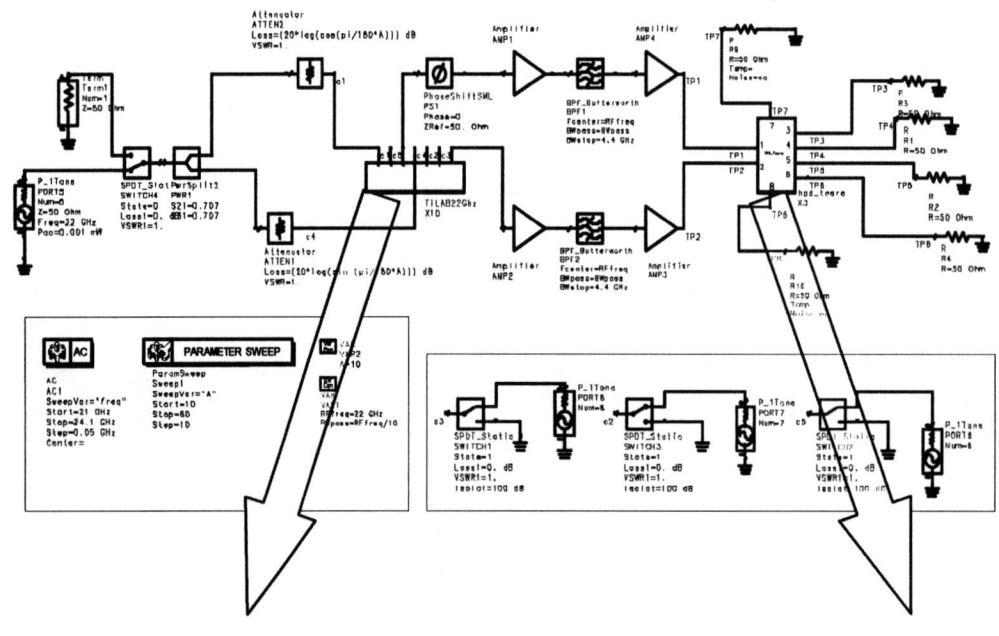

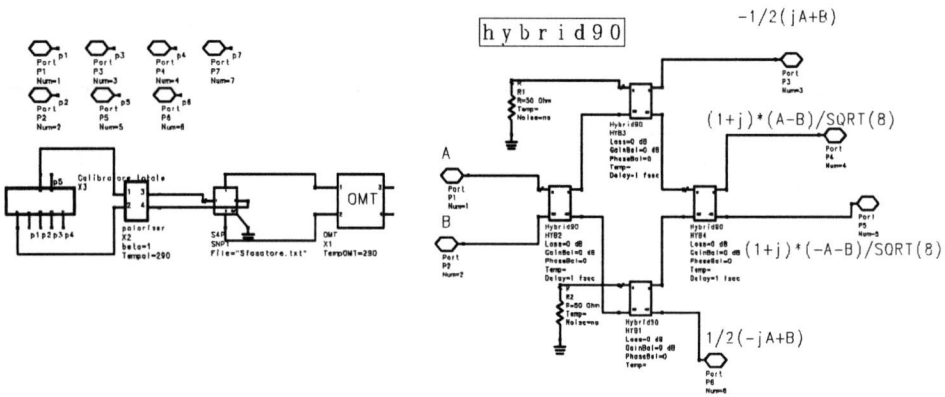

FIGURE 8. SPOrt Radiometer diagram defined with HP-ADS tool

Calibration Techniques and Devices for Correlation Radiometers Used in Polarization Measurements

M. Baralis*, O. A. Peverini*, R. Tascone*, D. Trinchero*, V. Niculae*, A. Olivieri*,
E. Carretti†, S. Cortiglioni†, C. Macculi†, J. Monari**, A. Orfei**, G. Sironi‡ and
M. Zannoni‡

*IRITI-CNR, Politecnico di Torino, Torino , Italy
†ITESRE-CNR, Bologna, Italy
**IRA-CNR, Bologna, Italy
‡Univ. Milano-Bicocca, Milano, Italy

Abstract. The polarized sky emission can be detected by measuring its Stokes parameters through correlation radiometers. In order to compensate for the variation of the characteristics of the various components of the radiometers and of the environmental parameters, an on line calibration is required. In fact, the evaluation of the proportional factor between the detected signal and the level of the radiation is not sufficient for a complete characterization of the radiometers, since, in polarization measurements, the detection occurs in a two-dimensional space, that is the Q and U Stokes parameters. In this case special procedures and devices are needed.

INTRODUCTION

The measurements of the polarized sky emission requires instrumentations with a very high level of sensitivity. Nowadays, various radiometer architectures are proposed, see for example [1]. The basic principle is to measure the Q and U Stokes parameters by comparing different observations obtained by different antenna orientations or by using a double polarized antenna. The Stokes parameters can be determined, in principle, by a differential or a correlation process. Because of the high level of the unpolarized radiation, the measurements based on the correlation process can be more promising. The astrophysical project SPOrt [2] will try to measure the polarized radiation of the sky by means of four radiometers, which will cover the frequency range 20-90 GHz and will be mounted on the International Space Station (ISS). An online calibration of the radiometers is highly recommended in order to compensate for the spurious effects induced by the variation of the environmental parameters and of the instrumentation characteristics [3].

In the SPOrt correlation radiometer, the polarized emission will be detected by measuring simultaneously the Q and U Stokes parameters which correspond to

$$Q = |E_x|^2 - |E_y|^2, \quad U = 2\Re\{E_x E_y^*\} \tag{1}$$

These parameters will be detected by means of correlation units (HPD) whose input signals come from an antenna with double circular polarization [2]. The antenna system consists of a corrugated horn, a polarizer which converts the circular polarizations into the linear ones and an ortho-mode transducer (OMT), which separates the two polarizations. Before entering the HPD, these two signals are amplified by means of a pair of amplification chains.

CALIBRATION PROCEDURE

In order to evaluate the nature of the degradations of the measurements, it is convenient to represent the entire radiometer in terms of a linear system with two input and output signals, as sketched in Fig. 1. The two inputs A and B are high frequency signals in the circular polarization base (LCP and RCP), whereas the two outputs Q_m and U_m are the low frequency signals corresponding to the Stokes parameters directly measured by the radiometer. With

CP609, *Astrophysical Polarized Backgrounds*, edited by S. Cecchini et al.

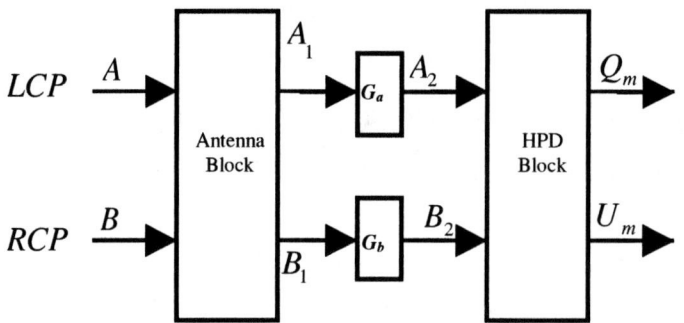

FIGURE 1. Block diagram of the correlation radiometer.

reference to Fig. 1, the whole radiometer system is represented in terms of four blocks. The antenna block, comprising the antenna, the polarizer and the OMT; two blocks describing the two chains of amplifiers and filters; the HPD block containing filters, the HPD itself, diodes, differential amplifiers and low pass filters. The two input signals A and B contain the polarized radiation (A_P and B_P) and the unpolarized one (A_N and B_N): $A = A_P + A_N$, $B = B_P + B_N$. The components A_N and B_N are uncorrelated and can be defined as to contain also the noise produced by the radiometer. According to this model, the output of the radiometer can be described by the following matrix expression:

$$\begin{bmatrix} Q_m \\ U_m \end{bmatrix} = \begin{bmatrix} H_{qq} & H_{qu} \\ H_{uq} & H_{uu} \end{bmatrix} \begin{bmatrix} Q \\ U \end{bmatrix} + \begin{bmatrix} T_{qa} & T_{qb} \\ T_{ua} & T_{ub} \end{bmatrix} \begin{bmatrix} |A|^2 \\ |B|^2 \end{bmatrix} \tag{2}$$

where the Q and U Stokes parameters of the polarized radiation can be expressed in terms of the signals A_P and B_P as follows:

$$\begin{bmatrix} Q \\ U \end{bmatrix} = 2 \begin{bmatrix} \Re\{A_P B_P^\star\} \\ \Im\{A_P B_P^\star\} \end{bmatrix} \tag{3}$$

In the case of a linearly polarized signal the Q and U Stokes parameters can be rewritten as:

$$\begin{bmatrix} Q \\ U \end{bmatrix} = \begin{bmatrix} |E|^2\cos(2\theta) \\ |E|^2\sin(2\theta) \end{bmatrix} \tag{4}$$

where θ is the angle between the incident electric field E and the principal direction of the polarizer. The matrices $\underline{\underline{H}}$ and $\underline{\underline{T}}$ in (2) are generic, but real, since they deal with quadratic quantities. Moreover, they transform the circle in the QU-plane, obtained by varying the angle θ, into a rotated and translated ellipse.

The calibration procedure, i.e. the evaluation of the matrices $\underline{\underline{H}}$ and $\underline{\underline{T}}$, can be accomplished by directly measuring the quantities Q_m and U_m in presence of predefined signals (markers). Under the reasonable assumption that the intensities of the total signals A and B are equal ($|A|^2 = |B|^2 \propto P$), equation (2) can be rewritten as:

$$\underline{Y} = \underline{\underline{H}}\,\underline{X} + \underline{C}P \tag{5}$$

where $\underline{C}$ is a two-element column vector. By injecting the marker M_1, corresponding to a linearly polarized field according to the angle θ_1, the following variation of the output is detected:

$$\Delta\underline{Y}_1 = \underline{\underline{H}}\,\Delta\underline{X}_1 + \underline{C}\Delta P_1 \tag{6}$$

where $\Delta\underline{X}_1$ and ΔP_1 are the variations of the input Stokes parameters and of the total power, respectively. The detection of the total power can be accomplished by the two auxiliary outputs of the HPD [2]. Injecting subsequently other two markers M_2 and M_3, one can measure the variation of both the Stokes parameters and of the total power and, hence, derive the following matrix equation:

$$\underline{\underline{H}} \left[\frac{\Delta X_3}{\Delta P_3} - \frac{\Delta X_1}{\Delta P_1}, \quad \frac{\Delta X_2}{\Delta P_2} - \frac{\Delta X_1}{\Delta P_1} \right] = \left[\frac{\Delta Y_3}{\Delta P_3} - \frac{\Delta Y_1}{\Delta P_1}, \quad \frac{\Delta Y_2}{\Delta P_2} - \frac{\Delta Y_1}{\Delta P_1} \right] \tag{7}$$

from which the matrix $\underline{\underline{H}}$ and the vector $\underline{C}$ can be easily derived. In order to obtain a well-conditioned matrix in equation (7) the columns of the matrix to be inverted should be orthogonal vectors. To this end, the markers M_1, M_2 and M_3 should correspond to signals with a relative rotation of $45°$.

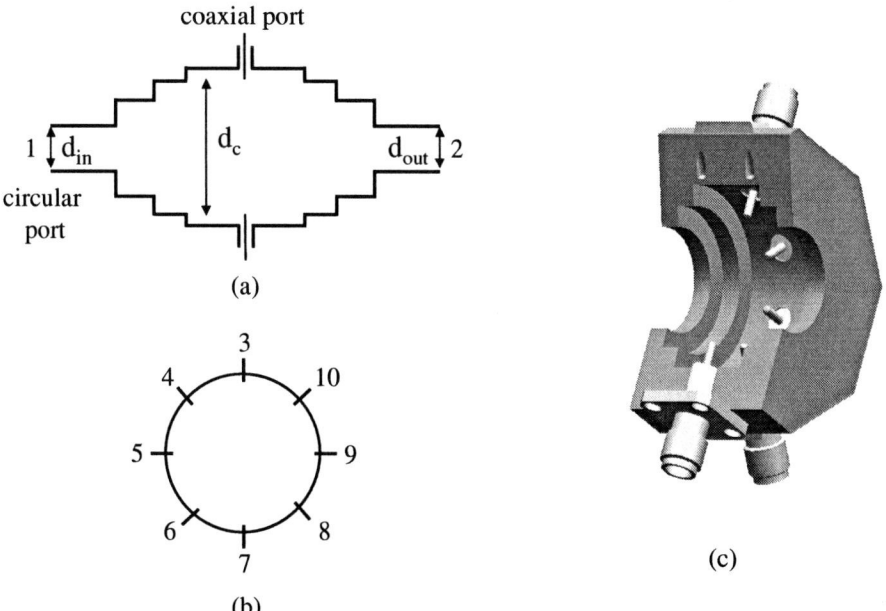

(a)

(b)

(c)

FIGURE 2. Scheme of the marker injector. (a) Longitudinal cut. (b) Cross-section of the central block. (c) Cutaway perspective, only five out of the eight SMA connectors are visible.

MARKER INJECTOR

The calibration procedure described in the previous section requires the injection of different markers in the radiometer, just behind the antenna. Unfortunately, this procedure must be carried out in the operative condition, that is in presence of the polarized and unpolarized radiation. Therefore, the marker injector must degrade the relevant signals as little as possible. In particular, this device has to exhibit a high return loss and to prevent depolarization of the incoming signals.

The design of the injector is depicted in Fig. 1, which consists of three blocks. The central block is formed by a circular waveguide in which eight SMA connectors are inserted, at 45° from each other. Although only three markers have to be injected inside the circular waveguide, all the eight connectors are necessary to preserve the azimuthal symmetry of the structure and, hence, to avoid depolarization of the incoming signals. The coupling level between coaxial and circular ports, i.e. the value of the scattering parameters $S_{1j} = S_{2j}$ with $j \neq 1,2$, can be controlled by adjusting the penetration length of the internal wire of the SMA connector. Due to the low value of the desired coupling between coaxial and circular ports ($\approx -35\,\mathrm{dB}$) and because of mechanical constrains, the diameter of the circular waveguide must be of the order of $d_c = 16 \div 20\,\mathrm{mm}$ for the device operating in the Ka-band. For this reason, the injector must include also two lateral blocks, which provide a matching structure between the input/ouput circular waveguides and the enlarged central waveguide. Because of its dimensions, the central waveguide operates in a multimode regime, the TE_{11}, TM_{11} and TE_{12} modes being propagating in the band of interest. In order to avoid spurious resonances, which can occur inside the central cavity of the injector due to modal conversion, the matching sections must be designed to minimize the level of excitation of higher modes. The matching section is formed by the cascade of several circular waveguide sections with increasing/decreasing dimensions. In order to optimize the heights of the steps and the lengths of the waveguides sections, we developed a simulation tool, which is based on the evaluation of the Generalized Scattering Matrices (GSM) of the steps by the Method of Moments (MoM). On the contrary, the GSM of the central block was evaluated by a Finite Element Method (FEM) code because of its geometrical complexity. In this case a design procedure similar to those adopted for waveguide filters is not readily applicable, since we are concerned with multimode cavities. Therefore, the optimization was carried out by means of an evolution-strategy method, which allows ones to control simultaneously the transmission level of all the modes of interest.

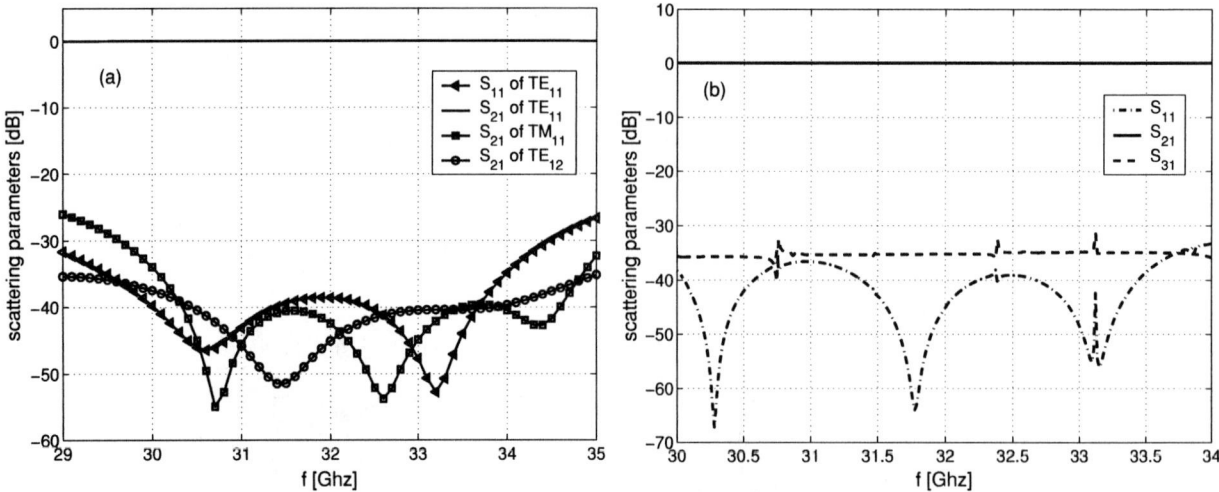

FIGURE 3. Scattering parameters versus frequency for the Ka-band device. (a) Matching structure with 8 sections ($d_{in} = 7.57$ mm, $d_c = 18$ mm). (b) Complete injector with a cavity length $l_c = 7.74$ mm.

Finally, the Generalized Scattering Matrices of the matching and central blocks are cascaded, yielding the complete electromagnetic characterization of the whole injector.

In Fig. 3 the computed scattering parameters versus frequency of the first matching section [(a)] and of the entire injector [(b)] are reported. The simulations were carried out by the full-wave method described above. The matching section consists of 8 waveguides sections in the case $d_{in} = 7.57$ mm and $d_c = 18$ mm. It has to be noted that the transmission levels relative to the higher modes (TM_{11} and TE_{12}) are lower than -40 dB, while the fundamental mode exhibits a reflection coefficient better than -35 dB in the band of interest. As for the whole injector, the coupling factor between the circular and coaxial ports is about -35 dB for a penetration length of the internal wire of ≈ 0.5 mm, whereas the reflection coefficient at the input/ouput circular ports remains at -35 dB. The small peaks which are visible in the scattering parameters of the injector correspond to resonances inside the central block due to modal conversion and are caused by the non zero excitation of the higher modes by the matching sections and by the internal wires.

CONCLUSIONS

In this paper a calibration procedure for correlation radiometers has been presented, which requires the injection of three polarized markers. A design of the marker injector, which exhibits high return loss and ideally zero depolarization of the incoming signals, has been reported as well. This work has been financially supported by the Agenzia Spaziale Italiana.

REFERENCES

1. G. Sironi, G. Boella, G. Monelli, L. Brunetti, F. Cavaliere, M. Gervasi, G. Giardino, A. Passerini, *New Astronomy*, Vol. **3**, pp.1-13 (1998).
2. S. Cortiglioni, S. Cecchini, M. Orsini, G. Boella, M. Gervasi, G. Sironi, R. Fabbri, J. Monari, A. Orfei, Ng. Kin-Wang, L. Nicastro, U. Pisani, R. Tascone, L. Popa, I. A. Strukov, " Sky Polarization Observatory (SPOrt): a Project for the International Space Station," *Proc. of ESA Workshop on Space Exploration and Resources Exploitation* , Cagliari, 1998, edited by Mauro Novara-ESTEC.
3. E.Carretti, R. Tascone, S. Cortiglioni, J. Monari, M.Orsini, *New Astronomy*, Vol. **6/3**, pp. 173-188 (2001).

Synchrotron Angular Spectra at Various Galactic Latitudes and Frequencies

M. Bruscoli[*], M. Tucci[†], V. Natale[**], E. Carretti[‡], R. Fabbri[§] and C. Sbarra[‡]

[*]*Dipartimento di Astronomia, Università di Firenze, Largo Fermi 5, I-50125 Firenze, Italy*
[†]*Dipartimento di Fisica, Università di Milano – Bicocca, Piazza della Scienza 3, I-20126 Milano, Italy*
[**]*C.A.I.S.M.I./C.N.R., Largo Fermi 5, I-50125 Firenze, Italy*
[‡]*I.Te.S.R.E./C.N.R, Via P. Gobetti 101, I-40129 Bologna, Italy*
[§]*Dipartimento di Fisica, Università di Firenze, Via Sansone 1, I-50019 Sesto Fiorentino, Italy*

Abstract. We study the angular power spectra of the polarized Galactic synchrotron emission in the range $10 \lesssim l \leq 800$, at several frequencies between 0.4 and 2.7 GHz and at several Galactic latitudes up to near the North Galactic Pole. We properly distinguish between the various polarization spectra that can be defined and measured, including the $C_{PI\ell}$ spectrum which is insensitive to the polarization direction.

INTRODUCTION AND MAIN RESULTS

Angular power spectra (APS) of the polarized Galactic synchrotron emission have been investigated in recent papers by Tucci et al. (2000, 2001a), Baccigalupi et al. (2001) and Giardino et al. (2001). These works were stimulated by the need for a clear separation of the cosmic microwave background (CMB) from the polarized radio foreground in future experiments, as it was the case for the modelling of the thermal dust foreground by Prunet et al. (1998). However, the study of synchrotron APS may also have a bearing on the knowledge of Galactic structure, in particular of the magnetic fields both in emission regions and in compact foreground screens (Tucci et al., 2001b).

The present work is motivated by the need for a verification of a conjecture by Tucci et al. (2000), based on the analysis of the 2.4 GHz Parkes survey (Duncan et al., 1997: D97) of the Southern Galactic Plane: Is the behaviour of electric- and magnetic-parity APS governed everywhere (with reasonable approximation) by power laws $C_{E,B} \propto \ell^{-\alpha_{E,B}}$ with $\alpha_{E,B} \simeq 1.4 - 1.5$ in the full range $\ell \lesssim 10^3$? Although such slopes are close to the values $\alpha_{E,B}^{dust} \simeq 1.3 - 1.4$ found for thermal dust (Prunet et al. 1998), later works cast doubts on the generality of this behaviour. For the APS of the scalar $PI \equiv \sqrt{Q^2 + U^2}$ somewhat steeper spectra, with α_{PI} ranging from 1.7 ± 0.2 to 1.9 ± 0.3, are found in the Galactic Plane by Baccigalupi et al. (2001) and Giardino et al. (2001); the former authors also find a much higher slope, $\alpha_{PI} = 2.9 \pm 0.2$, from the 1.4-GHz low-resolution survey of Brouw and Spoelstra (1976: BS76). Here we extend our previous analysis of APS in the Galactic Plane considering the 2.7-GHz Effelsberg survey (Duncan et al. 1999: D99); further, we analyse several patches at intermediate Galactic latitudes, $|b| \leq 20°$, from the 1.4-GHz Effelsberg survey (Uyaniker et al. 1999: U99), and finally three regions from BS76, covering latitudes up to near the North Galactic Pole at 5 frequencies between 408 and 1411 MHz. We carefully distinguish between those APS which provide a full description of the spin-2 polarization field (i.e., $C_{E\ell}$ and $C_{B\ell}$, from which $C_{P\ell} = C_{E\ell} + C_{B\ell}$ can be derived), and the $C_{PI\ell}$ spectrum which takes into account only the magnitude of the polarization pseudovector. Although this distinction was ignored by other authors, it is in fact mandatory, since Tucci et al. (2001b) find important differences between $C_{P\ell}$ and $C_{PI\ell}$, for both simulated CMB maps and real synchrotron maps.

The partial regularities that we find for polarization APS in this work do not support the usefulness of global (i.e., full sky) $C_{E,B\ell}$ for a satisfactory description of the spatial distribution of the synchrotron background. On the contrary, the local $C_{E,B\ell}$ based on Fourier analysis are much more suitable for a fuller description of angular structure. We can summarize our results as follows:

- In both the Northern and the Southern Galactic Plane, the slopes of $C_{E,B\ell}$ are quite moderate at 2.4 and 2.7 GHz, with averages $\alpha_E \simeq \alpha_B \simeq 1.4 - 1.5$ in the range $100 \leq \ell \leq 800$, although significant fluctuations exist around

CP609, *Astrophysical Polarized Backgrounds*, edited by S. Cecchini et al.

TABLE 1. The sky regions used for computation of APS

Ref.	ν (GHz)	FWHM	Galactic latitudes	Galactic longitudes
D97	2.4	10.4$'$	$-5° \leq b \leq 5°$	$-122° \leq l \leq 5°$
D99	2.7	5.1$'$	$-5° \leq b \leq 5°$	$5° \leq l \leq 74°$
U99	1.4	9.3$'$	$4° \leq b \leq 20°$ $5° \leq b \leq 15°$ $-15° \leq b \leq 5°$ $3.5° \leq b \leq 10°$ $3.8° \leq b \leq 15°$	$45° \leq l \leq 55°$ $65° \leq l \leq 95°$ $70° \leq l \leq 100°$ $140° \leq l \leq 153°$ $190° \leq l \leq 210°$
BS76	$0.41 - 1.4$	$2.3° - 0.6°$	$-10° \leq b \leq 20°$ $26.5° \leq b \leq 62.5°$ $60.5° \leq b \leq 84.5°$	$120° \leq l \leq 180°$ $l_c = 60.5°, \Delta\psi = \pm27°$ $l_c = 30°, \Delta\psi = \pm15°$

theses values in $10° \times 10°$ patches.

- Local fluctuations do not allow us to establish equally significant average slopes out of the Galactic Plane in the same angular range.
- At lower resolution, $\ell \leq 70$, large correlations between fit parameters cause large error bars; however, the best-fit slopes α_E and α_B stay in the range $0.5 - 2$ for almost all frequencies (within $\nu = 0.4 - 1.4$ GHz) and Galactic latitudes, and are quite inconsistent with values around 3.
- Our best value for α_{PI} lies in the interval $1.6 - 1.8$ for the full range $10 \lesssim \ell \leq 800$, for any Galactic latitudes and all frequencies $\nu \geq 1.4$ GHz. Although error bars again are large for $\ell \leq 70$, we find no evidence for steep spectra even in this case; on the contrary, at low frequencies $\nu \leq 0.8$ GHz we occasionally find rather flat $C_{PI\ell}$ spectra.
- The above behaviours of polarization APS should be attributed to Galactic synchrotron, with no appreciable contamination from point sources. On the other hand, the total intensity APS may be locally dominated by sources, when they exhibit small amplitude and slope α_I close to zero.

It is an intriguing fact that the Galactic Plane result, $\alpha_E \simeq \alpha_B \simeq 1.4 - 1.5$, is very close to the polarization APS of both thermal dust (Prunet et al., 1998) and starligth (Fosalba et al., 2001). However, the difference with α_{PI}, although generally small, has some statistical significance. It has been argued by Tucci et al. (2001b) that the scalar polarization APS, $C_{PI\ell}$, being less sensitive to Faraday rotation from foreground screens, can probably be extrapolated to the cosmological window with better confidence. This quantity however should be confronted with $C_{PI\ell}^{CMB}$, not with the $C_{E,B\ell}^{CMB}$ spectra which are popular among cosmologists.

DATA ANALYSIS

Polarized synchrotron surveys

Our first analysis of Galactic-Plane synchrotron polarization was performed on the 2.4 GHz Parkes survey (D97); this study, using twelve $10° \times 10°$ square patches (Tucci et al. 2000), is extended here to the 2.7 GHz Effelsberg survey (D99) covering the additional Galactic longitude range $5° \leq l \leq 74°$ (see Table 1) and providing six more $10° \times 10°$ patches. The nominal rms noise in D97 is 8 mK for total power and 5.3 mK for polarization (5.3 and 2.9 mK in some more sensitive areas), while in D99 the rms noise is 9 mK. We also analyse the maps of the Effelsberg 1.4 GHz survey (U99), consisting of five regions at moderate Galactic latitudes, $-15° \leq b \leq 20°$. The rms noise is there about 15 mK for total intensity and about 8 mK for linear polarization. From those regions we extracted different numbers of $10° \times 10°$ patches, namely, 1, 3, 3, 2 and 2 respectively for regions referred to by rows 3 to 7 in Table 1. Finally we analysed the low-resolution BS76 maps at five frequencies between 408 and 1411 MHz. The corresponding beamwidths range from $2.3°$ to $0.6°$, and noise levels from 0.34 to 0.06 K. In order to get good sampling and large signal-to-noise ratios we selected three rectangular regions, whose locations are reported in the last three rows of the Table.

The BS76 maps are affected by undersampling, and the grid spacing (larger than the FWHM) depends on the sky region and frequency. To make Fourier analysis feasible we constructed evenly spaced grids adopting a Gaussian smoothing function with dispersion $\sigma_{sm} = 3°/\sqrt{8\ln 2}$. Near the Galactic Plane, geodesics orthogonal to meridians could be identified with parallels with no appreciable errors. At high latitudes we constructed the grids projecting orthogonal geodesics, evenly spaced with $\Delta b = 1.5°$ at starting points, out of the central meridian of each patch, having longitude $l = l_c$ as reported in the last two rows of Table 1. On such geodesics, extended respectively up to proper lengths $\Delta\psi = \pm 27°$ and $\pm 15°$, we picked out the centers of the Gaussian beams, with spacing $\Delta\psi' = 1.5°$. (Since these centers do not lie on parallels, the latitude ranges in the Table refer to central longitudes l_c.) For the chosen $\Delta\psi$ the proper distance between corresponding points on *neighbour* geodesics remains sufficiently close to $\Delta\psi'$ even at the extremal longitudes $l = l_c \pm \Delta\psi$, and this provides an intrinsic check for the applicability of flat-space concepts to celestial-sphere patches.

Computation of angular spectra

Because of the limited sky coverages, power spectra can be suitably obtained from Fourier analysis instead of the standard spherical-harmonic approach (Seljak, 1997). The estimators for the APS of fields in a patch of size Ω are given by

$$C_{X\ell} = \frac{\Omega}{N_\ell} \sum_{\vec{\ell}} \left[X(\vec{\ell}) X^*(\vec{\ell}) - w_X^{-1} \right] b_\ell^{-2}, \tag{1}$$

with N_ℓ the number of Fourier modes in the interval around ℓ, w_X^{-1} the noise contribution and b_ℓ the beam function; $X(\vec{\ell})$ can be the discrete Fourier transform of any of the Stokes parameters I, Q and U, but also of the electric- and magnetic-parity polarization fields E and B, being

$$E(\vec{\ell}) = Q(\vec{\ell})\cos(2\phi_{\vec{\ell}}) + U(\vec{\ell})\sin(2\phi_{\vec{\ell}}), \qquad B(\vec{\ell}) = -Q(\vec{\ell})\sin(2\phi_{\vec{\ell}}) + U(\vec{\ell})\cos(2\phi_{\vec{\ell}}). \tag{2}$$

The APS describing the total power of the polarization field (taking into account spatial variations of both magnitude and direction) is

$$C_{P\ell} = C_{Q\ell} + C_{U\ell} = C_{E\ell} + C_{B\ell}, \qquad \ell \geq 2. \tag{3}$$

One can also consider the APS of the polarized intensity $PI = \sqrt{Q^2 + U^2}$, which is defined for any $\ell \geq 0$. Equation (1) still applies with $X = PI$. The technique based on Eqs. (1) and (2), already applied by Tucci et al. (2000) on D97 data, is implemented with a cosine apodization to suppress border effects, and with subtraction of mean values to suppress aliasing. However, a more careful analysis is required for BS76 maps because of frame-of-reference effects in Q and U fields. Such effects are especially important for our last patch under analysis, whose border almost touches the Galactic Pole. In order to smooth the Q and U fields we follow the following procedure: (i) At each point within a proper distance $\chi \leq 5\sigma_{sm}$ from a Gaussian beam center, we construct the polarization vector $\vec{p}$, (ii) we perform a parallel transport of $\vec{p}$ to the beam center, obtaining

$$p_\theta^{(PT)} = p_\theta \cos t + p_\phi \sin t, \qquad p_\phi^{(PT)} = -p_\theta \sin t + p_\phi \cos t, \tag{4}$$

with $t = \int \cos\theta \, d\phi$, and (iii) from $p_{\theta,\phi}^{(PT)}$ we recover the parallel-transported Stokes parameters $Q^{(PT)}$ and $U^{(PT)}$ and apply the weight function $\exp[-\chi^2/(2\sigma_{sm}^2)]$ to them. We checked that in practice the parallel transport could be made with negligible variations on any reasonable path close to geodesics, and a very convenient approximation is $t = \cos\theta_{ave}\Delta\phi$, with θ_{ave} the average of polar angles at the starting and end points. The smoothed fields $Q_{sm}^{(PT)}$ and $U_{sm}^{(PT)}$ are then used in Eq. (1) for the computation of the angular spectra of the BS76 maps.

BEST-FIT RESULTS

Intensity and polarization APS in the Southern Galactic Plane are reasonably approximated by power laws, $C_{X\ell} = A_X \ell^{-\alpha_X}$, for most of the 12 square patches investigated by Tucci et al. (2000), to which we refer for details. In Table 2 we report similar results of spectral fits with $X = I$, E and B, in the range $100 \leq \ell \leq 800$ for the 6 patches

TABLE 2. APS parameters for the D99 survey

$l(°)$	A_I (K^2)	α_I	A_E (K^2)	α_E	α_B
10	3.40	1.60 ± 0.12	0.21×10^{-4}	0.92 ± 0.10	$1.13^{+0.08}_{-0.12}$
20	124.4	2.18 ± 0.07	0.39×10^{-3}	$1.44^{+0.13}_{-0.07}$	$1.52^{+0.07}_{-0.15}$
30	11.5	1.79 ± 0.10	0.88×10^{-3}	1.50 ± 0.10	1.63 ± 0.10
40	0.10	1.21 ± 0.12	0.25×10^{-3}	$1.36^{+0.11}_{-0.06}$	$1.67^{+0.05}_{-0.10}$
50	0.31	$1.28^{+0.21}_{-0.15}$	0.80×10^{-4}	1.24 ± 0.09	1.52 ± 0.13
60	0.65×10^{-3}	$1.00^{+0.12}_{-0.15}$	0.71×10^{-3}	1.68 ± 0.07	1.82 ± 0.10

TABLE 3. Average parameters for Galactic Plane fits

Survey & method	A_I (K^2)	α_I	A_E (K^2)	α_E	α_B
D97, $\langle\alpha_X\rangle$	–	1.37 ± 0.44	–	1.44 ± 0.30	1.46 ± 0.29
D97, $\langle C_\ell\rangle$	2.2	1.60 ± 0.13	0.12×10^{-3}	1.53 ± 0.11	1.43 ± 0.12
D99, $\langle\alpha_X\rangle$	–	1.71 ± 0.43	–	1.40 ± 0.23	1.57 ± 0.19
D99, $\langle C_\ell\rangle$	10.3	1.82 ± 0.11	0.31×10^{-3}	1.39 ± 0.11	1.55 ± 0.12
D97+D99, $\langle\alpha_X\rangle$	–	1.48 ± 0.46	–	1.42 ± 0.30	1.51 ± 0.26
D97+D99, $\langle C_\ell\rangle$	6.9	1.72 ± 0.10	0.26×10^{-3}	1.40 ± 0.12	1.54 ± 0.10

from the Northern Galactic Plane D99 survey. While we found large variations in the normalization of the intensity spectra, the same feature is not found in polarization spectra. This can be explained by observing that, while the total power emission decreases fast when moving away from the Galactic center, the polarized component remains much more uniform when changing the Galactic longitude and the latitude: D97 noticed in fact a polarized "background component" of about 20 mK, nearly constant over the entire survey. The distributions of the indices α_X highlight some more differences between total intensity and polarization spectra: The values of α_I range between 1.0 (0.4 including D97 data) and 2.2, but those of $\alpha_{E,B}$ remain relatively close to the mean value $\simeq 1.4 - 1.5$ in all patches. Moreover, low–emission regions show very flat spectra in total intensity, while no meaningful differences are found between high– and low–polarized emission regions.

TABLE 4. APS parameters for the U99 survey

$\ell(°)$	$b(°)$	A_I (K^2)	α_I	A_E (K^2)	α_E	α_B
50	13	0.53×10^{-5}	$0.37^{+0.13}_{-0.10}$	0.16×10^{-3}	1.22 ± 0.08	1.19 ± 0.08
143	7	0.35×10^{-6}	$0.^{+0.02}$	0.22	$2.55^{+0.14}_{-0.17}$	$2.70^{+0.03}_{-0.25}$
150	7	0.50×10^{-6}	$0.^{+0.02}$	0.80×10^{-1}	$2.38^{+0.06}_{-0.17}$	$2.10^{+0.23}_{-0.16}$
195	10	0.45×10^{-6}	$0.^{+0.02}$	0.35×10^{-1}	2.28 ± 0.08	2.32 ± 0.08
205	10	0.72×10^{-6}	$0.^{+0.11}$	0.70×10^{-2}	$1.99^{+0.08}_{-0.12}$	1.98 ± 0.10
70	10	0.99×10^{-4}	0.82 ± 0.15	0.73×10^{-1}	2.41 ± 0.11	2.39 ± 0.12
80	10	0.89×10^{-3}	1.13 ± 0.13	0.90×10^{-3}	1.71 ± 0.15	1.79 ± 0.19
90	10	0.6×10^{-6}	$0.^{+0.13}$	0.11	$2.48^{+0.02}_{-0.12}$	2.23 ± 0.13
75	-10	0.16×10^{-5}	0.16 ± 0.13	0.23×10^{-5}	0.87 ± 0.08	1.50 ± 0.12
85	-10	0.12×10^{-4}	0.87 ± 0.12	0.36×10^{-5}	1.00 ± 0.11	1.17 ± 0.12
95	-10	0.20×10^{-4}	0.81 ± 0.11	0.47×10^{-4}	1.20 ± 0.09	$0.63^{+0.07}_{-0.02}$

In Table 3 we report the values of the average slopes (rows labelled with $\langle\alpha_X\rangle$), as well as the fit parameters from the average spectra (rows labelled by $\langle C_\ell\rangle$) for both D97 and D99 surveys, and also combining the two surveys. For the latter computation, the D99 data are rescaled to 2.4 GHz assuming a spectral index -3. The agreement is very satisfactory. On the other hand, the slope of C_{PII} turns out somewhat higher, $\alpha_{PI} \simeq 1.6 - 1.8$ (see Table 6), and agrees with values by other authors. Table 4 reports the results obtained from 5 intermediate–latitude regions of U99. The intensity spectra are found to be extremely flat ($\alpha_I < 1$) and low, indicating that the diffuse emission drops just out the Galactic plane. On the contrary, the polarization spectra do not show a decrease in amplitude with respect to the other two surveys; this means that the polarized background observed in D97 and D99 extends at least up to $l \sim 15°$. However, in U99 we observe large differences in the slope of polarization spectra from region to region: there are three patches with $\alpha_{E,B} > 2$ and two with $\alpha_{E,B} \sim 1$. However, some of these regions cannot be considered as typical; for example, the area $140° \le l \le 153°$, $4° \le b \le 10.4°$ lies within the so called "fan region", and the two regions centered at $l = 80°$ show rather complex structures. The dispersion in the results undoubtedly shows how local (rather than global) spectra will be important for foreground subtraction in CMB polarization measurements.

It is interesting to observe that the nearly flat intensity spectra, i.e. those with $\alpha_I \simeq 0$ found in the U99 patches, can be used to give upper limits on the contribution of extragalactic point sources to the polarized background. Since point sources do have flat APS as far as clustering can be neglected (Tegmark and Efstathiou 1996; Toffolatti et al. 1998), we can assume that they may at most dominate in the above U99 patches, and we thereby obtain the limit $C_{I\ell}^{(PS)} < 5 \times 10^{-7}$ K^2. From this number, adopting a radio-source polarization degree of 5% [in agreement with De Zotti et al. (1999)] we get $C_{X\ell}^{(PS)} < 1.3 \times 10^{-9}$ K^2 for $X = P, PI$ at 1.4 GHz. Assuming further a (frequency) spectral index $\beta = -2$ for radiogalaxies, we also get $C_{X\ell}^{(PS)} < 1.5 \times 10^{-10}$ K^2 at 2.4 GHz. We conclude that the contribution of point sources should be negligible in the whole range $\ell \le 800$ for all of our *polarization* APS. More stringent, but less safe limits on $C_{I\ell}^{(PS)}$ and $C_{P,PI\ell}^{(PS)}$) could be derived from estimates of Tegmark et al. (1996) and Toffolatti et al. (1998).

For the BS76 survey we analysed the maps at all of the frequencies, but due to the moderate resolution and the patch size, we could investigate only a limited spectral range. In particular, uncertainties in the beam angular function for $\ell > \pi/$FWHM must exist in the original experiment; further, adopting a Gaussian shape for our smearing function $\exp[-\chi^2/(2\sigma_{sm}^2)]$ with χ the bin center, we neglect the finite bin of the original measurements. Therefore, it is advisable to limit ourselves to the range such that $b_\ell^{-2} \lesssim e$. In fact, the spectra computed by means of Eq. (1) turned out to increase for $b_\ell^{-2} \ge e$, as can be expected for the difficulty of an accurate calculation of noise after beam smoothing. Since there is no hint of this increase for $\ell \le 70$, we performed fits of angular spectra only up to this limit, and we adopted the modified function

$$C_{X\ell} = A_X \ell^{-\alpha_X} + \tilde{C}_X b_\ell^{-2}. \tag{5}$$

Here $\tilde{C}_X$, a parameter ù to be determined by the fit, is not intended to describe the field X, but rather to account for the inaccurate a priori estimate of w_X^{-1}. The upper limits on point source APS given in the previous pharagraph, rescaled to the BS76 frequencies, exclude that the $\tilde{C}_X b_\ell^{-2}$ term may mask important contributions from the above sources. As a matter of facts, for $\beta = -2.3$ we find for instance $C_{X\ell}^{(PS)} < 4 \times 10^{-7}$ K^2 for $X = P, PI$ at 408 MHz, some orders of magnitude below the measured APS.

TABLE 5. APS parameters for 3 patches with increasing Galactic latitudes from the BS76 survey

ν (MHz)	$\log A_E$	α_E	$\log A_B$	α_B	$\log A_P$	α_P	$\log A_{PI}$	α_{PI}
408	$-0.9^{+0.4}_{-1.7}$	$1.0^{+0.5}_{-1.4}$	$-0.7^{+0.5}_{-1.3}$	$1.2^{+0.5}_{-1.2}$	$-0.7^{+0.1}_{-0.1}$	$0.8^{+0.7}_{-0.2}$	$-0.2^{+0.2}_{-0.3}$	$1.3^{+0.1}_{-0.3}$
465	$-2.1^{+0.8}_{-0.9}$	$0.4^{+0.3}_{-1.0}$	$-0.7^{+0.4}_{-0.8}$	$1.5^{+0.3}_{-0.9}$	$-0.9^{+0.4}_{-1.7}$	$1.2^{+0.5}_{-1.2}$	$-0.5^{+0.5}_{-0.1}$	$1.1^{+0.3}_{-0.2}$
610	$-1.2^{+0.4}_{-1.1}$	$1.2^{+0.4}_{-1.0}$	$-1.0^{+0.3}_{-0.5}$	$1.3^{+0.6}_{-0.5}$	$-0.7^{+0.2}_{-0.1}$	$1.1^{+0.1}_{-0.2}$	$-0.5^{+0.6}_{-0.1}$	$1.5^{+0.2}_{-0.3}$
820	$-1.3^{+0.6}_{-0.5}$	$1.6^{+0.7}_{-0.6}$	$-1.3^{+0.3}_{-0.4}$	$1.4^{+0.3}_{-0.3}$	$-0.9^{+0.3}_{-0.1}$	$1.2^{+0.6}_{-0.1}$	$-0.7^{+0.6}_{-0.7}$	$1.5^{+0.7}_{-0.5}$
1411	$-2.0^{+0.3}_{-1.2}$	$1.9^{+0.4}_{-0.2}$	$-2.8^{+0.3}_{-0.3}$	$1.2^{+0.4}_{-0.3}$	$-1.6^{+0.3}_{-0.7}$	$1.8^{+0.5}_{-0.3}$	$-2.0^{+0.5}_{-0.7}$	$1.8^{+0.7}_{-0.9}$
408	$-1.8^{+0.3}_{-0.7}$	$1.2^{+0.4}_{-0.7}$	$-1.5^{+0.4}_{-0.8}$	$0.6^{+0.6}_{-0.4}$	$-0.8^{+1.1}_{-1.0}$	$1.1^{+1.0}_{-0.8}$	$-1.1^{+0.9}_{-0.5}$	$1.3^{+1.1}_{-1.0}$
465	$-1.2^{+0.6}_{-0.5}$	$1.5^{+0.6}_{-0.8}$	$-1.7^{+0.6}_{-0.9}$	$1.3^{+0.8}_{-0.9}$	$-1.0^{+0.2}_{-0.8}$	$1.7^{+0.3}_{-1.1}$	$-1.0^{+1.0}_{-0.4}$	$1.3^{+1.2}_{-0.3}$
610	$-1.7^{+1.4}_{-1.2}$	$1.3^{+1.5}_{-0.6}$	$-2.5^{+0.5}_{-1.3}$	$1.1^{+0.4}_{-0.9}$	$-1.9^{+0.2}_{-1.2}$	$1.1^{+0.4}_{-0.9}$	$-1.2^{+1.6}_{-0.6}$	$1.6^{+1.0}_{-0.9}$
820	$-2.3^{+0.5}_{-2.1}$	$1.5^{+1.0}_{-1.8}$	$-2.0^{+0.8}_{-1.8}$	$1.2^{+1.0}_{-1.8}$	$-1.8^{+0.7}_{-1.5}$	$1.1^{+0.9}_{-0.8}$	$-1.9^{+1.0}_{-1.3}$	$1.7^{+1.0}_{-0.6}$
1411	$-2.5^{+0.3}_{-0.8}$	$1.4^{+0.5}_{-0.7}$	$-1.4^{+0.7}_{-0.6}$	$2.2^{+1.0}_{-0.8}$	$-1.5^{+0.9}_{-0.6}$	$1.9^{+1.0}_{-0.7}$	$-2.0^{+0.9}_{-0.7}$	$1.7^{+0.9}_{-0.7}$
408	$-2.0^{+0.7}_{-1.6}$	$0.8^{+0.4}_{-1.3}$	$-2.2^{+0.5}_{-1.2}$	$0.5^{+0.4}_{-0.9}$	$-1.4^{+0.4}_{-0.6}$	$0.4^{+0.3}_{-0.3}$	$-0.4^{+1.6}_{-0.7}$	$1.1^{+1.2}_{-0.3}$
465	$-1.7^{+1.4}_{-1.0}$	$0.6^{+1.4}_{-0.5}$	$-2.0^{+0.1}_{-0.2}$	$0.5^{+0.1}_{-0.1}$	$-1.7^{+0.3}_{-1.7}$	$0.3^{+0.3}_{-1.1}$	$-0.3^{+1.0}_{-0.1}$	$0.9^{+0.7}_{-0.5}$
610	$-3.7^{+1.2}_{-1.0}$	$-0.4^{+1.2}_{-0.8}$	$-1.8^{+3.5}_{-0.9}$	$1.0^{+3.0}_{-0.8}$	$-1.4^{+1.4}_{-1.8}$	$1.2^{+1.3}_{-1.9}$	$-1.5^{+2.4}_{-0.1}$	$1.2^{+2.2}_{-0.9}$
820	$-2.1^{+0.8}_{-1.8}$	$1.2^{+0.5}_{-2.0}$	$-1.7^{+1.1}_{-2.1}$	$1.5^{+0.9}_{-1.5}$	$-1.4^{+0.7}_{-0.8}$	$1.0^{+0.5}_{-0.6}$	$-1.5^{+1.2}_{-0.3}$	$1.2^{+0.8}_{-0.4}$
1411	$-3.0^{+0.5}_{-2.0}$	$1.4^{+0.6}_{-2.1}$	$-1.3^{+1.0}_{-1.7}$	$2.1^{+1.0}_{-1.4}$	$-1.5^{+1.4}_{-1.5}$	$1.9^{+1.1}_{-1.3}$	$-2.0^{+0.5}_{-1.8}$	$1.8^{+1.0}_{-1.5}$

The results of the fits are given in Table 5. The declared uncertainties, which are 1-sigma error on individual parameters computed from chi-square fields, are large because of correlations in 3-parameter fits. In particular, we find

TABLE 6. Average slopes of $C_{P\ell}$ and $C_{PI\ell}$ spectra

ν (MHz)	ℓ-range	α_P	α_{PI}
408	≤ 70	0.56 ± 0.24	1.29 ± 0.19
465	≤ 70	1.05 ± 0.43	1.09 ± 0.22
610	≤ 70	1.10 ± 0.14	1.50 ± 0.24
820	≤ 70	1.14 ± 0.28	1.43 ± 0.37
1411	≤ 70	1.82 ± 0.34	1.76 ± 0.51
2417	$100 - 800$	1.48 ± 0.12	1.68 ± 0.30
2695	$100 - 800$	1.50 ± 0.11	1.59 ± 0.12

strong, positive correlations between A_X and α_X; we thereby evaluated errors from inspection of chi-square contours in (A_X, α_X) space after minimization with respect to C_X. In spite of such uncertainties, we find moderate slopes for α_E and α_B, generally in the range $1 \div 2$. At low frequencies, $\nu \leq 610$ MHz, we occasionally find some slopes < 1, and in one case (near the Galactic Pole in the 610 MHz map) a best value $\alpha_E < 0$. There is sufficient evidence that polarization APS are somewhat flatter at frequencies where Faraday rotation is large. Slopes as large as ~ 3 are ruled out everywhere.

The angular spectral behaviour does not exhibit any clear dependence on Galactic latitude at these moderate resolutions. It makes sense therefore to average over our three patches from the BS76 survey. Table 6 gives the average slopes α_P and α_{PI} for each BS76 frequency. At low frequencies, where Faraday rotation becomes more and more important, we observe a gradual flattening of both spectra; the deficit of steepness is more apparent for α_P, in agreement with the considerations of Tucci et al. (2001). The result quoted for α_{PI} at 1411 MHz is quite consistent with those found at 2.4 and 2.7 GHz, with higher resolution in the Galactic Plane (cfr. the last 3 rows in the Table), but not with the results of Baccigalupi et al. (2001), which however refer to BS76 regions with a smaller signal-to-noise ratio.

ACKNOWLEDGMENTS

This work was performed within the SPOrt Collaboration, and is supported by Agenzia Spaziale Italiana (ASI).

REFERENCES

1. Baccigalupi, C., et al., 2001, A&A, **372**, 8.
2. Brouw, W.N., & Spoelstra, T.A., 1976, A&AS, **26**, 129 (BS76).
3. De Zotti, G., et al., 1999, NewA, **4**, 481.
4. Duncan, A.R., Haynes, R.F., Jones, K.L., & Stewart, R.T., 1997, MNRAS, **291**, 279 (D97).
5. Duncan, A.R., Reich, P., Reich, W. & Fürst, E., 1999, A&A, **350**, 447 (D99).
6. Fosalba, P., Lazarian, A., Prunet, S. & Tauber, J., 2001, this volume.
7. Giardino, G., et al., 2001, in: Mining the Sky, Proc. MPA/ESO/MPE Conference, Springer-Verlag Series "ESO Astrophysics Symposia", to be published
8. Prunet, S., Sethi, S.K., Bouchet, F.R. & Miville-Deschenes, M.-A. 1998, A&A **339**, 187.
9. Seljak, U., 1997, ApJ, **482**, 6.
10. Tegmark, M., & Efstathiou, G., 1996, MNRAS, **281**, 1297.
11. Toffolatti, L., et al., 1998, MNRAS, **297**, 117
12. Tucci, M., Carretti, E., Cecchini, S., Fabbri, R., Orsini, M., & Pierpaoli, E., 2000, NewA, **5**, 181.
13. Tucci, M., Carretti, E., Cecchini, S., Cortiglioni, S., Fabbri, R., & Pierpaoli, E., 2001a, Proc. 20th Texas Symposium on Relativistic Astrophysics, in press.
14. Tucci, M., et al., 2001b, this volume.
15. Uyaniker, B., et al., 1999, A&AS, **138**, 31.

Submillimetre Polarization of M82 and the Galactic Centre: Implications for CMB Polarimetry

J.S. Greaves & W.S. Holland

UK Astronomy Technology Centre, Royal Observatory, Blackford Hill, Edinburgh EH9 3HJ, UK

Abstract. Polarized foreground galaxies may be a contaminant in CMB polarimetry experiments. We use 850 μm maps of polarized dust emission in M82 and towards the Galactic Centre to estimate the possible level of contamination. The dust grains are aligned by magnetic fields which in both cases have complex structure over large scales, and this field 'tangling' reduces p_{net} to only 0.4%. If this is typical of luminous local galaxies, we estimate that the polarized signal in a 5 arcmin pixel centred on the galaxy would be of the order of 0.5 μK.

INTRODUCTION

Various experiments exist or are proposed to map polarization of the cosmic microwave background. Since the effect is very small, contamination by foreground sources is a concern, and it is important to estimate the effects of likely contaminants. For experiments at wavelengths of about a millimetre, the foreground will be dominated by cold dust clouds, since the dust grains can be aligned by magnetic fields and hence their thermal emission will be slightly linearly polarized. Typically, observations at wavelengths of 0.35–1.3 mm have measured degrees of polarization of 1–10% in Galactic star-forming clouds [1, 2, 3].

CMB polarimetry experiments can of course avoid the Galactic Plane where most star-forming activity is concentrated, which leaves high-latitude (cirrus) clouds and extragalactic sources as possible contaminants. Little is known about the detailed structure of the millimetre-wavelength dust emission from cirrus clouds, particularly on scales as small as arcminutes. This is because sensitive submillimetre cameras have been in use for only about five years and wide-field imaging is still in the future. However, it is just possible to map the submillimetre polarization of a few nearby galaxies, and in this paper we describe the first such observations of the nearby starburst M82. Some simple calculations are presented of the effects which galaxies such as M82 and our own Milky Way might have on searches for CMB polarization.

M82 AND THE GALAXY

M82 is the nearest starburst galaxy, at a distance of about 3.25 Mpc. In the submillimetre, it has a bright dusty 'torus' about 1 kpc across which we see edge-on [5], and inside which are at least 100 stellar superclusters [6]. The powerful winds from the galactic nucleus blow dust grains at least 1 kpc up into the galactic halo [7]. Nearby normal spiral galaxies are much fainter in dust emission, but we can observe the centre of our own Galaxy at a distance of 8.5 kpc. Here there is a comparable 'Central Molecular Zone' of dense clouds [8] which has a diameter of about 0.4 kpc and constitutes roughly 10% of the Galaxy's interstellar medium by mass.

We observed the submillimetre polarization of M82 in 1998/9 [4], using the SCUBA camera on the James Clerk Maxwell Telescope. With a 15m single dish, the resolution was 15$''$ (about 250 pc) and the flux densities at 850 μm were up to 1.4 Jy/beam. The current flux limit of the camera and telescope is about 0.1 Jy/beam for polarization experiments, so magnetically aligned dust grains could be detected out to a moderate distance in the galactic halo.

The centre of the Milky Way was observed in 850 μm polarization in late 1999 [9], and included 8 maps of overlapping fields of view each covering 2.4 arcminutes. The final mosaic covers about 25 pc along the Galactic Plane and includes over 600 polarization vectors. For the purposes of the this paper, we calculate the net polarization

CP609, *Astrophysical Polarized Backgrounds*, edited by S. Cecchini et al.

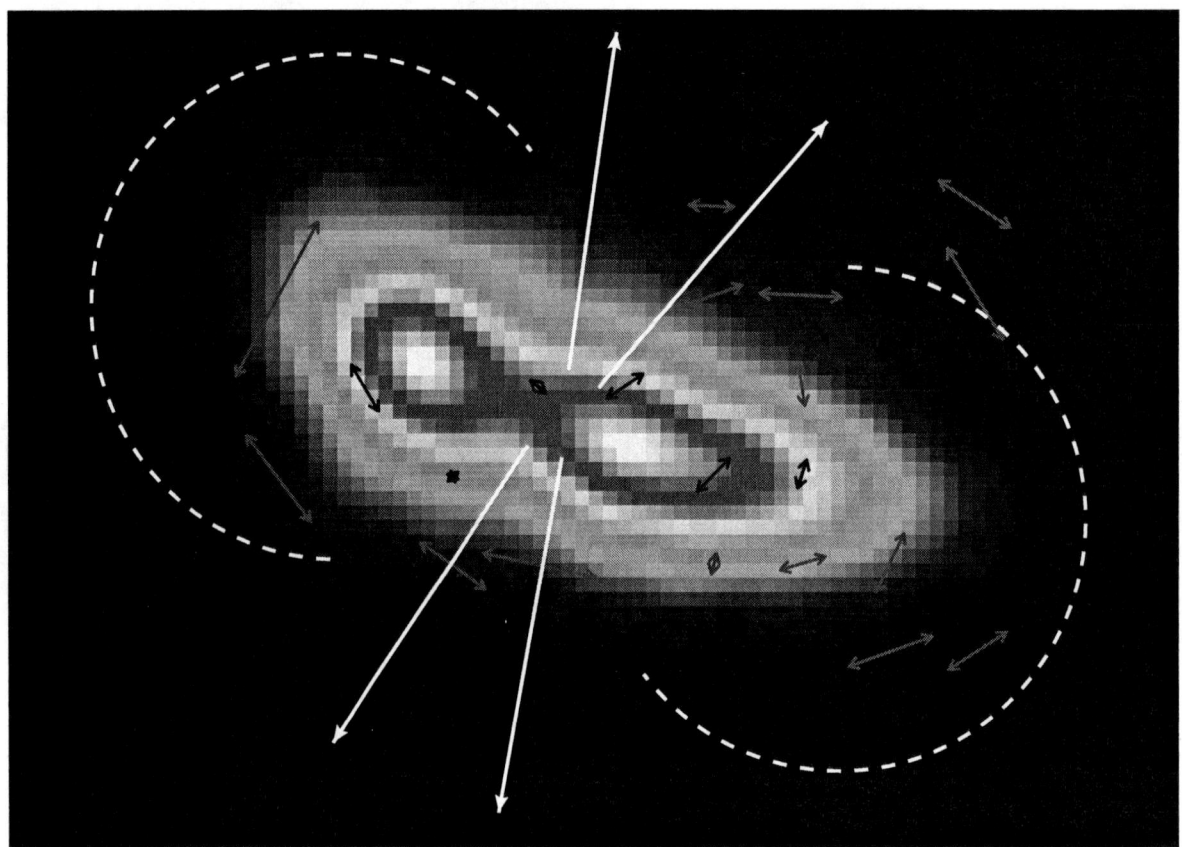

FIGURE 1. 850 μm polarization results for M82, overlaid on an image of the 450 μm dust emission [4]. The observed E-plane vectors have been rotated 90° to show the directions of the magnetic fields that align the dust grains. The vector lengths are proportional to percentage polarization in the range 0.7–9.1%, and all the vectors plotted are $\geq 3\sigma$ so that the uncertainty in vector direction is less than $\pm 10°$. The data were independently sampled every 6 arcseconds, although the telescope beam size is 15 arcseconds full-width half-maximum. The white arrows show the location of the galactic wind and the white dashed lines sketch out the 'magnetic bubble' inferred from the magnetic vectors.

of this region and assume it is typical of the 400 pc-wide Central Molecular Zone.

Polarization of M82

Two main magnetic field structures were seen in M82 (Figure 1). Towards the brightest regions of the torus, magnetic field vectors point roughly inwards towards the nucleus, and so this field could promote gas inflow to the starburst core. The degree of polarization is small, so the field may have sub-structure not seen with our 250 pc resolution. In the fainter halo region, the vector directions are very diverse, but appear to trace out a 'bubble' structure, as sketched in Figure 1. This bubble could have been blown out by the galactic 'superwind'. The magnetic structure is surprising as most other galaxies have either flat fields in the galactic plane or poloidal fields. This is in fact the case in the close vicinity of the M82 nucleus, where near-infrared polarimetry shows fields aligned with the wind [10].

Although individual points are up to 9% polarized, the variations in direction result in a much lower *net* effect. Linear polarization can be described in terms of two orthogonal components called the Stokes parameters, and given by $Q = p\,cos(2\theta)$ and $U = p\,sin(2\theta)$ where p and θ are the polarization magnitude and direction; the factor of 2 arises because directions 180° apart are equivalent. With a large angle-dispersion, $\overline{cos(2\theta)}, \overline{sin(2\theta)} \to 0$ so the polarization becomes very small. In the case of M82, averaging the Stokes parameters over a circular region of diameter 1.4 kpc gives p_{net} of $0.37 \pm 0.07\%$ (although the error excludes possible systematic effects of $\sim 0.1\%$). The net magnetic field direction is $82 \pm 5°$ (east-of-north), which is roughly along the major axis of the torus. More importantly in the CMB

Declination

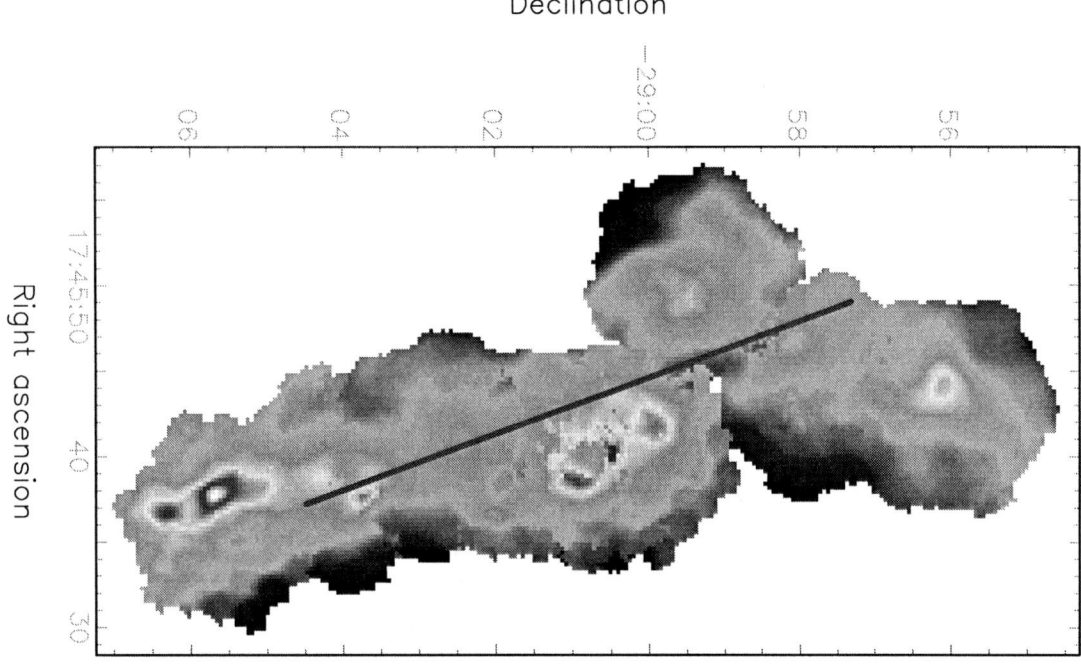

FIGURE 2. Net magnetic field direction from 850 μm polarization measurements, overlaid on an 850 μm image of the central 10 arcminutes of the Galaxy (data courtesy A. Chrysostomou). The vector length corresponds to 0.4%. The bright emission near the map centre is the highly-inclined '2 parsec ring', at the centre of which is the point source Sgr A*. Co-ordinates are RJ; the figure is oriented with RA running vertically.

context, the net polarization is small – only half the magnitude of the smallest vector shown in Figure 1.

Polarization of the Galactic Centre

The Galactic Centre mosaic (Figure 2) includes over 600 high signal-to-noise detections, covering several giant molecular clouds and the '2 parsec ring' in the centre of the map, which surrounds the black hole candidate Sgr A*. The synchrotron emission from Sgr A* is quite highly polarized [11] but the source is very compact so the extended dust clouds are much more important for the net polarization. Figure 2 shows the result of co-adding all the vectors: p_{net} is $0.40 \pm 0.01\%$ with a magnetic field direction of $20 \pm 1°$. This is well aligned with the group of clouds lying along the Galactic Plane (position angle of 32° in RA-Dec. co-ordinates); the single vector is plotted very large to demonstrate this.

A BACK-OF-THE-ENVELOPE CALCULATION

The relevant quantity for the question of contamination of CMB experiments is the foreground polarized flux, i.e. the product of flux and fractional polarization. We use submillimetre fluxes from the SCUBA Local Universe Galaxy Survey [12], which is based on the IRAS Bright Galaxy Sample. We then assume that a 0.4% net polarization is typical and will be identical at any wavelength of observation; the latter should be true if the dust emission is optically thin (i.e. longwards of the far-infrared) and the grain population is reasonably uniform.

— The mean 850 μm flux of the bright nearby galaxies is 175 mJy [12], which at the 15m JCMT corresponds to a brightness temperature of 6 mK. With p_{net} of 0.4%, the polarized signal is therefore $0.004 \times 6000 = 24 \,\mu K$.

— The local galaxies are larger than the JCMT beam, but smaller than planned pixel sizes in CMB polarimetry experiments. If the typical galaxy diameter is $\sim 0.75'$, then the beam dilution factor in a $5'$ pixel, for example, would be $0.75^2/(0.75^2 + 5^2)$ or 0.02.

— Thus the typical polarized signal would be $0.02 \times 24 = 0.5\,\mu K$. The maximum signal would be from M82 itself (unless any other nearby galaxies are significantly more polarized), and would be $55\,\mu K$. Since there is only 1 bright galaxy per $10 \times 10°$ area of sky, out to distances ~ 200 Mpc, the probability of one of these sources lying in a pixel is only 0.005%.

— For the Milky Way, the $850\,\mu m$ flux of the Central Molecular Zone has recently been measured as 10^5 Jy [8], and it is perhaps an order of magnitude higher for the Galaxy as a whole. At a distance of ~ 200 Mpc instead of 8.5 kpc, this flux would be reduced to 2 mJy, which is negligible compared to the luminous IRAS galaxies.

CONCLUSIONS

We conclude that a local luminous galaxy might be expected to contribute $\sim 0.5\mu K$ of contamination to a CMB polarimetry experiment with a pixel size of a few arcminutes. This would be significant if the cosmic signal were only of the order of one to a few μK. The filling factor of such galaxies on the sky is small, and known local objects can of course be avoided; a much more detailed calculation is required to take into account the galaxy foreground beyond about 200 Mpc. Further polarimetry observations of local galaxies are also needed, as the degree of polarization depends on inclination and we see both M82 and our Galaxy edge-on, so the 0.4% net polarization may be untypical. However, a *worst-case scenario* can be considered. If we assume that galaxies could have very ordered magnetic fields with p_{net} of 10%, as seen in some Galactic clouds [2], then the polarized signal would be $12\,\mu K$. This is equivalent to the $0.5\,\mu K$ calculated above but from a galaxy at 1 Gpc, using the square-law decrease in signal. If we then also consider that a $0.2\,\mu K$ contamination might be significant, this distance limit extends out to 1.5 Gpc. Finally, the number of galaxies in a volume out to 1.5 Gpc instead of 200 Mpc is increased by a factor of 400, so instead of a 0.005% chance of a 'bad' pixel we have a 2% chance. This worst-case scenario would produce a small but non-trivial number of bad pixels in a wide-field CMB polarimetry experiment.

ACKNOWLEDGMENTS

We wish to thank Tim Jenness and Tim Hawarden for their invaluable contributions to the work on M82, Antonio Chrysostomou for allowing us to use the Galactic Centre data prior to publication, David Berry for much software assistance, and Gianfranco De Zotti for bringing the idea to our attention. The JCMT is operated by the Joint Astronomy Centre on behalf of the UK Particle Physics and Astronomy Research Council, the Netherlands Organisation for Pure Research and the National Research Council of Canada.

REFERENCES

1. Hildebrand, R. H., Dotson, J. L., Dowell, C. D., Schleuning, D. A., and Vaillancourt, J. E., *ApJ*, **516**, 834–842 (1999).
2. Coppin, K. E. K., Greaves, J. S., Jenness, T., and Holland, W. S., *A&A*, **356**, 1031–1038 (2000).
3. Kane, B. D., Clemens, D. P., Barvainis, R., and Leach, R. W., *ApJ*, **411**, 708–719 (1993).
4. Greaves, J. S., Holland, W. S., Jenness, T., and Hawarden, T. G., *Nature*, **404**, 732–733 (2000).
5. Hughes, D. H., Gear, W. K., and Robson, E. I., *MNRAS*, **270**, 641–649 (1994).
6. O'Connell, R. W., Gallagher, J. S., Hunter, D. A., and Colley, W. N., *ApJ*, **446**, L1–L4 (1995).
7. Alton, P. B., Davies, J. I., and Bianchi, S., *A&A*, **343**, 51–63 (1999).
8. Pierce-Price, D., Richer, J. S., Greaves, J. S., Holland, W. S., Jenness, T., Lasenby, A. N., White, G. J., Matthews, H. E., Ward-Thompson, D., Dent, W. R. F., Zylka, R., Mezger, P., Hasegawa, T., Oka, T., Omont, A., and Gilmore, G., *ApJ*, **545**, L121–L125 (2000).
9. Chrysostomou, A., *in prep.* (2002).
10. Jones, T. J., *AJ*, **120**, 2920–2927 (2000).
11. Aitken, D. K., Greaves, J., Chrysostomou, A., Jenness, T., Holland, W., Hough, J. H., Pierce-Price, D., and Richer, J., *ApJ*, **534**, L173–L176 (2000).
12. Dunne, L., Eales, S., Edmunds, M., Ivison, R., Alexander, P., and Clements, D. L., *MNRAS*, **315**, 115–139 (2000).

Polarization of the Cosmic Microwave Background from Non-Uniform Reionization

Guo-Chin Liu*, Naoshi Sugiyama†, Andrew J. Benson**, C. G. Lacey‡ and Adi Nusser§

*Institute of Astronomy and Astrophysics, Academia Sinica, Taiwan, R.O.C.
†Division of Theoretical Astrophysics, National Astronomical Observatory Japan, Mitaka, 181-8588, Japan
**California Institute of Technology, MC 105-24, Pasadena, CA 91125, U.S.A.
‡SISSA, via Beirut 2-4, 34014 Trieste, Italy
§The Physics Department, The Technion-Israel Institute of Technology, Technion City, Haifa 32000, Israel

Abstract. The secondary anisotropies and polarization of the Cosmic Microwave Background (CMB) provide a laboratory for the study of the epoch of reionization in the Universe. Here, we concentrate on the CMB *polarization* in models with inhomogeneous reionization. Although the amplitude of the polarization anisotropy is estimated to be much smaller than that in the temperature, it is advantageous to consider the polarization signal since it is generated when photons and electrons scatter for the last time. Detection of these signals will place important contraints on the reionization history of the Universe.

INTRODUCTION

Reionization produces interesting effects on CMB temperature and polarization anisotropies at both first and second order in the perturbations. At first order, the effects of reionization are the same as for an IGM with spatially uniform density and ionized fraction. Density fluctuations in the free electrons around the reionization epoch produce CMB anisotropies and polarization only at second order, as in the Vishniac effect [5]. These second-order anisotropies and polarization are small in amplitude, but nonetheless dominate over the first-order anisotropies on small angular scales of order of arc minutes. They are thus cosmologically interesting as a probe of structure present at reionization.

In this work, we concentrate on the second order polarization, which is generated by coupling of the temperature quadrupole with density fluctuations of the free electrons. In general, density fluctuations in the free electrons can be considered from two parts: one is the case of a homogeneous reionization of the IGM, with the fluctuations in the free electron density being assumed to follow the variations in the total matter density, the so-called "density modulation model". Another case is that the reionization is expected to be patchy or inhomogeneous, with some regions already being fully ionized while others are still neutral, and the ionized regions growing until they encompass the whole IGM, the so-called "patchy reionization model". Here, we consider a realistic reionization process by combining a semi-analytic model of galaxy formation with an N-body simulation of the distribution of dark matter in the universe to determine the distribution of ionized regions.

GALAXY FORMATION

The reionization history of the universe is determined by a semi-analytic model of galaxy formation. The semi-analytic model is that of Cole et al. (2000), which includes the following processes: formation and merging of dark matter halos through hierarchical clustering; shock-heating and radiative cooling of gas within these halos; collapse of cooled gas to form galactic disks; star formation in disks and feedback from supernova explosions; galaxy mergers; chemical evolution of the stars and gas; and luminosity evolution of stellar populations based on stellar evolution codes and model stellar atmospheres. The model has been shown by Cole et al. to agree fairly well with a wide range of observed properties of galaxies in the local universe. We used this model to calculate the ionizing luminosities of galaxies at

CP609, *Astrophysical Polarized Backgrounds*, edited by S. Cecchini et al.

different redshifts, including the effects of absorption by interstellar gas and dust on the fraction of ionizing photons escaping, and followed the propagation of the ionization fronts around each galaxy. To find the ionizing luminosity, we first calculate the rate at which ionizing photons are being produced by stars in the galaxy, then apply an attenuation due to dust, and finally allow a fraction f_{esc} of the remaining photons to escape into the IGM. The mass of ionized hydrogen in each spherical ionization front is found by integrating the equation [4]

$$\frac{1}{m_H}\frac{dM}{dt} = S(t) - \alpha_H^{(2)} a^{-3} f_{clump} n_H \frac{M}{m_H},$$

(1)

where n_H is the comoving mean number density of hydrogen atoms (total, HI and HII) in the IGM, m_H is the mass of a hydrogen atom, $a(t)$ is the scale factor of the universe normalized to unity at $z = 0$, t is time (related to the conformal time by $dt = d\tau/a$), $S(t)$ is the rate at which ionizing photons are being emitted and $\alpha_H^{(2)}$ is the recombination coefficient to levels $n \geq 2$. The clumping factor $f_{clump} \equiv \langle \rho_{IGM}^2 \rangle / \bar{\rho}_{IGM}^2$ gives the effect of clumping on the recombination rate of hydrogen in the IGM. A larger f_{clump} increases the recombination rate resulting in a delay of the reionization epoch. We use the clumping factor $f_{clump}^{(halos)}$ as defined in Benson et al 2000. By summing over the ionized volumes due to all galaxies, we can calculate the fraction of the IGM which has been reionized at any redshift.

SECOND ORDER POWER SPECTRUM OF POLARIZATION

The second order polarization is generated by the coupling of *primary* temperature quadrupole with the density fluctuation of the free electrons. The density field of the gas is obtained by the simpler approach of Benson et al, in which the semi-analytic galaxy formation model is combined with a high-resolution N-body simulation of the dark matter. The simulation volume, which is a box of length $141.3h^{-1}$Mpc and contains 256^3 dark matter particles each of mass $1.4 \times 10^{10}h^{-1}M_\odot$, is divided into 256^3 cubic cells. Then we determine which regions of the simulation box become ionized by using one of the five toy models A-E listed below, which span the likely range of possible behaviour.

Model A (Growing front model) Ionize a spherical volume around each halo with a radius equal to the ionization front radius for that halo calculated assuming a large-scale uniform IGM. Since in the simulation the IGM is *not* uniform, but is assumed to trace the dark matter, and also because some spheres will overlap, the ionized volume calculated in this way will not contain the correct total ionized mass. We therefore scale the radius of each sphere by a constant factor, f, and repeat the procedure. This process is repeated, with a new value of f each time, until the correct total mass of hydrogen has been ionized.

Model B (High density model) In this model, we ignore the positions of halos in the simulation. Instead, we simply rank the cells in the simulation volume by their density. We then completely ionize the gas in the densest cell. If this has not ionized enough HI then we ionize the second densest cell. This process is repeated until the correct total mass of HI has been ionized.

Model C (Low density model) This is like model B, except that we begin by ionizing the least dense cell, and work our way up to cells of greater and greater density.

Model D (Random spheres model) As Model A, except that the ionized spheres are placed at completely random positions in the simulation volume, rather than on the dark matter halos to which they belong. By comparing to Model A this model allows us to estimate the importance of the spatial clustering of dark matter halos.

Model E (Boundary model) Ionize a spherical region around each halo with a radius equal to the ionization front radius for that halo. This may ionize too much or not enough HI depending on the density of gas around each source. We therefore begin adding or removing cells at random from the boundaries of the already ionized regions until the required mass of HI is ionized.

In left panel of Fig. 1, we plot the second-order power spectrum of the polarization in the five toy models with fixed extreme escape fraction $f_{esc} = 1$ and $\Omega_b = 0.02$. The cosmological parameters are $\Omega_0 = 0.3$, $\Lambda_0 = 0.7$, Hubble constant $H_0 = 70$ km/s/Mpc and $\sigma_8 = 0.9$. We find reionization occurs at $z \sim 9$ (corresponding to the optical depth to reionization is 0.034) from semi-analytic model. Although the shapes of the curves are all very similar, their amplitudes are different. Note that the reduction in power above $l \sim 10,000$ is artificial and due to the limited resolution of the N-body simulation we use (the density field of the ionized gas is calculated on a grid with cell size $0.55h^{-1}$Mpc, corresponding to $l \sim 10^4$). On the other hand, the finite size of the simulation box ($256h^{-1}$Mpc) affects the power spectrum for l below a few hundred. We see that the amplitude of the power spectrum around the peak ($l \approx 10,000$) varies by a factor ≈ 2.5, depending on which of the models A–E is used. The amplitude of the curves is affected by the

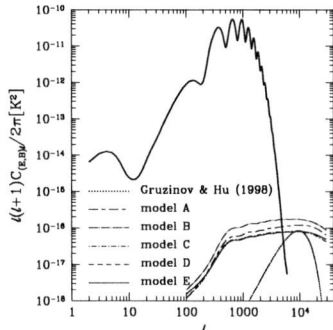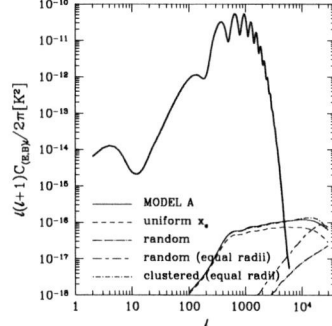

FIGURE 1. Power spectra of the second order effect for our models (left panel), and effect on the second-order anisotropy of varying the assumed geometry of the re-ionized regions (right panel).

strength of the correlations of δ_e present in each model. As a result, the "high density" model (B) is the most strongly correlated and has the highest amplitude, and conversely the "low density" model (C) has the lowest amplitude.

In left panel of Fig. 1, we also compare our results to the analytical toy model of Gruzinov & Hu (1998), in which the reionization is described by three free parameters. In their model, each luminous source is assumed to ionize a spherical region with fixed comoving radius R, the first source appears at redshift z_i, and new sources turn on at a constant rate until reionization is complete after an interval δz. An artificial assumption is made that luminous sources appear at random locations in space, so there are no correlations between the positions of the ionized spheres. Assuming that the spheres remain ionized forever, the fractional ionization increases with increasing number density of ionized spheres during δz until the universe is completely ionized. We chose $R = 0.85h^{-1}$ Mpc, $z_i=11$ and $\delta z = 5$ in the Gruzinov & Hu model to match the peak in the power spectrum of secondary *temperature* anisotropies predicted for our model E. For small l, little power is generated in the Gruzinov & Hu model because, by design, the patches are uncorrelated.

To further clarify what determines the *shape* of the second-order anisotropy spectrum in our models, we carried out the following additional tests. The first test was to force the ionized fraction x_e to be uniform and equal to the same mean value as before, so that fluctuations δ_e in the free electron density are then simply equal to fluctuations δ in the *total* density. In this case, the angular power spectrum has an almost identical shape to the model A. In the next two tests (labelled "random" in right panel of Fig. 1), the *total* gas density was forced to be uniform (i.e. we set $\delta = 0$), and put bubbles down at random positions, so that fluctuations in δ_e resulted only from the patchiness of the reionization. In one case, the bubble radii were chosen from the size distribution predicted by our galaxy formation model. In the other case, all bubbles were given the same comoving radius of $0.62h^{-1}$Mpc, which corresponds to the mean bubble radius (weighted by bubble volume) predicted by the galaxy model, at the redshift corresponding to the peak of the visibility curve. Both of these cases give power spectra with shapes (at large scales) similar to the analytical Gruzinov & Hu model, and completely different from when fluctuations in δ are included. In the final test (labelled "clustered" in right panel of Fig. 1), we again forced the bubble radii to be equal at a given redshift, but placed them on random halos, and included the fluctuations in the total gas density. The starting value for the radii in this last case was again $0.62h^{-1}$Mpc, but the spheres were then grown by a uniform factor at each redshift to produce the correct mean ionized fraction, as in the model A. The power spectrum in this case is almost the same as in model A, showing that the distribution in bubble sizes in the latter case does not have much effect.

We conclude that in our model A, the *shape* of the power spectrum on scales large compared to the typical size of the ionized bubbles is determined primarily by the correlations in total gas density. However, in the case of *patchy* reionization, the *amplitude* depends on the spatial distribution of these patches, which produces *biasing* for the correlations in the ionized gas density relative to those in the total gas density, which in turn boosts the amplitude of the polarization fluctuations.

Finally, the second order power spectrum of polarization provides a very important constraint on the galaxy formation because their amplitude depends on models. A signal of this amplitude is below the detectability limits of the Planck Surveyor mission, which is the most accurate experiment in the near future. Therefore, detection of this signal should be a key aim of a post-Planck experiment with increased sensitivity and resolution in the next decade.

ACKNOWLEDGMENTS

We thank the Virgo Consortium for making available the GIF N-body simulations used here, and Shaun Cole, Carlos Frenk and Carlton Baugh for allowing us to use their model of galaxy formation. GCL thanks N. Seto for useful discussion. NS, AN and CGL acknowledge kind hospitality of Carlos Frenk and the physics department of University of Durham during the TMR network meeting. NS is supported by the Sumitomo fundation. CGL acknowledges support at SISSA from COFIN funds from MURST and from ASI. AN and AJB acknowledge the support of the EC RTN network "The Physics of the Intergalactic Medium".

REFERENCES

1. Benson, A.J., Nusser, A., Sugiyama, N. & Lacey, C. G., 2001 MNRAS, 320, 153
2. Cole, S., Lacey, C. G., Baugh, C. M. and Frenk, C. S., 2000, MNRAS, 319, 168
3. Gruzinov, A. & Hu, W., 1998, ApJ, 508, 435
4. Shapiro, P. R., & Giroux, M. L., 1987, ApJ, 321, L107
5. Vishniac, E. T.,1987, ApJ, 322, 597

Thermal design and preliminary performance evaluation of the cooling system for BaR–SPOrt

Macculi C.* and Zannoni M.†

*I.Te.S.R.E./C.N.R., via P. Gobetti 101, I–40129 Bologna
†Dip. di Fisica, Univ. di Milano - Bicocca, P.za della Scienza 3, I–20126 Milano

Abstract. BaR-SPOrt is an experiment to measure the linearly polarised emission of 20°x20° sky patches at 32 GHz and 90 GHz from a stratospheric balloon. It consists of correlation polarimeters for direct measurements of the Q and U Stokes parameters, coupled to an optics providing a beam of 0°.5 (32 GHz) and 0°.2 (90 GHz). Its aim is the study of the polarisation of the Diffused Galactic Background as well as the Cosmic Microwave Background Radiation (CMBR). The instrument thermal design and the preliminary performance evaluation of the cooling system are described.

INTRODUCTION

BaR-SPOrt (Balloon-borne Radiometer for Sky Polarisation Observations) is a balloon experiment housing correlation microwave polarimeters (32 & 90 GHz) for the direct measurement of the Q and U Stokes parameters with HPBW=0°.5 & 0°.2, sharing most of the SPOrt–ISS know-how [1, 2]. Usually, to maximize the sensitivity of a radio receiver, it is necessary to minimize the noise injected by its front-end. In the case of a polarimeter such as the one used for BaR-SPOrt, this means cooling to cryogenic temperatures both the system Polariser - OMT (Orthomode transducer) and the first stage of the low noise amplifiers (HEMT). In order to reach and keep constant their temperature, a proper cooler as well as a cryostat housing the cold components in a vacuum chamber are necessary. Drifts and instabilities of the temperature of the warm and cold parts generate thermal noise (offset and its fluctuations), leading to degradation of the sensitivity. In order to control and remove efficiently such offset fluctuations and to guarantee the goal sensitivity, the maximum thermal instability inside the cryostat has to be ≤ 0.1 K. A shield, whose temperature is actively controlled, as well as a closed loop cryocooler (CLC), with high temperature stability (0.1 K), have thus been adopted. Since the expected level of the signal is very low (the expected polarised component of the CMB is ≤ 1 μK) we have to maximize the signal–to–noise ratio of the instrument. Starting from the correlation radiometer sensitivity equation:

$$\Delta T_{rms} = \sqrt{\frac{k^2 T_{sys}^2}{B\tau} + T_{offset}^2 \left(\frac{\Delta G}{G}\right)^2 + \Delta T_{offset}^2} \tag{1}$$

care has been taken to reduce the system noise temperature (T_{sys}), the gain fluctuations ($\Delta G/G$), the instrumental offsets (T_{offset}) and the offset fluctuations (ΔT_{offset}) by means of correlation and modulation techniques, and employing thermally controlled devices. Considering the polarimeter design [3] and the instrumental thermal specification (maximum environment temperature: COLD $\rightarrow$ 80.0 $\pm$ 0.1 K, WARM $\rightarrow$ 300.0 $\pm$ 0.1 K), the main features of the BaR-SPOrt experiment are summarized in Table 1.

TABLE 1. BaR-SPOrt main characteristics: σ_{1s} is the instantaneous sensitivity, σ_{PX} and σ_{FP} are the final per pixel and the full patch rms sensitivities for a flight of two weeks.

Frequencies (GHz)	Bandwidth	Beam	$\sigma_{1s}[\text{mKs}^{1/2}]$	$\sigma_{PX}[\mu\text{K}]$	$\sigma_{FP}[\mu\text{K}]$
32	10%	0.5°	0.5	18	0.4
90	10%	0.2°	0.7	64	0.6

CP609, *Astrophysical Polarized Backgrounds*, edited by S. Cecchini et al.
© 2002 American Institute of Physics 0-7354-0055-5/02/$19.00

BAR-SPORT THERMAL INSTRUMENT DESIGN

In order to cool a box to a fixed temperature T by a CLC, we must guarantee that the thermal input on the CLC cold finger is lower than the refrigeration power of the mechanical cryogenerator at the same temperature. It is possible to estimate the thermal input to the cryocooler according to the following budget:

$$\dot{Q} = \dot{Q}_R + \dot{Q}_C + \dot{Q}_J \qquad (2)$$

where the total thermal input equals the radiative (R) and the conductive (C) heat plus the heat dissipated by Joule effect (J) in the active components. The convective exchange is negligible because the pressure level in the vacuum chamber is 10^{-5}–10^{-6} mbar. The radiative thermal input is due to the emission of the inner surface of the cryostat "seen" from the cold box. The conductive thermal input is due to the waveguide (WG) (fiberglass with internal silver coating) between the feed (300 K) and the polariser (80K), the two WG (stainless steel) from the first stage of the cold HEMT to the warm box, the four spacers (fiberglass) between the cold and the warm box and the cryogenic wires for the LNAs power supply. In order to understand how the total heat could vary as a function of some technical parameters (see Figure 1), we explore two possibilities: one with maximum (UPPER CASE) and one with minimum (LOWER CASE) total thermal input to the cold finger.
We consider the following designs:

UPPER CASE

- the two WG (L_A = 0.681 cm, L_B = 0.325 cm, t = 0.02 cm, l = 10 cm) from the HEMT to the warm box are internally silver coated (5 μm);
- the four spacers are solid (Φ = 1 cm, l = 8 cm);
- the lenght of the cryogenic wires is four times shorter than in the minimum case (total number of wires = 38: nine for each cold HEMT and two for each thermometer (total number of thermometers = 10));

LOWER CASE

- the two WG from the HEMT to the warm box are not internally silver coated;
- the four spacers are hollow with a fixed thickness (t = 3 mm);
- the lenght of the cryogenic wires is l = 40 cm

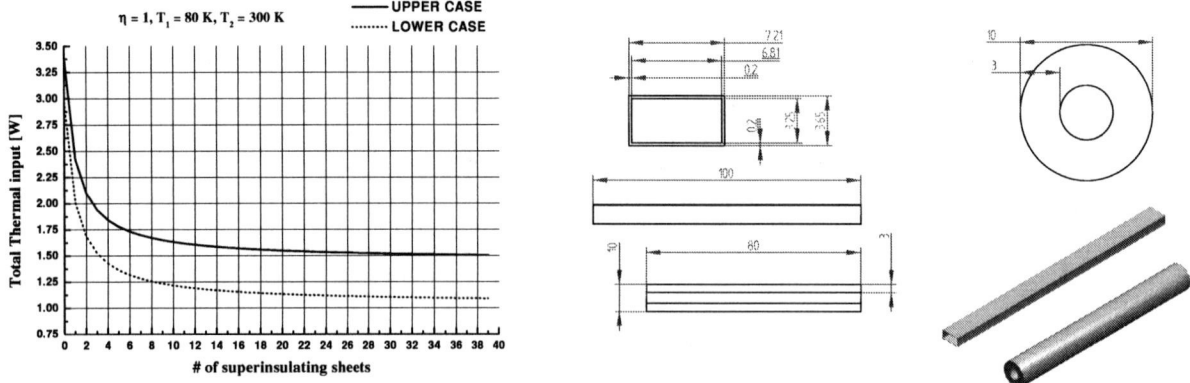

FIGURE 1. Expected thermal input (left). Instrument design parameters [mm] for the LOWER CASE: waveguide and spacer (right).

In both cases, assuming the maximum efficiency ($\eta = 1$) for the heat transfer due to Joule effect and using superinsulation techniques to reduce the total thermal input, the total heat insisting on the cold finger versus the number of superinsulating sheets is shown in the Figure 1. Combining the UPPER CASE and the specifications of the Stirling CLC (6,4W @ 80 K, 180 W Power drawn) provided by the manufacturer (see section below), the goal of 80 K results achievable.

PRELIMINARY EVALUATION OF THE ACTIVE COOLING SYSTEM

First measurements related to the cryocooler have been made in order to verify both its power refrigeration (Stand Alone Test) and its temperature stability. During these measurements, a P_{MAX} (maximum motor power) of 180 W has been set and 1 W has been applied by a heater directly linked to the cold finger. The cold finger temperature has been lowered down to 55 K by 1 K steps in order to determine the minimum temperature the cryocooler is able to keep constant with a heat input of about 3 W (see later). It is possible to see from Figure 2 that the temperature stability, connected to that of the power consumption, is guaranteed when the latter is lower than P_{MAX}. Since a sudden decrease of the cold finger temperature implies a sudden spike of the motor power absorption, such a spike has to be lower than P_{MAX} in order to avoid production of overcurrents. This is necessary because overcurrent lasting a few seconds imply automatically the switch off of the cryocooler (failure $\rightarrow$ safety mode). The lowest stable temperature we reached is 55.0 ± 0.1 K. Lower temperatures imply a power consumption exceeding the limit we set (180 W). From Figure 4

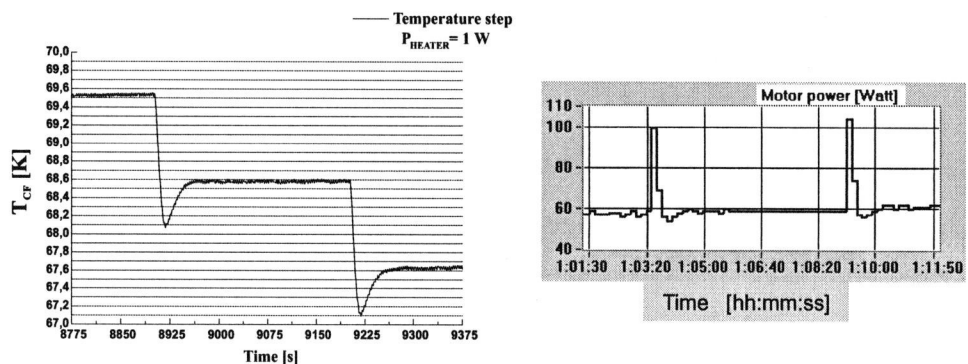

FIGURE 2. Example of cold finger temperature (left) and motor power stabilities (right).

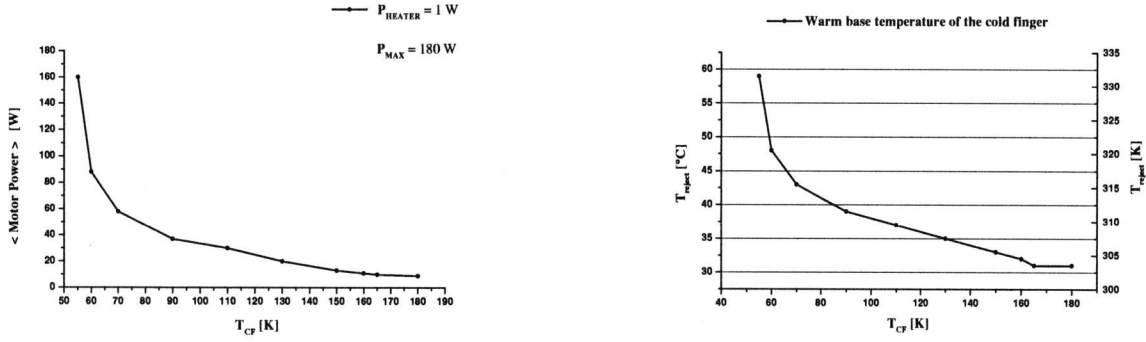

FIGURE 3. Average motor power compressor (left). Reject temperature of the cold finger (right).

we find a cooling power of about 3.2 W for a minimum stable temperature of 55.0 ± 0.1 K. At 80 K such a cooling capacity requires a power consumption of ~ 45 W (see Figure 3). Considering the "UPPER CASE" of the thermal budget (~ 1.75 W for 10 sheets of superinsulator) and the thermal specifications, the performances of our cryocooler, both in terms of low temperature and stability, meet the thermal requirements for an active cooling system for the BaR-SPOrt polarimeter. Since during our test 1 W had been applied by a heater on the cold finger, we can estimate a background heat input of the test vacuum chamber of ~ 2.2 W. The results of this analysis suggest this cryocooler is suitable for our purposes.

Other tests are being planned in order to verify, by passive loads, the dynamics of the heat transfer [4, 5] toward the cold finger, in terms of temperature working point and stability of the loads.

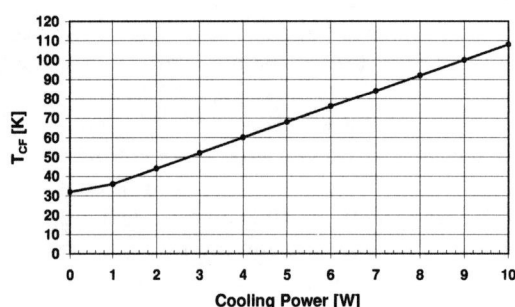

FIGURE 4. Nominal cold finger temperature (50 Hz operation, Power drawn: 180 W, Ambient temperature: 25°C)

ACKNOWLEDGMENTS

Authors wish to thank all the BaR-SPOrt collaboration for useful discussion. This work has been financially supported by ASI.

REFERENCES

1. Cortiglioni S. et al., The SPOrt Project: an Experimental Overview, in proceedings of the *International Conference on 3K Cosmology EC-TMR, Roma 1998*, edited by L. Maiani, F. Melchiorri and N. Vittorio, AIP Conference Proceedings 476, 1999, pp.186-193.
2. Fabbri R. et al.,The SPOrt Project: Cosmological and Astrophysical Goals, in proceedings of the *International Conference on 3K Cosmology EC-TMR, Roma 1998*, edited by L. Maiani, F. Melchiorri and N. Vittorio, AIP Conference Proceedings 476, 1999, pp.194-203.
3. Zannoni M. et al., The BaR–SPOrt experiment, these proceedings.
4. Macculi C., Thermal fluctuations induced by cryocooler temperature fluctuations, *BaR-SPOrt Int. Tech. memo*, 01–01, 2001.
5. Macculi C. et al., BaR–SPOrt: a technical overview, in proceedings of the *Experimental Cosmology at Millimeter Wavelenghts – 2K1BC Workshop, Breuil–Cervinia 2001*, edited by De Petris M. and Gervasi M., AIP Conference Proceedings, in press.

SDAMS: SPOrt Data Archiving and Management System

Luciano Nicastro* and Giorgio Calderone*

*IFCAI-CNR, Via U. La Malfa 153, 90146 Palermo, Italy

Abstract. SDAMS is the ensemble of database + software packages aimed to the archiving, quick-look analysis, off-line analysis, network accessibility and plotting of the SPOrt produced data. Many of the aspects related to data archiving, analysis and distribution are common to almost all the astronomical experiments. SDAMS ambition is to face and solve problems like accessibility and portability of the data on any hardware/software platform in a way as simpler as possible, though effective. The system is conceived in a way to be used either by the scientific community interested in background radiation studies or by a wider public with low or null knowledge of the subject. The user authentication system allows us to apply different levels of access, analysis and data retrieving. SDAMS will be accessible through any Web browser though the most efficient way to use it is by writing simple programs. Graphics and images useful for outreach purposes will be produced and put on the Web on a regular basis.

INTRODUCTION

SPOrt ([1]) on ISS is an experiment which will produce a limited amount of data with respect to other space experiments. The foreseen total (i.e. housekeepings and scientific data) bit rate is ~ 2.5 kbit s^{-1}. This means that in 3 years of operation time it will collect ~ 30 Gbytes of data. In addition SPOrt has not pointing capabilities and it is expected to work in a stable and semi-automatic way. As a consequence the output data stream will be simply identified by *time* and *orbital parameters* of the Space Station. The data management system (i.e. quick look, archive and scientific analysis) can be build to be both simple and highly automatic.

The SDAMS system has two fundamental parts: *database server* and *application server*. The former performs the data storing and allows a simple (low level) fast access to them. To this aim we chose MySQL^{TM}, a very easy to use data management system capable to deal with a relatively large amount of data. On the Web there are several examples of intensive usage of MySQL with application tools and there are extensions to use it within several programming languages (e.g. PHP). It has also the advantage to be an Open Source package, that means one can modify the source code to match any particular need and no license is required to use it. This is crucial to allow a standard (system) to be developed.

The SDAMS application server is a multithreaded server which uses a simple C-written communication protocol to interact with its clients. A task-oriented module library allows us to easily add further capabilities to those so far foreseen. The fundamental task of this component is to accept client connections (through socket) and invoke the appropriate modules to perform the required tasks.

News about SDAMS will be posted on `http://sport.tesre.bo.cnr.it`.

HOW TO USE IT

There are several ways to use SDAMS, the simplest being through a Web interface. Such kind of interface allows users to setup a simple query in order to request data retrieval and visualization. This is done, for example, selecting a time or sky coordinates range. It will also be possible to view the sky portion seen by SPOrt (pointing at the zenith of the ISS) in any given time interval. In addition to Sun, Moon and some Planets, also "interesting" strong radio sources will be highlighted in the sky maps. Access to the data will be granted only to registered users. However, this is the less flexible way to use SDAMS as it only offers a limited set of choices; it will be likely used to perform data "quick-look" or by users not directly involved in the SPOrt project or by the general public. The communication protocol is so easy that it is possible to connect to SDAMS through a telnet session and typing in the commands. The output will be ASCII formatted. The most flexible way to access SDAMS is instead through a dedicated, user written (in any language) program. The only requirements are

- capability to open TCP/IP sockets,
- capability to read ASCII and binary files.

In this case the output can be formatted as a FITS binary (or ASCII) table. For these "high-level" users, a reference manual will be available reporting characteristics of the protocol and a full description of commands and message codes. Once again we note how SDAMS does not put any restriction to the hardware/software platform of the client.

HOW IT WORKS

Figures 1 and 2 show how SDAMS works: for each client successfully connected to the socket a new thread is created; it is a sort of sub-process running on the server. It waits for queries from the client and executes them independently of what the other clients are doing. To each query, written by using high level commands, corresponds a spawned process which passes a reformatted low-level query to the database server. The output data are processed either by C or IDL^{TM} routines (depending on the request) and then sent to the client in the requested format. This process is repeated until a "disconnect" request is received. The "language" used by the client to put its query is very simple, being an ASCII string containing the name of the procedure to use and its parameters. For example, to select the data for the Stokes parameter Q in the time interval (MJD) 52810–52813, in the 32 GHz frequency channel, averaged over 30 s, and perform a polynomial fit (third order) with automatic graphical output, the command is:

```
sel, par='q', dt=[52810.0,52813.0], fr=32, av=30, fit='poly 3'
```

By default the graphics goes into a PostScript file, but other formats can be chosen.

It is worth to note how, in spite the application server talks to the database server using the SQL (Structured Query Language) syntax, it is not necessary to write the queries using this low-level format (but can be done starting the query with SQL). A dedicated module, the *SQL wrapper*, takes care of translating the input command string (after a validity check). When the client queries require complex operations (e.g. correlation analysis or sky-map production), the application server first extracts the data into a local directory then the appropriate (IDL) routine (one or more) is invoked for the analysis and finally the output is sent to the client (e.g. in FITS format). The clients do not need an IDL license as the application software only runs on the server. Figure 3 shows an example of sky coverage map extracted using the command "map, /coverage". It assumes only about 20 days are elapsed since the start of the mission. Adding a new feature to the system is extremely easy: it is enough to add the IDL routine (accessing the data using the SDAMS standards) to the library and adding its name to the list of allowed commands.

ACKNOWLEDGMENTS

SDAMS is part of the SPOrt project and it is supported by Agenzia Spaziale Italiana (ASI).

REFERENCES

1. Carretti, E., "The SPOrt project", in *Astrophysical Polarized Backgrounds*, edited by e. a. S. Cecchini, AIP Conference Proceedings 0, American Institute of Physics, New York, 2001, pp. xx–yy.

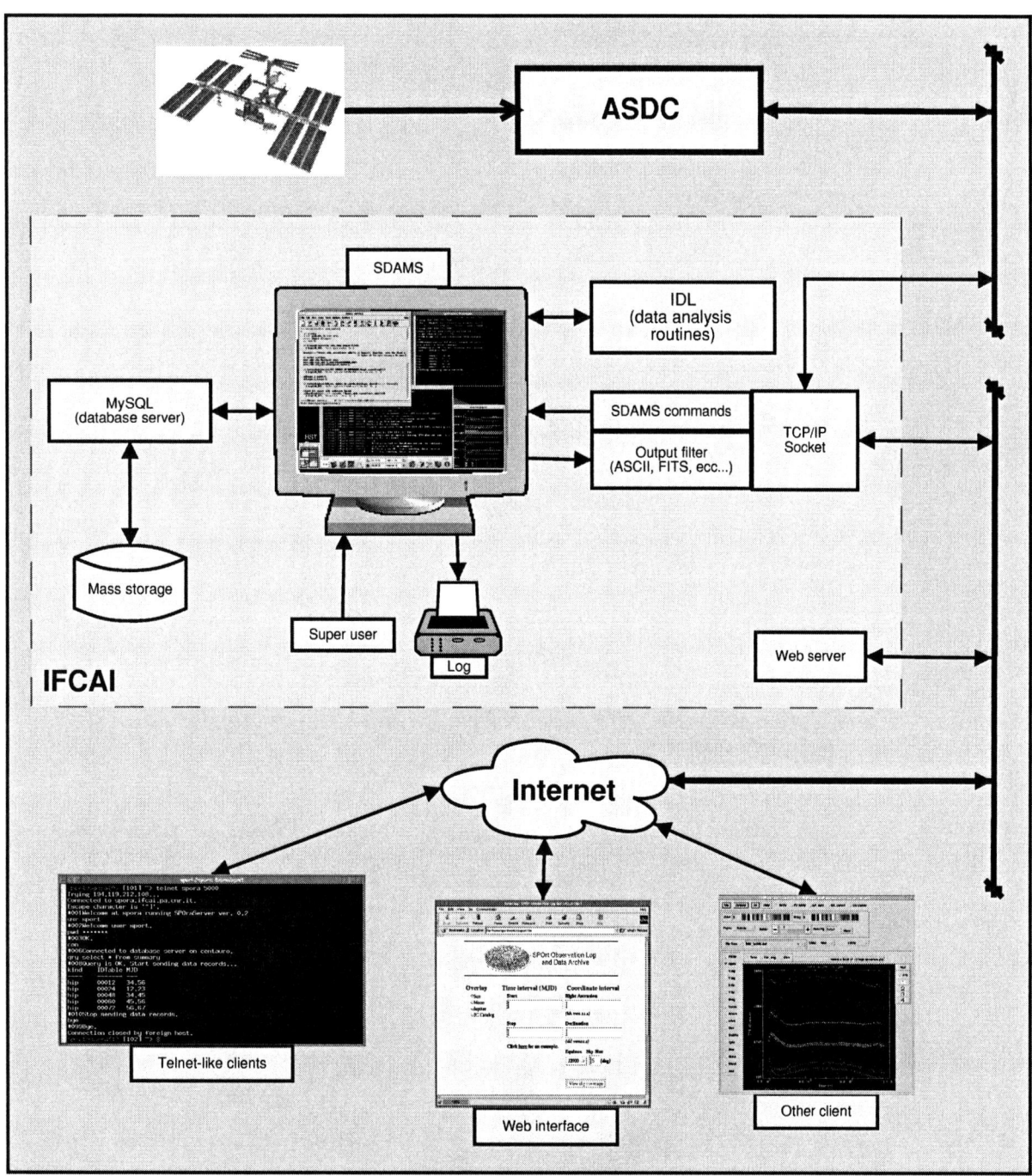

FIGURE 1. This diagram shows the main components of SDAMS (dashed inset) and the way it can interact with external systems (clients). The ASDC (ASI Science Data Center) network gives access to the raw data downloaded from the International Space Station. Here we assume the database resides at IFCAI Institute (Palermo)

281

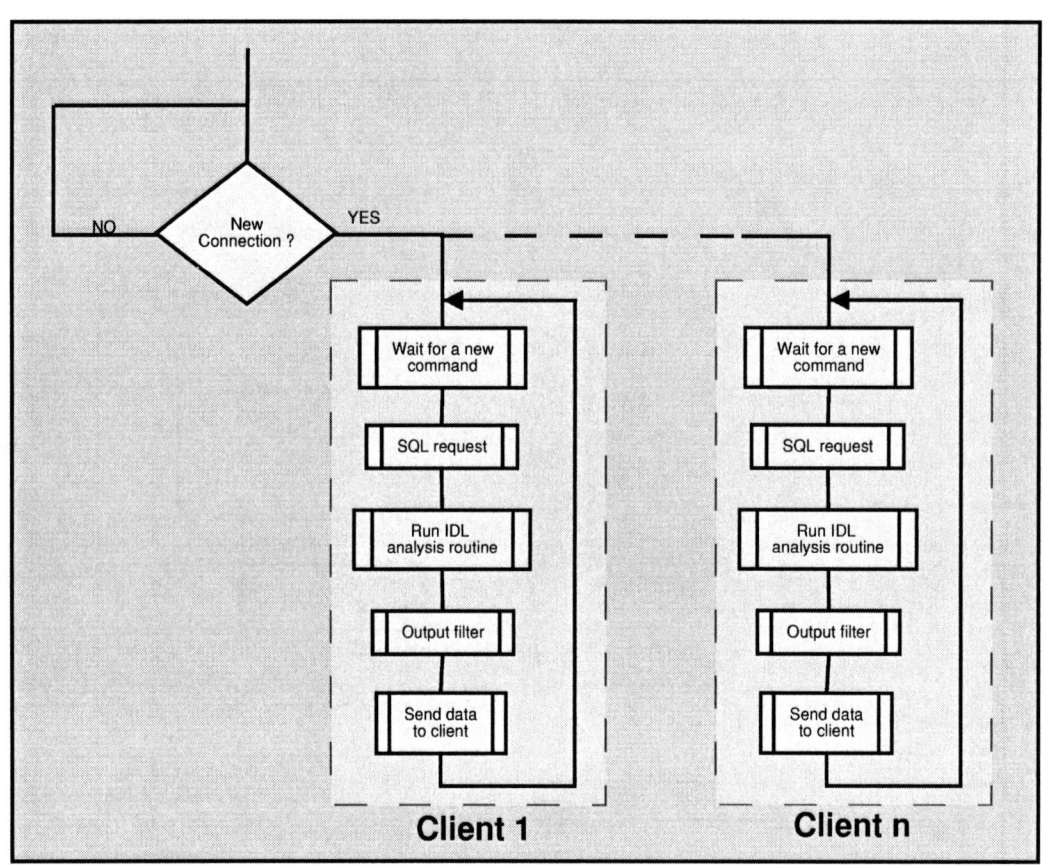

FIGURE 2. The scheme above shows the parallel running of the threads (dashed insets). Each thread is a sub-process (on the server) performing the client queries and data output

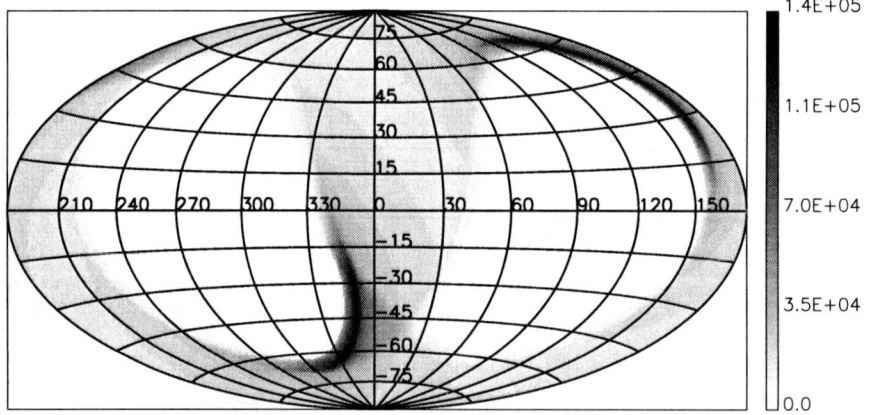

FIGURE 3. In this image (Galactic coordinates) the SPOrt sky coverage after about 20 days of operative life is reported. Grey scale is proportional to the integration time

A millimeter polarimeter for the 45m telescope at Nobeyama [1]

Hiroko Shinnaga[†*], Masato Tsuboi[†] and Takashi Kasuga[**]

*Harvard-Smithsonian Center for Astrophysics, SAO / SMA Project, 82 Pu'uhonu Pl. Hilo, HI, 96720 USA
[†]Institute of Astrophysics and Planetary Science, Ibaraki University, Bunkyo, Mito, Ibaraki, 310-8512 Japan
[**]Department of System and Control Engineering, Hosei University, Kajino, Koganei, Tokyo, 184-8584 Japan

Abstract. We have designed and constructed a tunable polarimeter system to cover frequencies from 35 GHz to 250 GHz (8.6 mm and 1.2 mm in wavelength) for the 45m telescope at Nobeyama Radio Observatory (NRO). Both circular and linear polarizations can be measured by the polarimeter. The performance of the polarimeter in astronomical observations was tested by simultaneously measuring the linear polarization of the $J = 2 - 1$ transition of SiO in the $v=0$ and 1 states at 86 GHz toward VY CMa. The design, construction, and tests are presented briefly.

INTRODUCTION

Astronomical polarimetry is a powerful tool for investigating the structures and strengths of interstellar and circumstellar magnetic fields, which are keys to understanding the physical processes in star formation and the evolution of interstellar medium. Since the first clear detection of linear polarization had been made [1], a number of radio polarimeter systems have been developed for various radio telescopes and interferometers. We have designed and constructed a millimeter wave polarimeter with a reflection-type wave plate to cover the frequency range from 35 GHz to 250 GHz and with low insertion loss for the Nobeyama 45 m telescope [2]. Using this polarimeter system, we have made several astronomical observations [2, 3, 4].

In this paper, the design, construction, control system, and performance tests are reported briefly. Detailed description of the instrument can be found in Shinnaga et al. (1999) [2].

DESCRIPTION OF THE INSTRUMENT

The reflection-type wave plate is assembled by mounting a free-standing wire grid in front of the reflecting surface of a plane mirror with both surfaces parallel to each other (Howard et al. (1986) [5], Prigent et al. (1988) [6]). The grid is made of a plane array of parallel wires with circular cross section. Figure 1 shows two pictures of the instrument. Figure 2 is a schematic diagram of the wave plate. The separation c between the wire grid and the plane mirror gives rise to an optical-path difference, and hence a phase shift between the two orthogonal linear-polarized components. When the optical path difference d between $E_{//}$ and $E_\perp$ is $c/\cos\varphi$, the corresponding phase shift δ can be written as $\delta° = 360d/\lambda$, where λ is an observational wavelength. It can be seen from the equation that a reflection-type wave plate can be used as a half-wave plate (HWP) and a quarterwave plate (QWP) by adjusting c so that $\delta = 180°$ or $90°$, respectively. A wave plate of diameter 400 mm is used in the optical path of the beam guiding system of the telescope at $\varphi = 45°$. With a and b equal to 20 μm and 50 μm, respectively, the cross polarization of the grid from the ratio is estimated to be -46 dB at 35 GHz, -38 dB at 100 GHz, and -30 dB at 250 GHz, respectively.

The polarimeter comprises a reflection-type wave plate and it's rotator. The wave plate, fixed on a rotational stage,is driven by a stepping motor. The separation c is hanged by replacing eight spacers with a certain thickness for an

---

[1] Nobeyama Radio Observatory is a branch of the National Astronomical Observatory, operated by the Ministry of Education, Science, Sports and Culture.

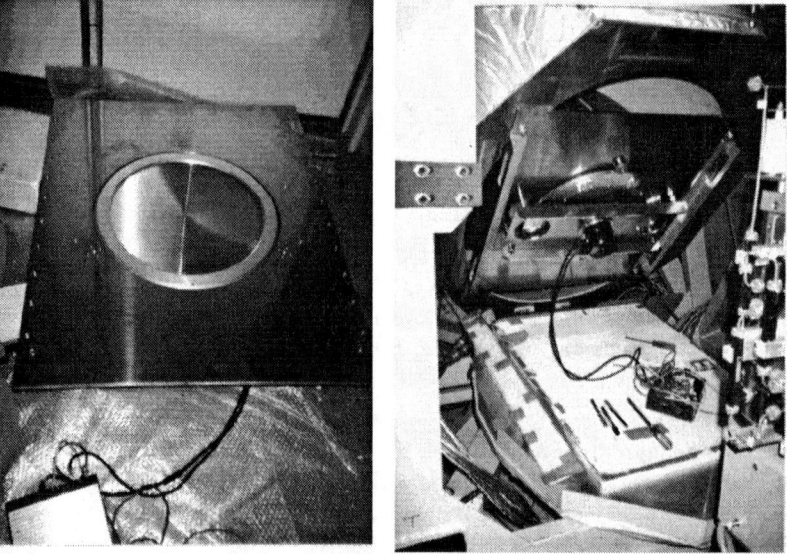

FIGURE 1. Left: Top view of the millimeter polarimeter. The wire grid with a circular frame is seen at the center. Right: The instrument installed on the beam guiding sysmte at the telescope.

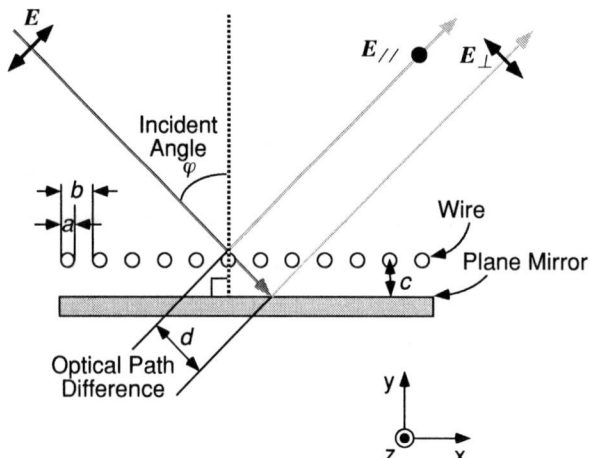

FIGURE 2. Schematic diagram of the reflection-type wave plate [2]. The grid wires along the z directions are shown in open circles. The electric field E associated with the incident beam reaches the wave plate at an incident angle of φ. The two components of E with orthogonal linear polarization, $E_{//}$, parallel to the grid (along the z direction), and $E_{\perp}$, perpendicular to the plane defined by the wire and the direction of the beam propagation have an optical path dirence of d after being reflected from the wave plate.

observational frequency. Both linear and circular polarization could be studied with the wave plate set, respectively, as HWP and QWP.

For sensitive polarimetry, most of the power of a beam in the guiding system should be reflected by the wave plate. Assuming a Gaussian beam, the large diameter of the wave plate allowed us to collect about 99.7%of the total power of the beam at 35 GHz, the lowest detectable frequency of the available receivers. The large radius of curvature of the beam at the polarimeter (for example, 1269 mm at 35 GHz, 872 mm at 120 GHz) reduces the difference between the ideal value of δ and its measured value. The millimeter polarimeter is capable, in principle, of covering a much wider frequency range; the the frequency coverage of the available receivers limits its operational range from 35 GHz to 250 GHz. Additionally, the surface accuracy of the telescope also limits the highest frequency to 250GHz.

The instantaneous bandwidth of the polarimeter set as HWP, $\Delta f/f$ (f is the observation freqnency), depends on the repeatability of the measurements with the system. It was found that there was an error of 1% in the relative intensity for repeated measurements of the same point of a source using the telescope system. In observing linear polarization, this error introduced a 2% uncertainty in measuring the degree of polarization P, which means 100% linearly polarized radiation is observed as $P = 98\%$, using the system. The degration in P of 2% is also estimated to be caused by the phase-shift error at the band edge, $f \pm \Delta f \sim 1 \pm 0.08$. Thus the bandwidth of the polarimeter $\Delta f / f$ is considered to be $\sim \pm 0.08$.

When the polarimeter was implemented with the telescope for astronomical observations, we used the position-switching method for data acquisition.The rotation of the wave plate was controlled by a personal computer, which received a signal from a spectrometer at the end of each integration period,and then sent a start signal of rotation to the controller of the stepping motor. The rotation was completed before the start of the subsequent integration period.

PERFORMANCE

The performance of our polarimeter has been tested in two aspects, namely, the insertion loss and the uncertainty in linear-polarization measurements. The latter is referred to as instrumental polarization. Here, only linear polarization is considered very bridfly.

A low insertion loss is essential for the good performance of a polarimeter. The insertion loss of the polarimeter is due to the surface resistance and irregularities of the wave plate. The loss at 110 GHz was measured to be 0.14 ± 0.05 dB, in good agreement with the estimated value of 0.14 dB at 110 GHz, calculated from the (0.08 dB). The losses at other frequencies estimated using the same assumption were 0.11 dB at 35 GHz and 0.17 dB at 250 GHz. The insertion loss of the system is low compared to a transmission-type polarimeter made of Teflon stack with a QWP configuration that was described by Watanabe 1981 [7], for example. The calculated loss of our polarimeter at 90 GHz was 0.13 dB, about half of the value reported by Watanabe (1981 [7], 0.3 dB in the same frequency band). For a system at a noise temperature of 200 K, a loss of 0.14 dB corresponds to an increase of noise by ~ 15 K, compared to an increase of ~ 35 K at a loss of 0.3 dB.

The three sources contributing to the overall instrumental polarization are considered to be (i) the wave plate, which makes the phase shift δ ; (ii) the beam guiding system of the telescope, which is composed of many plane mirrors, beam splitters and focusing mirrors; and (iii) the main- and sub- reflectors of the telescope, which are deformed by gravity. The instrumental polarization due to these three sources has been tested separately. The detail of each measurement is described in Shinnaga et al. 1999 [2].

Here we briefly mention the overall instrumental polarization of the system, including the telescope, by observing the non-polarized celestial sources Jupiter, Uranus,and Saturn at 86 GHz, by using orthogonal dual-channel SIS receivers. Some of the results of these observations are shown in the figure 6 in the paper by Shinnaga et al. 1999 [2]. Recent measurements show linear correlations between the elevation angle and the polarization angle of the polarimeter as well as the polarization degree (Miyahara et al. 2001, private communication). This is presumably caused by a gravitational deformation of the main reflector and the rotation of the Nasmyth focus. A systematic change in the size of the beam width at half power (HPBW) with the elevation angle was not found in the measurement.

ASTRONOMICAL OBSERVATIONS

Following the various tests, the polarimeter was also used for an observation towards VY CMa, which is one of the brightest red supergiant stars and is known to be one of the strongest SiO sources, to study the polarization phenomenon of the $J = 2 - 1$ transitions of SiO in the $v = 0$ (86.846998 GHz) and 1 (86.243442 GHz) states. The observation took place on 1998 March 30. The emission of two transitions was simultaneously measured. The beam size of the telescope at 86 GHz was 18". The beam and the aperture efficiencies were 49 % and 42%, respectively. The sensitivity of the antenna is 3.5 Jy K^{-1}. A detailed description of the observation can be found in Shinnaga et al [2]. The figures 3a and 3b show the spectra of the two SiO transitions of VY CMa sampled at 8 different polarization angle of the polarimeter. The figure 3a clearly shows that the two bright velocity components are highly polarized(up to $P \sim 50\%$). In contrast, as shown in figure 3b, the $v = 1$, $J = 2 - 1$ maser emission of SiO is strong, but their intensities and shapes hardly change at different PA's, whose P is smaller than $\sim 3\%$ at the LSR velocity of 18.5 kms $^{-1}$, for example. The polarization properties of the $v = 1$, $J = 2 - 1$ emission is consistent with previous observations (e.g.,McIntosh et al. 1994 [8];Troland et al. 1979 [9]). Since both transitions were observed simultaneously, they are expected to suffer from the same experimental error, and therefore the observed polarization should be genuine.

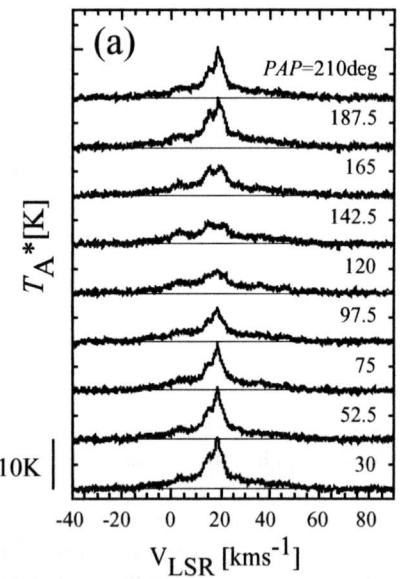 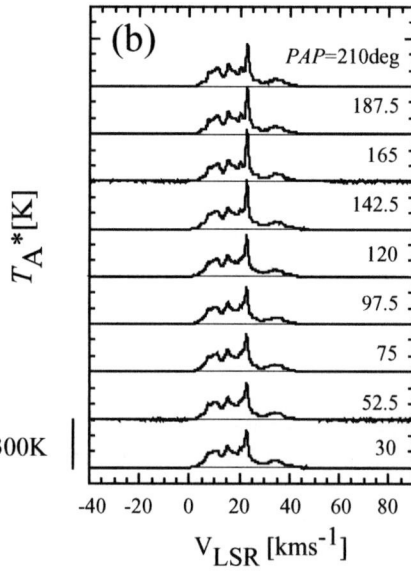

FIGURE 3. Spectra of SiO $J = 2 - 1$ emission of VY CMa in (a)$v = 0$ and (b)$v=1$ states simultaneously sampled at 8 different polarimeter polarization angles [2].

The high degree of polarization is a strong evidence that the $v = 0$ emission originates from maser action, even if the brightness temperatures were low and their line widths were not narrow. So far, we found several cases of the $v = 0$ maser emission (Tsuboi et al. 1996 [10], Shinnaga et al. 1999, [2], Shinnaga et al. 2001 [4]). In order to infer the mechanism of the maser emission further, polarization measurements with higher angular resolution are needed.

In addition, using the polarimeter system, we also carried out CCS Zeeman observations towards a dense molecular core in Taurus Molecular Cloud[3]. The detail of the study will be presented elsewhere.

ACKNOWLEDGMENTS

This work was supported by Yamada Science Foundation, Foundation for Promotion of Astronomy, and a grant-in-aid (No.1044061) from the Ministry of Education, Science, Sports and Culture. The authors thank N. Ukita of NRO for useful comments and the members of the 45 m telescope group for supporting the measurements. Thanks are also due to T. Ohno, A. Miyazaki, S. Imaizumi, and H. Kanno of Ibaraki University for their extensive help during the course of this work. The authors thank H. Miyahara for allowing us to refer their unpublished measurements for the instrumental polarization of the polarimeter system.

REFERENCES

1. Mayer, C.H., McCullough, T. P., and Sloanaker, R. M. 1957, ApJ, 126, 468
2. Shinnaga, H., Tsuboi, M., and Kasuga, T. 1999, PASJ, 51, 175
3. Shinnaga, H., Tsuboi, M., and Kasuga, T. 1999, in Proceedings of Star Formation 1999, 175, June 1999, Japan
4. Shinnaga, H., Tsuboi, M., and Kasuga, T. 2001, in Proceedings of IAU symposium vol.206, "Cosmic Masers: from protostars to blackholes", Brazil
5. Howard J., Peebles W.A., and Luhmann N.C.Jr. 1986, Int. J. Infrared Millimeter Waves 7, 1591
6. Prigent C., Abba P., and Cheudin M. 1988, Int.J.Infrared Millimeter Waves 9, 477
7. Watanabe R.1981,PhD Thesis,Hokkaido University
8. McIntosh G.C., Predmore C.R., and Patel N.A. 1994, ApJ 428,
9. Troland T.H., Heiles C., Johnson D.R., and Clark F.O. 1979, ApJ 232,143
10. Tsuboi M., Ohta E., Kasuga T., Murata Y., and Handa T.1996, ApJ 461,L107

Observations of Linear Polarization of Background Galactic Radio Emission in Selected Directions at 8.3 GHz

E.N.Vinyajkin[1], E.Carretti[2], S.Cortiglioni[2], S.Poppi[3]

[1]Radiophysical Research Institute (NIRFI), 25, B.Pecherskaya ul., 603950, Nizhnij Novgorod, Russia
[2]ITeSRE/CNR,Via P.Gobetti 101, I-40129 Bologna, Italy
[3]IRA/CNR, Via P.Gobetti 101, I-40129 Bologna, Italy

Abstract. Polarization observations of the Galactic radio emission at 8.3 GHz were made by the 32-m Medicina (Italy) radio telescope in four pixels (HPBW=4.'8). A method of tracking around the upper culmination was used in order to use the rotation of the parallactic angle for detecting the weak linearly polarized Galactic radio emission against the background of relatively strong and variable spurious and instrumental polarization. The well known source 3C 286 was used as calibrator. As a result the brightness temperatures of linearly polarized component of the Galactic radio emission and positions angles were measured for all pixels. Comparison was made for the pixels in the first Galactic quadrant with Duncan et al. 2695 MHz polarization measurements and as a result spectral indexes and rotation measures were determined.

1. INTRODUCTION

By now the linearly polarized component of the Galactic diffuse synchrotron radio emission has been partly investigated at frequencies of meter (e.g. [1]) and decimeter (e.g. [2,3]) wavebands up to 2.695 GHz [4]. Current estimates of the Galactic polarized contributions to the microwave region are obtained by extrapolating available low frequency data and are affected by uncertainties because of Faraday depolarization and the use of spectral indexes from low frequencies. To study the spectrum and the angular distribution of the Galactic polarization at centimeter wavelengths new observations are urgently needed [5]. A knowledge of the polarized radio emission spectral index for high enough frequencies to avoid Faraday depolarization would be useful also for the separation of the synchrotron and free-free components of the total microwave Galactic diffuse radio emission. Besides that, higher frequency polarization observations will bring information on the interstellar magnetic fields at larger distances. To make a first step before observations of more extended areas of the Galactic plane we have observed four pixels at 8.3 GHz by the 32-m Medicina (Italy) radio telescope (HPBW=4.'8). The centers of these pixels are the points with Galactic coordinates l=141°09', b=7°53'; l=145°20', b=4°00'; l=61°36', b=3°43' and l=65°38', b=3°43'. The first point (α_{1950}=57°.0, δ_{1950}=64°.0) is a well known calibrator for polarization measurements (e.g. [2,6]). The 3rd and the 4th points are in the longitude and latitude ranges examined by Duncan et al. [4] at 2.695 GHz. The areas with a radius ~5 HPBW around each of these 4 points are free of discrete sources with flux densities higher than 30-40 mJy at 6 cm and free of Galactic supernova remnants.

As a result we have obtained values of brightness temperature of the linearly polarized component of the Galactic radio emission and position angle at 8.3 GHz for all four pixels.

This paper is organized as follows. In section 2 our observations are described. In section 3 we describe the data reduction. Results and discussion are given in section 4. In section 5 conclusions are formulated.

2. OBSERVATIONS

Observations were made in December 2000 and March-April 2001 at the Medicina (Italy) radio observatory by a 32-m radio telescope. A method of tracking the selected point on the sky around its upper culmination was used.

CP609, *Astrophysical Polarized Backgrounds*, edited by S. Cecchini et al.
© 2002 American Institute of Physics 0-7354-0055-5/02/$19.00

Observations of a point in the second Galactic quadrant consist of an hour angle (*t*) interval tracking between the point's East elongation and the West one. Observations of a point in the first Galactic quadrant consist of an hour angle interval tracking between the minimum and the maximum value of the equatorial parallactic angle $q(t)$ of the point. We used a correlation polarimeter to measure Stokes parameters *I*, *Q* and *U*. Some of the telescope and observing parameters are listed in Table 1. The characteristics of 3C286 were taken from [7,8].

TABLE 1. Polarimeter data, telescope parameters and assumed calibrator 3C286 values

Centre frequency (MHz)	Receiver bandwidth (MHz)	Beamwidth (HPBW)	Q,U RMS noise (mK s$^{1/2}$)	3C286 assumed flux density (Jy)	3C286 degree of polarization (%)	3C286 position angle
8300	80	4'.8	10	(5.0 ± 0.1)	(11.8 ± 0.2)	34°.0 ± 1°.5

For each point several observations were made. Calibration source 3C286 was observed daily during observations. One observation of 3C286 consists of an hour angle interval tracking between the minimum and the maximum value of the equatorial parallactic angle of the source using the "on-off" technique. Fig. 1 shows an example of such observation of the Stokes parameters *Q* and *U*. The 3C286 observations were used for calibrating both the brightness temperature and the position angle of the linearly polarized Galactic radio emission. Instrumental linear polarization was also measured from 3C286 observations. Values of the radio telescope Müller matrix elements as $M_{21}=-1.8\%$ (conversion $I \rightarrow Q$) and $M_{31}=1.7\%$ (conversion $I \rightarrow U$) were measured.

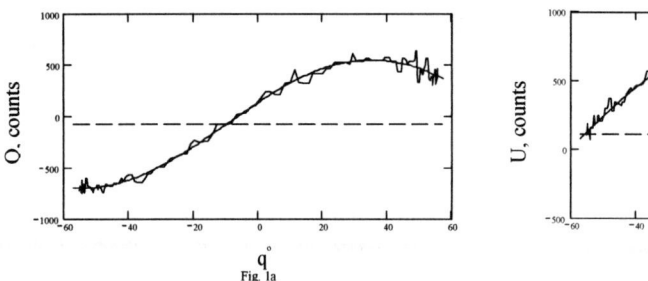 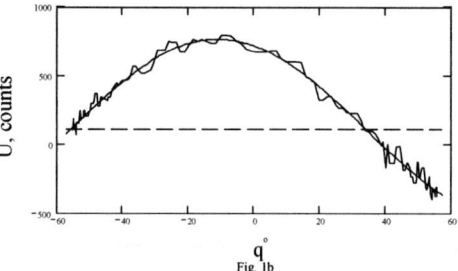

FIGURE 1. The observation of 3C286 made in Dec 3rd, 2000: a) Measured values of the Stokes parameter *Q* as a function of the parallactic angle *q* and the best fit sinusoid are shown. The horizontal dashed line shows the value of the instrumental polarization. b) The same for the Stokes parameter *U*.

Measured values of the Stokes parameters *Q*, *U* of an incoming radio emission are the sums of the true very weak signals of the Galactic linearly polarized radio emission Q_g, U_g, spurious Q_s, U_s and instrumental Q_i, U_i polarization: $Q=Q_g+Q_s+Q_i$, $U=U_g+U_s+U_i$. The main source of spurious polarization is a partly linearly polarized ground radio emission received by the side lobes of the radio telescope. It depends mainly on the elevation angle *h* and the time and it is vertically polarized. During the observation $h=h(t)$, so that spurious polarization depends only on time in fact. Besides that, the total ground radio emission converts into Q_i, U_i via instrumental polarization. Contributions from the atmosphere radio emission and the CMBR into Q_i, U_i are much smaller than the contribution of the ground radio emission. The method of tracking gives the possibility to use specific time dependence of Q_g, U_g due to the rotation of parallactic angle for detecting the weak linearly polarized Galactic radio emission against the background of relatively strong and variable spurious and instrumental polarization.

3. DATA REDUCTION

The data reduction procedure is as follows. In order to take into account a bulk of spurious and instrumental polarization effects, the measured values *Q*, *U* are fitted to a polynomial of degree 2 as a function of the parallactic angle *q*. It is convenient to use *q* as argument instead of the time because around culmination the ratio *q* to *t* is approximately constant. Fig. 2 shows an example of observation (Dec 2nd, 2000) of the point with coordinates l=145°20', b=4°00' and the approximation of Q, U by a polynomial of degree 2.

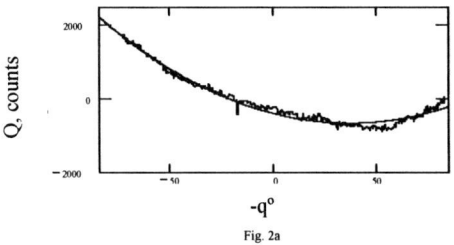

Fig. 2a

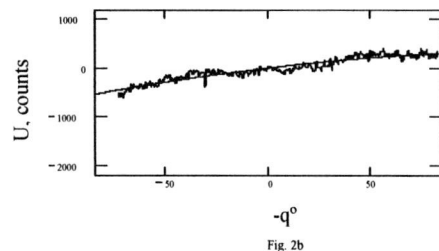

Fig. 2b

FIGURE 2. Results of fitting the measured values of Q, U by a polynomial of degree 2 for observations of the point with coordinates l=145° 20', b=4° 00' made in the Dec 2nd, 2000: a) Stokes parameter Q, b) Stokes parameter U.

After that, the best fit polinomials $Q_2(q)$, $U_2(q)$ are subtracted from the observed Stokes parameters to obtain Q_r =Q-Q_2, U_r =U-U_2. The values Q_r, U_r consist of the true signals, Q_g, U_g, and fluctuations of spurious and instrumental polarization. Finally $Q_r(q)$ is fitted to $Q_g(q)$ and $U_r(q)$ is fitted to $U_g(q)$, where

$$Q_g(q) = Q_0 cos(2q) + U_0 sin(2q)$$
$$U_g(q) = -Q_0 sin(2q) + U_0 cos(2q)$$

(1)

in order to obtain the values of Q_0 and U_0. Fig. 3 shows the results of these fits for the same observations of Fig. 2. For each observation two values of Q_0 and two values of U_0 are obtained. Q_0 and U_0 values were averaged over N observations to obtain the mean values Q_{0m} and U_{0m}, they were calibrated and the values of the polarized brightness

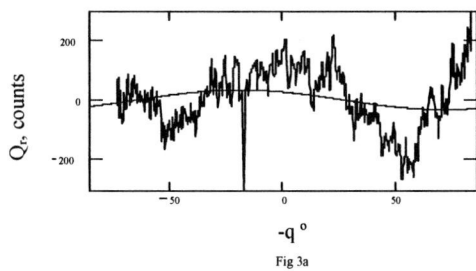

Fig 3a

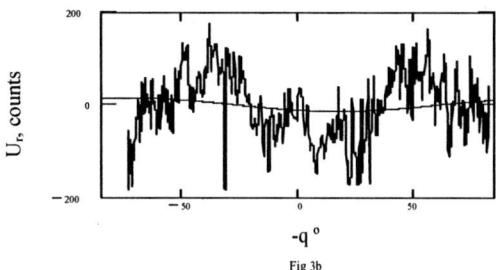

Fig 3b

FIGURE 3. Results of the fits of Q_r and U_r to Q_g and U_g, respectively, for the same observations of Fig. 2: a) Stokes parameter Q_r, b) Stokes parameter U_r.

temperature T_b^P=$(Q_{0m}^2 + U_{0m}^2)^{1/2}$ and of the position angle χ, obtained from formulae $tan(2\chi)$=U_{0m}/Q_{0m}, were calculated for each pixel.

4. RESULTS AND DISCUSSION

Table 2 shows the values of polarized brightness temperature and position angle we obtained. It is of interest to compare polarization characteristics of the 3rd and 4th pixels with 2695 MHz polarization data [4] obtained with similar angular resolution (map resolution (HPBW) is 5.'1 in [4]). Table 3 shows the values of T_b^P and the Galactic position angle χ_g for the 3rd and 4th pixels at 2695 MHz [4] and 8300 MHz (this paper). Values of temperature spectral index of linearly polarized radio emission β_p ($T_b^P\infty\nu^{-\beta_p}$) and rotation measures RM were calculated by comparing 2695 MHz and 8300 MHz data (see also Tab. 3). One can see that in the 3rd pixel *RM* is positive and in the 4th it is rather positive too. This is in agreement with the sign of rotation measures of extragalactic radio sources and distant pulsars and also with values for the positive low latitudes of the first Galactic quadrant [9].

TABLE 2. Results of brightness temperature and position angle measurements at 8300 MHz of the linearly polarized component of the Galactic radio emission for four pixels. *N* is the number of observations.

Galactic longitude	Galactic latitude	Polarized brightness temperature, mK	Galactic position angle	N
141°09'	7°53'	3.3 ± 1.2	143° ± 11°	7
145°20'	4°00'	4.4 ± 1.9	126° ± 11°	4
61°36'	3°43'	1.4 ± 0.8	99° ± 7°	6
65°38'	3°43'	2.7 ± 1.2	86° ± 13°	5

TABLE 3. The values of polarized brightness temperature and position angle at 2695 MHz [4] and 8300 MHz for the 3rd and 4th pixels and values of polarization brightness temperature spectral index β_p and rotation measure *RM*.

l	b	T_b^p (2695 MHz) mK	T_b^p (8300 MHz) mK	χ_g (2695 MHz)	χ_g (8300 MHz)	β_p (2695-8300 MHz)	RM rad m^{-2}
61°36'	3°43'	37.4 ± 7.3	1.4 ± 0.8	132°.9 ± 7°.2	99° ± 7°	2.91 ± 0.53	53.3±15.8
65°38'	3°43'	39.1 ± 7.3	2.7 ± 1.2	92°.7 ± 6°.9	86° ± 13°	2.37 ± 0.43	10.5±23.2

5. CONCLUSION

As a first step before future observations of more extended areas of the Galactic plane, polarization observations of the pixels in the first Galactic quadrant and two pixels in the second quadrant were made at 8.3 GHz. The main difficulty of these observations is the detection of a very weak true signal against the background of much stronger and variable spurious ground polarization. Rotation of the parallactic angle during the tracking of an observed pixel (space filtering), repeating observations and averaging were used for the detection of the signal. Values of polarized brightness temperature and position angle were obtained for all four pixel. By comparison with Effelsberg 2695 MHz brightness polarized temperatures and position angles, obtained with nearly the same angular resolution, values of temperature polarization spectral index of linearly polarized radio emission β_p(2695-8300 MHz) and rotation measures were determined for two pixels from the first Galactic quadrant. Both the absolute values and the sign of our *RM*s are in an agreement with the configuration of the magnetic field inside the Solar circle of the Galaxy revealed by extragalactic source and pulsar polarization observations.

ACKNOWLEDGMENTS

Thanks are due to prof. R. Wielebinski and dr. W. Reich who have generously made available their polarization maps through the Web. Thanks to dr. S. Trushkin for availability of CATS data base. E.N.V. was supported by CNR-RAS agreement during his three visits to Bologna, and partly by Russian grant of the leading scientific schools (grant 00-15-96591).

REFERENCES

1. Vinyaikin, E.N., Kuznetsova, I.P., Paseka, A.M., Razin, V.A., and Teplykh, A.I., Astronomy Letters 22, 582-588 (1996).
2. Brouw, W.N., and Spoelstra, T.A.Th., Astron. and Astrophys. Suppl. 26, 129-146 (1976).
3. Uyaniker, B., Fürst, E., Reich, W., Reich, P., and Wielebinski, R., Astron. Astrophys. Suppl. 138, 31-45 (1999).
4. Duncan, A.R., Reich, P., Reich, W., and Fürst, E., Astron. Astrophys. 350, 447-456 (1999).
5. Cortiglioni S. et al. AMiBA 2001: High-z Clusters, Missing Baryons, and CMB Polarization. ASP Conference Series (in press).
6. Razin, V.A., Khrulev, V.V., Fedorov, V.T., et.al., Radiophysics and Quantum Ellectronics (translated from Russian) 11, 824-829 (1968).
7. Kühr, H., Witzel, A., Pauliny-Toth, I.I.K., Nauber, U., Astron. and Astrophys. Suppl. 45, 367-430 (1981).
8. Tabara, H., Inoue, M., Astron. and Astrophys. Suppl. 39, 379-393 (1980).
9. Han, J.L., Manchester, R.N., Berkhuijsen E.M., Beck, R., Astron. Astrophys. 322, 98-102 (1997).

Radio Polarimetry: A Historical Development at Effelsberg

R. Wielebinski, O. Lochner, W. Reich, H. Mattes

Max-Planck-Institut für Radioastronomie, Auf dem Hügel 69, 53121 Bonn, Germany

Abstract. To measure radio polarization one needs to analyse the incoming wave by means of a polarimeter. This is the crucial device that allows to determine the Stokes parameters after abstraction of the incoming wave in two orthogonal planes. The development of polarimeters in Effelsberg over the past 30 years is described in the present paper.

THE BACKGROUND

The need to measure polarization in radio astronomy comes from the wish to understand the emission mechanisms and to delineate magnetic fields in the interstellar regions. The radio continuum emission due to synchrotron radiation should be up to 75% linearly polarized. This linearly polarized emission suffers a Faraday rotation on passage through the interstellar medium. This is illustrated in Figure 1. Radio astronomy methods have been crucial in giving us information that is not available in other spectral ranges.

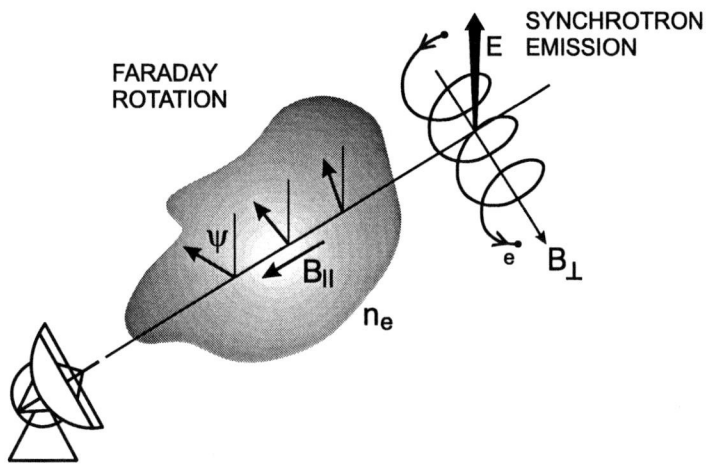

FIGURE 1. Situation in the interstellar medium

THE HISTORY

Radio polarization of the Crab nebula was detected by Mayer et al. [1] implying that the synchrotron emission process was responsible for the radio continuum. Radio polarization of the Galactic emission has been detected first by Westerhout et al. [2] and Wielebinski et al. [3]. Ionospheric Faraday rotation was reported by Wielebinski & Shakeshaft [4]. Galactic Faraday rotation was measured by Muller et al. [5]. Radio source polarization, detected by Mayer et al. [6], lead to studies of Rotation Measures. In those early days several methods of measurement of polarization were developed. The early method of switching between orthogonal dipoles or using two different bandwidths were superseded by correlation methods. These correlation methods are still in use in present-day polarimeters.

CP609, *Astrophysical Polarized Backgrounds*, edited by S. Cecchini et al.

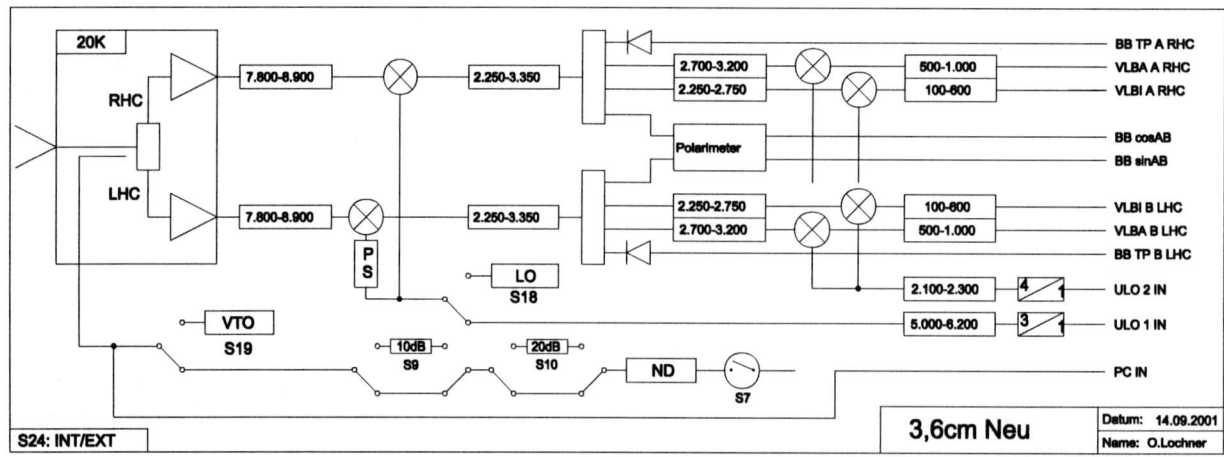

FIGURE 2. Block diagram of a polarization front-end

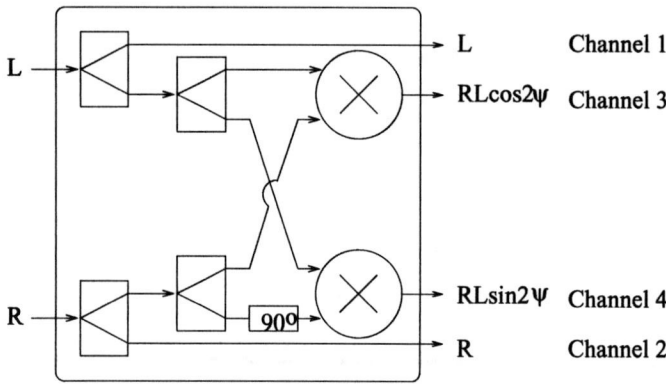

FIGURE 3. Block diagram of a multiplying polarimeter

THE BASIC POLARIMETERS

From an optimized polarization feed either two orthogonal linear polarization signals or two circular polarizations (RHC and LHC) can be coupled out. Great care is needed in the feed design to ensure minimal polarization cross-coupling. In Effelsberg all polarimeters use the circular polarization scheme (Figure 2) leading, after two Low Noise Amplifiers (LNA's), to an IF polarimeter (Figure 3). The reason for the use of the circular polarizations is the fact that this way the Stokes parameters Q and U, that define linear polarization, are derived from identical multiplying correlators and hence give minimal spurious signals. Before the Orthogonal Mode Transducer a calibration signal must be coupled in at 45° to allow equal matching of the two signal paths. A phase shifter is used in one of the Local Oscillator paths to allow phase equalization. The IF polarimeters vary in type depending on frequency. The wish to make sensitive polarization observations led to the development of polarimeters with wide frequency bands. In particular the development of the 90° phase-shift circuits with sufficient bandwidth was a problem of the early developments. The Effelsberg radio telescope, with its inherent axial symmetry, is one of the best polarization measuring systems.

THE EFFELSBERG POLARIMETERS

There have been 30 years of development of polarimeters in Effelsberg. The first polarimeters were narrow-band units matched to the then available front ends. The bandwidth has been increased with time to allow sensitive measurements at high radio frequencies. More recent developments have followed the direction of increasing the bandwidth for the

TABLE 1. Polarimeters at Effelsberg.

Centre frequency [MHz]	Bandwidth [MHz]	Channel	Type*	Developed
150	160	2	mult	1974
350	500	2	mult	1980
150	200	2	mult	1981
3000	2000	2	ph dis	1983
150	4	2×8	mult	2001
150	10	2×8	mult	2002
750	500	2	ph dis	2001
32000	4000	2	pdmic	2002

*mult: multiplier; ph dis: phase discriminator; phase discriminator in MIC technology

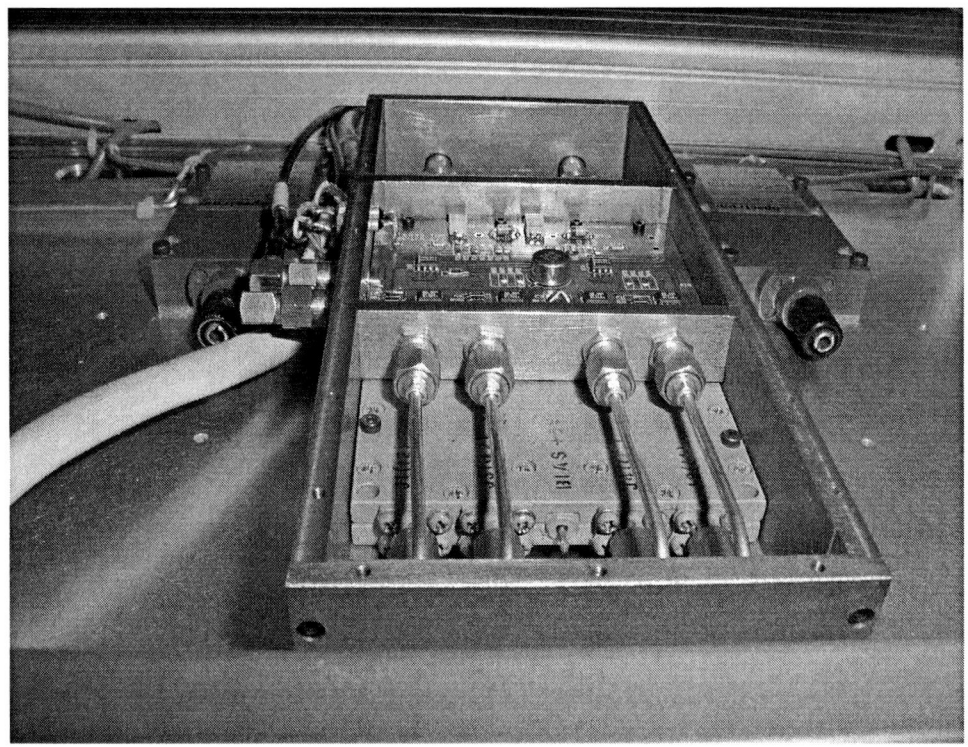

FIGURE 4. A broadband ($\Delta f = 2000$ MHz) polarimeter (see Table 1)

highest radio frequencies but also of providing multi-band polarimeters. These multi-band devices are designed to measure Rotation Measure but also to allow interference rejection that may occur in individual channels. Table 1 gives an overview of all the systems available in Effelsberg.

OBSERVATIONAL RESULTS

A large number of papers showing the polarization of the radio continuum emission have been published. Further current projects are nearing completion. A short description of results follows.

21 cm   a medium latitude survey of the Galaxy with $b = \pm 20°$
          selected areas, also at high Galactic latitudes (Uyaniker et al. [7])
11 cm   a survey of the Galactic plane with $b = \pm 5°$ and $74° > \ell > 5°$ (Duncan et al. [8])
          selected areas, SNRs, nearby galaxies

FIGURE 5. An 8-channel polarimeter ($\Delta f = 4$ MHz) (see Table 1)

6 cm   selected areas, SNRs, nearby galaxies, clusters of galaxies
3 cm   SNRs, nearby galaxies, radio galaxies, clusters of galaxies, Galactic centre (Beck [9], Wielebinski [10])
9 mm   SNRs, Galactic centre (Reich et al. [11], Lesch & Reich [12])

REFERENCES

1. Mayer, C. H., McCullough, T. P., and Sloanaker, R. M., *Astrophys. J.*, **126**, 468 (1957).
2. Westerhout, G., Seeger, Ch. L., Brouw, W. N., and Tinbergen, J., *Bull. Astron. Inst. Netherlands*, **16**, 187 (1962).
3. Wielebinski, R., Shakeshaft, J. R., and Pauliny-Toth, I. I. K., *The Observatory*, **82**, 158–164 (1962).
4. Wielebinski, R., and Shakeshaft, J. R., *Nature*, **195**, 982 (1962).
5. Muller, C. A., Berkhuijsen, E. M., Brouw, W. N., and Tinbergen, J., *Nature*, **200**, 155 (1963).
6. Mayer, C. H., McCullough, T. P., and Sloanaker, R. M., *Astrophys. J.*, **135**, 656 (1962).
7. Uyanıker, B., Fürst, E., Reich, W., Reich, P., and Wielebinski, R., *Astron. Astrophys. Suppl.*, **138**, 31 (1999).
8. Duncan, A. R., Reich, P., Reich, W., and Fürst, E., *Astron. Astrophys.*, **350**, 447 (1999).
9. Beck, R., *Phil. Trans. R. Soc. Lond. A*, **385**, 777 (2000).
10. Wielebinski, R., in *Encyclopedia of Astronomy and Astrophysics*, Vol. 3, edited by P. Murdin, Institute of Physics Publishing, Bristol, 2001, pp. 1947–1955.
11. Reich, W., Fürst, E., and Kothes, R., *Mem. Soc. Astron. Italiana*, **69**, 933 (1998).
12. Lesch, H., and Reich, W., *Astron. Astrophys.*, **264**, 493 (1992).

Workshop Summary and the Future

Gianfranco De Zotti

Osservatorio Astronomico di Padova, Vicolo dell'Osservatorio 5, I-35122 Padova, Italy

Abstract. We present a tentative summary of the many very interesting issues that have been addresses at this workshop, focussing in particular on the perspectives for measuring the polarization power spectra of the Cosmic Microwave Background produced by scalar and tensor perturbations, in the presence of foregrounds.

INTRODUCTION

The astonishing advances in our understanding of the basic properties of the Universe and in precision determinations of its fundamental parameters made possible by the recent accurate measurements of acoustic peaks of the Cosmic Microwave Background (CMB) anisotropy power spectrum by the TOCO [38], Boomerang [7, 8], Maxima [21], DASI [20], and CBI [40] experiments, have strongly highlighted the extraordinary wealth of cosmological information carried by the CMB. As the ongoing MAP mission and the forthcoming PLANCK satellite will provide high sensitivity, high resolution all sky CMB temperature maps the new frontier has become CMB polarization, first investigated by [46] and [26].

The information content of the CMB polarization field is in fact richer than that of the temperature field since, in addition to amplitude it also has an orientation. Recent analyses [47, 30, 58] have shown that, rather than with the conventional Stokes parameters Q and U (the Stokes parameter V describing circular polarization can generally be neglected because it cannot be generated through Thomson scattering), it is more convenient to work with the two rotationally invariant fields E and B, which are linear, but non-local, combinations of Q and U. This decomposition of the 2×2 symmetric trace-free tensor describing the linear polarization state is analogous to that of a vector field into a curl and a curl-free (gradient) component (e.g. [27]).

If the CMB fluctuations are Gaussian, they can be fully characterized by four power spectra, $C_\ell^{TT}, C_\ell^{EE}, C_\ell^{BB}, C_\ell^{TE}$. Under reflections (parity transformations) E transforms as a scalar (like T), while B transforms as a pseudo-scalar; therefore the cross-correlations TB and EB vanish, as discussed by *Sahzin*. The only fact that the temperature field yields just the first of those power spectra indicates how CMB polarimetry increases the cosmological information. But there is much more than that: scalar perturbations, held responsible for the acoustic peaks in the temperature power spectrum, do not produce, to first order, B-mode signal. The B-mode component may contain the signature of tensor perturbations (gravity waves) produced during the inflationary epoch and allow to measure the inflaton potential height[1] (see, e.g. [29]). Its detection would therefore give us a glimpse of the Universe at 10^{-38} seconds after the initial singularity, at energy scales of $\sim 10^{16}$ GeV, many orders of magnitude above those accessible at accelerators, and would therefore have profound implications not only for cosmology but also for particle physics [41].

Measurements of CMB polarization are however extremely challenging because the signal is very weak (at several μK level on small angular scales and much less on large angular scales) and liable to be strongly contaminated by polarized foreground emissions. The definition of suitable strategies therefore requires a close collaboration between experiment builders, theorists, experts on the various relevant Galactic and extra-galactic foregrounds. This very timely workshop has offered a badly needed opportunity for experts in the different fields to meet.

---

[1] A B component can also be generated by vector perturbations that may be excited by topological-defect models but are not present in inflationary models.

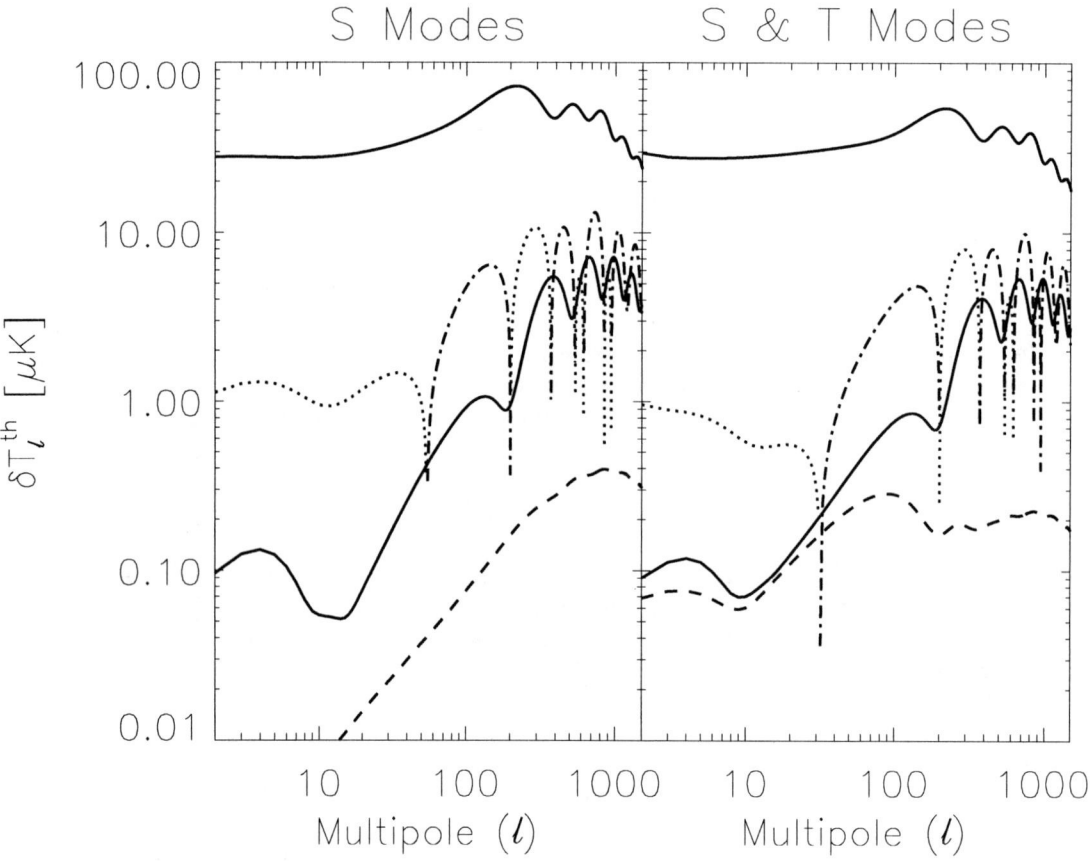

FIGURE 1. Temperature (upper solid line) and polarization power spectra for the cosmological model specified in the text. The panel on the left corresponds to pure scalar perturbations, the one on the right to scalar plus tensor perturbations yielding equal contributions to the temperature quadrupole. The partly dotted, partly dot-dashed lines represents the TE cross-correlation; the dot-dashed portion shows the absolute value of the cross-correlation when it is negative. The lower solid lines and the dashed lines represent E and B power spectra, respectively. In the case of scalar perturbations, a B contribution is originated by gravitational lensing. In the case of scalar plus tensor perturbations, gravitational lensing adds a contribution showing up at large ℓ.

CMB POLARIZATION POWER SPECTRA

CMB polarization is induced by Thomson scattering of anisotropic radiation with a quadrupole pattern in the rest frame of the electron (see e.g. [24]). Before recombination, anisotropies were strongly damped out by the tight coupling between photons and baryons, so that the polarization that could be generated was correspondingly depressed. To generate a quadrupole, a gradient in the velocity of the photon-baryon fluid across the photon mean free path, λ_p, is necessary: only perturbations on scales small enough to produce anisotropies on scales $< \lambda_p$ can give rise to polarization. But on small scales perturbations are damped by photon diffusion. As recombination proceeds, λ_p increases rapidly and polarization is produced. The polarization degree increases with decreasing angular scale (increasing ℓ) reaching a maximum of $\sim 10\%$ on the minimum scales that survived damping prior to recombination [23].

Figure 1 shows simulated temperature, polarization and TE correlation power spectra (discussed by *Balbi*), in terms of thermodynamic temperature fluctuations per logarithmic interval in ℓ, $\delta T = [\ell(2\ell+1)C_\ell/4\pi]^{1/2}$, for a flat ΛCDM cosmology with $\Omega_{\text{baryon}} = 0.03$, $\Omega_{\text{CDM}} = 0.3$, $\Omega_\Lambda = 0.67$, $H_0 = 65\,\text{km}\,\text{s}^{-1}\,\text{Mpc}^{-1}$, 3 species of massless neutrinos, $Y_{\text{He}} = 0.24$ and a re-ionization optical depth $\tau = 0.04$. Calculations have been made using the CMBFAST code by [48]. The panels show the cases of purely scalar, scale invariant, isentropic primordial perturbations (left), and of scalar plus tensor perturbations with a unit ratio of tensor (T) to scalar (S) contributions to the temperature quadrupole (right). A $T/S \simeq 1$ is probably an upper limit, once all the relevant constraints are taken into account [61].

As illustrated by Fig. 1, the (E-mode) polarization fluctuations associated to acoustic peaks have relatively more power on small scales than temperature fluctuations. Also, since polarization is related to peculiar velocities on the last scattering surface, the peaks of its power spectrum are out of phase by $\pi/2$ with those of temperature anisotropies (velocities are out of phase with density perturbations, like velocity and position of a harmonic oscillator). The TE power spectrum, being the product of the two, oscillates at twice the frequency. The relative position of the temperature and polarization peaks is thus the signature of coherent perturbations (as opposed to causal mechanisms, e.g. those due to topological defects), hence of inflation, as the origin of structure. If large scale structure was produced by some causal mechanism, the first peak would have to occur at smaller angular scales in order to be within the causal horizon at last scattering [49]. Therefore, just the detection of the peaks of the strongest CMB polarization component, induced by scalar perturbations, will allow an unambiguous test on the nature of primordial perturbations.

A lot more can be learned from an accurate determination of the power spectrum of the E-mode on small angular scales, and, to some extent, to the T-E correlation power spectrum [60, 32, 13, 14, 4]. For example, polarization measurements: identify contributions from isocurvature perturbations that can be confused with tensor perturbations or early re-ionization effects on temperature data; constrain scalar-tensor theories of gravity which may generate scalar, vector and tensor modes, leaving distinctive signatures in the CMB; are sensitive to effects of quintessence models; increase significantly the precision of the determination of most cosmological parameters; are directly informative on the details of the recombination and, therefore, on processes governing it [39].

As mentioned above, the amplitude of the polarization power spectrum on large angular scales originated at recombination is expected to be very small because the photon mean free path is small and multiple scattering erase polarization. On the other hand, re-ionization strongly increases the polarization signal at low ℓ [57, 28] producing a characteristic bump (see Fig. 1) at $\ell \sim \sqrt{z_{ri}}$, z_{ri} being the redshift of re-ionization, with amplitude roughly proportional to the optical depth, τ for Thomson scattering:

$$\tau = 0.041 \frac{h\Omega_b}{\Omega_m} \left\{ \left[1 - \Omega_m + \Omega_m (1+z_{ri})^3 \right]^{1/2} - 1 \right\}, \tag{1}$$

Ω_b and Ω_m are the baryon and the matter density, respectively, in units of the critical density. The polarization power spectrum at low ℓ is therefore a sensitive probe of re-ionization.

The lack of an observed Lyα (Gunn-Peterson) trough in the spectrum of the SDSS quasar at $z = 5.8$ [17] implies, for the above choice of cosmological parameters, $\tau > 0.022$. Re-ionization, however, suppresses temperature fluctuations on small scales: the lack of observed suppression of the acoustic peaks gives $\tau \leq 0.2$ [56], implying $z_{ri} \leq 20$. Very recently [2] reported compelling evidence of a Gunn-Peterson trough in a $z = 6.28$ quasar discovered by the Sloan survey (SDSS), suggesting that $z_{ri} \simeq 6$ (see also [16]), implying τ close to the lower limit. Keck spectroscopy by [10] have shown evidence of dark portions of the spectrum of an SDSS quasar at $z = 5.73$, indicating a dramatic increase of the Lyα opacity at $z \geq 5.2$, as expected if we are approaching (from lower redshifts) the re-ionization era of a non-uniform medium, with ionized regions intermingled with islands of neutral gas. Clearly, it may be premature to draw very definite conclusions. Only a tiny fraction ($\sim 2 \times 10^{-5}$) of baryons need to be in the form of neutral hydrogen to account for the Gunn-Peterson optical depth implied by the re-ionization may be non-uniform, so that the line-of-sight where the effect was observed may be not representative.

On the other hand, given the rapid evolution of the optical depth that is observed, it seems unlikely that the effective re-ionization redshift is $\gg 6$. If so, the re-ionization bump shown in Fig. 1, which corresponds to $\tau = 0.04$, is probably an upper limit. CMB polarization fluctuations on large angular scales are thus probably at the $0.1\,\mu$K level and their detection is therefore very hard. Since the SDSS is expected to find ~ 20 quasars at $z \geq 6$ [15], more detailed insight into the re-ionization process is expected in the next few years.

The CMB polarization power spectrum provides additional information compared to observations of the Gunn-Peterson trough. While the latter measures just z_{ri}, the former also measures τ, in addition to allowing an estimate of z_{ri} (from the position of the peak), thus providing a determination of the baryon fraction in the intergalactic medium, hence, e.g., of the galaxy formation efficiency, a piece of information very difficult to derive with other means.

The specific signature of tensor and vector perturbations is B-type polarization, which cannot be produced by scalar perturbations because of their symmetry properties. This opens the exciting possibility of a direct investigation of tensor components, in spite of the fact that their contribution to the CMB polarization is expected to be much smaller than that of scalar perturbations. Since the amplitude of the B-mode component due to gravity waves is proportional to the square of the inflationary energy scale [58, 30], its detection amounts to a measurement of the energy scale of inflation!

For roughly scale invariant tensor perturbations the B-mode power spectrum peaks at $\ell \simeq 100$ (see Fig. 1). As pointed out by [59], gravitational lensing by the matter distribution, in addition to smoothing the acoustic peaks of both temperature and polarization, distorts the polarization pattern on the sky, mixing E and B modes. B-type polarization

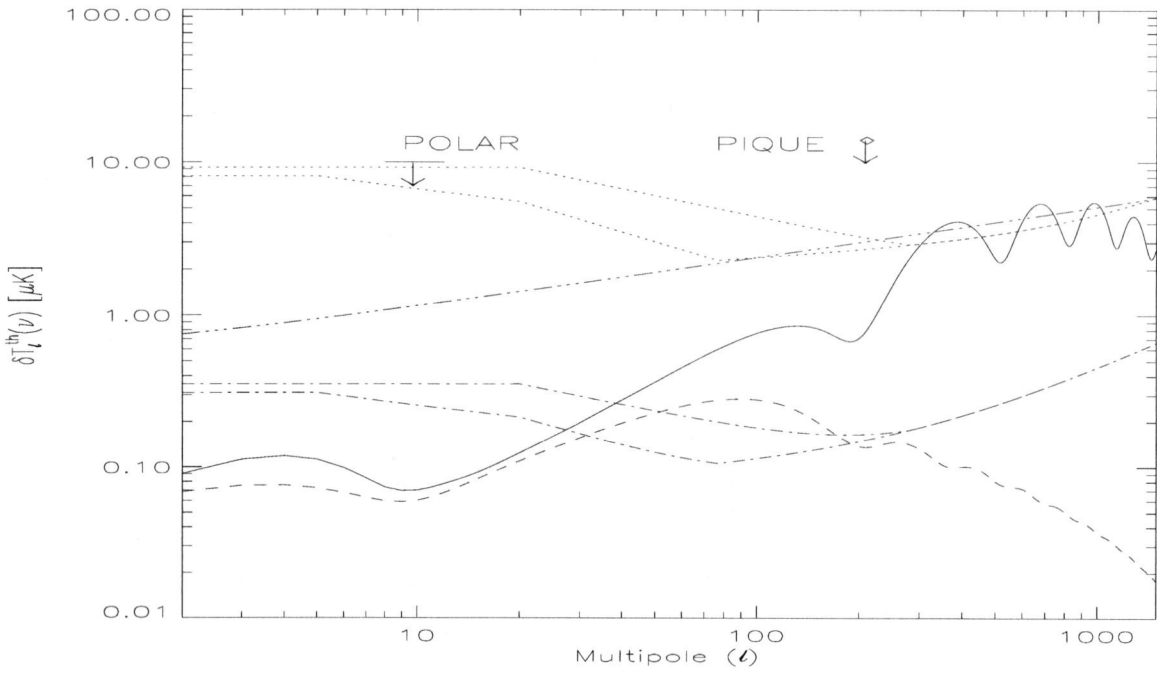

FIGURE 2. CMB versus foreground polarization power spectra. For CMB we have plotted the scalar plus tensor model of Fig. 1, but without gravitational lensing. The foreground power spectra are shown for 3 frequencies: 30 (dots), 100 (dot-dash), and 217 (three dots - dash) GHz. They comprise the contributions of Galactic synchrotron and thermal dust emission, and of extragalactic sources (radio sources and dusty galaxies); polarization fluctuations of spinning or magnetized grains are not included. At 30 and 100 GHz, the dominant polarized foregrounds are synchrotron for $\ell \leq 1000$ (the limit actually decreases with increasing frequency) and extragalactic radio sources at higher ℓ. The two lines at these frequencies for $\ell \leq 200$ bound the range of synchrotron power spectra reported by *Burigana* at this meeting; the synchrotron power spectrum for higher ℓ is from [1]. For an independent estimate of the synchrotron power spectrum see [54] and the contribution by *Tucci*. At higher frequencies, the dominant polarized foreground is expected to be thermal dust emission, as illustrated by the line at 217 GHz. We have assumed that the global net polarization degree of dusty galaxies is 0.4%, as found by *Greaves* (these proceedings) for M82; if so their contribution to polarization fluctuations can be significant only on very small angular scales. The E and B mode power spectra for these foreground components are found to be almost equal; only one of them is plotted. The recent upper limits set by the POLAR [31] and the PIQUE [22] experiments are also shown.

is thus generated even in the case of pure scalar perturbations. The effect peaks on small angular scales ($\ell \sim 1000$) and does not interfere with the possibility of measuring the B-mode power spectrum from gravity waves.

Of course, gravity waves are not the only primary sources of B-mode polarization. Primordial magnetic fields (*Wielebinski*) would also produce tensor and vector metric perturbations, resulting in further CMB fluctuations [37, 25]. A distinctive feature of such fields is that, for a range of power spectra of perturbations produced by them, the polarization fluctuations are comparable to, or larger than, the corresponding temperature fluctuations. The amplitude of the CMB power spectra vary as the square of the energy density, i.e. as the 4th power of the magnetic field amplitude; the effects on the CMB of primordial stochastic magnetic fields with comoving amplitude much below 10^{-9} Gauss are essentially undetectable. Cosmological magnetic fields may also be detectable through the (frequency dependent) Faraday rotation of CMB polarization. Faraday rotation also mixes E and B modes. These effects are however small for realistic magnetic field strengths [34, 25].

POLARIZED FOREGROUNDS

Foregrounds may be a more serious hindrance for CMB polarization measurements than for temperature measurements because at least some of them are more polarized than the CMB. At the moment they are very poorly understood, so that it is still unclear to what extent they will limit our ability to measure CMB polarization. Particularly uncertain is

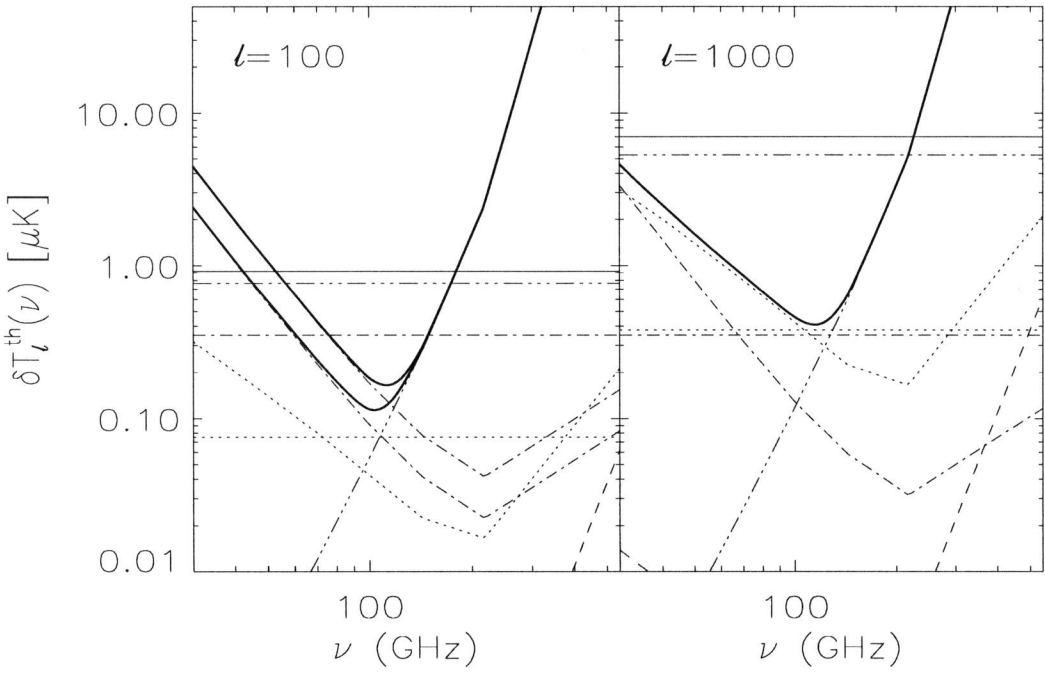

FIGURE 3. Frequency dependence of foreground polarization fluctuations for $\ell = 100$ (left) and $\ell = 1000$ (right). The dot-dashed curves represent Galactic synchrotron. The two such curves shown in the left-hand panel bracket the estimates reported by *Burigana* (see caption to Fig. 2). The synchrotron curve on the right-hand panel is from [1]. The three-dots/dash curve is for thermal dust emission [45], while the dotted curve is for extragalactic radiosources [9], which, for $\nu \leq 100\,\mathrm{GHz}$, dominate foreground fluctuations on small angular scales, as illustrated by the $\ell = 1000$ panel. The dashed line in the lower right-hand corner of each panel is for dusty galaxies, assumed to have a net polarization degree of 0.4%. The solid curves show the sum of the various foreground contributions. For $\ell = 100$, and in general for $\ell \leq 200$, and $\nu \leq 100\,\mathrm{GHz}$, the two solid lines reflect the range of estimates for the dominant synchrotron contributions. The horizontal lines show the CMB polarization fluctuations for the cases shown in Fig. 1, namely: *E*-mode for pure scalar fluctuations (solid), *E*-mode for pure scalar plus tensor fluctuations (three dots/dash), *B*-mode from the tensor component in the case of scalar plus tensor fluctuations (dot/dash), *B*-mode due to gravitational lensing for pure scalar fluctuations (dotted).

the polarized emission from thermal [45], spinning and ferro- or ferri-magnetic dust grains [11]. Much more work is needed in this area.

Recent analyses of polarization fluctuations due to the various foregrounds have been presented and discussed by several speakers (*Baccigalupi, Burigana, Fosalba, Lazarian, Ponthieu, Prunet, Sazhin, Tucci*). Figures 2 and 3 attempt to summarize the main conclusions. At frequencies $\leq 100\,\mathrm{GHz}$ foreground polarization fluctuations are likely dominated by Galactic synchrotron emission on large angular scales and up to $\ell \sim 1000$ (the limiting value of ℓ actually decreases with increasing frequency) and by extragalactic radio sources on small scales. Note that synchrotron emission seems to have much more small-scale structure in polarization than in total intensity, at least in the relatively low-frequency maps currently available. Such structure may be due, to some extent, to differential Faraday rotation, and might therefore not be present in high frequency maps. Spinning dust grains may be important contributors particularly at 20–30 GHz, while their polarized emission is probably negligible above 40 GHz. Polarized microwave emission from magnetic grains may be important even if they do not dominate the overall emission.

At higher frequencies, the main polarized foreground is expected to be, on all scales, thermal dust emission [45]. Polarization surveys at (sub)-mm wavelengths, however, cover very small regions, mostly on the Galactic plane. On the other hand, very interesting statistical information on polarized dust emission can be derived from data on starlight polarization due to selective absorption by magnetically aligned dust grains [18]. *Fosalba* estimates a typical polarization degree of Galactic dust of about 2% and a power spectrum $C_\ell \propto \ell^{-1.5}$.

The (still scanty) data on sub-mm polarization of dusty galaxies (*Greaves, Scott*) indicate that their global net polarization, is likely < 1%. Polarization maps of the nearest starburst galaxy, M82 [19] show values from 0.7% to

9.1%. However, when the contributions of the differently oriented magnetic vectors are summed together, the global net polarization is found to be $\simeq 0.4\%$ (*Greaves*).

Figure 2 also suggests that the POLAR experiment [31], at 30 GHz, is already close to detecting synchrotron polarization fluctuations. There is apparently little hope of detecting large scale CMB polarization fluctuations at this frequency, and even at higher frequencies, a delicate subtraction of the synchrotron emission will be probably necessary. Such subtraction is further complicated by the substantial variations of the synchrotron spectral index across the sky [43].

As illustrated by Fig. 3, the foreground contamination is expected to be minimum at frequencies $\simeq 120\,\mathrm{GHz}$. In the optimal frequency window, foregrounds should not seriously limit our ability to measure the E-mode power spectrum except for large ($\ell \leq 40$) and very small ($\ell \geq 2000$) angular scales.

As pointed out by [47], in most cases foregrounds yield essentially equal contributions to the E and B power spectra (see also the contribution by *Sazhin*). This is obviously the case for uncorrelated point sources. In the case of Galactic synchrotron and dust emission, the alignment is primarily determined by magnetic fields, which are not scalar in nature. The analysis of low-frequency Galactic polarization surveys and of extragalactic radio sources, presented by *Baccigalupi*, indeed yields essentially equal E- and B-mode power spectra in both cases. Since, the CMB polarization is, on small angular scales, predominantly E-mode, the difference between E and B power spectra may be a direct measure of the CMB signal.

The situation is clearly much more difficult for the B-mode, which, even in the optimal frequency range, appears to dominate foreground fluctuations only if the amplitude of the tensor component is close to current upper limits, and only over a rather narrow range of multipoles. Ongoing measurements and the forthcoming multifrequency maps provided by MAP and PLANCK will be essential to design future high sensitivity CMB polarization experiments.

Astrophysical information from polarization surveys

Clearly, the study of the foreground sources mentioned above is of great astrophysical interest per se, since it will provide crucial information on their physical properties. Still in the framework of CMB experiments, as noted above, some polarized astrophysical sources (Galactic synchrotron, extragalactic radio sources) may be more easily extracted from polarization than from temperature maps because their polarization degree may be higher than that of the CMB. This is particularly true for B-mode maps: CMB fluctuations are very small even in the "cosmological window" (50–200 GHz) where they dominate by far temperature maps.

High resolution, high sensitivity polarization surveys of the Galaxy are becoming available (see contributions by *Reich, Gaensler, Landecker, Sault, Vinyakin*), with particularly detailed imaging of the Galactic plane. These maps effectively amount to a tomography of the ISM, to quote *Wielebinski*, and allow to investigate the ecosystem of the Galaxy, in *Landecker*'s words. In fact these observations are informative on the large scale structure of the Galactic magnetic field, its structure in the solar neighborhood in relation with ISM features (*Haverkorn*) (cloud complexes, bubbles, loops, shells, SN remnants, etc.), its z-height structure and the relation between the magnetic field within the thin ISM disk and the thicker synchrotron emitting disk, the relation between the magnetic field within interstellar clouds and their density and velocity structure, the role of the magnetic field in regulating the star formation efficiency, dust properties and dust alignment mechanisms (from polarization of dust emission). Our current understanding of the large scale structure of the Galactic magnetic fields has been reviewed by *Han*.

The available information on polarization properties of extragalactic sources is still very limited particularly at sub-mm wavelengths (*Scott, Greaves*), but not much is available also for radio sources at $\geq 10\,\mathrm{GHz}$. There is obviously a lot of astrophysics to be learned from polarization measurements, particularly at mm/sub-mm wavelengths where the effects of internal synchrotron self-absorption and of Faraday rotation can be (with rare exceptions) ignored so that we can reliably assume that the magnetic field direction lies perpendicular to the observed polarization position angle.

Another tiny signal with a large information content is the polarized component of the Sunyaev-Zeldovich effect [51]. It will be important to assess the effect of other polarized sources either in the cluster (radio sources, radio halos – μGauss magnetic fields and relativistic particles seem to be ubiquitous in clusters, as discussed by *Wielebinski* –, dust) or along the line of sight.

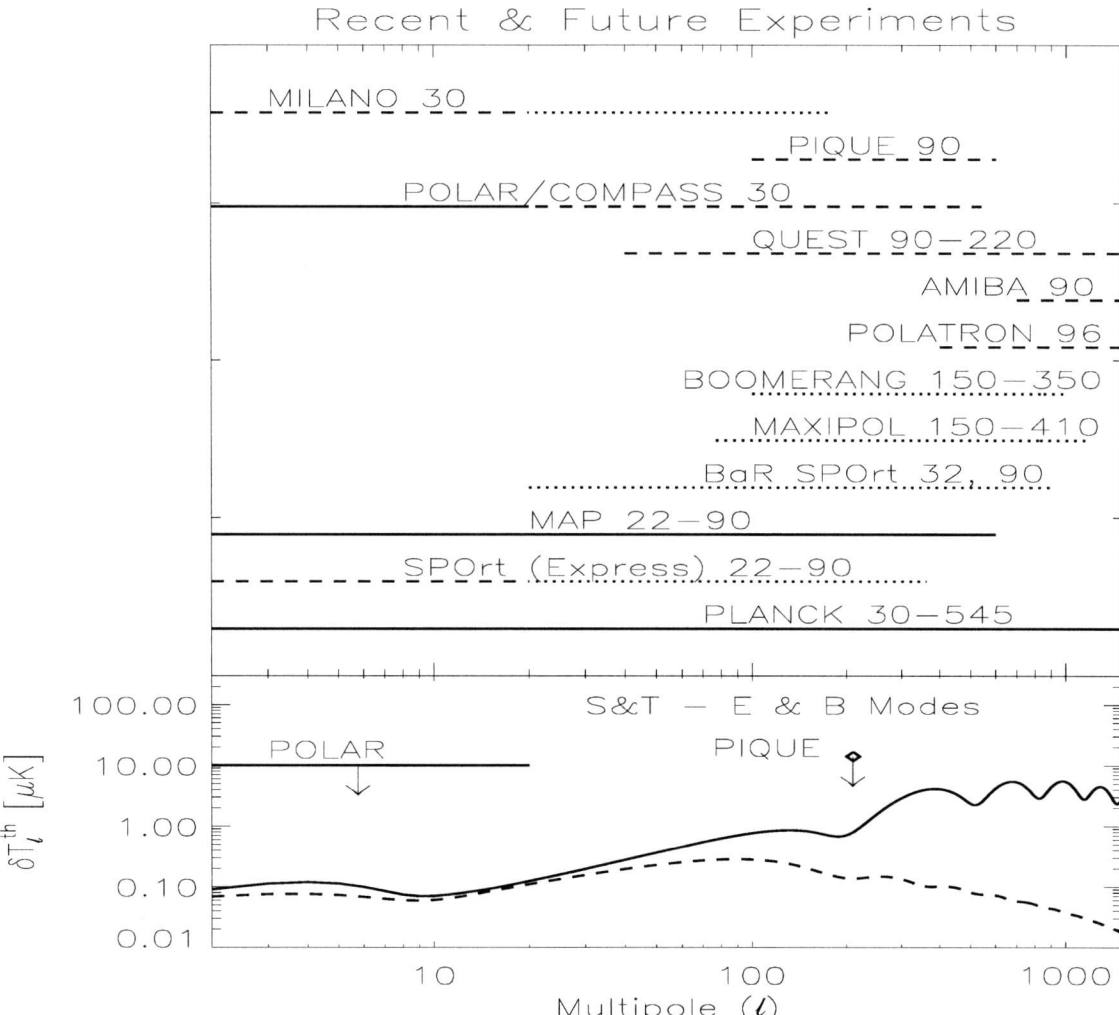

FIGURE 4. Summary of CMB polarization experiments discussed at this meeting. The straight lines show the multipole coverage. Note that the y-coordinate of such lines has nothing to do with the sensitivity of the instruments. Ground-based experiments are on top, balloon experiments in the middle and space-borne experiments on the bottom. The figures close to each experiment give the observing frequency or the frequency interval covered. The E- (solid) and B-mode CMB power spectra (shown for reference) are for the same model as in Fig. 2.

CMB POLARIZATION EXPERIMENTS

An excellent review of ongoing and planned CMB polarization experiments is given by [50]. Figure 4 summarizes those discussed at this meeting, indicating the multipole range they cover and the frequency bands they probe.

Although recent experiments (PIQUE [22] operating at 90 GHz, and POLAR [31] at 30 GHz) have considerably improved on previous results, CMB polarization is still undetected. *Timbie*, the upgrade of POLAR, called COMPASS, which can reach angular scales of $20'$, has the potential of detecting the CMB polarization if its sensitivity per pixel will be similar to POLAR. CMB polarization measurements are expected in the next few years from several other experiments. The NASA Microwave Anisotropy Probe (MAP) is expected to obtain a highly significant detection of the CMB temperature-polarization cross-correlation, which is less affected by foregrounds than polarization maps [33]. Detection of the E-mode polarization is also expected from the new flights of the balloon-borne experiments, the bolometer-based, now with polarization capabilities, BOOMERANG (*de Bernardis*) and MAXIMA (MAXIMAPOL), and BAR-SPort (*Zannoni, Macculi*), using radiometers. Very promising ground based experiments include QUEST (*Piccirillo*) and Polatron [42], both using bolometric receivers, and the interferometer array AMiBA (*Kesteven*, [36]).

Measuring polarization on large angular scales (cf. the Milano experiment, presented by *Gervasi*, and the SPOrt project, due to fly on the International Space Station, presented by *Carretti* and *Nicastro*) appears to be much more difficult. The SPOrt-Express project, proposed for reuse of the Mars Express platform, adds to the SPOrt payload a second instrument, HARI, equipped with a 50cm antenna, giving an angular resolution $\simeq 0.5°$ at 90 GHz.

PLANCK (see presentations by *Delabrouille* and *Villa*) should provide accurate measurements of the E-mode power spectrum up to $\ell \simeq 1500$–2000; its sensitivity would allow to detect B-mode polarization if the ratio of tensor to scalar contributions to the quadrupole anisotropy is $T/S \geq 0.045$. Clearly, even if foreground contamination will not prevent the experiment from reaching these levels, detection of B-mode polarization is by no means guaranteed! Much higher sensitivities are needed to explore the inflationary parameter space [27].

As discussed by *Kaplan* and *Leahy*, systematic errors may be most critical. There are a number of potential sources of systematic errors that could affect the various experiments: foreground subtraction, beam asymmetries, straylight, electronic interference, thermal variations, etc.. Experiments measuring polarization by differencing the outputs from radiometers sensitive to orthogonal polarization have common systematics for intensity and polarization measurements, although some cancellation can be expected in polarization (*Leahy*). The need of combining data from different detectors introduces effects of calibration, pointing, beam shape mismatches (*Kaplan*). Issues related to the removal of SPOrt systematics have been discussed by *Amisano*. Since the number of physically interesting parameters is much smaller than the number of multipoles measured in the power spectrum (typically ~ 10 parameters, compared with ~ 2000 multipoles in the case of PLANCK), very small correlated systematic errors which mimic the ℓ-dependence of a parameter, can produce a large error in that parameter [12].

The sky coverage is also an issue. Large area maps are important to maximize the information from the power spectrum (maps of a patch of angular size θ lose the information from modes with $\ell < 180°/\theta$), to minimize the sample variance (which scales inversely with the sky coverage), to minimize the mixing between B and E caused by incomplete sky coverage (due to the non-local nature of these quantities), which may cause the B component to be swamped by the, typically much larger, E component [52, 5]: to detect the B component a survey size much larger than the coherence length of this component is required. On the other hand, given the extreme sensitivity required, very long exposures are needed to decrease the noise per pixel: with realistic integration times only relatively small patches can be covered. Optimal survey sizes may be of ~ 20–30°, with $\sigma_{beam} \simeq 0.3$–0.5° [27, 5, 35].

The flood of data expected from ongoing or forthcoming CMB experiments is facing us with huge computational challenges. *Balbi* has described an efficient method for constructing polarization maps. *Sbarra* has described a new destriping technique and discussed its applications with reference to the SPOrt experiment. The application of a multi-frequency Wiener filtering technique to polarized maps was investigated by [3] and [44]. The method, generalized by [53], requires the knowledge of the average frequency and angular dependence of the foreground emission. Other possibilities include pseudo-C_ℓ estimates [55] and harmonic analysis methods that bypass traditional map-based methods [6].

ACKNOWLEDGMENTS

I'm deeply indebted to C. Burigana who made all the figures and offered many very useful ideas and comments, and to C. Baccigalupi whose contributions substantially improved this paper. I'm also grateful to S. Cecchini and C. Sbarra for a careful reading of the manuscript and to the organizers of this very timely, useful and stimulating meeting. Work supported in part by ASI and MIUR.

REFERENCES

1. Baccigalupi, C., Burigana, C., Perrotta, F., et al., *A&A*, **372**, 8–21 (2001).
2. Becker, R.H., Fan, X., White, R.L., et al., *A. J.*, in press, astro-ph/0108097 (2001).
3. Bouchet, F.R., Prunet, S., Sethi, S.K., *MNRAS*, **302**, 663–676 (1999).
4. Bucher, M., Moodley, K., Turok, N., astro-ph/0012141 (2000).
5. Bunn, E.F., *Phys. Rev. D*, submitted, astro-ph/0108209 (2001).
6. Challinor, A., et al., in preparation (2001).
7. de Bernardis, P., Ade, P., Bock, J.J., Bond, J.R., Borrill, J,, et al., *Nature*, **404**, 955–959 (2000).
8. de Bernardis, P., Ade, P., Bock, J., Bond, J.R., Borrill, J., et al., *Ap. J.*, in press, astro-ph/0105296 (2001).
9. De Zotti, G., Gruppioni, C., Ciliegi, P., Burigana, C., Danese, L., *New Astr.*, **4**, 481–488 (1999).
10. Djorgovski, S.G., Castro, S., Stern, D., Mahabal, A.A., *Ap. J.*, **560**, L5–L8 (2001).
11. Draine, B.T., Lazarian, A., in *Microwave Foregrounds*, A. de Oliveira-Costa & M. Tegmark eds., ASP, **181**, 133–149 (1999).
12. Efstathiou, G., Bond, J.R., *MNRAS*, **304**, 75–97 (1999).

13. Eisenstein, D.J., Hu, W., Tegmark, M., *Ap. J.*, **518**, 2–23 (1999).
14. Enqvist, K., Kurki-Suonio, H., *Phys. Rev. D*, **61**, 043002 (2000).
15. Fan, X., Narayanan, V.K., Lupton, R.H., et al., *A. J.*, in press, astro-ph/0108063 (2001).
16. Fan, X., Narayanan, V.K., Strauss, M.A., White, R.L., Becker, R.H., Pentericci, L., Rix, H.-W., *A. J.*, submitted, astro-ph/0111184 (2001).
17. Fan, X., White, R.L., Davis, M., et al., *A. J.*, **120**, 1167–1174 (2000).
18. Fosalba, P., Lazarian, A., Prunet, S., Tauber, J.A., *Ap. J.*, in press, astro-ph/0105023 (2001).
19. Greaves, J.S., Holland, W.S., Jenness, T., Hawarden, T.G., *Nature*, **404**, 732–733 (2000).
20. Halverson, N.W., et al.,*Ap. J.*, in press, astro-ph/0104489 (2001).
21. Hanany, S., Ade, P., Balbi, A., Bock, J., Borrill, J., et al., *Ap. J.*, **545**, L5–L9 (2000)
22. Hedman, M.M., Barkats, D., Gundersen, J.O., Staggs, S.T., Winstein, B., *Ap. J.*, **548**, L111–L114 (2001).
23. Hu, W., Dodelson, S., *Ann. Rev. Astron. Ap. 2002*, in press, astro-ph/0110414 (2001).
24. Hu, W., White, M., *Ap. J.*, **479**, 568–579 (1997).
25. Kahniashvili, T., Kosowsky, A., Mack, A., Durrer, R., in proc. *Cosmology and Particle Physics*, eds. J. Garcia-Pulido et al., astro-ph/0011095 (2000).
26. Kaiser, N., *MNRAS*, **202**, 1169–1180 (1983).
27. Kamionkowski, M., Jaffe, A.H., proc. *DPF2000*, Columbus, in press, astro-ph/0011329 (2000).
28. Kamionkowski, M., Kosowsky, A., *Phys. Rev. D*, **67**, 685–691 (1998).
29. Kamionkowski, M., Kosowsky, A., *Ann. Rev. Nucl. Part. Sci.*, **49**, 77–123 (1999).
30. Kamionkowski, M., Kosowsky, A., Stebbins, A., *Phys. Rev. D*, **55**, 7368–7388 (1997).
31. Keating, B.G., O'Dell, C.W., de Oliveira-Costa, A., et al., *Ap. J.*, **560**, L1–L4 (2001).
32. Kinney, W.H., *Phys. Rev. D*, **58**, 123506 (1998).
33. Kogut, A., and Hinshaw, G., *Ap. J.*, **543**, 530–534 (2000).
34. Kosowsky, A., Loeb, A., *Ap. J.*, **469**, 1–6 (1996).
35. Lewis, A., Challinor, A., Turok, N., *Phys. Rev. D*, in press, astro-ph/0106536 (2001).
36. Lo, K.Y., Chiueh, T., Liang, H., et al., in *New Cosmological Data and Values of the Fundamental Parameters*, IAU Symp. 201
37. Mack, A., Kahniashvili, T., Kosowsky, A., *Phys. Rev. D*, submitted, astro-ph/0105504 (2001).
38. Miller, A.D., Caldwell, R., Devlin, M.J., Dorwart, W.B., Herbig, T., et al., *Ap. J.*, **524**, L1–L4 (1999).
39. Naselsky, P., Schmalzing, J., Sommer-Larsen, J., Hannestad, S., *MNRAS*, submitted, astro-ph/0102378 (2001).
40. Padin, S., Cartwright, J.K., Mason, B.S., Pearson, T.J., Readhead, A.C.S., et al., *Ap. J.*, **549**, L1–L5 (2001).
41. Peterson, J.B., Calstrom, J.E., Cheng, E.S., et al., astro-ph/9907276 (1999).
42. Philhour, B.J., Keating, B.G., Ade, P.A.R., et al., astro-ph/0106543 (2001).
43. Platania, P., Bensadoun, M., Bersanelli, M., de Amici, G., Kogut, A., Levin, S., Maino, D., Smoot, G.F., *Ap. J.*, **505**, 473–483 (1998).
44. Prunet, S., Sethi, S.K., Bouchet, F.R., *MNRAS*, **314**, 348–353 (2000).
45. Prunet, S., Sethi, S.K., Bouchet, F.R., Miville-Deschenes, M.-A., *A&A*, **339**, 187–193 (1998).
46. Rees, M., *Ap. J.*, **153**, L1–L5 (2001).
47. Seljak, U., *Ap. J.*, **482**, 6–16 (1997).
48. Seljak, U., Zaldarriaga, M., *Ap. J.*, **469**, 437–444 (1996).
49. Spergel, D.N., Zaldarriaga, M., *Phys. Rev. Lett.*, **79**, 2180-2183 (1997).
50. Staggs, S.T., Gundersen, J.O., Church, S.E., in *Microwave Foregrounds*, A. de Oliveira-Costa & M. Tegmark eds., ASP, **181**, 299–309 (1999).
51. Sunyaev, R.A., Zeldovich, Ya.B., *MNRAS*, **190**, 413–420 (1980).
52. Tegmark, M., and de Oliveira-Costa, A., *Phys. Rev. D*, **64**, 063001–063015 (2001).
53. Tegmark, M., Eisenstein, D.J., Hu, W., de Oliveira-Costa, A., 2000, *Ap. J.*, **530**, 133–165 (2000).
54. Tucci, M., Carretti, E., Cecchini, S., Fabbri, R., Orsini, M., Pierpaoli, E., *New Astr.*, **5**, 181–190 (2000). MPA/ESO/MPE conf. *Mining the Sky*, (2000).
55. Wandelt, B.D., Gorski, K.M., Hivon, E., in *Energy Densities in the Universe*, XXXV Rencontres de Moriond, astro-ph/0004178 (2000).
56. Wang, X., Tegmark, M., Zaldarriaga., M.*Phys. Rev. D*, in press, astro-ph/0105091 (2001).
57. Zaldarriaga, M., *Phys. Rev. D*, **55**, 1822–1829 (1997).
58. Zaldarriaga, M., Seljak, U., *Phys. Rev. D*, **55**, 1830–1840 (1997).
59. Zaldarriaga, M., Seljak, U., *Phys. Rev. D*, **58**, 023003 (1998).
60. Zaldarriaga, M., Spergel, D.N., Seljak, U., *Ap. J.*, **488**, 1–13 (1997).
61. Zibin, J.P., Scott, D., White, M., *Phys. Rev. D*, **60**, 123513 (1999).

LIST OF PARTICIPANTS

Amisano Franco
Alenia Spazio S.p.A.
Strada Antica di Collegno, 253
10146 Torino - Italy

Balbi Amedeo
Dip. Fisica Universita' Tor Vergata
Via della Ricerca, 1
00133 Roma - Italy

Baralis Massimo
IRITI - CNR
Corso Duca degli Abruzzi, 24
10129 Torino - Italy

Bignami Giovanni Francesco
Agenzia Spaziale Italiana
Via di Villa Grazioli, 23
00198 Roma - Italy

Bonometto Silvio A.
Dip. Fisica Univ. Milano Bicocca
Piazza della Scienza, 3
20126 Milano - Italy

Carretti Ettore
ITeSRE - CNR
Via Gobetti, 101
40129 Bologna -Italy

Cecchini Stefano
ITeSRE - CNR
Via Gobetti, 101
40129 Bologna - Italy

Cosentino Orazio
Base Lancio Palloni Stratosferici
Agenzia Spaziale Italiana
S.S. 113 Nr 174 C.da Milo
91100 Trapani - Italy

Baccigalupi Carlo
SISSA/ISAS
Via Beirut, 2-4
34014 Trieste - Italy

Baldemar Per
SSC Esrange
P.O. Box 802
SE-98128 Kiruna - Sweden

Bernardi Gianni
ITeSRE - CNR
Via Gobetti, 101
40129 Bologna - Italy

Boen Kjell
Andoya Rocket Range
P.O. Box 54
NO-8483 Andenes - Norway

Burigana Carlo
ITeSRE - CNR
Via Gobetti, 101
40129 Bologna - Italy

Cecchini Gian Piero
Agenzia Spaziale Italiana
Via di Villa Grazioli, 23
00198 Roma - Italy

Cortiglioni Stefano
ITeSRE - CNR
Via Gobetti, 101
40129 Bologna - Italy

De Bernardis Paolo
Dip. Fisica Univ. La Sapienza
Piazzale A. Moro, 2
00185 Roma - Italy

de Gasperis Giancarlo
Universita' di Roma Tor Vergata
Via della Ricerca Scientifica, 1
00133 Roma - Italy

De Zotti Gianfranco
Osservatorio Astronomico di Padova
Vicolo dell'Osservatorio, 5
35122 Padova - Italy

Fabbri Roberto
Dip. Fisica Univ. Firenze
Via Sansone, 1
50019 Sesto Fiorentino - Italy

Gervasi Massimo
Dip. Fisica Univ. Milano Bicocca
P.zza della Scienza, 3
20126 Milano - Italy

Han Jinlin
Chinese Academy of Science
Jia-20 DaTun Road
ChaoYang District
Bejing 100012, China

Ibba Roberto
Agenzia Spaziale Italiana
Via di Villa Grazioli, 23
00198 Roma - Italy

Kesteven Michael
Australia Telescope, CSIRO
P.O. Box 76
Epping, NSW, 1710
Australia

La Porta Laura
ITeSRE - CNR
Via Gobetti, 101
40129 Bologna - Italy

De Troia Grazia
Universita' di Roma La Sapienza
Piazzale A. Moro, 2
00185 Roma - Italy

Delabrouille Jacques
College de France
11, Place Marcelin Berthelot
75231 Paris Cedex 05 - France

Fosalba Pablo
Institut d'Astrophysique de Paris
98bis Boulevard Arago
75014 Paris - France

Greaves Jane S.
U.K. Astronomy technology Centre
Blackford Hill
EH9 3HJ Edinburgh - UK

Haverkorn Marijke
Leiden Observatory
P.O. Box 9513
2300 RA Leiden
The Netherlands

Kaplan Jean
College de France
11, Place Marcelin Berthelot
75231 Paris cedex 05 - France

Kees van't Klooster
ESTEC
Keplerlaan 1 - P.O. Box 299
2200 AG Noordwijk ZH
The Netherlands

Landecker Tom
Dominion Radio Astrophysical Observatory
P.O. Box 248
Pentincton, B.C., V2A 6K3
Canada

Leahy Patrick J.
Jodrell Bank Observatory
SK11 9DL Macclesfield
Cheshire - UK

Macculi Claudio
ITeSRE - CNR
Via Gobetti, 101
40129 Bologna - Italy

Masi Silvia
Dip. di Fisica, Univ. La Sapenza
Piazzale A. Moro, 2
00185 Roma - Italy

Morelli Ennio
ITeSRE - CNR
Via Gobetti, 101
40129 Bologna - Italy

Musso Carlo
Agenzia Spaziale Italiana
Via di Villa Grazioli, 23
I-00198 Roma - Italy

Natoli Paolo
Universita' di Roma Tor Vergata
Via della Ricerca Scientifica, 1
00133 Roma - Italy

Olthof Henk
ESA/ESTEC, P.O. Box 299
2200 AG Noordwijk
The Netherlands

Palumbo Giorgio
Agenzia Spaziale Italiana
Via di Villa Grazioli, 23
00198 Roma - Italy

Liu Guo Chin
Institute of Astronomy & Astrophysics
Academia Sinica
106 Taipei -Taiwan

Mandolesi Nazareno
ITeSRE - CNR
Via Gobetti, 101
40129 Bologna - Italy

Monari Jader
IRA - CNR
Via Gobetti, 101
40129 Bologna - Italy

Muhonen Vesa
Helsinki Institute of Physics
PO Box 64 (Nils Hasselblomin katu 2)
FIN-00014 Helsinki - Finland

Musso Ivano
Istituto C.N.U.C.E. - CNR
Via Giuseppe Moruzzi, 1
56124 Pisa - Italy

Nicastro Luciano
IFCAI - CNR
Via U. La Malfa, 153
90146 Palermo - Italy

Orfei Alessandro
IRA - CNR
Via Gobetti 101
40129 Bologna - Italy

Peverini Oscar Antonio
IRITI - CNR
Corso duca degli Abruzzi, 24
10129 Torino - Italy

Piccirillo Lucio
Cardiff University
5 The Parade
Cardiff, CF24 3YB - U.K.

Poppi Sergio
IRA - CNR
Via Fiorentina
40060 Villafontana (BO) - Italy

Reich Wolfgang
MPI für Radioastronomie
Auf dem Hügel, 69
53121 Bonn - Germany

Rubano Claudio
Univ. Federico II di Napoli
Via Cintia, Ed N
80126 Napoli - Italy

Sazhin Mikhail V.
Sternberg Astronomical Institute
Universitetsky pr. 13
119899 Moscow - Russia

Scott Douglas
University of British Columbia
6224 Agricultural Road
Vancouver, BC V6T 1Z1 - Canada

Setti Giancarlo
Istituto Nazionale per l' Astrofisica
Viale del Parco Mellini, 84
00136 Roma- Italy

Sironi Giorgio
Dip. Fisica Univ. Milano Bicocca
Piazza della Scienza, 3
20126 Milano - Italy

Pontieu Nicolas P.
ISN
53, Avenue des Martyrs
38026 Grenoble - France

Prunet Simon
Institut d'Astrophysique de Paris
94bis boulevard Arago
75014 Paris - France

Rosset Cyrille
PCC-College de France
11, Place Marcellin Berthelot
75231 Paris Cedex 05 - France

Sault Bob
Australia Telescope, CSIRO
P.O. Box 76
Epping, NSW, 1710 - Australia

Sbarra Carla
ITeSRE - CNR
Via Gobetti, 101
40129 Bologna - Italy

Scudellaro Paolo
Univ. Federico II di Napoli
Via Cintia, Ed. N
80126 Napoli - Italy

Shinnaga Hiroko
Harvard Smithsonian Center
for Astrophysics
60 Garden St., MS 42
Cambridge MA 02138 - USA

Tascone Riccardo
Politecnico di Torino
Corso Duca degli Abruzzi, 24
10129 Torino - Italy

Tauber Jan
Space Science Dpt., ESA
Keplerlaan, 1
2201AZ Noordwijk -The Netherlands

Tucci Marco
Univ. di Milano Bicocca
Piazza della Scienza, 3
20216 Milano - Italy

Villa Fabrizio
ITeSRE - CNR
Via Gobetti, 101
40129 Bologna - Italy

Widell Ola
SSC Esrange
P.O. Box 802
SE-98128 Kiruna - Sweden

Zannoni Mario
Universita' di Milano Bicocca
Piazza della Scienza, 3
20216 Milano - Italy

Timbie Peter
University of Wisconsin
University Avenue 1150
Madison, Wisconsin 53706 - USA

Valiviita Jussi
University of Helsinky, Dep. of Physics
PO Box 64 (Nils Hasselblomin katu 2)
FIN-00014 Helsinki - Finland

Vinyajkin Evgeny N.
NIRFI, 25 B.Pecherskaya st.
603950 Nizhnij Novgorod
Russia

Wielebinski Richard
MPI für Radioastronomie
Auf dem Hügel, 69
53121 Bonn - Germany